80 Notes on Numerical Fluid Mechanics and Multidisciplinary Design (NNFM)

Springer
Berlin
Heidelberg
New York
Barcelona
Hong Kong
London
Milan
Paris
Tokyo

Drag Reduction by Shock and Boundary Layer Control

Results of the Project EUROSHOCK II Supported by the European Union 1996-1999

Egon Stanewsky, Jean Délery, John Fulker
and Paolo de Matteis (Editors)

Springer

Dr. Egon Stanewsky
DLR Institute of Aerodynamics
and Flow Technology
Bunsenstr. 10
D - 37073 Göttingen
Germany

Prof. Jean Délery
ONERA Fundamenal/Experimental
Aerodynamics Department
8, rue des Vertugadins
F - 92190 Meudon
France

Prof. John Fulker
QinetiQ Ltd.
Centre of Aerospace Technology
Bedford MK41 6AE
U.K.

Dr. Paolo de Matteis
CIRA Experimental
Aerodynamics Laboratories
Via Maiorise
I - 81043 Capua (CE)
Italy

Library of Congress Cataloging-in-Publication Data applied for

Die Deutsche Bibliothek - Cip-Einheitsaufnahme

Drag reduction by shock and boundary layer control : results of the project EUROSHOCK II / Egon Stanewsky ... (ed.). - Berlin ; Heidelberg ; New York ; Barcelona ; Hong Kong ; London ; Milan ; Paris ; Tokyo : Springer, 2002
(Notes on numerical fluid mechanics and multidisciplinary design (NNFM) ; 80)
(Engineering online library)
(ISBN 3-540-43317-1)

ISSN 0179-9614
ISBN 3-540-43317-1 Springer-Verlag Berlin Heidelberg New York

Springer-Verlag Berlin Heidelberg New York
a member of BertelsmannSpringer Science+Business Media GmbH

http://www.springer.de

Printed in Germany

Typesetting: camera-ready by authors
Cover design: de'blik, Berlin
Printed on acid-free paper SPIN: 10851364 62/3020/M - 5 4 3 2 1 0

NNFM Editor Addresses

Prof. Dr. Ernst Heinrich Hirschel
(General editor)
Herzog-Heinrich-Weg 6
D-85604 Zorneding
Germany
E-mail: e.h.hirschel@t-online.de

Prof. Dr. Kozo Fujii
Space Transportation Research Division
The Institute of Space
and Astronautical Science
3-1-1, Yoshinodai, Sagamihara,
Kanagawa, 229-8510
Japan
E-mail: fujii@flab.eng.isas.ac.jp

Dr. Werner Haase
Höhenkirchener Str. 19d
D-85662 Hohenbrunn
Germany
E-mail: wug.haase@t-online.de

Prof. Dr. Bram van Leer
Department of Aerospace Engineering
The University of Michigan
Ann Arbor, MI 48109-2140
USA
E-mail: bram@engin.umich.edu

Prof. Dr. Michael A. Leschziner
Department of Engineering
Queen Mary & Westfield College (QMW)
University of London
Mile End Road
London E1 4NS
U.K.
E-mail: mike.leschziner@qmw.ac.uk

Prof. Dr. Maurizio Pandolfi
Politecnico di Torino
Dipartimento di Ingegneria Aeronautica
e Spaziale
Corso Duca degli Abruzzi, 24
I - 10129 Torino
Italy
E-mail: pandolfi@polito.it

Prof. Dr. Jaques Periaux
Dassault Aviation
78, Quai Marcel Dassault
F-92552 St. Cloud Cedex
France
E-mail: periaux@rascasse.inria.fr

Prof. Dr. Arthur Rizzi
Department of Aeronautics
KTH Royal Institute of Technology
Teknikringen 8
S-10044 Stockholm
Sweden
E-mail: rizzi@aero.kth.se

Dr. Bernard Roux
IRPHE-IMT
Technopole de Chateau-Gombert
F-13451 Marseille Cedex 20
France
E-mail: broux@irphe.univ-mrs.fr

Preface

The survival of the Aeronautical Industries of Europe in the highly competitive World Aviation Market is strongly dependent on such factors as time-to-market of a new or derivative aircraft and on its manufacturing costs but also on the achievement of a competitive technological advantage by which an increased market share can be gained. Recognizing this, cooperative research is continuously encouraged and co-financed by the European Union in order to strengthen the scientific and technological base of the Aeronautical Industries thus providing – among others – the technological edge needed for survival. Corresponding targets of research within Area 3, Technologies for Transport Means, and here in particular Area 3A, Aeronautics Technologies, of the Industrial and Materials Technologies Program (Brite - EuRam III, 1994 - 1998) have been identified to be aircraft efficiency, cost effectiveness and environmental impact. Concerning aircraft efficiency – relevant to the present research – a reduction in aircraft drag of 10%, a reduction in aircraft fuel consumption of 30%, and a reduction in airframe, engine and system weight of 20% are envisaged. Meeting these objectives has, of course, also a strong positive impact on the environment.

In order to further technology, it is prudent to concentrate on the feasibility demonstration of a limited number of technologies of high economic and industrial impact. Examples of such technologies are, for instance, with regard to aircraft efficiency, the application of laminar flow and drag reduction technologies, technologies related to advanced large primary structures, and propulsion technologies. A general prerequisite for technology development is, of course, also the continuous improvement of the theoretical / numerical and experimental tools and, particularly in the case of aeronautical fluid dynamics, which is of interest here, the understanding of complex viscous compressible flow phenomena such as turbulence, transition, shock boundary layer interaction and separation.

The fundamental research program described here is related to drag reduction and separation control; it is based on the following consideration: the development of the boundary layer and the interaction of the wing-upper-surface shock wave with the boundary layer essentially establish the flight performance of transonic transport aircraft at cruise as well as at high-speed off-design conditions. Consequently, employing shock and boundary layer control can be assumed to have a large potential for improving flight performance in terms of cruise drag, hence speed and/or fuel consumption, and with respect to the drag-rise and buffet boundaries. Control can also be utilized to design simpler-geometry wings, allowing to reduce weight and increase pay load, without the penalty of reduced aerodynamic performance. Based on the experience gained during the EUROSHOCK I project, where it was found that passive shock control by a perforated-surface/cavity arrangement generally leads to an increase or, at best, to marginal reductions in drag, the specific objective of the research performed here was to study the various aspects of ***active*** shock and boundary layer control, to develop and improve the computational and experimental tools needed to incorporate control concepts into the design of advanced transonic wings and to determine the aerodynamic merits of control up to flight Reynolds numbers, but also to assess the penalties associated with incorporating potential control methods into existing and new wing designs.

The work was carried out by five research organizations, viz., Deutsches Zentrum für Luft- und Raumfahrt e.V. (DLR), Centro Italiano Ricerche Aerospaziali S.C.p.A. (CIRA), Instituto Nacional de Técnica Aerospacial (INTA), Office National d'Etudes et de Recherches Aérospatiales (ONERA) and Defense Evaluation Research Agency (DERA), three universities, viz., the Universities of Cambridge and Karlsruhe and the Universita´di Napoli "Federico II", and four industrial partners, viz., Alenia Aeronautica, EADS-Airbus (Airbus-D), BAE SYSTEMS-Airbus (Airbus-UK Ltd.), and Dassault Aviation.

The present book is, similarly to the EUROSHOCK I book, structured as follows: Firstly, the scientific and economical reasons leading to this investigation and the approach taken are outlined. This is followed by a comprehensive and critical account of the research and the results obtained – without going into excessive detail. Finally, the individual contributions of the partners are presented in the form of papers giving appropriate details of the fundamental, numerical and experimental research performed.

The editors would like to thank all partners for their contribution to the success of EUROSHOCK II and for the effort they put into the preparation of the present book. The work was performed in a very harmonious way which is reflected in the high quality of the results. On behalf of the entire team, we would also like to thank the European Commission for its support. Finally, thanks are due to E.H. Hirschel, the general editor of the Notes on Numerical Fluid Mechanics and to the Springer Verlag for making this publication possible.

September 2001

Egon Stanewsky	Göttingen
Jean Délery	Paris
Paolo de Matteis	Capua
John Fulker	Bedford

Table of Contents

A. SYNOPSIS OF THE PROJECT EUROSHOCK II

by

Egon Stanewsky
DLR, Institute of Aerodynamics and Flow Technology
Bunsenstraße 10, D-37073 Göttingen

Jean Délery
ONERA, Fundamental / Experimental Aerodynamics Department
8, rue des Vertugadins, 92190 Meudon, France

John Fulker
DERA Center for Aerospace Technology
Bedford MK44 1JP, UK

Paolo de Matteis
CIRA, Experimental Aerodynamics Laboratories
Via Maiorise, 81043 Capua (CE), Italy

Summary

The development of the boundary layer and the interaction of the boundary layer with the wing-upper-surface shock wave play an essential role in determining the design and off-design performance of transonic transport aircraft in the case of a turbulent wing but more so for laminar wings where moderate to strong shock waves may already be present at cruise conditions in order to take full advantage of the potential of laminarization. Based on the experience gained during the EUROSHOCK I project, where it was found that passive shock control by a perforated surface / cavity arrangement always lead, for laminar wings, to an increase in total drag, active control by contour bumps, discrete slot suction, a perforated surface / cavity arrangement with part-suction, and by hybrid control, i.e., a combination of control schemes, was investigated. The study consisted of four elements: basic experiments with the objective of improving the physical models associated with control, the extension of numerical prediction methods to properly treat shock and boundary layer control and the performance of parametric control effectiveness studies, the performance of airfoil and sheared-wing tests to provide data for the validation of the computational methods and to determine — in conjunction with the computational results — the aerodynamic merits of active shock and boundary layer control, and the assessment of benefits and penalties associated with incorporating potential control methods into existing and/or new wing designs. The results have shown that active shock control by a perforated surface / cavity arrangement with part-suction and similarly hybrid control, consisting of a passive cavity arrangement upstream followed by active suction downstream, always lead to an increase in total drag for the airfoils and the sheared wing considered here, while discrete suction resulted in a noticeable decrease in drag, even when accounting for "pump" drag. The most effective device, however, was found to be an adaptive contour bump placed in the shock region which lead to drag reductions of up to 23%. A further reduction in drag was achieved when combining the contour bump with discrete suction upstream of the bump. The implementation studies have shown, accordingly, that by incorporating an adaptive, variable-height bump into a laminar-wing aircraft, fuel reductions of up to 2.11% can be achieved on typical long-range flight missions. All devices investigated had a positive effect on the buffet boundary.

1 Introduction

The development of the boundary layer and the interaction of the wing-upper-surface shock with the boundary layer essentially establish the flight performance of transonic transport aircraft at cruise as well as at high-speed off-design conditions. Consequently, employing shock and boundary layer control can be assumed to have a large potential for improving flight performance in terms of cruise drag, hence speed and/or fuel consumption, and with respect to the drag-rise and buffet boundaries. Control can also be utilized to design wings of simpler geometry, e.g., thicker wings, without the penalty of performance degradation, allowing to reduce weight and increase pay load.

Before continuing, it is deemed worthwhile to consider briefly the transonic flow development whose control seems of such benefit: as, for instance, the Mach number for a given lift coefficient is increased — and similar considerations hold for increasing lift at a constant Mach number — shock waves develop on the airfoil upper surface, resulting in an increase in drag mainly due to the occurrence of wave drag, Figure 1 [1.1]. Subsequently, in addition to the further increase in wave drag, viscous drag increases essentially due to a thickening of the boundary layer — initially without separation present — caused by the shock and the sustained rear adverse pressure gradients. At a later stage flow separation develops which leads, finally, to the buffet process. Shock and boundary layer control may delay this development as hypothetically exemplified in Figure 2 for a turbulent (T) and a laminar (L) airfoil or wing, respectively, showing the effect of control, the latter denoted by the subscript "SB", on drag and the drag-rise and buffet boundaries. Also shown in Figure 2 (a) are characteristic pressure distributions for a laminar and a turbulent airfoil, respectively, which will be addressed below when considering control effectiveness for these airfoils.

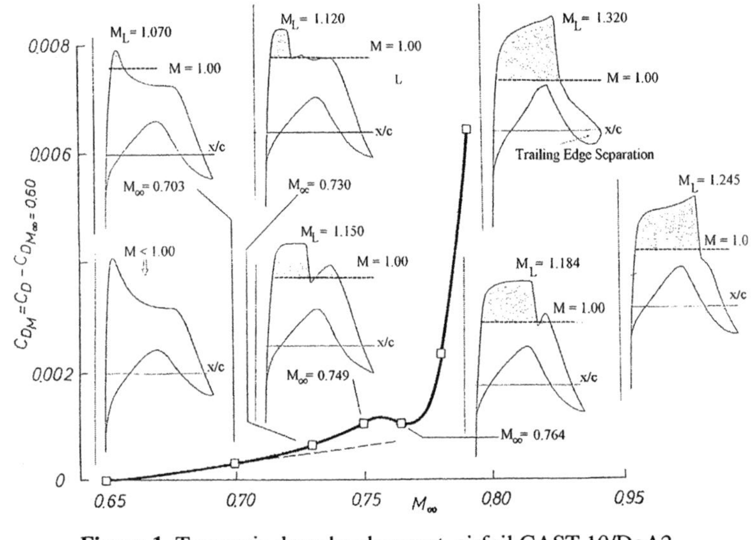

Figure 1 Transonic drag development, airfoil CAST 10/DoA2, $C_L = 0.50$, $Re_c = 30 \times 10^6$ [1.1]

Early experiments have revealed that the detrimental effect of strong shock waves and sustained rear adverse pressure gradients on the *off-design* performance of airfoils and wings can be reduced by some means of shock and boundary layer control, such as vortex generators and single-slot suction and blowing [1.2]. In later experiments with double-slots and perforated surfaces with underlying cavities in areas of strong shocks in conjunction with suction from the cavity, it was shown that for a turbulent airfoil — besides suppressing shock-induced and rear separation, hence shifting the buffet boundary to higher Mach numbers and/or lift coefficients — drag could be reduced over a considerable range of freestream conditions. The effect was, qualitatively, also achieved when no external suction, C_Q, was applied, Figure 3 [1.1, 1.3]. The initial success of shock control has initiated experimental and numerical research activities at various European universities and research organizations with similar research also having been carried out in the US and in Japan. Corresponding references are enumerated in [1.4].

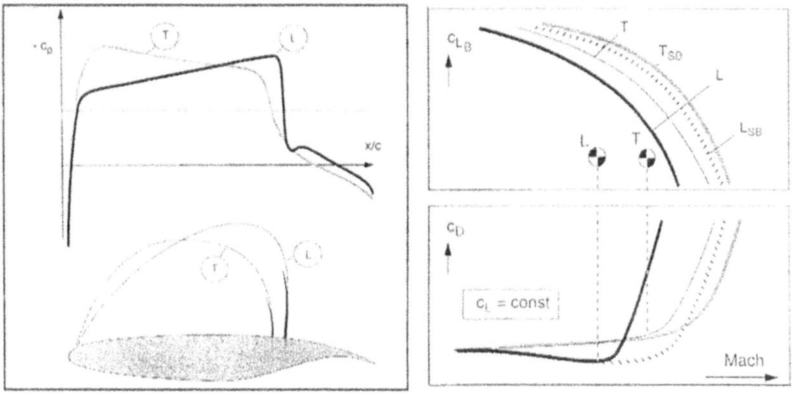

a. Pressure distribution for a turbulent (T) and a laminar (L) airfoil at equal lift

b. Design points and drag-rise and buffet boundaries for the two airfoils without and with (SB) shock control

Figure 2 Characteristic transonic features for a laminar and a turbulent airfoil

The advance of laminar airfoil and wing technology and the positive effect of shock control on the flow development about transonic (turbulent) airfoils led to the proposal and the subsequent execution of the EC Research Project EUROSHOCK (I) — Drag Reduction by Passive Shock Control — mainly concerned with the detailed investigation of **passive** shock control on laminar-type airfoils previously not considered. To repeat: passive shock control means here control via a perforated surface with underlying cavity without applying suction as indicated in Figure 3. Laminar-type airfoils were considered to be more susceptible to shock control, hence of higher potential for control, since the inherent acceleration of the flow on the upper surface of an airfoil or wing may lead for a prescribed lift coefficient already at the design condition to stronger shock waves, Figure 2a, thus higher wave drag. The flow development in the vicinity of the drag-rise and buffet boundary is, due to the stronger shocks, also much more sensitive to small changes in the freestream conditions and shock control can reduce this sensitivity by reducing the otherwise existing strong pressure gradients and their inherent effect on (sudden) boundary layer separation.

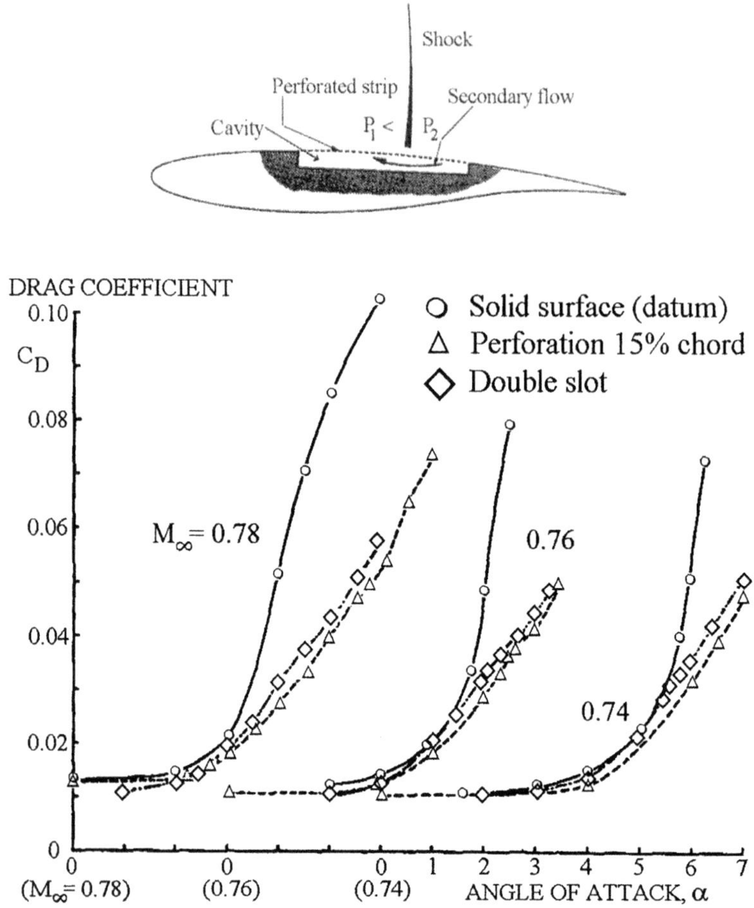

Figure 3 Mechanism and effect of passive shock control ($C_Q = 0$) on drag $Re_c = 2.5 \times 10^6$, airfoil VFW VA-2 [1.1, 1.3]

The specific objective of the EUROSHOCK (I) research was to investigate all aspects of passive shock control aiming at the reduction of aircraft drag and the improvement of aircraft off-design performance. The results of the three-year very thorough investigation of passive shock control, fully described in [1.5], can be summarized as follows:

- The basic mechanism associated with passive shock control has clearly been identified. The blowing taking place in the upstream part of the control region provokes a rapid thickening of the boundary layer with the resulting increase in the displacement thickness being felt by the outer inviscid flow as a ramp with an almost continuous (isentropic) compression replacing the single normal shock present in the interaction without control, thus considerably reducing wave drag, Figure 4 [1.6, 1.7].

- The re-circulating flow in the cavity region results in an increase in the boundary layer thickness parameters, e.g., momentum and displacement thicknesses, downstream of the interaction/control region. Nevertheless, a reduction in total drag of approximately 4% is obtained for the control region due to the dominating effect of the wave-drag reduction [1.6].
- For the laminar-type airfoils investigated, an **increase in total drag** due to passive control was determined, experimentally and numerically, Figure 5 [1.8]. This drag increase was traced to the fact that the increase in the boundary layer thickness parameters over the control region — also observed in the basic experiments — was being amplified over the rear part of the airfoils due to the strong adverse pressure gradients prevailing there, Figure 6 [1.9], generating viscous drag which overcompensated the reduction in wave drag.
- Passive shock control strongly dampened the shock oscillations thus shifting the buffet boundary to higher Mach numbers and/or lift coefficients [1.10].

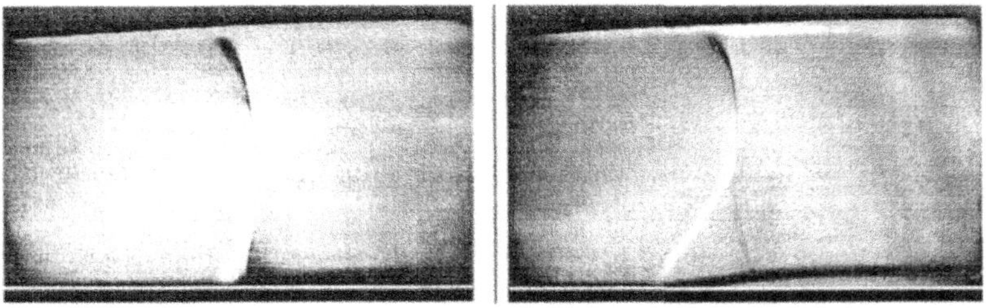

Solid wall — Inclined holes, diameter 0.15 mm

a. Schlieren photograph of the flow field w/o and with control, ONERA [1.6]

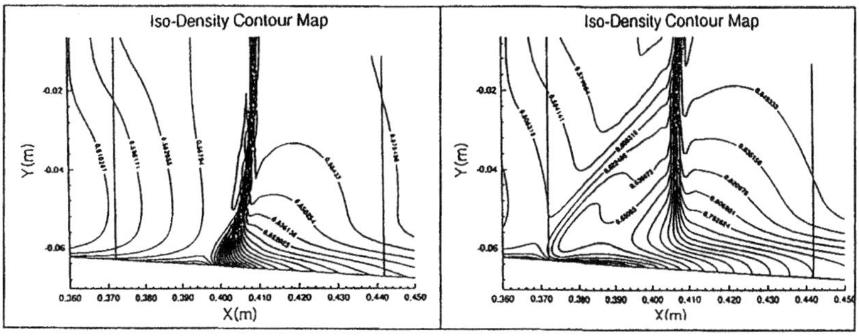

Solid surface — Perforated surface, porosity 5.3%

b. Karlsruhe University Navier-Stokes Simulation [1.7]

Figure 4 Flow field observations obtained by ONERA and Karlsruhe University

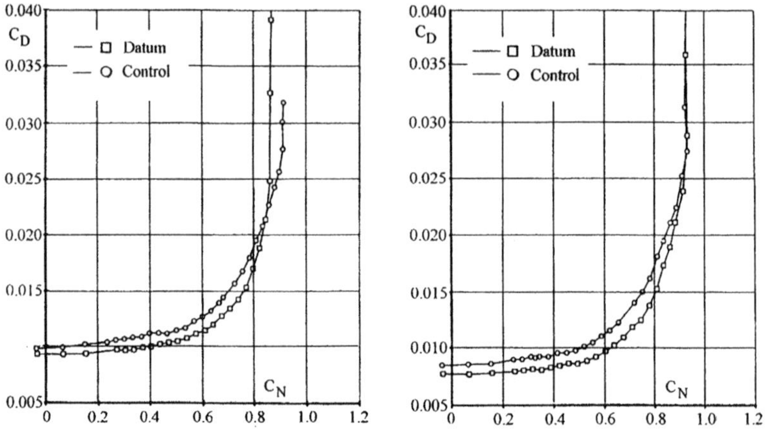

a. $Re_c = 6x10^6$ b. $Re_c = 19x10^6$

Figure 5 Drag- versus normal-force coefficient at $M\infty = 0.68$, DERA-experiments [1.8]

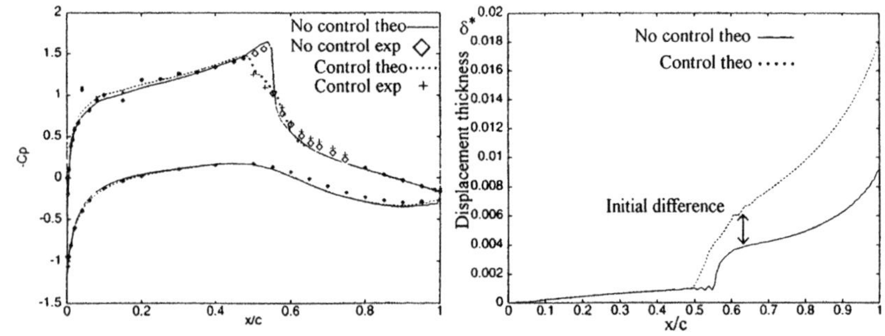

a. Surface pressure distributions with and w/o control

b. Boundary layer development with and w/o control

Figure 6 Experimental and computational results for the laminar-type airfoil DRA-2303, $M\infty = 0.680$, $Re_c = 19x10^6$, $C_L = 0.81$, CIRA computations, DERA experiments [1.9, 1.8]

Based on the experience made during the EUROSHOCK (I) project, it may be stated that *passive shock control can now be ruled out as an effective means of reducing drag of laminar wings.* Even for turbulent wings, the sensitivity of the effectiveness of passive control to changes in the flow and boundary layer parameters makes this type of control rather impracticable at conditions where drag reduction is of prime interest, especially since the benefits are marginal. If one is concerned with reducing drag, other control techniques must be considered, the latter constituting the main — but not the sole — objective of the research described and discussed below. Of course, in applications where drag reduction is not the main driver, passive control may still be of use, such as, for example, in supersonic intakes where shock waves are utilized to compress and slow down the flow at the engine face and where shock-induced separation must be avoided and the best possible flow uniformity assured.

2 The EUROSHOCK II – Project

The specific objective of the research described here was to study, experimentally and numerically, the effect of various means of active shock and boundary layer control on cruise performance in terms of cruise drag and/or speed and on off-design performance, i.e., essentially the drag-rise and buffet boundaries, of transonic airfoils and wings. It was, furthermore, aimed at clearly defining the benefits and penalties associated with incorporating potential control methods into existing and new wing designs for typical transonic long-range transport aircraft and, similarly, for regional-jet aircraft, considering characteristic aircraft missions. Also to be provided were the tools needed for the design of transonic wings with control, i.e., essentially extending the numerical codes to be able to accurately treat control, which requires a clear understanding of the flow phenomena associated with control and the establishment of corresponding control laws and boundary conditions.

The active control mechanisms proposed — based on the experience gained during the EUROSHOCK (I) project [1.5] — and subsequently investigated included, Figure 7:

- A local contour modification (distensible bump) in the shock region mainly designed to reduce shock strength, hence wave drag, possibly also reducing viscous drag and delaying the development of separation due to the lesser load on the boundary layer.
- A perforation / cavity arrangement placed in the shock region, similar to EUROSHOCK I, however, with part-suction from the cavity intended to reduce wave drag while, at the same time, keeping viscous drag low.
- Discrete slot suction intended to reduce the boundary layer thickness in the shock region thus reducing viscous drag by reducing the growth in displacement thickness throughout and downstream of the shock boundary layer interaction region, and delaying the onset of separation.
- Hybrid control, e.g., a passive cavity in the shock region with active suction downstream or a contour bump in conjunction with upstream discrete suction, a mechanisms likely to reduce both wave drag and viscous drag.

In order to reach the overall objectives, the project was divided into four major, interrelated tasks, Figure 8, with the individual task objectives being as follows:

Task 1 *- Modeling of Active Control Phenomena:* ① Execution of basic experiments to study the influence of fundamental parameters on steady shock and boundary layer control and to optimize control devices. ② Improvement and validation of physical models for the control region based on the experiments.

The results shall be used to extend and improve the numerical methods to be treated under Task 2 and to recommend potential control devices to be investigated in Tasks 3 and 4. Involved in this task were ONERA and the Universities of Karlsruhe and Cambridge (also see Chapters 11 to 13).

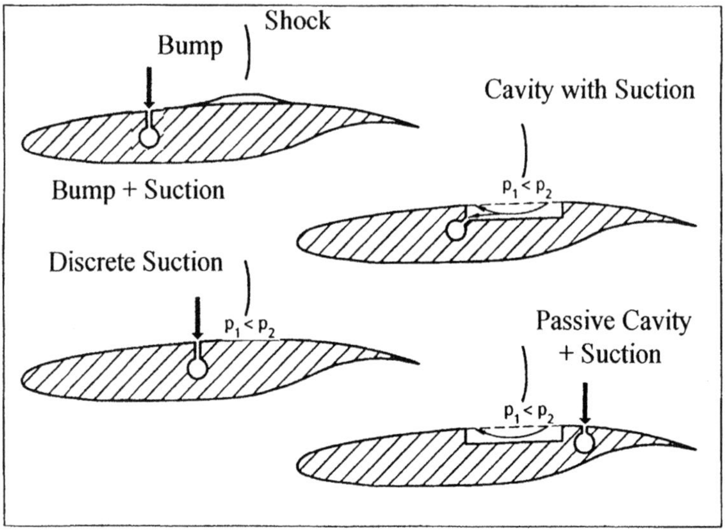

Figure 7 Active control mechanisms investigated

Task 2 - *Prediction of Transonic Airfoil / Wing Flow with Control:* ① Extension of available steady and unsteady two-dimensional and steady three-dimensional computational methods to predict flows with shock and boundary layer control. ② Performance of first parametric studies to assess control concepts and validation of the methods employing, among others, results obtained within Task 3.

Involved in this task were CIRA, the University of Naples, INTA, ONERA, and DLR, and EADS-Airbus and Alenia for the validation of the codes to be used in Task 4 (also see Chapters 14 to 18 and 21 and 23).

Task 3 - *Wind Tunnel Experiments on Airfoils and Wings with Control:* ① Performance of detailed measurements on airfoils and an infinitely-swept (sheared) wing up to flight Reynolds numbers to assess the improvements to be gained by active shock and boundary layer control and to determine possible scaling effects. ② Provision of results to validate the computational methods of Task 2.

Involved in this task were DERA and DLR and initially ONERA (also see Chapters 19 and 20).

Task 4 - *Control Application Aspects:* ① Determine benefits and assess penalties associated with aircraft implementation of control. ② Study ways and

means of incorporating potential control methods identified here into existing or new wing designs.
Airbus-UK and Airbus-D considered the application of control to a large A340-type aircraft with a turbulent and a laminar wing design, respectively, while Alenia and Dassault, in a combined effort, studied the application of boundary layer and shock control to regional-jet aircraft. (also see Chapters 21 to 24).

The interrelation between the tasks is summarized in Figure 8: first one has to consider the basic control phenomena and the associated flow developments and the critical parameters involved, e. g., shock strength, boundary layer conditions and surface geometries. Physical models of the viscous-inviscid interactions with control must be established and/or confirmed and the rather complex boundary conditions determined. The physical models must be incorporated into computational methods which, in turn, are to be validated by the results of realistic experiments on airfoils and wings. Both experiments and computations must be employed to assess the aerodynamic merits of the shock and boundary layer control concepts. Finally, ways must be found to incorporate potential control methods into existing and/or new wing designs and benefits and penalties that can be expected from an incorporation of control into aircraft must be assessed.

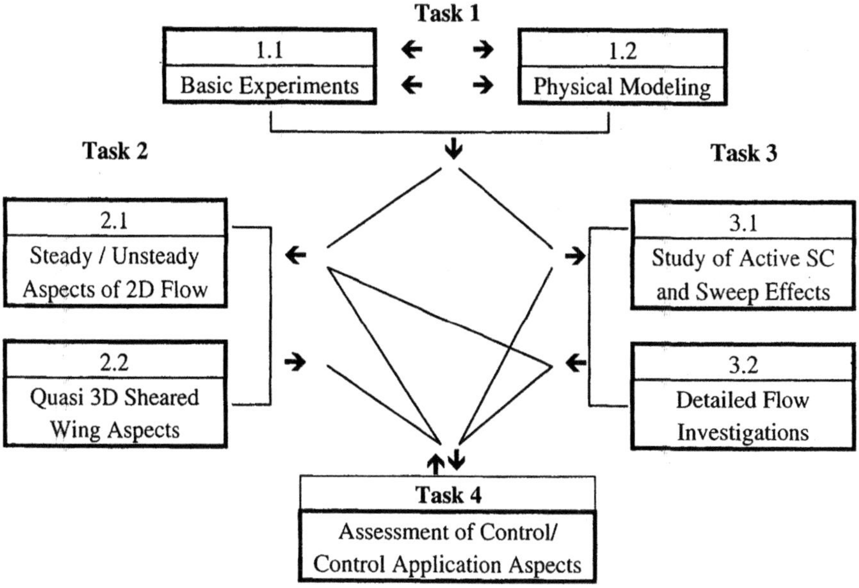

Task 1: Modeling of Active Control Phenomena **Task 2:** Prediction of Transonic Airfoil / Wing Flow with Control **Task 3:** Wind Tunnel Experiments on Airfoils and Sheared Wings with Control **Task 4:** Control Application Aspects

Figure 8 General task flow chart and interrelation between tasks

3 Modeling of Active Control Phenomena (Task 1)

The objective of the work performed here was to further the understanding and modeling of the physical phenomena involved in transonic shock boundary layer interactions under active control conditions. Considering control, as envisaged here, one may distinguish between control applied to the incoming boundary layer before it enters the interaction region, thus increasing the resistance of the boundary layer to the destabilizing action of the shock and sustained adverse pressure gradients, or actively applied in the interaction region with the positive effect of preventing separation and/or restricting the thickening of the boundary layer. Applying suction upstream or in the interaction region tends, however, to increase the shock strength compared to the no-control and passive control cases, respectively, provoking an increase in wave drag, while, at the same time, viscous (or pressure) drag is likely to be reduced. Thus it can be advantageous to associate passive cavity control in the interaction region with active control by suction downstream, i.e., exert hybrid control. A contour bump in the shock region tends to mainly reduce shock strength, hence wave drag, without a major effect on the boundary layer development downstream, hence viscous drag.

The details of the physics of these control schemes and their effect on the flow development, especially the gains in performance that may be expected, must be established. The work was, accordingly, divided into two supplementing parts:

(i) Execution of detailed experiments on simplified (channel flow) configurations aiming at a detailed description of the controlled flow field dependent on major flow and geometrical parameters.

(ii) Exploitation of the results with regard to the derivation, assessment and improvement of the physical models used to represent the effect of shock and boundary layer control, also placing emphasis on turbulence effects and turbulence modeling.

Three institutions were involved in the basic investigations, namely, the Fundamental / Experimental Aerodynamics Department of ONERA, the University of Karlsruhe Institute of Fluid Mechanics and the Cambridge University Department of Engineering. These partners worked in close contact investigating essential and complementary aspects of the various control mechanisms:

ONERA focused its experimental work on a local analysis of various controlled interactions in order to establish the flow field, including both mean and turbulent properties, and to study the control effectiveness. The experiments were supplemented by Navier-Stokes computations to evaluate the physical models and boundary conditions prescribed utilizing several turbulence models. The control mechanisms considered were: a combination of several cavities through which hybrid control was applied, discrete slot suction, and a contour bump in the shock region.

The **University of Karlsruhe** investigated active control by part-suction through a perforated plate / cavity arrangement and hybrid control placing emphasis on a careful examination of the influence of the size (length) and location of the downstream active control cavity. In further work, the important problem of the mass-flow determination through a porous plate in the presence of an outer transonic stream was examined. Supplementing the experiments, 2D and 3D Navier-Stokes computations, simulating the channel-flow experiments, were performed.

Cambridge University considered active shock and boundary layer control in three-dimensional flows with the control mechanisms considered being a perforated plate / cavity arrangement with part-suction, and discrete slot suction with emphasis on the effect of the slot location with respect to the shock. Again, Navier-Stokes computations, using different turbulence models, supplemented the swept-shock channel-flow experiments.

A detailed description of the three investigations is given in Chapters 11, 12 and 13, respectively.

3.1 The Test Arrangements

3.1.1 Two-dimensional channel-flow experiments

Two-dimensional basic experiments were performed by ONERA [3.1] and the University of Karlsruhe [3.2], respectively, utilizing the channel-flow experimental set-ups depicted in Figures 9 and 10.

The ONERA experiments have been executed in a continuous transonic wind tunnel supplied with desiccated atmospheric air with the stagnation conditions being close to ambient conditions. The test set-up, Figure 9a, consisted of a transonic channel having a height of 100 mm and a span of 120 mm in the test section. The lower wall was flat, the upper wall consisted of a contoured profile (nozzle) designed to produce a uniform supersonic flow at a (maximum) Mach number close to $M = 1.4$. An adjustable second throat was placed at the test section outlet to produce a shock wave in the desired position with respect to the control mechanism / control region. The present test section set-up, Figure 9a, corresponds to control by discrete slot suction with the air to be removed from the cavity metered by a sonic throat in the suction duct. Figures 9 b and c exhibit the set-up for hybrid control with a 70 mm long passive cavity and a 15 mm long active cavity downstream. Furthermore investigated was, as a common test case with Karlsruhe University, a similar arrangement with the passive cavity being only 50 mm long, and a contour-bump in the shock region. The interacting boundary layer and the flow field were probed by using a two-component LDV system which allowed to determine the mean velocity vectors and the Reynolds-tensor components in the interacting flows.

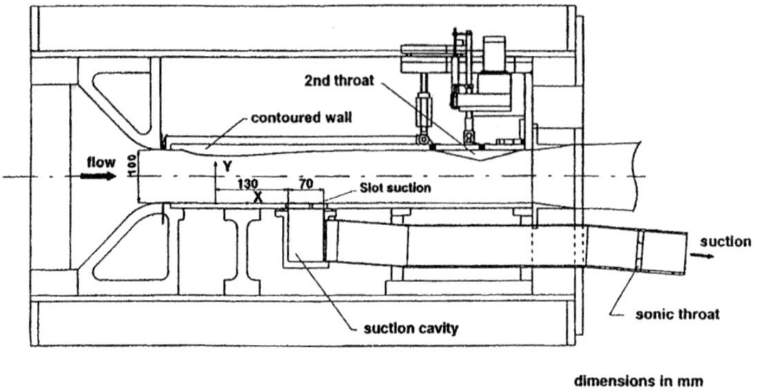

a. Overall test arrangement and suction-slot installation

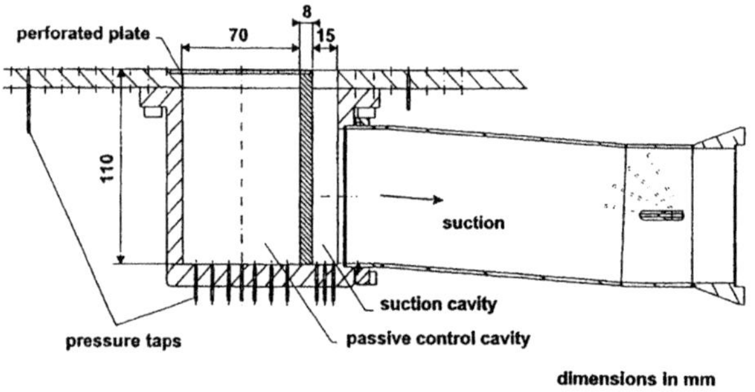

b. Details of the hybrid-control installation

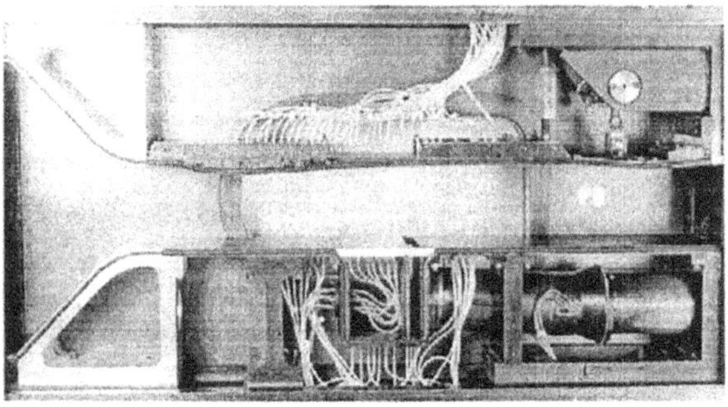

c. Photograph of the test installation for hybrid control

Figure 9 Active shock and boundary layer control set-up in the ONERA S8 wind tunnel [3.1]

At the University of Karlsruhe, the experiments have been performed in an atmospheric blow-down wind tunnel with a test section having a cross section of about 50 mm x 200 mm. The test section proper, located in the slightly diverging part of a Laval-type nozzle in order to achieve typical transonic shock-upstream Mach numbers, consisted of an interchangeable lower-wall region allowing the active and hybrid control device, respectively, to be installed. The lower-wall test set-up, here for hybrid control by a passive/active cavity combination and a freestream Mach number of M = 1.3, is shown in Figure 10. One of the main variables of the investigation was the length of the downstream active cavity which has been changed in steps between 15 mm and 45 mm. The measurements consisted of wall pressure measurements, boundary layer profile surveys by means of pressure probes, and Mach-Zehnder flow-field observations.

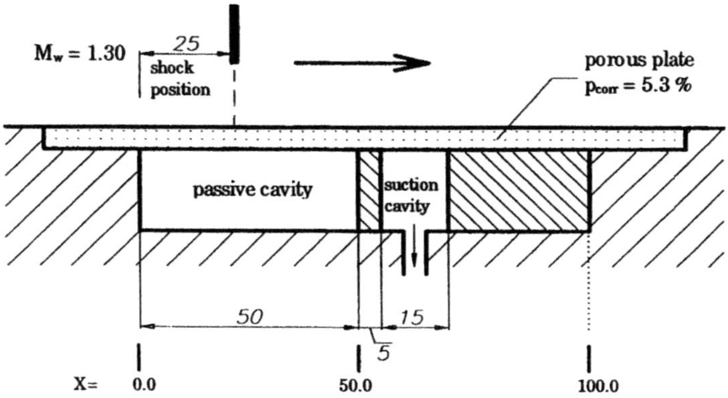

(all dimensions are in mm)

Figure 10 Test arrangement for the Karlsruhe University shock control experiments [3.2]

3.1.2 Three-dimensional channel-flow experiments

The work on three-dimensional oblique shock wave boundary layer interaction with and without control was carried out in one of the Cambridge University Department of Engineering's supersonic wind tunnels capable of simulating full-scale Reynolds number conditions [3.3]. The wind tunnel utilized, having a test section of 150 mm x 110 mm, operates at a freestream Mach number of $M_\infty = 1.85$; the shock wave was formed by a 6° wedge placed on the wind tunnel ceiling — with the shock boundary layer interaction to be investigated taking place on the side wall — providing a Mach number normal to the shock of $M_{\infty N} = 1.15$ which is comparable to that observed on transonic aircraft wings, Figure 11a. A plenum chamber, inserted into the tunnel side wall, was placed underneath the swept shock, at an angle of 40° to the freestream direction. The plenum can be covered with either a porous surface or various solid plates containing individual slots for discrete suction; mass-flow removal is provided by an ejector-driven suction system.

The arrangement of the perforated surface and the slot positions investigated are depicted in Figure 11b and c, respectively. Also indicated here are the locations of the surface pressure orifices and the positions where boundary layer probe traverses were carried out. Investigated were active control by a porous surface/cavity arrangement with part-suction, and discrete slot suction in three positions with respect to the shock, viz., upstream, downstream and at the foot of the shock; in all instances, the suction rate was varied.

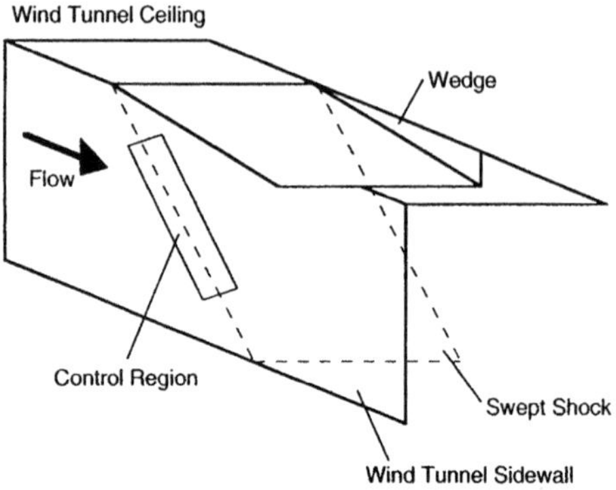

a. Sketch of experimental configuration

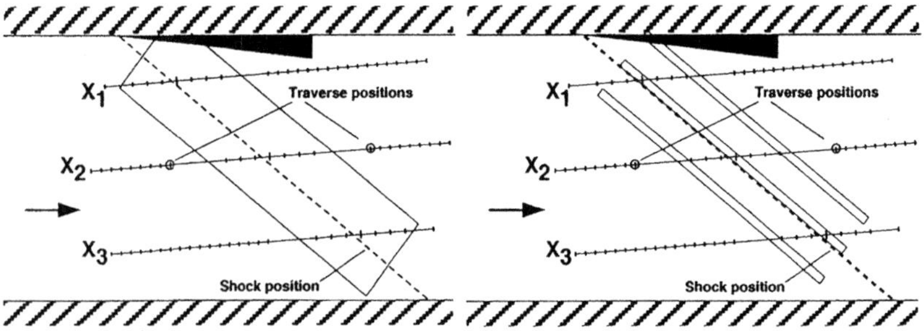

b. Porous surface (8% porosity) c. Suction-slot positions

Figure 11 Test set-up details for the Cambridge shock control experiments [3.3]

3.2 Numerical Codes Employed

All channel-flow basic experiments were supplemented by numerical simulations. Characteristic features of the computational procedures were as follows:

ONERA implicitly solved the Reynolds-averaged Navier-Stokes equations formulated, respectively, to incorporate the Baldwin-Lomax algebraic turbulence model, adapted to apply to separated flows, and the [k- ε] transport-equation model of Chien [3.1]. The wall vertical velocity distribution was obtained by either using Poll's law [3.4] or the law developed by Bohning and Doerffer at the University of Karlsruhe [3.2]; both laws will be further addressed in subsequent chapters. The ONERA computational domain as part of the experimental channel extended from a well chosen upstream section, where the experimental velocity profiles were imposed, to the end of the channel, where the experimental pressure was prescribed.

The **University of Karlsruhe** employed for their channel-flow computations the Navier-Stokes code KAPPA using the Baldwin-Lomax turbulence model [3.2]. The code consists of a finite-volume multi-grid/multi-level algorithm. Generally, the normal velocity distribution in the control region is obtained by the Bohning and Doerffer control law described below. The computational domain starts at the entrance to the convergent-divergent nozzle and ends at the second throat downstream of the test section.

Cambridge University performed all computations with a finite-volume implicit algorithm which solves the full three-dimensional Reynolds-averaged Navier-Stokes equations [3.3]. In these computations the energy equation is not solved and the total enthalpy is assumed constant. This results in considerable savings in computer time without impairing the overall accuracy. The physical domain of the computations was 235 mm x 76.2 mm x 240 mm with grid refinements at the walls and throughout the interaction region. A relation for the velocity normal to the wall within the control region was derived from a number of experiments in a calibration rig, without, however, accounting for an outer tangential flow as in the case of the Karlsruhe law, something that should be considered in future work. In the calculations three turbulence models, namely, Cebeci-Smith, Baldwin-Lomax and Johnson-King, were used and compared.

For more details on the computational methods and procedures, the reader is referred to the individual technical reports [3.1, 3.2, 3.3] and Chapters 11, 12, and 13, respectively.

3.3 Analysis of Results

The control concepts investigated included active control by a perforated plate/cavity arrangement with part-suction from the cavity mainly in three-dimensional (3D) interactions, hybrid control, essentially consisting of a passive cavity upstream and active control by suction downstream of the passive cavity, mainly in 2D interactions, discrete slot suction in various locations with respect to the shock for 2D and 3D flows, and, again only in 2D flow, control by a contour bump in the shock region. Representative results will be presented in the following sequence: single-cavity control with part-suction, hybrid control, control by discrete suction and control by a contour bump. The analysis will start, however, with perforated-plate calibration experiments which resulted in a new control law.

3.3.1 The new control law of Karlsruhe University

Since it was found during the EUROSHOCK (I) exercise that the establishment of the characteristics of a perforated plate — be it for passive or active shock control by perforated-plate/cavity arrangements — may require the presence of an outer transonic stream, a special test-rig was built and experiments carried out to derive these characteristics at corresponding conditions [3.2]. The test-rig employed consisted of a channel with adjustable walls allowing the generation of accelerating or decelerating flow thus providing weakly supersonic or subsonic streams. One of the walls, Figure 12, was fitted with a cavity covered by a porous plate through which, depending on cavity pressure, either blowing or suction was applied.

Representative results of the measurements, with and without an external stream, are depicted in Figure 12: one observes that without external flow (M = 0), the results agree quite well with the general function already established during the EUROSHOCK (I) investigation [1.7], viz.,

$$\frac{dP}{P_0} = M_{hole}^{1/0.55} \left(\frac{1}{1.2}\right)^{1/0.55},$$

where the subscript "hole" refers to the average conditions within the holes of the perforation and dP is the pressure difference across the perforated plate. However, in the presence of an outer tangential flow (M > 0), there is, in the case of suction, a strong dependence of the transpiration characteristics on the outer-stream Mach number. Note that in the case of blowing the external flow has no effect on the transpiration characteristics.

It can be surmised that the wall shear stress is the driving mechanism for blocking the holes in the perforated plate with the effect of the shear depending on the magnitude of the transpiration flow. A factor was, accordingly, introduced, comprised of the ratio of τ_{Wall} and the momentum of the flow in the hole, ½ $\rho_{hole} u_{hole}^2$, and a relation of the form

$$\frac{dP}{P_0} = M_{hole}^{1/0.55} \left[\left(\frac{1}{1.2}\right)^{1/0.55} + b(BM_{hole})^{1/a}\right] \text{ with } B = \frac{\tau_{Wall}}{\frac{\rho_{hole} u_{hole}^2}{2}},$$

derived. The best fit of the proposed formula to the measured results was obtained for a = 1.52 and b = 25 with the character of the curves and the qualitative behavior, as shown in Figure 12, well represented. The above formula, applicable in the case of active and/or passive control, can be used to determine the velocity distribution in the cavity region by means of the continuity equation. For more details one may refer here to [1.7] and [3.2].

3.3.2 Active control by perforated plate/cavity in 3D flow

Active control by a perforated plate/cavity arrangement with part-suction was mainly investigated by Cambridge University and here for three-dimensional swept-shock interactions [3.3]. The conditions were a cavity with a porous surface of 8% porosity and a Mach number upstream of and normal to the shock of $M_{\infty N} = 1.15$, generated by a 6° wedge at a freestream Mach number of $M_{\infty} = 1.8$, Figure 11 (also consult Figure 7). The Reynolds number was $Re = 3.3 \times 10^7$/m with the boundary layer being fully turbulent upstream of the interaction. The suction rate varied from 0 g/s (passive control) to 10.9 g/s, the latter corresponding to a suction coefficient of $C_\mu = q / (\rho_\infty U_\infty \delta^*) = 0.143$.

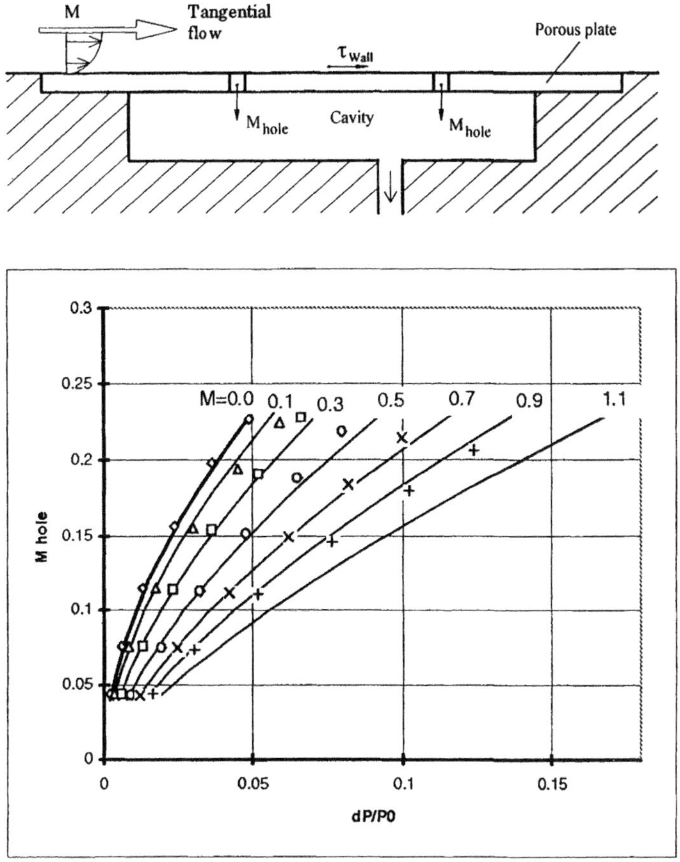

Figure 12 Results of the perforated-plate calibration and of computations according to the new Karlsruhe (Bohning-Doerffer) transpiration law [3.2]

The surface pressure distributions obtained at these conditions are compared in Figure 13 with the no-control reference case. Here, it must first be noted that the overall pressure rise is distinctly less than expected from inviscid theory which is believed to be partly due to upstream disturbances caused by the experimental set-up; however, the study of control effects is considered not to be impaired by this shortcoming.

This figure otherwise shows that passive control effectively smears the shock induced pressure rise exhibiting a distinct plateau in the streamwise pressure distribution, indicative of a λ-shock system, with a nearly isentropic compression following the forward shock (also see Figure 4). The application of suction to the cavity lowers the plateau pressure and therefore removes some of the benefits of passive control such as the reduction in wave drag. Large values of suction completely eradicate any shock-smearing, even causing an expansion. The pressure rise in these cases is seen to be steeper than in the uncontrolled test case, thus most likely increasing wave drag.

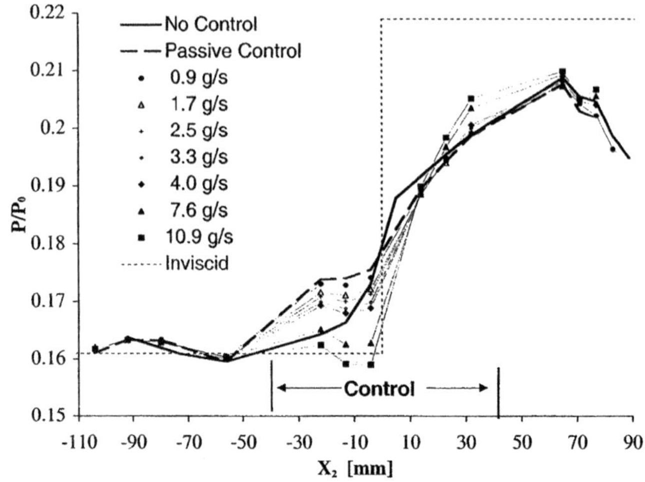

Figure 13 Surface pressure distributions for active control at various suction levels, centerline distributions [3.3]

Corresponding boundary layer profiles downstream of the interaction, with and without cavity control, show, Figure 14, that suction, as expected, increases the fullness of the boundary layer profiles near the surface. Large amounts of suction, however, affect the entire velocity profile rendering it even fuller than in the solid-wall case. The fuller profiles indicate a reduction in viscous drag in the case of airfoil flow so that suction, as seen, seems to increase wave drag but reduce viscous drag. This strongly suggests that a favorable balance must be struck in applying suction considering the change in wave drag and the change in viscous drag both, but also accounting for "pump" drag which will be addressed later.

The numerical simulation showed results quite similar to the experimental pressure distributions except that the level of the plateau pressure was over-predicted and, even for a suction mass flux well in excess of the experimental maximum, the plateau itself remained. The effect of suction, i.e., the reduction in the level of the plateau, was, however, well predicted. Considering the boundary layer development, again agreement between computation and experiment is good as far as the effect of suction is concerned: for instance, a similar reduction in the momentum and displacement thicknesses downstream of the interaction/control region with increasing mass flux was predicted by computation and experiment, although the level in the boundary layer thickness parameters is quite different, Figure 15.

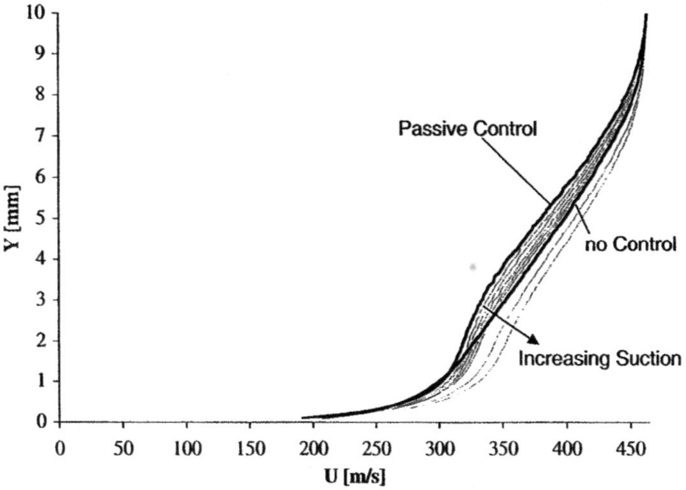

Figure 14 Velocity profiles downstream of the interaction for active control at various suction levels [3.3]

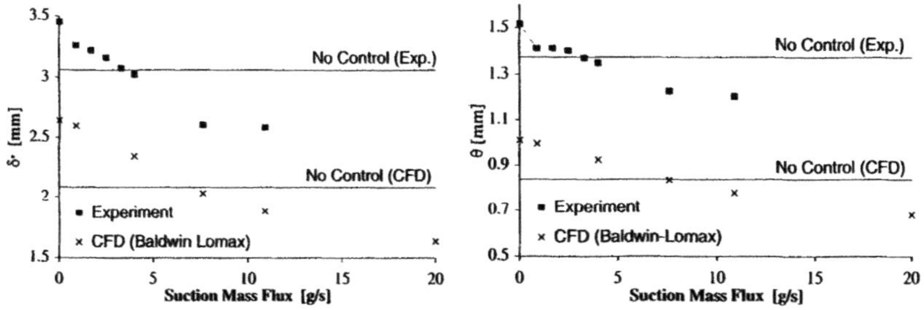

a. Displacement thickness

b. Momentum thickness

Figure 15 Boundary layer thickness downstream of the interaction for various levels of suction [3.3]

There is a further interesting observation to be made in Figure 15, reflecting on the balance between wave and viscous drag addressed above: considering, for example, the experimental results, the displacement and momentum thicknesses reach in the case of control the level of the no-control condition at a suction mass flux of q = 4 g/s, i.e., viscous drag should not be affected by control. Examining the corresponding pressure distribution in Figure 13, one observes that there is still a noticeable plateau with a considerable spreading of the shock at this suction rate indicating that wave drag is still being reduced; this implies, of course, that at this suction level also total drag is being reduced.

Concerning the issue of turbulence models — here considered were the Baldwin-Lomax, Cebeci-Smith and Johnson-King models, respectively — there was very little effect on the pressure distributions; however, for an actively controlled interaction there is a pronounced difference in the boundary layer profiles downstream of the interaction between the Baldwin-Lomax model (BL), which fits the experimental data quite well, and the other two models considered, Figure 16. This lead to the conclusion that, since no model seems to give consistently better results, the Baldwin-Lomax model may be preferable since it is relatively simple and well established.

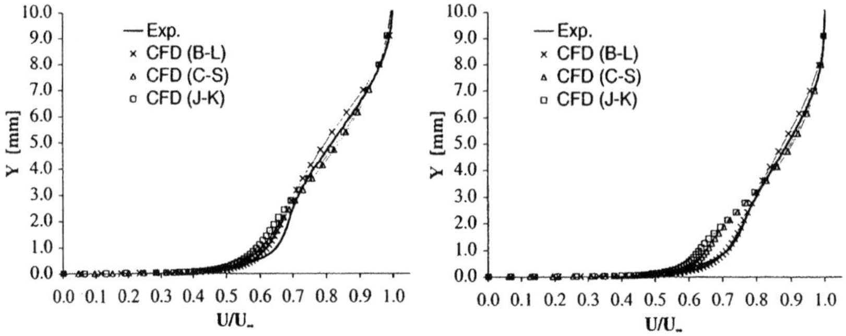

a. Passive control

b. Active control, mass flux 11 g/s

Figure 16 Velocity profiles downstream of the interaction region determined with different turbulence models [3.3]

3.3.3 Hybrid control by passive/active cavity in 2D flow

Hybrid control, as defined here, means control by a passive cavity, i.e., a perforated plate/cavity arrangement without suction, located in the shock region combined with active suction through a second, less extended cavity or a discrete slot downstream of the passive cavity. The effect of this type of control is thought to be similar to the one just described, except that the reduction in wave drag by means of the re-circulating flow in the passive-cavity region and the thinning of the boundary layer downstream are here achieved by separate control entities and therefore independently controllable.

Studies at the University of Karlsruhe

Typical boundary layer profiles downstream of the interaction/control region (X = 100), determined for $M_\infty = 1.30$ and a fully developed turbulent boundary layer upstream of the interaction, are depicted in Figure 17 together with the experimental setup [3.2]. Variables are here the suction intensity and the length of the second, active cavity as shown in Table 1. One observes that, while passive control causes an obviously separated flow downstream of the control region, suction through the downstream active cavity eventually reestablishes the fully attached boundary layer profile present at the no-control conditions. This implies that a considerable reduction in wave drag can be achieved without increasing viscous drag; pump-drag must, of course, still be taken into account.

Table 1 Mass-flows and cavity-lengths corresponding to Figure 17 and boundary layer parameters at X = 100 (also see Figure 18)

M_{hole}	Suction-cavity length	Mass-flow rate [g/s]	$C\mu$ 1)	δ [mm]	δ^* [mm]	H_{12}
0.19	15 mm	1.51	0.503	12.51	5.32	3.45
0.415	27 mm	5.52	1.840	11.23	4.29	2.96
0.57	27 mm	6.97	2.313	9.83	3.25	2.39
0.57	45 mm	12.07	4.023	8.85	2.69	2.2

1) $C\mu = q/(\rho_\infty U_\infty \delta^*)$, ρU taken at $M_\infty = 1.30$, δ^* taken upstream of the interaction region

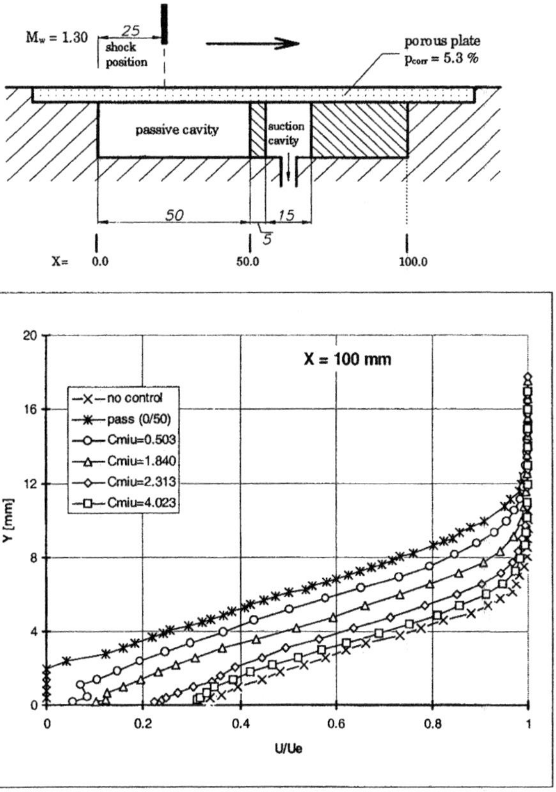

Figure 17 Boundary layer profiles downstream of the interaction for different mass-flow rates [3.2]

The results obtained are essentially similar to the active-single-cavity results described in the preceding chapter and the question arises as to the more effective control arrangement, for instance, for drag reduction. Although only a very detailed drag balance can answer this question, a first attempt is being made in Chapter 3.3.4.

One of the objectives of the present investigation was to study the effect of the suction-cavity length on control efficiency. However, when changing the length of the downstream active cavity, the hole Mach number, i.e., the suction intensity, was kept constant thus always increasing the mass-flow rate with increasing cavity length as indicated in Table 1. The boundary layer thickness parameters downstream of the interaction/control region were therefore plotted versus the suction coefficient for all suction rates/cavity lengths investigated, Figure 18. One observes that all data points for a given boundary layer parameter form a single curve, indicating that the important parameter of the interaction control is the suction mass-flow rate with the cavity length only having a secondary effect. Note that the boundary layer parameters for the no-control case, viz., δ = 7.75 mm, δ^* = 2.36 mm and H_{12} = 2.14, are close to the ones for the high-suction-rate case as already indicated by the boundary layer profiles in Figure 17.

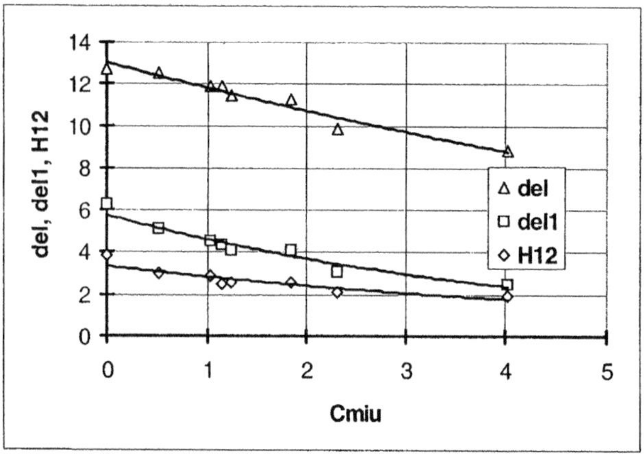

Figure 18 Hybrid-control effect on the boundary layer integral parameters downstream of the interaction-control region [3.2]

Two- and three-dimensional computations were carried out by Karlsruhe University employing, as described above, the Navier-Stokes code KAPPA. The 3D calculations, modeling the complete channel and test area, were performed in response to discrepancies observed previously between experimental and 2D computational results. The 3D computations for the case without control given here as example, Figure 19, show, indeed, that large three-dimensional effects exist in the experiments downstream of the initial interaction/control region which are thought to be due to the rather narrow channel of the experimental setup. It is, however, believed that this will not impair the conclusions drawn from the present investigation.

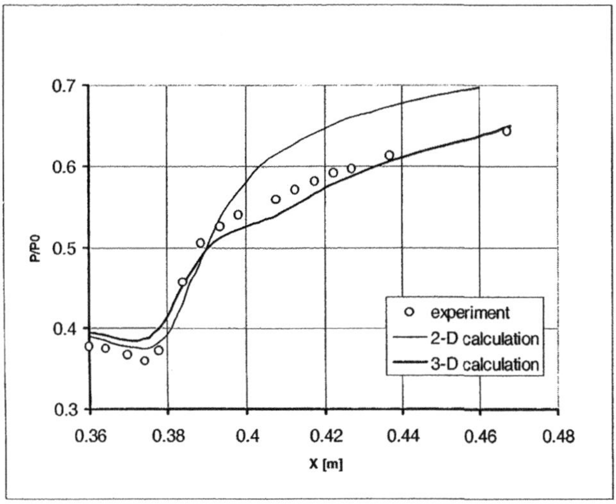

Figure 19 Comparison of experimental and computational wall static pressure distributions in the interaction region [3.2]

Studies at ONERA

ONERA has performed channel-flow experiments concerned with hybrid control for the configurations outlined in Chapter 3.1.1, Figure 9, i.e., 50 mm long and 70 mm long passive cavities in the shock region with 15 mm long active cavities downstream [3.1]. In the case of the former, the active cavity was covered by a perforation, while in case of the latter this cavity was open acting more like a slot. Due to the utilization of the LDV-system, detailed flow field information — besides surface pressure distributions — has been obtained.

Considering first the flow field, the Schlieren picture for an undisturbed shock-upstream Mach number of M = 1.4, Figure 20a, indicates that in the no-control case the shock thickens the boundary layer moderately with the shock strength in the vicinity of the wall being somewhat reduced due to a weak forward compression. For hybrid control, Figure 20b, the upstream passive control cavity causes a considerable increase in boundary layer thickness due to the outflow of air from the cavity resulting in the formation of an oblique shock wave followed by a nearly isentropic compression. The boundary layer thickness is, at the suction rate considered, somewhat reduced by the downstream active cavity. Note that the suction coefficient is here defined as $C'_Q = q / (\rho^* U^* L)$ with ρ and U taken at sonic conditions and L taken as 1 meter, a definition similar to the one — also giving similar coefficients — used for airfoil flow.

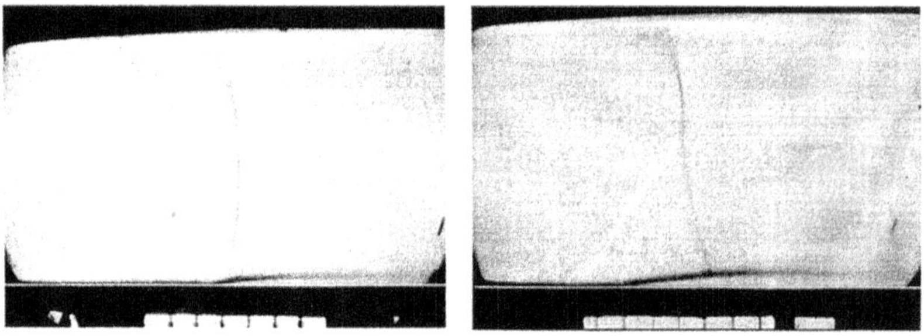

a. No-control reference case b. Hybrid control, $C'_Q = 8.2 \times 10^{-4}$

Figure 20 Schlieren photographs of the flow field in the interaction region, M = 1.4 [3.1]

A more quantitative resolution of the flow field is provided by the Mach number contour maps derived from the LDV-measurements, Figure 21: in the case of hybrid control, the leading shock provokes in the outer inviscid flow a first compression from an upstream Mach number of about M = 1.30 to M = 1.20 with a subsequent compression reducing the Mach number further to M = 1.10. The transition to subsonic flow occurs through the weak normal shock located near the downstream end of the active cavity. The boundary layer downstream of the interaction is considerably thickened compared to the reference case but reduced again by the subsequent suction to nearly no-control conditions.

A better indication of the effect of hybrid control on the boundary layer development is, of course, provided by the distribution of characteristic boundary layer integral parameters — here represented by the displacement thickness — within and downstream of the interaction/control region, Figure 22. It can be seen that at low suction rates, the boundary layer displacement thickness keeps growing downstream of the passive cavity to values much higher than the ones for the no-control case. Increasing suction reduces the thickness parameter which reaches, at a suction coefficient of $C'_Q = 8.2 \times 10^{-4}$, the no-control level downstream of the active cavity; increasing suction to $C'_Q = 30 \times 10^{-4}$ seems to almost eliminate the boundary layer downstream of the interaction, Figure 22b, confirming the trend observed in the University of Karlsruhe experiments.

One aspect of control, important to turbulence modeling, concerns the turbulence behavior in interactions with control. Staying with hybrid control, it is indicated in Figure 23, where the local maxima of the turbulent kinetic energy k (see Chapter 11 for definition) within and downstream of the interaction region are plotted, that the turbulence level in the interaction region drops below the one for the reference case, while downstream it is higher than or equal to the no-control level when moderate to high suction rates are applied; turbulence only reduces below the no-control level when using extreme mass-flow removal, a behavior similar to the one of the boundary layer thickness parameters.

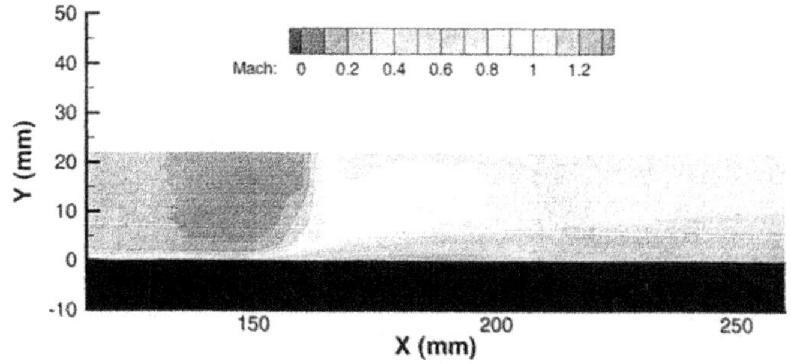

a. Solid-wall reference case, $x_{shock} = 165$ mm

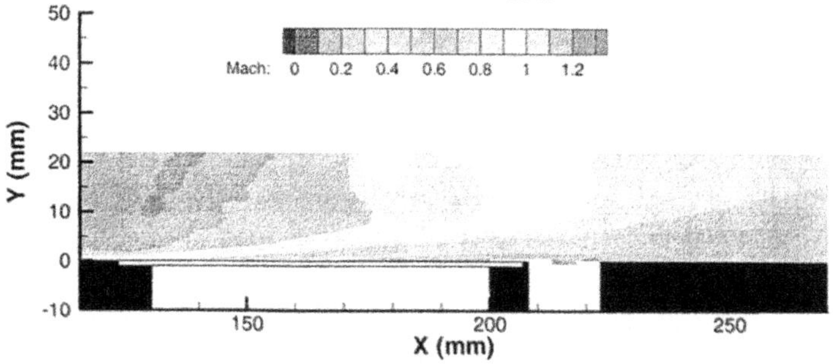

b. Hybrid control, $C'_Q = 8.2 \times 10^{-4}$

Figure 21 Mach number contour lines for the solid-surface reference case and for hybrid control [3.1]

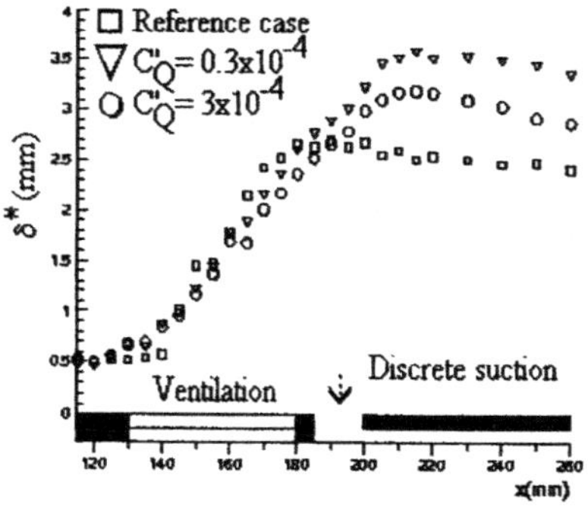

a. Displacement thickness, low C'_Q

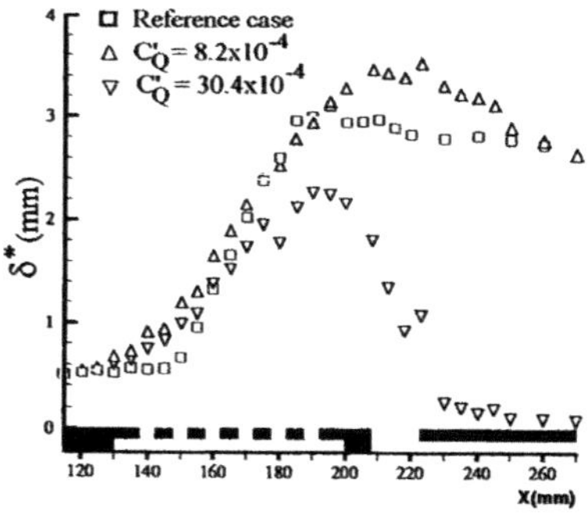

b. Displacement thickness, high C'_Q

Figure 22 Boundary layer development for hybrid control at various suction rates [3.1]

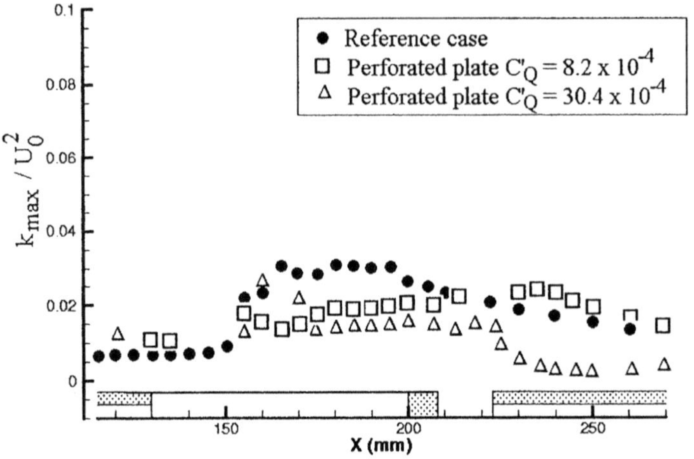

Figure 23 Streamwise variation of the maximum turbulent kinetic energy at no-control and hybrid-control conditions [3.1]

Supplementing the measurements, Navier-Stokes computations have been performed with and without control employing initially the Baldwin-Lomax and the [k-ε] Chien turbulence models, and using for the determination of the vertical velocity component at the wall within the control region the calibration law of Poll and the law of Bohning-Doerffer described in Chapter 3.3.1, respectively. In the passive cavity the experimental pressure was prescribed, while at the downstream active slot the ρv – value at each grid point within the slot region, derived at by dividing the measured suction mass-flow rate by the number of grid points, was prescribed. Since the [k-ε] Chien transport equation turbulence model showed difficulties in correctly reproducing the viscous flow in the long constant channel section preceding the test area proper, which lead to "viscous" choking of the channel flow in the sonic throat even in the smooth-wall no-control case, the computations with control were only carried out with the Baldwin-Lomax turbulence model. Computations included the no-control case, passive control, and hybrid control by the 70 mm long passive-cavity/slot arrangement.

As reflected in the computed wall pressure distributions, Figure 24, essential features of the controlled interactions are well represented in the computations. In detail there are, however, some discrepancies: while for the reference case the calculated wall pressures are in good agreement with the measured data, predictions in the case of passive control, using either Poll's or Bohning and Doerffer's law, show a steep pressure rise at the leading edge of the passive cavity, followed by an expansion, which does not exist in reality. This discrepancy is thought to be due to an insufficient mesh resolution. In the case of hybrid control, larger differences between experiment and computation also exist in the slot region where computations under-predict the pressure recovery, again considered due to grid resolution deficiencies.

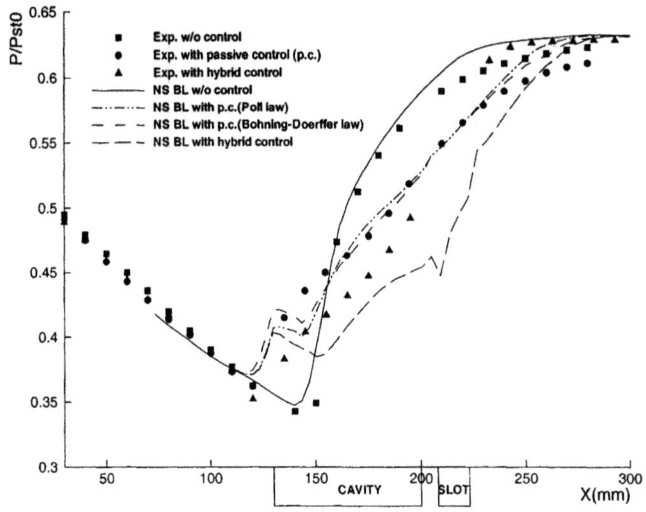

Figure 24 Measured and computed pressure distributions for the reference condition and passive and hybrid control [3.1]

3.3.4 Cavity-ventilation control efficiency

As was repeatedly stated, essential drag components of an airfoil or wing at transonic speeds are the wave drag, generally associated with the upper-surface shock wave, and viscous drag, associated with the boundary layer development throughout the interaction and control region and downstream thereof where sustained adverse pressure gradients prevail. Wave drag and viscous drag account, together with wing friction drag, for roughly 30% of aircraft total drag [1.2]. It is therefore of interest to consider for potential control arrangements the boundary layer and outer-flow-field conditions downstream of the interaction/control region representative of these drag components.

For this purpose, we examine first the boundary layer and flow field total pressure losses, P_o/P_{oo}, as function of the distance from the wall downstream of the control region for the conditions: ① no control, ② passive control, ③ hybrid control, and ④ active (part-suction) single cavity control, the latter two with maximum suction applied ($C_\mu = 4.023$).

The results for these conditions are compared to the profile upstream of the interaction, i.e., the profile without shock losses, in Figure 25. By comparison with the undisturbed profile it can be seen that ① the no-control case is associated with moderate viscous losses but strong shock losses, except close to the surface due to the boundary thickening by the shock, while ② passive control generates strong viscous losses but over a considerable distance into the flow field no shock losses. ③ Hybrid control shows moderate viscous losses, similar to the no-control case, and no shock losses, similar to passive control. ④ Active control through a single cavity exhibits essentially no viscous losses — compared to the upstream conditions — but the highest shock losses. It should be noted that the control mechanisms investigated are only effective up to a distance from the wall of Y = 50 mm; at Y > 50 mm, the shock is no longer reached by control.

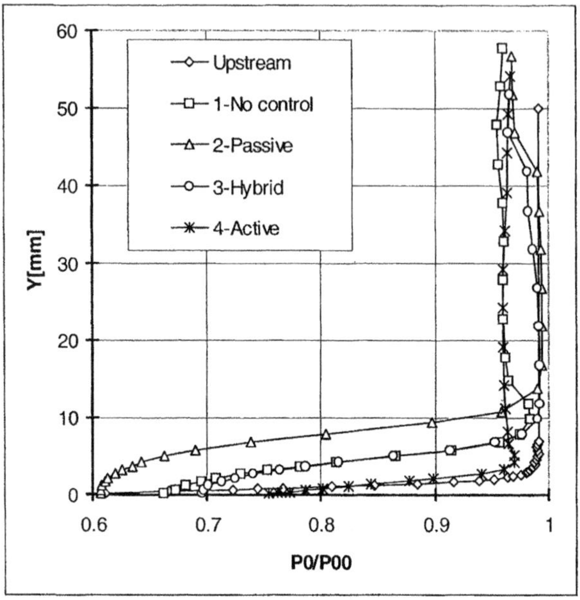

Figure 25 Stagnation pressure profiles for different control schemes downstream of the interaction/control region [3.2]

A direct indication of the efficiency of a control mechanism is given by plotting the integral form of the efficiency, defined as $\eta = U^2 / U^2_{is}$ — where the subscript "is" denotes isentropic conditions — viz.,

$$\eta_Y = \frac{1}{Y}\int_0^Y \eta dY$$

as function of the distance from the wall Y, Figure 26. It is seen that passive control has the lowest overall efficiency, while hybrid and active control exhibit the highest efficiencies. However, considering the "near-wall" region, one observes that the losses in the boundary layer in case of the single-cavity active control are less which makes this type of control also less sensitive to the sustained rear adverse pressure gradients prevailing on airfoils or wings so that — with the overall efficiency being the same — active control by single cavity seems preferable if drag reduction is of main concern.

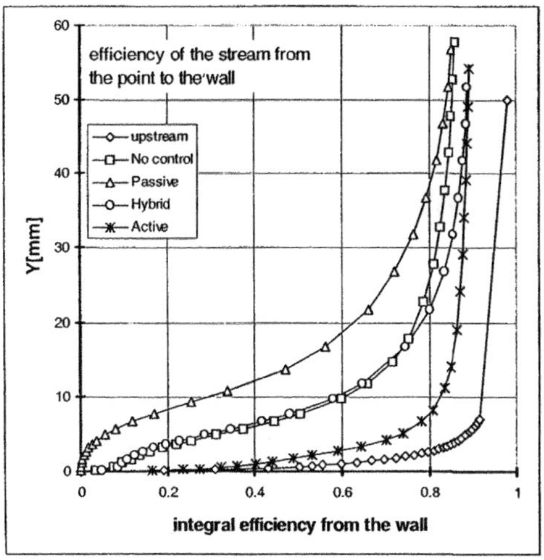

Figure 26 Integral efficiency of several control schemes determined downstream of the interaction/control region [3.2]

3.3.5 Control by discrete slot suction

Following the airfoil experiments of DERA within Task 3, Chapter 5, where it was found that, concerning drag reduction, discrete slot suction is one of the more viable mechanisms of shock and boundary layer control, it was decided to consider discrete slot suction also within the basic (channel) flow experiments. Accordingly, ONERA studied the effect slot suction in 2D flow with emphasis on the effect of slot geometry [3.1], while Cambridge University in their swept-shock experiments placed emphasis on the effect of the slot location on control effectiveness [3.3]; the suction rate was varied in both investigations.

ONERA 2D studies

The experimental set-up has been introduced in Figure 9. The slot, 1.5 initial boundary layer thicknesses (5 mm) wide, was located about one boundary layer thickness downstream of the (fixed) shock location. Geometric variables included the geometry of the slot leading edge, i.e., blunt or sharp, and the inclination of the slot, viz., 90° and 60°, respectively, relative to the flow direction.

The purpose of slot suction is, of course, to render the boundary layer more resistant to the detrimental effects arising from the interaction with the shock wave and — in the case of airfoil and wing flow — to the subsequent sustained rear adverse pressure gradients. Suction has, at the slot location considered here, the direct effect of eliminating or at least strongly reducing the extent of shock-induced separation, and the associated λ-shock region, thus increasing the shock strength and correspondingly wave drag. This is demonstrated in Figure 27 by Schlieren photographs and, in a more quantitative way, by the Mach number contour lines, derived from the LDV flow-field measurements, in Figure 28; both indicate the extreme thinning of the boundary layer despite the increase in shock strength. For the Mach number contours corresponding to the no-control case, needed for comparison, please consult Figure 21a.

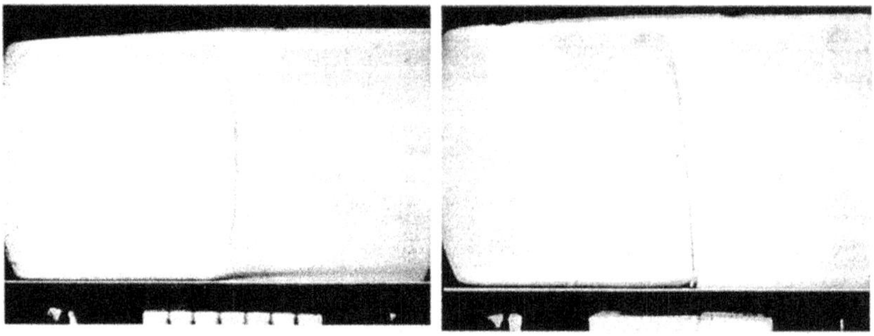

a. No-control reference case   b. Discrete slot suction, $C'_Q = 8.5\times10^{-4}$

Figure 27 Schlieren photographs of the flow field in the interaction region for the solid surface reference case and for control by discrete slot suction [3.1]

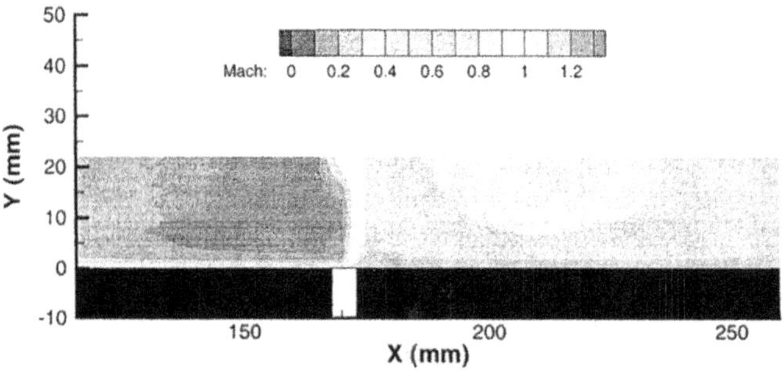

(For the solid-surface reference case see Figure 21a)

Figure 28 Mach number contours for control by discrete slot suction; $C'_Q = 8.5\times10^{-4}$ [3.1]

The effectiveness of discrete slot suction on the boundary layer development and, consequently, on viscous drag can, as before, best be judged by examining the development of relevant boundary layer thickness parameters, here the displacement and momentum thicknesses, in and downstream of the interaction region, Figure 29: suction reduces both boundary layer parameters well below the level of the no-control case. This means, of course, that viscous drag — if discrete suction is applied to airfoil or wing flow — will be considerably reduced, while wave drag is, as shown above, increased. It should be noted here that the geometry variations considered, viz., the rounding of the slot leading edge and the inclination of the slot, had only a minor effect on control effectiveness [3.1].

Considering the turbulence behavior, it was found that the turbulence level downstream of the interaction decreases below the level of the no-control case with increasing suction mass-flow rate, similar to the case of hybrid control at extreme suction rates.

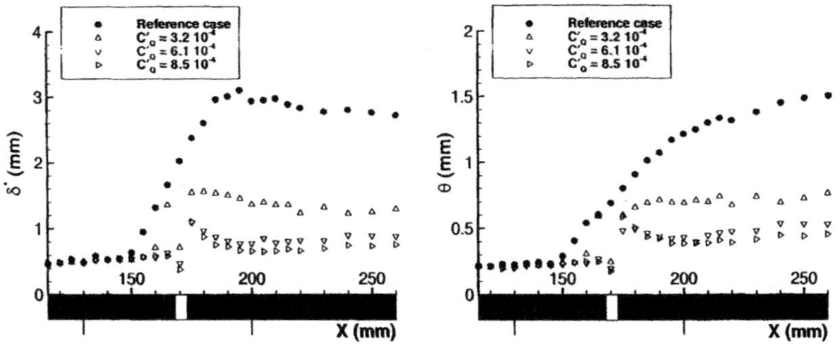

a. Displacement thickness

b. Momentum thickness

Figure 29 Boundary layer thickness characteristics without and with slot suction [3.1]

It should be noted that ONERA, in supplementary tests, also considered slots located further upstream (see [3.5] and Chapter 11), however, we shall study the effect of slot location here by referring to the Cambridge investigations.

Studies of Cambridge University

The experimental set-up for the investigation of the effect of discrete slot suction on swept-shock boundary layer interactions, including the slot locations studied, was introduced in Chapter 3.1, Figure 11. The slot width corresponded to one initial boundary layer thickness ($\approx$ 7 mm). The suction rate was varied between 0 g/s and 13 g/s.

Turning first to the surface pressure distributions, Figure 30, it should be noted that the slots at the locations considered here, viz., underneath and downstream of the (inviscid) shock, cause a severe disturbance of the downstream pressure distributions even without suction applied. This seems to indicate that, even without active suction applied, the suction system is possibly being filled during the initial phase of a test. In that case, no effect on the results for active suction is to be expected.

The effect of active suction on the pressure distributions is dependent on slot location: upstream suction has hardly any influence (plot not shown), while suction at the foot of the shock may, dependent on the suction rate, considerably reduce the pressure upstream of the shock and steepen the subsequent pressure gradient, similar to the 2D case, indicating that wave drag may be considerably increased. Downstream suction causes, at the lower suction rates, a further increase in pressure above the initial no-control pressure rise, while at higher suction rates these pressures are reduced following more the no-control distribution. The pressure gradients in the shock region are less severe than the ones for suction at the foot of the shock or even the no-control case.

For all slot locations considered, the boundary layer profiles downstream of the interaction/control region became fuller as the suction level was increased, similar to active control by the perforated plate/cavity arrangement (Figure 14).

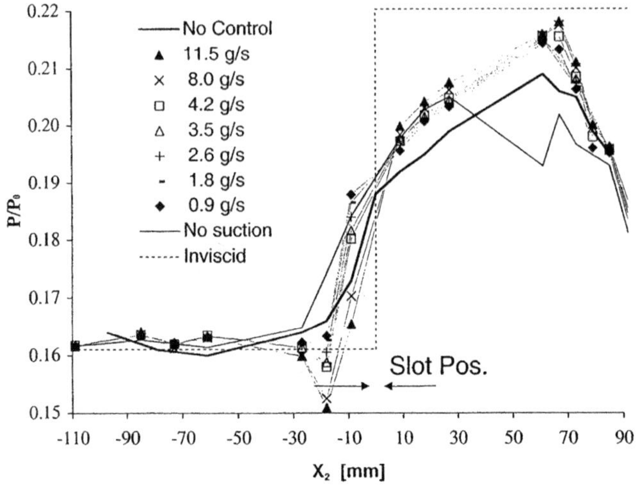

a. Central slot position

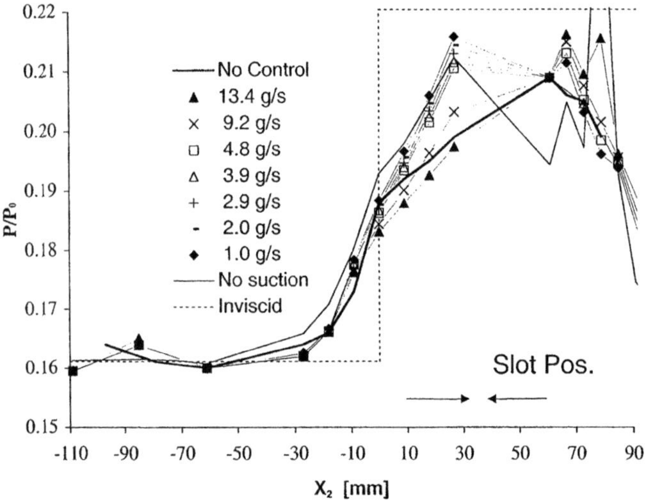

b. Downstream slot position

Figure 30 Surface pressure distributions for slot suction dependent on slot location and suction level [3.3]

The boundary layer displacement thickness downstream of the interaction/control region as essential parameter for the contribution of viscous drag to total drag in the case of airfoils and wings, is, for *all* control arrangements studied by Cambridge University, plotted in Figure 31 dependent on the suction mass flux: in all instances is the displacement thickness downstream of the interaction region reduced by suction. The highest δ^*-level is associated with active cavity control by part-suction due to the initial zero-mass-flux passive cavity effect. The slot-suction performance in reducing the displacement thickness is nearly independent of the slot location with the differences in the δ^*-level mainly being due to disturbances caused by the slots themselves. Normalizing the displacement thickness by the initial zero-mass-flux value, Figure 31b, shows, however, the downstream slot location to be somewhat less effective in reducing displacement thickness, while suction through the porous surface has the largest effect. The latter is thought to be due to the fact that suction through the porous surface not only improves the boundary layer condition directly — as does slots suction — but also reduces the detrimental effect of blowing from the cavity upstream of the shock.

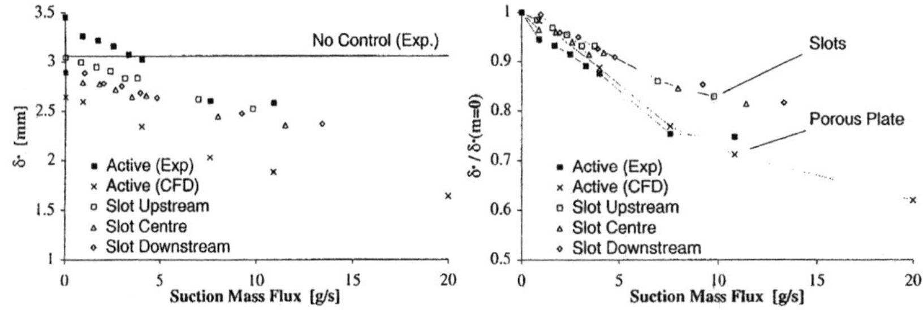

a. Displacement thickness b. Normalized displacement thickness

Figure 31 Displacement thickness downstream of the interaction/control region for various control mechanisms [3.3]

3.3.6 Control by a contour bump in the shock region

As in the case of slot suction, computations and airfoil tests within Tasks 2 and 3, following investigations of [3.6], revealed that a contour bump in the shock region was most effective in reducing wave drag. Bump contours were, therefore, also investigated in supplements to the ONERA channel flow experiments [3.5] in order to establish the details of the local flow development associated with bump control. Originally, a bump for shock control was selected since it constitutes a shape similar to the "effective" contour generated by the passive cavity/perforated-plate arrangement.

The bump investigated here had a length of 80 mm — corresponding to 20% chord in case of an airfoil — extending in the cannel, Figure 9a, from X = 130 mm to 210 mm, i.e., essentially covering the area previously taken up by the cavity arrangements. The bump was asymmetrical with the highest point of the bump (crest) being located at X = 186 mm, i.e., at 70% of the bump length, which was derived at in the bump optimization process described in Chapters 4.3.2 and 6.2.2. Three shock locations with respect to the bump were investigated, viz., X = 165, 186, and 210 mm, corresponding to 44%, 70%, and 100% of the bump length.

Representative results for the effect of the bump on the local flow development, here for the (optimum) reference shock position of 44% of the bump length, are presented in Figure 32: the Schlieren photograph, Figure 32a (for the no-control reference condition see Figure 20a), indicates that the bump produces a λ-shock structure as in the case of hybrid control, Figure 20 b, replacing the strong shock by an initial oblique shock, an isentropic compression and a weak second (normal) shock. The displacement distribution in the bump region, Figure 32b, indicates ① the smearing of the shock and ② a displacement thickness downstream of the interaction/control region that is only marginally higher than the no-control value — similar to the hybrid control case at high suction rates but without power addition — indicating the drag-reduction potential of the bump.

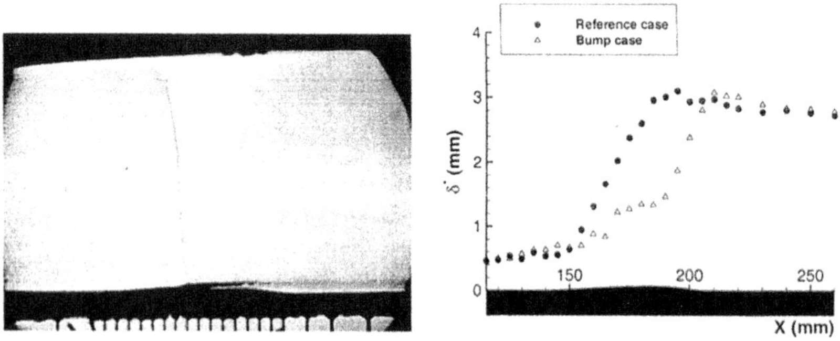

a. Schlieren photograph of bump region   b. Displacement thickness distribution

Figure 32 Effect of a contour bump on shock boundary layer interaction [3.5]

3.4 Conclusions and Future Work

Basic channel-flow experiments of shock and shock boundary layer interaction control at typical transonic shock-upstream Mach numbers were performed by ONERA and the University of Karlsruhe for two-dimensional interactions and by Cambridge University for three-dimensional swept-shock interactions. The basic area of investigation ranged from just upstream of the influence sphere of the shock to sufficiently downstream of the interaction/control region to judge the subsequent boundary layer development. We have hence considered local flow developments as affected by shock and boundary layer control which, of course, strongly shape the global developments associated with airfoil and wing flow. The control mechanisms studied were:

- Active control by means of a perforated plate / cavity arrangement with part-suction including the "zero-suction" passive control case.
- Hybrid control, consisting of a passive cavity in the shock region and suction, either through a cavity covered by a perforated plate or a slot, downstream.
- Discrete slot suction at various locations with respect to the shock.
- Control by a local contour bump.

The results of the basic experiments have shown, for two- as well as for three-dimensional interactions, that by *active single-cavity control* with part suction the boundary layer thickness parameters downstream of the interaction/control region can be considerably reduced — compared to the passive control case — and that, at moderate to large suction rates, the no-control boundary layer parameters downstream of the interaction can be reestablished while some shock-spreading still remains. This means that wave drag may be reduced without increasing viscous drag. *Hybrid control* has the same integral efficiency; however, it seems that single-cavity part-suction generates, at the same suction rate, a thinner boundary layer downstream of the control region. This is attributed to the fact that the boundary layer thickness is reduced — as in the case of hybrid control — while at the same time the detrimental effect of blowing from the cavity upstream of the shock is weakened.

Slot suction considerably reduces the boundary layer thickness parameters downstream of the interaction/control region. It seems, therefore, a viable tool for reducing viscous drag in the case of airfoil and wing flow; however, for slot locations upstream of and at the foot of the shock, wave drag may be strongly increased and care must be taken in determining the amount of suction — also considering pump drag — that still provides a total drag reduction. The rear slot location is somewhat less effective in reducing boundary layer thickness due to the thicker boundary layer approaching the slot, but has, on the other hand, the advantage that wave drag is somewhat reduced due to a certain degree of shock smearing by the thicker boundary layer.

A *contour bump* in the shock region, if placed correctly with respect to the shock, is likely to reduce wave drag by shock smearing without increasing viscous drag and, of course, without needing additional energy input as in the case of suction.

Concerning the numerical simulations accompanying the computations by Reynolds Averaged Navier-Stokes (RANS) codes, employing various turbulence models, generally predicted the effect of control quite well. However, details of the pressure distributions and absolute levels of the global boundary layer parameters were not always satisfactorily determined. This was considered to be due to, respectively, an insufficient grid resolution in the control region and deficiencies in the turbulence models employed, an issue still unresolved. A new control law has been established and introduced by Bohning and Doerffer which takes the presence of an outer transonic tangential stream into account and allows a more accurate determination of the vertical velocity in the control region in the case of passive or active ventilation.

Future work is envisaged to include the investigation of a wider range of boundary layer and flow control mechanisms at both, low- and high-speed flow conditions, including, e.g., sub-boundary-layer devices, such as vortex generators, mass-less air jets, reversed-flow flaps, and streamwise slots, without, however, neglecting the mechanisms investigated up to now where many unresolved issues still exist. Especially a combination of control mechanisms is worth considering. Emphasis must again be placed on the understanding of the physics of the controlled flow and on the establishment of the proper boundary conditions. Especially turbulence remains an issue where continued research is urgently needed.

4 Numerical Simulation of Airfoil and Wing Flow with Control (Task 2)

The main objective of the work discussed here was to extend, where needed, available steady and unsteady two-dimensional and steady three-dimensional computational methods to predict airfoil and wing flows with shock and boundary layer control, to assess these methods, and to perform first parametric studies to determine the effectiveness of control schemes in reducing drag and improving the drag-rise and buffet boundaries. The control mechanisms considered included local contour bumps in the shock region, active shock control by a perforated plate/cavity arrangement, hybrid control consisting of a passive cavity in the shock region coupled with active suction downstream, and discrete slot suction, pure and in combination with a contour bump. The basic computational methods employed included coupled Viscous/Inviscid Interaction (VII) procedures and Navier-Stokes solvers.

The partners involved were **CIRA** Transport Aircraft Aerodynamics Division, the Institute of Aerodynamics and Fluid Mechanics of **DLR**, **INTA** Fluid Dynamics Department, the Computational Fluid Dynamics and Aeroacoustics Department of **ONERA**, and the **University of Naples** Aerospace Design Department; the Aeronautics Division of **ALENIA** and **EADS Airbus-D** (DASA-Airbus) participated here in order to validate the computational methods to be used for their control application assessment to be discussed in Chapter 6.

4.1 Numerical Methods

Most of the basic numerical methods used here were already extended and utilized to treat flows with shock control by passive/active cavity ventilation within the EUROSHOCK (I) project [1.5]. Since they are described in detail in the individual contributions to [1.5], only a brief account of the basic methods will be given here. During the present project, the codes were generally improved by introducing, for instance, the new control law of Bohning and Doerffer, described in

Chapter 3.3.1 [3.2], and by grid refinements in the control region whose necessity became evident also during the course of the present computations. Other common issues treated concerned the prediction of the various components comprising total drag, including pump-drag in the case of suction, specifics of the numerical simulation of discrete slot suction, especially minimum grid resolution in the slot region, and the establishment of effective freestream conditions, particularly in relation to the present experiments. Some of these issues are addressed in Chapter 4.1.3 and, when relevant to the understanding of the results, in the corresponding discussion.

4.1.1 Basic numerical methods

The basic numerical methods employed to treat airfoil and wing flow with control are briefly described below. More details are given in the respective chapters of Part B, Individual Contributions, and the literature as indicated.

ALENIA *ALN Navier-Stokes Code [4.1]:* Solved are the Reynolds-averaged full Navier-Stokes equations for 2D steady/unsteady flow with a finite-difference technique. Centered space discretization is used with added non-linear second- and fourth-order damping. A two-equation, fully point-wise k-Rt turbulence model is implemented. Control laws were not introduced since only control by a contour bump was considered. Computations regarding the application of bump control to a regional-jet aircraft were performed with a 3D coupled full potential/ boundary layer code.

CIRA *EUBL2D Code and Qusi-3D-Extension [4.2, 4.3]:* The code is based on the semi-inverse coupling of the Euler- and the integral boundary layer equations. The former are solved using a standard cell-centered finite-volume discretization technique. The calculation of the laminar boundary layer is based on the method of Cohen-Reshotko, the turbulent boundary layer calculation on the direct and inverse formulation, respectively, of Green's integral method. Lighthill's surface source model is used for the viscous/inviscid interaction. The extension to wall transpiration has been accomplished by adding a natural or a forced ventilation term to the contribution of the body thickening. The method has been extended to treat infinitely swept-wing configurations by de-coupling the streamwise flow and the cross flow, employing 2D Euler computations in the direction normal to the leading edge and boundary layer calculations along 3D streamlines.

EADS Airbus-D *VII Code [4.4, 4.5]:* The method, applicable to 2D and infinite-swept-wing flow, is based on the inviscid/viscous coupling of a 2D full-potential flow solver with an interactive 3D boundary-layer finite difference method. Using LeBalleur's defect formulation concept, the real viscous flow is split into a viscous defect flow and an equivalent inviscid flow extending to the airfoil/wing surface and the wake streamline as the location of the viscous boundary condition for the inviscid outer flow. The inviscid flow is solved in a transformed domain; a shock operator performs the entropy correction in order to satisfy the Hugoniot-Rankine condition. The viscous solution is obtained by solving a stream function formulation of the compressible boundary layer equations with an algebraic eddy-viscosity turbulence model based on the Cebeci-Smith formulation. The introduction of the wall mass-flow transfer was accomplished by a slight modification of the wall boundary condition in the inviscid and viscous flow solvers.

***DLR/INTA** 2D Time-Accurate Navier-Stokes Code [4.6, 4.7, 4.8]:* The (DLR) code is based on the Beam/Warming approximate factorization implicit methodology using central differences in the space coordinates. Due to the latter, numerical damping terms had to be added to avoid numerical instabilities. To represent turbulent flow, the Baldwin-Lomax algebraic turbulence model is generally being used; however, two further turbulence models have been introduced, viz., the Johnson-King model and the Spalart-Allmaras model, respectively. Provisions to treat the control mechanisms considered here have been implemented by INTA. The code can be applied to steady conditions, i.e., below the buffet boundary, and will automatically provide unsteady oscillatory results as soon as the buffet process commences; it can therefore be employed to directly determine the buffet boundary.

***ONERA** VIS Codes [4.9, 4.10]:* The Viscous-Inviscid Interaction (VII) codes, namely the steady code VIS05c and the unsteady code VIS15, have been utilized to compute airfoil flow with shock control. Both codes have a common methodology and similar coding, including adaptive grids. The approach is based on LeBalleur's defect-formulation theory for the full Navier-Stokes equations that replaces the single-field Navier-Stokes domain by a double viscous-inviscid-interaction field. The steady code VIS05c solves the full potential equation for the inviscid field. The viscous method is a hybrid-field/integral method solved in a marching thin-layer 2D numerical technique in direct/inverse modes. The time-consistent code VIS15 uses the same viscous methodology as the steady code VIS05c, however, for the outer inviscid flow, the unsteady Transonic Small Perturbation equation (TSP) is solved. A semi-implicit time-consistent coupling algorithm is used in the code. The code is able to distinguish between a steady and an unsteady solution and can, therefore, be used to determine the buffet boundary. For the computation of the quasi-3D infinite swept-wing flow, the unsteady code VIS25 has been employed which is a direct extension of the 2D code VIS15 adapted to sheared-wing conditions.

***University of Naples** VII Code [4.11]:* The inviscid flow field is computed by solving the Euler equations on a structured grid following the scheme of Schmidt, Jameson and Turkel. The method is based on a central space discretization with explicit adaptive artificial dissipation. Steady state calculations are performed by using a pseudo-transient approach. The unsteady Euler equations are integrated until steady state is reached by using an explicit multi-stage scheme with local time stepping. One of the features of the present solver is the possibility of handling complex configurations by using a multi-block structured approach which allows local grid refinement for improved accuracy at low costs. The viscous solver is based for the laminar part on the method of Cohen-Reshotko, for the turbulent part on a modification of Green's method, similar to the CIRA code.

4.1.2 Control laws and control-law simulation

All codes have adopted a common procedure for the treatment of control by cavity ventilation, based on developments during the EUROSHOCK (I) project where different control laws for the calculation of the transpiration velocity as function of the pressure difference over the perforated plate have been compared [1.5]. Since Poll's law was shown to provide the best results compared to experimental data, this law — besides the new law of Bohning and Doerffer at a later stage — has been adopted by the present codes for all computations involving control by suction/ventilation. In order to rule out any discrepancies possibly arising from the introduction of this law into the individual computer codes, a common (Task 2) exercise was set up to ensure the correct, or at least a compatible, implementation. The results of this exercise will briefly be discussed below.

Poll's law [3.4] is based on the experimental investigation of a large number of samples of laser-drilled titanium plates carried out to establish a relation between the mass-flow rate and the pressure drop through a porous surface. As a final result, this relation can be expressed by the equation

$$Y = \frac{1}{K}[40.7X + 1.95X^2]$$

with X proportional to the mass flux $\dot{m}$ and Y proportional to the pressure difference across the porous plate as follows:

$$X = \frac{\dot{m}}{\mu e} \quad \text{with} \quad \dot{m} = \rho\pi\frac{d^2}{4}v$$

$$Y = \frac{(p_c - p(x))d^2}{\rho v^2}\cdot\frac{d^2}{e^2}.$$

Here, d is the nominal determined hole diameter, e is the perforated-plate thickness, v is the velocity through the holes of the perforation and ρ, μ, ν are the density, the dynamic viscosity and the kinematic viscosity, respectively, in the hole; p_c is the cavity pressure and p(x) the pressure distribution along the perforated plate exposed to the outer flow. Since the laser drilling of a porous sheet does not produce perfect holes, it is necessary to specify an effective diameter, d_{eff} = K d, which must be determined for individual porous plates by calibration.

The common **control-law simulation** was carried out for prescribed airfoil pressure distributions measured on the DA LVA-1A laminar-type airfoil during EUROSHOCK (I) tests at a freestream Mach number of $M_\infty = 0.77$ and a Reynolds number of $Re_c = 4.64 \times 10^6$ [4.13, 4.14]. Furthermore prescribed were the cavity pressure coefficient C_{pcav} = -0.72 and the suction coefficients C_Q = 0 and –0.0007, respectively. Figure 33 shows as an example of the results of the simulation the mass-flow distribution over the control region determined by the various participating codes employing Poll's formula. One observes that the agreement in the mass-flow distribution is quite good. However, some differences occurred in the average transpiration velocities. These were attributed to the use of different approaches to the calculation of the density and viscosity within the holes of the perforation needed to determine the transpiration velocity rather than an incorrect implementation of the control law. This deficiency was remedied.

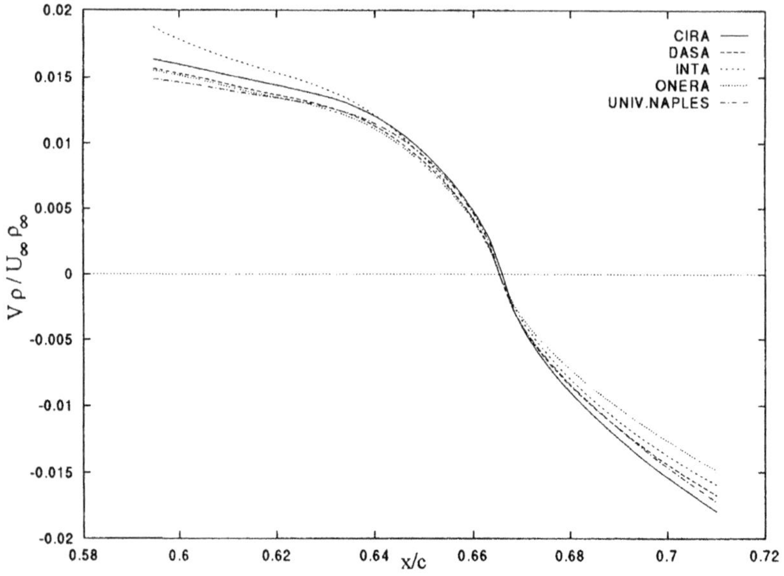

Figure 33 Comparison of mass flows computed in the shock region using Poll's formula for a prescribed pressure distribution

4.1.3 Drag determination and the simulation of control by suction

Generally two approaches have been used to determine total drag. One approach calculates drag as the sum of pressure drag and friction drag, obtained by integrating the pressure and skin friction distributions, respectively, along the airfoil surface, i.e.,

$$C_D = C_{DP} + C_{DF}.$$

This method has been used by CIRA, the University of Naples and by the partners using Navier-Stokes codes, namely, ALENIA, DLR and INTA. The second approach is based on determining wave drag from the entropy jump across the shock and viscous drag from the momentum thickness in the far wake, i.e.,

$$C_D = C_{DW} + C_{DV}.$$

This method, found to be less sensitive to numerical errors, has been used by EADS-Airbus-D and ONERA.

A further drag component, viz., excrescence drag due to the perforated surface, has been determined by comparison with experimental results and approximated by

$$C_{Dexcr} = 2.5\,[C_{DF}]_{S1/c}^{S2/c},$$

where $[C_{DF}]_{S1/c}^{S2/c}$ is the friction drag of the clean airfoil obtained by integrating skin friction over the control region. Finally, there is "pump" drag to be accounted for when suction is applied. Under certain assumptions, e.g., that duct losses are negligible and that there is a single plenum to receive the suction mass flow, pump drag can be expressed as

$$C_{Dpump} = C_{pcav}\, C_Q,$$

where C_Q is the total mass-flux and C_{pcav} the cavity pressure coefficient [3.6].

Another issue of common concern related to the numerical simulation of slot suction, and specifically to the effect of grid spacing in the slot region. Here, it was found that at least two grid points should be located within the slot region but about 7 points over a 1%-chord slot width was recommended for a sufficiently accurate simulation. Furthermore of concern were differences in the results for control by suction or ventilation accruing due to prescribing either cavity pressure or the suction mass-flux coefficient, or due to applying either Poll's law or the "non-linear" relation of Bohning and Doerffer, respectively. As mentioned above, these issues will be addressed, if relevant to the understanding of the results, during the ensuing analysis of CFD capabilities and the parametric studies of control concepts.

4.2 CFD Capabilities and Preliminary Control Concept Assessment

For the initial validation of the participating computational methods and to ensure common procedures, mandatory and optional test cases were selected for which, in part, experimental results were available. The test cases comprised the following configurations (also see Figure 7):

- The laminar-type airfoil DRA-2303 without and with control by a local contour bump, by discrete slot suction, and by the latter in conjunction with a contour bump. The DRA-2303 bump-configuration was not part of any EUROSHOCK project, but experimental results were made available by DERA [3.6].
- The turbulent airfoil RAE-5225 with control mechanisms as above; this airfoil was also not part of any EUROSHOCK project, but experimental results were again provided by DERA [3.6].
- The turbulent airfoil VA-2 with control by passive and active cavity ventilation; this configuration was investigated during the EUROSHOCK (I) project [4.15].

In the assessment of CFD capabilities we shall mainly discuss the DRA-2303 airfoil with control by a contour bump and with the latter in combination with discrete suction since ① these configurations are very sensitive to changes in the freestream conditions and to the predictive capabilities, and ② the airfoil was extensively investigated within the present project in conjunction with, besides discrete slot suction, active single-cavity control, and hybrid control thus allowing a comparison of a wide range of control mechanisms applied to the same airfoil. For further and more detailed results, especially concerning the other test cases considered, the reader is referred to the individual contributions in Chapters 14 to 18 and to the detailed account of the Task 2 activities in Annex B of [1.5].

DRA-2303 airfoil with contour bump

The **DRA-2303** airfoil is characterized by strongly accelerating flow on the suction side and a relatively fixed shock position (see Chapter 5). It was tested in the DERA 8Ft x 8 Ft wind tunnel without and with control by a surface contour bump in the shock region. Typical pressure distributions at the nominal condition $M_\infty = 0.68$, $Re_c = 19 \times 10^6$, $C_L = 0.747$ — selected for the purpose of validation — are presented in Figure 34 for the no-control datum airfoil (a) and for the airfoil with control by an asymmetric bump with a relative height of 0.25% chord consisting of two arcs placed between x/c = 0.485 and 0.685 with the junction of the arcs at x/c = 0.57 (b); for a typical bump arrangement on an airfoil, please turn briefly to Figures 7 and 56. Computed displacement thickness distributions corresponding to these conditions are shown in Figure 35.

In the case without control, Figure 34a, the computed pressure distributions, here generally obtained with a grid spacing of 0.5%-chord in the shock/control region, agree quite well with the experimental distributions except for the INTA results which show a more forward shock location associated with a stronger spreading of the shock — possibly due to a reduced grid resolution — and the distribution determined by ONERA that slightly under-predicts the absolute pressures on the upper surface which was attributed to the use of the transonic-small-perturbation method for the flow-field calculations.

Similar agreement is obtained in the case of the airfoil with bump, Figure 34b, except that a larger scatter between computations occurs in the shock region. Also noteworthy is that some codes predict a nearly isentropic compression while others show a pronounced plateau in the control region. Generally, small differences in the effective freestream conditions and in the predicted development of the boundary layer, especially in the shock region and downstream of the shock, Figure 35, resulting in differences in the location of the shock with respect to the contour bump, might be the reason for the discrepancies observed here. Nevertheless, all codes determine the spreading of the shock indicative of a reduction in wave drag.

The experimental and numerical pressure distributions shown in Figures 34 a and b, respectively, were obtained at slightly different freestream Mach numbers, viz., $M_\infty = 0.683$ (a) and 0.681 (b), which, of course, has no consequences for the comparison between experiment and computation. For the evaluation of the drag reduction potential of the bump, however, there is likely to be an influence due to the relatively strong dependence of drag on Mach number, so the computations with and without control have been repeated at the same Mach number and a grid spacing in the interaction/control region of 0.5% chord. The results are presented in Table 2.

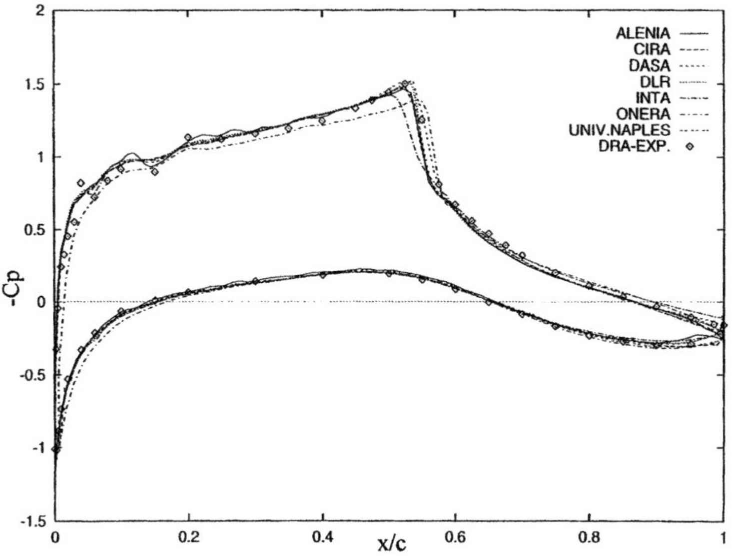

a. Airfoil DRA-2303, datum configuration, $M_\infty = 0.683$, $Re_c = 19x10^6$, $C_L = 0.74$

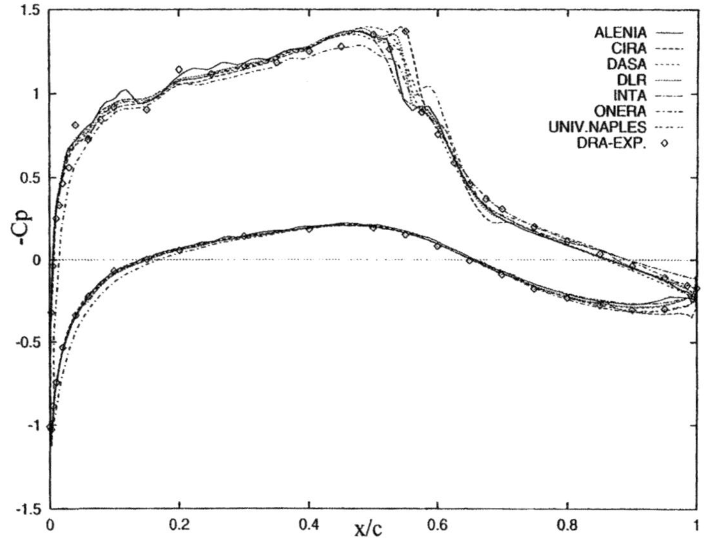

b. Airfoil DRA-2303 with bump control, $M_\infty = 0.681$, $Re_c = 19x10^6$, $C_L = 0.74$

Figure 34 Comparison of experimental and computational pressure distributions for the airfoil DRA-2303 with and without control by contour bump

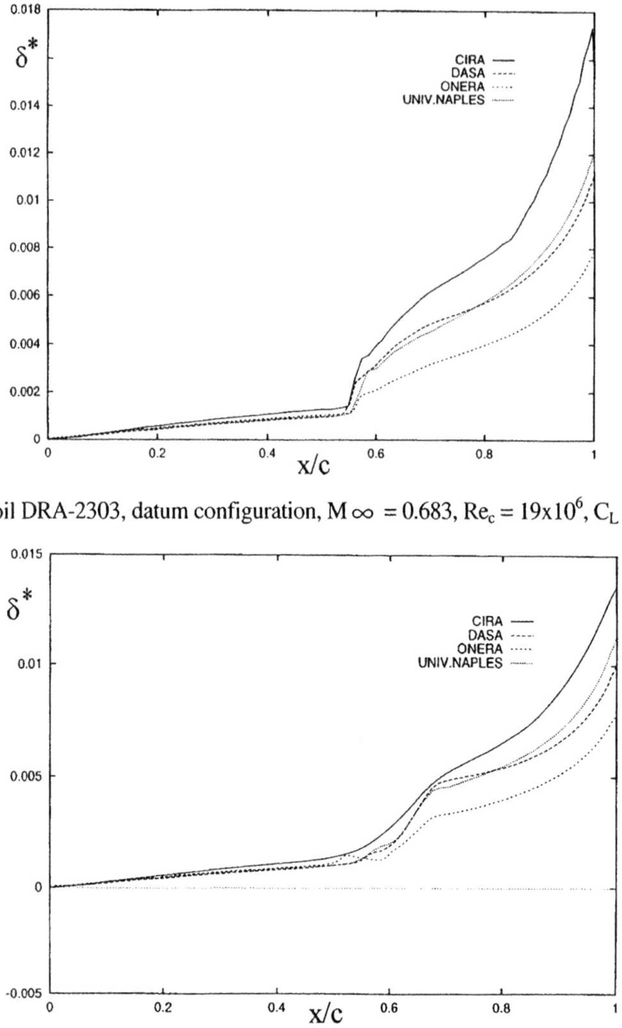

a. Airfoil DRA-2303, datum configuration, $M_\infty = 0.683$, $Re_c = 19\text{x}10^6$, $C_L = 0.74$

b. Airfoil DRA-2303 with bump control, $M_\infty = 0.681$, $Re_c = 19\text{x}10^6$, $C_L = 0.74$

Figure 35 Computational displacement-thickness distributions for DRA-2303 airfoil with and without control by contour bump

Referring to Table 2, at nearly identical freestream conditions for the datum airfoil and the airfoil with control, all codes predict a drag reductions due to the contour bump; there are, however, considerable differences in the predicted drag reductions as well as in the absolute drag levels with the former ranging from 12% in the case of ALENIA and the University of Naples to 6.6% for the INTA computations.

Differences in the predicted drag reductions are partly due to differences in the location of the bump relative to the computed shock locations, i.e., they result partly from differences in the wave drag reduction. However, as is indicated in Figure 35 by the development of the displacement thicknesses for the datum airfoil and the airfoil with control, viscous drag is generally also reduced. Since there are considerable deviations in the curves predicted by the various codes, which seem to originate in or even upstream of the shock region and then increase downstream due to the influence of the sustained rear adverse pressure gradients prevailing on the airfoil, one can assume that differences in viscous drag are also partly responsible for the observed differences in total drag and, correspondingly, in the drag reductions. This reflects a major drawback of the codes. The difference in the boundary layer development is, as already mentioned, also responsible for differences in the predicted shock locations.

Table 2 Aerodynamic coefficients for the DRA-2303 airfoil with and without control
$M\infty \approx 0.681$, $Re_c = 19x10^6$, $C_L \approx 0.74$

		$M\infty$	α *(deg)*	C_L	C_m	C_D	ΔC_D *(%)*
Experiment	Datum	0.6829	2.097	0.74721	-0.1006	0.01346	
	Bump	0.6801	2.098	0.73988	-0.0989	0.01138	-15.4
ALENIA	Datum	0.6800	2.500	0.73216	-0.0861	0.01323	
	Bump	0.6800	2.500	0.73123	-0.0906	0.01165	-11.9
CIRA	Datum	0.6801	2.228	0.73978	-0.0897	0.01133	
	Bump	0.6801	1.898	0.73991	-0.1024	0.01017	-10.1
EADS-	Datum	0.6800	2.050	0.74200	n.a.	0.01131	
AIRBUS-D	Bump	0.6800	1.990	0.74100	n.a.	0.01029	-9.0
DLR	Datum	0.6800	2.250	0.75130	-0.0919	0.01226	
	Bump	0.6800	2.150	0.74540	-0.0937	0.01113	-9.2
INTA	Datum	0.6800	2.000	0.7343	-0.1035	0.0135	
	Bump	0.6800	2.000	0.7443	-0.1054	0.0126	-6.6
ONERA	Datum	0.6873	1.998	0.7467	-0.1010	0.01173	
	Bump	0.6845	1.998	0.7393	-0.1005	0.01069	-8.8
U. Naples	Datum	0.6800	2.100	0.7397	-0.0939	0.01079	
	Bump	0.6800	2.070	0.7393	-0.0943	0.00950	-11.9

The experimentally predicted drag reduction is with 15.4% higher than the computed ones, however, this might be due, at least in part, to the difference in the freestream Mach number with and without control. Differences in the measured and computed drag coefficients in the case with control at nominally the same Mach number ranged from ± 2% to ± 10% with the scatter in the computational results due to the reasons outlined above.

The second airfoil considered was the turbulent airfoil RAE-5225. The results of the comparison between experiment and computations and between the individual computations are quite similar to the ones observed above. Here too, the contour bump resulted, at $M_\infty \approx 0.73$, $Re_c = 19 \times 10^6$, $C_L \approx 0.756$, in a reduction in drag which varied in the computations between 2% and 9% with the majority, however, being closer to the latter. The experimentally determined reduction was to 13.7%. It should be noted that the lower effectiveness of the bump in reducing drag is here mainly due to the weaker shocks prevailing on this airfoil.

DRA-2303 airfoil with contour bump and slot suction

Further test-case configurations concerned the airfoil DRA-2303 with discrete slot suction upstream of the shock and the latter in conjunction with a contour bump in the shock region as sketched in Figure 7. For pure slot suction at the nominal condition $M_\infty = 0.680$, $Re_c = 19 \times 10^6$, $C_L = 0.74$, $C_Q = -0.00007$, drag reductions ranging from about 2% to 7% were predicted; the corresponding experimentally determined drag reduction was 7.5%, indicating both that slot suction seems less effective in reducing drag than the contour bump.

The more interesting control scheme — in conjunction with the DRA-2303 airfoil within EUROSHOCK II only treated numerically — is the control by a combination of a bump contour in the shock region and slot suction upstream of the bump. Computations were again carried out for the nominal conditions $M_\infty = 0.680$, $Re_c = 19 \times 10^6$, $C_L = 0.74$. The slot was located here between x/c = 0.475 and 0.485 (1%-chord slot width), the suction coefficient was $C_Q = -0.0003$. The bump had the same location and shape as described above.

Considering first the global aerodynamic coefficients, Table 3, one observes that an increase in drag reduction due to the addition of slot suction above the gain achieved by the bump alone (compare Table 2) is predicted by EADS-Airbus D and the University of Naples, while the calculations of CIRA, INTA, and ONERA show a negative effect with the total drag reduction determined by ONERA only being 2%, although at a different but consistent Mach number.

Table 3 Aerodynamic coefficients for the DRA-2303 airfoil with bump and slot suction $M = 0.681$, $Re_c = 19 \times 10^6$, $C_L \approx 0.74$, $C_Q = -0.0003$

	α 1) (deg)	C_L w/o control	C_L w control	C_D w/o control	C_D w control	ΔC_D%
CIRA	2.196	0.73978	0.74015	0.011329	0.010602	-6.42
EADS-A.	1.990	0.74180	0.74104	0.01121	0.00991	-11.6
INTA	2.300	0.74843	0.77630	0.01220	0.01147	-5.98
ONERA 2)	1.964	0.74050	0.74022	0.01127	0.01103	-2.13
U.NAPLES	2.015	0.74697	0.73982	0.01202	0.00948	-12.14

1) Average between test cases with and w/o control 2) $M_\infty = 0.687$

The differences in the drag reductions shown in Table 3 are, of course, due to differences in the drag components as affected by control. It seems, therefore, worthwhile to analyze, for representative computations, the effect of control on these components more closely, Table 4 (also see the specification of the drag components in Chapter 4.1.3). The subject was already repeatedly addressed when considering the boundary layer development within and just downstream of the control region in the basic experiments.

Table 4 Effect of control on drag components for the DRA-2303 airfoil
$M = 0.681$, $Re_c = 19x10^6$, $C_L \approx 0.74$, $C_Q = -0.0003$; grid spacing 0.5%

(1)	C_{DP} / C_{DW} w/o contr.	C_{DF} / C_{DV} w/o contr.	C_{DP} / C_{DW} bump	C_{DF} / C_{DV} bump	C_{DP} / C_{DW} bump+suc	C_{DF} / C_{DV} bump+suc
CIRA	0.00715 (p)	0.00418 (f)	0.00620 (p)	0.00400 (f)	0.00595 (p)	0.00465 (f)
EADS-Airbus	0.00202 (w)	0.00919 (v)	0.00120 (w)	0.00900 (v)	0.00122 (w)	0.00869 (v)
ONERA (2)	0.00227 (w)	0.00901 (v)	0.00172 (w)	0.00915 (v)	0.00244 (w)	0.00848 (v)

1) Indices: p = pressure drag, w = wave drag, f = friction drag, v = viscous drag
2) 2) $M\infty = 0.687$

Table 4 indicates that, when applying pure bump control, a strong reduction in wave drag, associated with the spreading of the shock, is predicted by the EADS-Airbus D and ONERA computations. Viscous drag is in the case of the former also reduced, while ONERA predicts an increase in viscous drag, which might — similar to the lesser decrease in wave drag — be due to the higher freestream Mach number. In the EADS-Airbus D computations, applying suction increases wave drag somewhat due to the thinner boundary layer — the bump is, however, still very effective — but viscous drag is, as expected, further reduced resulting in the 11.6% reduction in total drag indicated in Table 3. In the case of the ONERA computations, adding suction raises wave drag above its no-control level while viscous drag is noticeably reduced still leading, in spite of the increase in wave drag, to a total drag reduction of 2.13%. The physical mechanism behind the ineffectiveness of the bump seems here mainly a mismatch between shock location and bump position as a result of adding suction which seems to lead to a strong double-shock system associated with higher wave drag (see Figure 36).

The CIRA computations show that pressure drag is considerably reduced by the bump, accounting for the main contribution to drag reduction, with an additional small reduction in friction drag. Adding suction, further decreases pressure drag, while friction drag is, however, considerably increased due to the fuller boundary layer profiles associated with suction which is here the dominating effect in reducing control benefits compared to pure bump control.

Computed pressure distributions for the case of control by the combination of bump and discrete suction are compared in Figure 36; the agreement between codes is essentially as described for the bump-alone control case considered above (Figure 34b). Note that the spikes in the pressure distributions upstream of the shock are caused by the suction slot modeling.

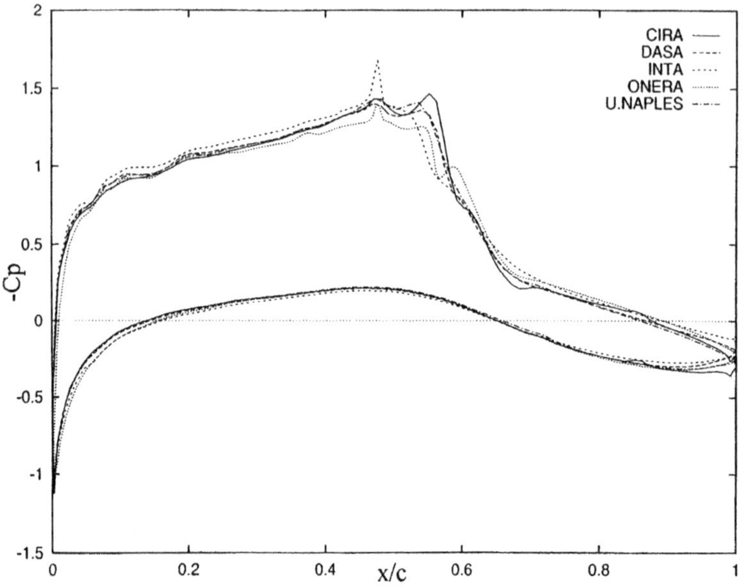

Figure 36 Pressure distributions for the airfoil DRA-2303 with contour bump plus suction, $M_\infty = 0.681$ (0.687 for ONERA), $Re_c = 19 \times 10^6$, $C_L = 0.74$, $C_Q = -0.0003$

Resume: The test cases considered placed relatively high demands on computational codes, procedures, and the definition of boundary conditions since the control by bump is very sensitive to changes in the freestream conditions, especially to the associated changes in shock location with respect to the bump position. This situation is even more severe when suction is applied in conjunction with the bump since suction will change flow conditions at the trailing edge of an airfoil or wing, hence circulation, which will, in turn, change the shock location. For comparisons between codes it is, for these reasons, extremely important that the effective freestream conditions, the grid resolution and procedures related to control closely match; similar considerations hold, of course, also for comparisons between computation and experiment.

The code-assessment exercise led to several improvements in the computational procedures, in grid resolution, and in the definition of the boundary conditions. Computations were, in part, repeated and agreement improved. All codes were, thereafter, able to predict the effect of control qualitatively quite correct, although the absolute levels, especially concerning drag, still deviated and further improvements, for instance, in turbulence modeling for flows with shock waves and/or separation, are needed. Concerning the present exercise, a very detailed account of the comparison between codes for all test cases is given in Annex B of [1.5].

4.3 Parametric Study of Shock and Boundary Layer Control Concepts

Since the comparison of the results generated by the various computational methods, described above, allows some judgement of the capabilities of the codes and needed improvements, the following discussion will primarily concentrate on the assessment of control mechanisms and gains in the design and off-design performance of airfoils and wings possible due to control. We shall, therefore, also not consider all results obtained for a specific control scheme by all codes involved but rather select results best suited for an analysis, trying, nevertheless, to strike a fair balance between the individual contributions sometimes also duplicating evidence in support of certain conclusions.

4.3.1 Active control by ventilation and suction schemes

Control schemes considered here comprise discrete slot suction, single cavity ventilation with part-suction, and hybrid control consisting of passive cavity ventilation in the shock region and suction downstream of the passive cavity. Furthermore briefly considered will be distributed blowing representing a "pneumatic bump".

4.3.1.1 Discrete slot suction

In the experiments with the DRA-2303 airfoil it was, as already mentioned, found that discrete slot suction is one of the more viable tools if drag reduction is the main driver [5.1]. A rather extensive computational study was, therefore, carried out especially with respect to possible improvements in cruise drag and in the buffet boundary with variables being — besides the freestream conditions — the suction mass flux and the slot location.

In our discussion, we will first consider a complete drag polar for the DRA-2303 airfoil without and with discrete suction applied upstream of the shock region, Figure 37. The computations were carried out by ONERA with the unsteady code VIS15 at a Mach number slightly higher than the experimental one ($M_{\infty comp} = 0.69$; $M_{\infty exp} = 0.68$), using a fine grid in the shock region and about 7 grid points covering the slot, located here between x/c = 0.45 and 0.46 [4.10]. A suction coefficient of $C_Q = -0.00007$, considered reasonable for aircraft installation, was applied; Poll's law was employed to determine the normal velocity in the slot area. In the computations, the angle of attack was progressively increased, starting at $\alpha = -1.8°$, up to about $\alpha = 4.0°$.

Figure 37 shows that the overall drag reduction, present in the entire incidence range investigated, i.e., also at sub-critical conditions, is mainly due to a reduction in viscous drag, while wave drag is, at $C_L > 0.60$, actually increased due to suction confirming observations discussed above and in Chapter 3.3.5, e.g., Figure 29. An effect of discrete suction on the angle of attack for buffet onset was not observed at the present mass-flow rate, however, maximum lift is slightly increased due to suction.

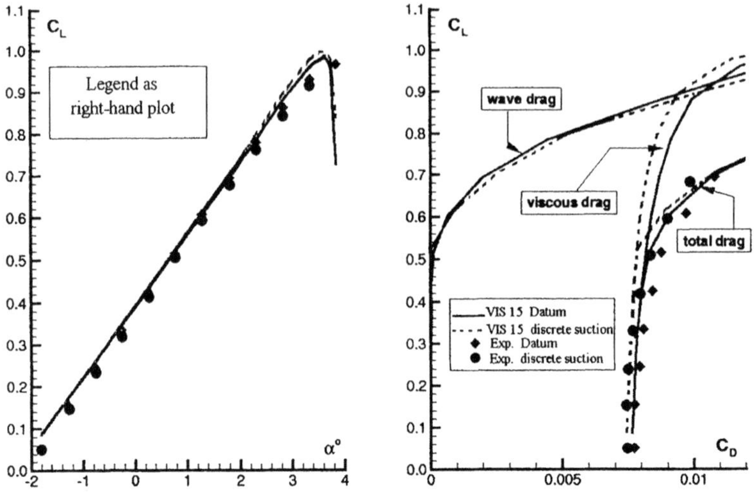

Figure 37 Influence of discrete suction, DRA-2303 airfoil, $M_\infty = 0.69$, $Re_c = 19 \times 10^6$ $C_Q = -0.00007$ [4.10, 5.1]

As already evident from the basic experiments and also indicated in the preceding chapter, shock-upstream suction or suction at the foot of the shock reduces the thickness of the boundary layer entering the shock region or interacting with the shock thereby reducing the shock spread and increasing wave drag; however, the boundary layer leaving the shock region is still thinner than the one without control (see, e.g., Figures 28 to 31). In the case of airfoils and wings this initial difference, say, in the displacement thicknesses may be enlarged by the rear adverse pressure gradients so that an even thinner boundary layer reaches the trailing edge in the case of control resulting in the dominating viscous drag reduction. This development is demonstrated by the pressure and displacement and momentum thickness distributions, determined by EADS-Airbus D at a lift coefficient of $C_L = 0.608$, Figure 38, and by the corresponding total drag reduction, which amounts, as in the ONERA computations, to about 4.4% [4.5]. We have already seen in the preceding chapter that the increased wave drag due to suction may be softened by a correctly positioned contour bump.

The effect of slot location on the efficiency of slot suction was studied by ONERA [4.10] for the freestream parameters and the suction rate corresponding to Figure 37 at a constant lift coefficient of $C_L = 0.6075$, a condition where wave drag starts, as indicated in Figure 37, to be negatively affected by suction. The pressure distributions corresponding to this lift coefficient (and the upstream slot location) are depicted in Figure 39 which also shows an example of the suction velocity distribution determined by the law of Bohning and Doerffer prescribing the suction coefficient and the cavity pressure, respectively. The figure also allows to relate the suction slot positions to be considered below to the relevant pressure distribution.

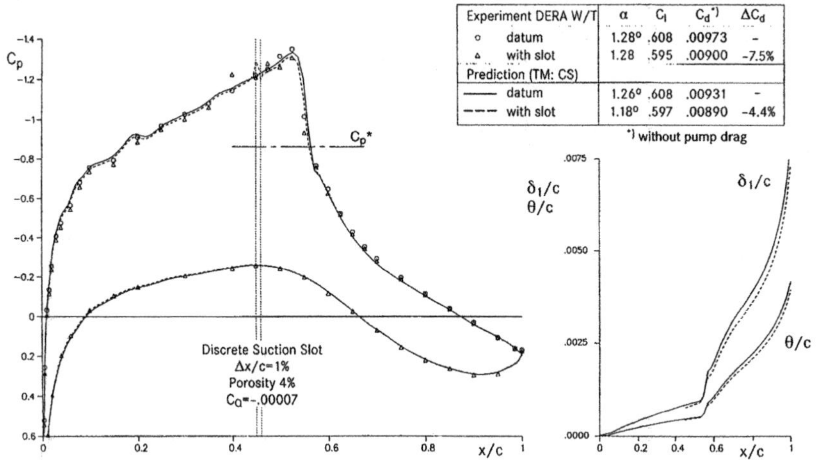

Figure 39 Influence of discrete suction on pressure distribution and boundary layer development, DRA-2303 airfoil, $M_\infty = 0.68$, $Re_c = 19 \times 10^6$, $C_Q = -0.00007$ [4.5, 5.1]

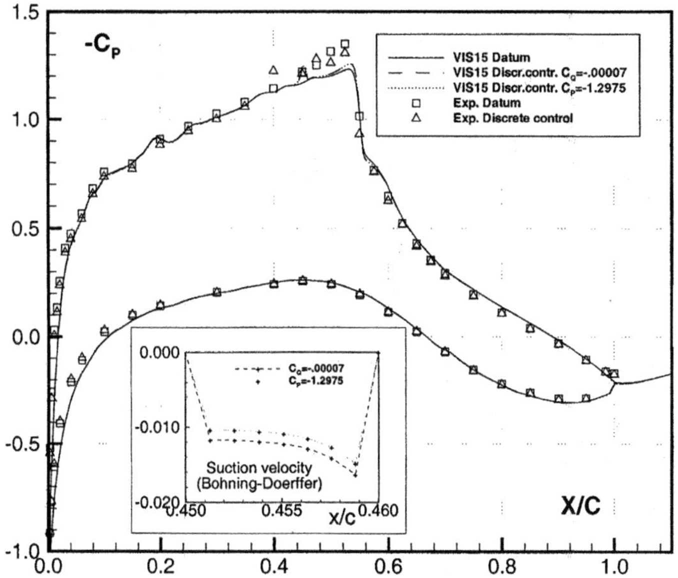

Figure 39 Influence of discrete suction on the pressure distribution for the airfoil DRA-2303, $M_\infty = 0.69$, $Re_c = 19 \times 10^6$, $C_L = 0.6075$, $C_Q = -0.00007$ [4.10, 5.1]

Referring again first to the basic experiments, it was found that rear slot suction was less effective in reducing displacement thickness while central and upstream suction exhibited about the same effectiveness, Figure 31. This is also reflected in the ONERA computations for slot locations varying between x/c = 0.45 and 0.66, Figure 40 [4.10]: the drag reduction is highest for the most-upstream slot location with the reduction amounting to about 4%. When the slot is moved closer to the interaction region, wave drag increases due to the thinner boundary layer and reduces the gain in total drag. When locating the slot downstream of the shock, wave drag again decreases but the increase in boundary layer thickness due to the interaction with the shock and the reduced effectiveness of slot suction almost halve the initial drag reduction.

The second important airfoil performance parameter to be considered is the buffet boundary, here determined by increasing the angle of attack until the time-accurate code employed indicated the onset of unsteady conditions, Figure 40b [4.10]. For the most forward slot position, an angle of attack of $\alpha = 3.7^{\circ}$ was estimated for buffet onset with and without suction applied. Moving the slot downstream reduces this angle slightly, obviously due to an increase in shock strength which furthers separation; however, placing the slot downstream of the shock increases the angle of attack for the onset of unsteadiness to $\alpha = 3.94^{\circ}$, corresponding to an increase in lift at buffet onset of about 4%. The effect seems due to delaying the development of separation when suction is applied downstream.

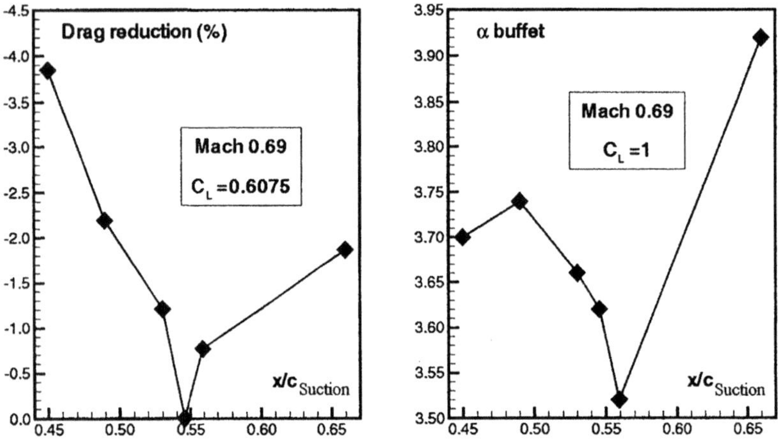

a) Drag reduction at C_L=0.6075 b) Onset of buffet

Figure 40 Effect of slot location on drag reduction and buffet onset for the airfoil DRA-2303, $M_{\infty} = 0.69$, $Re_c = 19 \times 10^6$, $C_L = 0.6075$, $C_Q = -0.00007$ [4.10]

INTA has similarly computed the effect of slot location and mass flux on the buffet boundary employing the DLR time-accurate Navier-Stokes code [4.7]. The slot location was varied between $x/c = 0.45$ and 0.54 at suction rates of $C_Q = -0.00007$ and -0.0007, the latter corresponding to a 10-fold increase. At the freestream conditions for the airfoil DRA-2303 considered, i.e., $M_\infty = 0.68$, $Re_c = 19 \times 10^6$, the angle of attack for the onset of unsteadiness was in the no-control case determined to be $\alpha = 4.5°$. Since suction at the lower suction rate was found to have no effect on buffet onset, independent of the slot location, a different approach was taken: first the angle of attack was increased to $\alpha = 4.8°$ resulting for the datum airfoil in fully developed buffet with large oscillations in shock location, hence lift, represented in Figure 41 by the forward part of the curves. The corresponding iso-Mach contours, Figure 42a, indicate that at maximum lift the shock is in a mid-chord position while at the lower lift coefficient of the cycle the shock is far forward with a large area of separation present.

Once the buffet process was completely developed, slot suction was activated while the shock was at the position of maximum lift. At the lower suction rate there was again little effect. However, at $C_Q = -0.0007$ the process changed considerably, essentially dependent on the slot location, Figure 41: suction at the forward slot location $x/c = 0.45 - 0.46$ (1%-chord slot width) caused a higher frequency coupled with a lower amplitude in the shock movement; suction at $x/c = 0{,}49 - 0.50$, i.e., closer to the shock, further reduced the amplitude with, as can also be seen in Figure 42b, very little separation present. Placing the slot at the foot of the shock completely eliminated shock-induced separation and the buffet process ceased.

It is obvious that a very high suction rate had to be applied at the foot of the shock to stop the buffet process. At the lower suction rate with the slot positioned at the foot of the shock — and also shown by the ONERA results in Figure 40 — the buffet process was not affected or even commenced at a lower angle of attack. A slot position further downstream seems, as far as buffet is concerned, considerably more effective.

The effect of discrete slot suction — varying slot location — on drag and buffet onset for the DRA-2303 airfoil at $M_\infty = 0.681$ (nominal), $Re_c = 19 \times 10^6$, $C_L = 0.6075$, and $C_Q = -0.00007$ was also determined by other participants in Task 2 without the results generally showing any differences in the observed trends due to suction. Differences similar to the ones discussed in Chapter 4.2 occurred, however, in the absolute levels of drag and drag reductions.

A further issue studied in conjunction with slot suction was the application of Poll's law and the new law of derived at the University of Karlsruhe by Bohning and Doerffer, the latter accounting, as described in Chapter 3.3.1, for the presence of an outer tangential stream. In the computations it was generally found that the new law gave, in agreement with Poll's law, answers independent of whether prescribing the suction coefficient or the cavity pressure, a characteristic that lead, when applying the law of Bohning and Doerffer in its original form, i.e., not accounting for the outer tangential stream, to large discrepancies in the results (also see Annex B of [1.5]).

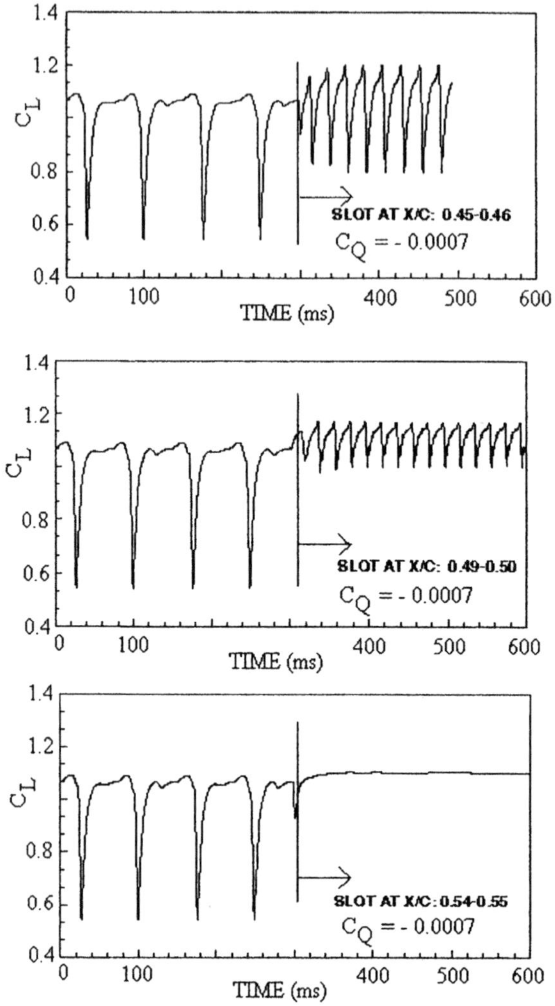

Figure 41 Lift oscillations at fully developed buffet with and w/o control by slot suction DRA-2303 airfoil, $M_\infty = 0.68$, $Re = 19 \times 10^6$, $\alpha = 4.8^O$ [4.7]

4.3.1.2 Cavity ventilation and hybrid control

The control mechanisms considered here are a perforated plate/single-cavity arrangement with part suction and hybrid control comprised of an upstream passive cavity in the shock region and slot suction downstream of the passive cavity as sketched in Figure. This type of control was applied to the airfoil DRA-2303 and investigated numerically as well as experimentally. The freestream conditions considered in the computations were, as above, $M_\infty = 0.68$, $Re = 19 \times 10^6$, $C_L = 0.608$. Iso-Mach lines for the datum airfoil at these freestream conditions, as computed by INTA [4.7], are shown in Figure 43.

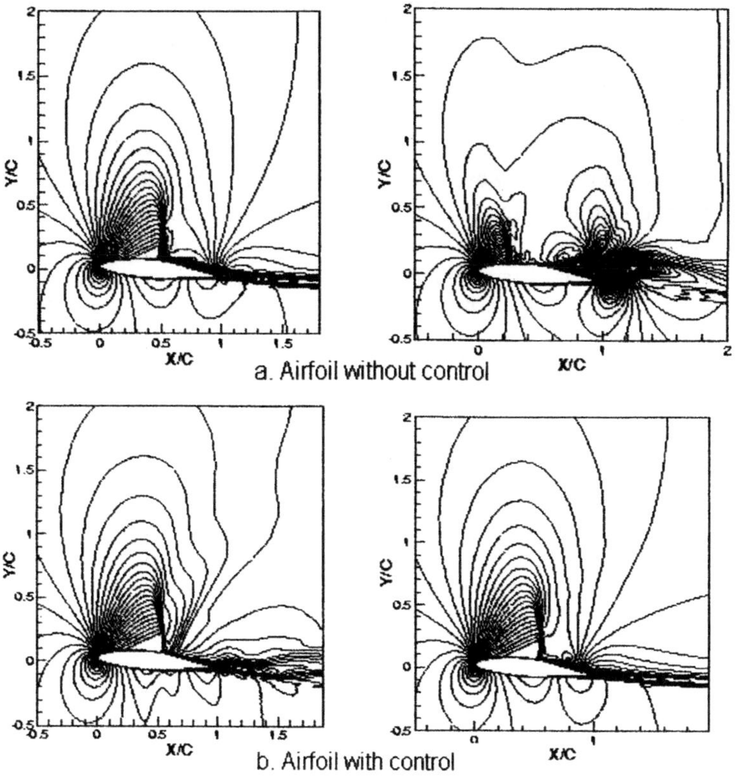

Figure 42 Iso-Mach lines corresponding to the extrema in Figure 41, DRA-2303 airfoil $(x/c)_{Slot} = 0.49 - 0.50$, $M_\infty = 0.68$, $Re = 19 x 10^6$, $\alpha = 4.8^O$ [4.7]

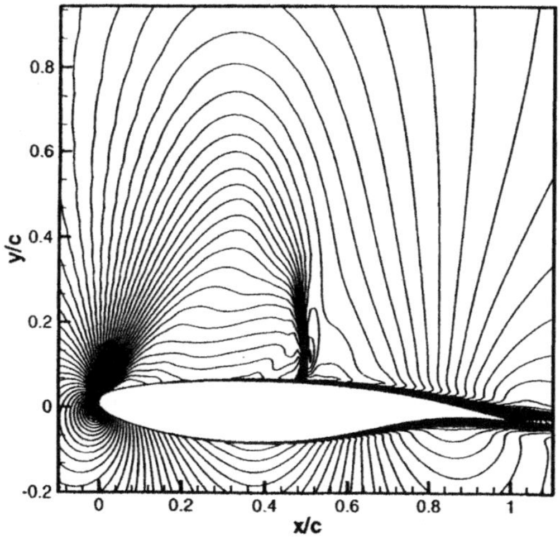

Figure 43 Computed iso-Mach contours for the DRA-2303 airfoil without control $M_\infty = 0.68$, $Re = 19 \times 10^6$, $C_L = 0.608$ [4.7]

Computations for the DRA-2303 airfoil with single-cavity control were — among others — carried out by EADS-Airbus D at the freestream conditions given above assuming zero mass flux (passive control) and a, for aircraft applications reasonable and realizable, mass-flux coefficient of $C_Q = -0.00009$, Figure 44 [4.5]. Quite obvious is the spread of the shock due to passive as well as active control, hence the likely reduction of wave drag. However, as can be seen in the right part of the figure, upper-surface displacement and momentum thicknesses are considerably increased due to control with the initial difference, i.e., the difference immediately downstream of the control region, amplified by the sustained rear adverse pressure gradients prevailing on the airfoil. Table 5 shows that, correspondingly, drag is increased due to passive control by 11.8% which is reduced to 5.7% as suction is applied. A similar trend in drag behavior is exhibited by the experiments with experiment and computation both demonstrating the dominating effect of the increase in viscous drag which has its origin in the boundary layer development within the control region as already shown by the basic experiments.

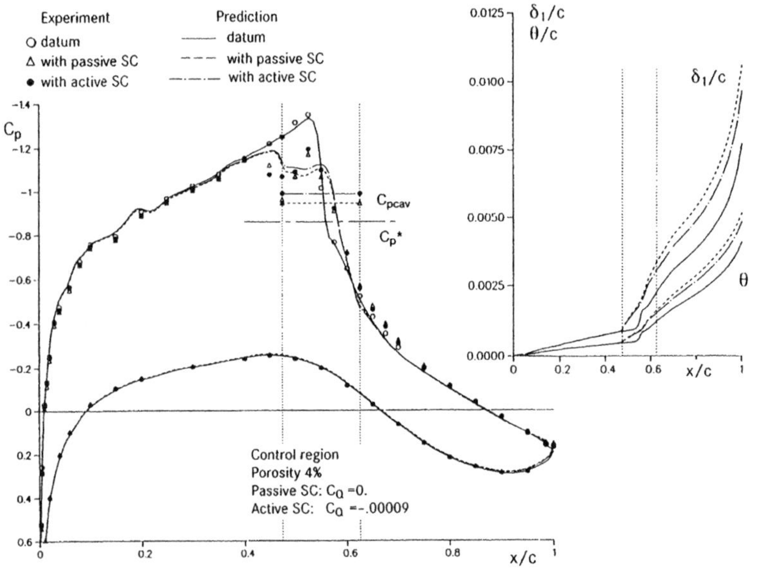

Figure 44 Active control effects on the flow about the DRA-2303 airfoil $M_\infty = 0.68$, $Re = 19 \times 10^6$, $C_L = 0.608$, $C_Q = 0$ and -0.00009 [4.5, 5.1]

Hybrid control was, e.g., numerically considered by INTA with the control mechanism consisting of the upstream passive cavity coupled with suction downstream of the passive cavity [4.7]. Again, a spreading of the shock due to the passive cavity action is predicted in agreement with experiment indicating the reduction in wave drag, Figure 45. However, as in the case of active single-cavity control, total drag is increased, here by about 7.2% if no suction is applied downstream, while applying suction reduces the drag increase to 6.2% at the lower suction rate and to 1.3% as the suction coefficient is raised to $C_Q = -0.0001$, Table 5.

Since the wave drag reduction is essentially not affected by suction, as in the case of single-cavity control, the drag increase is here entirely due to an increase in viscous drag. As shown in Table 5, this drag development is also indicated by the experiments, although the levels of influence differ as before. Please note that we will return to the experimental results in Chapter 5.3.

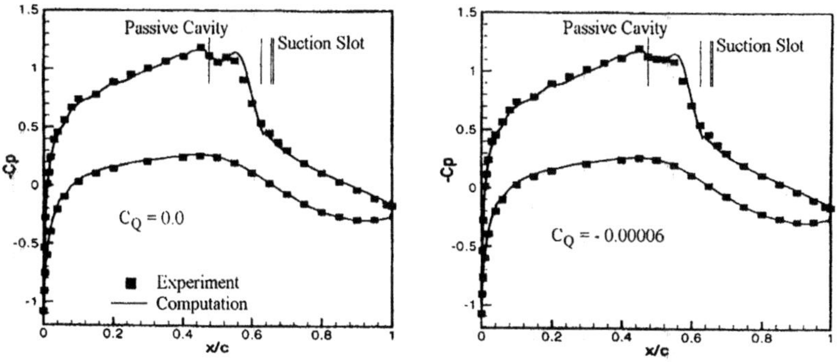

Figure 45 Hybrid control effect on the flow about the DRA-2303 airfoil $M_\infty = 0.68$, $Re_c = 19\text{x}10^6$, $C_L = 0.608$, $C_Q = 0$ and -0.00006 [4.7, 5.1]

Table 5 Aerodynamic coefficients for the DRA-2303 airfoil with active and hybrid control $M\infty = 0.681$, $Re_c = 19\text{x}10^6$, $C_L = 0.608$ (nominal)

	C_Q	C_L 2) w/o contr.	C_L w contr.	C_D 2) w/o contr.	C_D w contr.	ΔC_D 3)%
EADS-A-pass. SC 1	0	0.6070	0.5860	0.00932	0.01042	11.8
EADS-A-active SC	-0.00009	0.6070	0.5950	0.00932	0.00985	5.7
INTA-hybrid SC	0	0.6125	0.5841	0.00970	0.01040	7.2
INTA-hybrid SC	-0.00006	0.6125	0.5871	0.00970	0.01030	6.2
INTA-hybrid SC	-0.00010	0.6125	0.5971	0.00970	0.00970	1.3
Exp.-passive SC	0	0.6080	0.5860	0.00973	0.01196	22.9
Exp.-active SC	-0.00009	0.6080	0.5930	0.00973	0.01121	15.2
Exp.-hybrid SC	0	0.6080	0.5905	0.00973	0.01040	7.2
Exp.-hybrid SC	-0.00006	0.6080	0.5941	0.00973	0.01030	6.2
Exp.-hybrid SC	-0.00010	0.6080	0.5960	0.00973	0.01020	5.1

1) SC = Shock Control 2) Corresponds to datum (clean) airfoil 3) Pump drag is not accounted for

It is worthwhile to recall that the basic experiments already indicated that, for the control mechanisms considered here, mass fluxes in the order of $C_Q' = -0.0008$, i.e., mass fluxes about eight times higher than the ones considered in the present computations, are required to reduce the global boundary layer parameters downstream of the interaction/control region to the no-control level (see, e.g., Figure 23), thereby eliminating the viscous drag increase. These suction rates are, however, as already mentioned, not realistic for aircraft application. One may, therefore, conclude that, at feasible suction rates, active control by perforated plate/cavity arrangement and hybrid control seem to be generally associated with an increase in total drag — at least in the case of the laminar-type airfoil investigated here — so that this type of control does not seem a viable tool if drag reduction is of main interest. However, in other applications where the avoidance of separation and/or the weakening and stabilizing of the shock is of primary concern, as in supersonic air intakes, or, generally, in internal flows, this type of control seems well suited.

4.3.2 Active control by contour modifications (Bumps)

Single-point computations for the laminar airfoil DRA-2303 and the turbulent airfoil RAE-5225 with and without a contour bump in the shock region were already discussed above indicating that ① considerable drag reductions can be achieved by bump control and ② bump effectiveness seems very sensitive to the freestream conditions determining, e.g., the relative location of the bump with respect to the shock. It is, therefore, worthwhile to consider in the assessment of bump control complete aerodynamic polars and, of course, the effectiveness of a bump dependent on characteristic geometric features such as the chordwise extent of the bump, its height and its shape.

Within EUROSHOCK II the A340-type ADIF airfoil was investigated numerically and experimentally as basic airfoil and in the constant-chord infinite-swept-wing (sheared-wing) configuration without and with bump control (see Figures 56 and 57 for the airfoil/sheared-wing geometry and Figure 47 for a typical pressure distribution). The airfoil is at transonic conditions characterized by a moderate acceleration of the flow on the upper surface. Also investigated without and with shock control by contour bump was the laminar-type DA LVA-1A airfoil, already studied in EUROSHOCK (I) in conjunction with passive ventilation [1.5]. Compared to the ADIF airfoil, the latter exhibits a stronger upper-surface flow acceleration. Varied were, besides the freestream conditions, for both airfoils and the sheared wing the bump geometry and especially the bump height. We will also return here to the DRA-2303 airfoil when considering the effect of a bump on buffet.

4.3.2.1 Steady flow conditions: airfoil studies

Turning first to the **ADIF-airfoil** drag polars, computations were carried out, for instance, by the ONERA VIS15 code considering the airfoil with two asymmetric bumps of different height, namely, h_{bump} = 0.175%-chord and 0.350%-chord, respectively [4.10]. The bumps were optimized in preceding studies by EADS-Airbus D [4.5] and DLR [4.15]; they were located between x/c = 0.640 and 0.840 with the bump crest at x/c = 0.720 (also see Figures 59 and 56). The Mach numbers for the computations were selected close to the design Mach number of the airfoil, viz., M_∞ = 0.765. The computed and experimental polars, the latter obtained in the DLR

Cryogenic Ludwieg-tube Wind Tunnel [4.15], are presented in Figure 46; characteristic ADIF-airfoil pressure distributions are shown in Figure 47.

Figure 46 exhibits very characteristic features of control by contour bumps, some already discussed: the higher bump is more effective at higher lift coefficients, say, at $C_L > 0.60$, but shows at lower lift coefficients a very strong increase in drag. The smaller bump is less effective at higher lift, but stays effective over a considerably larger lift range starting at relatively low lift coefficients. Drag reductions of up to 17% are achieved at $C_L < 0.60$ while reductions of up to 23% are attainable above this lift coefficient. The figure shows — besides the sensitivity of the bump effectiveness to lift — also the sensitivity of the bump to Mach number changes indicated, for instance, by the shift in the cross-over lift coefficient from effective to ineffective when decreasing the Mach number from $M_\infty = 0.762$ to 0.757, Figure 46, left diagrams.

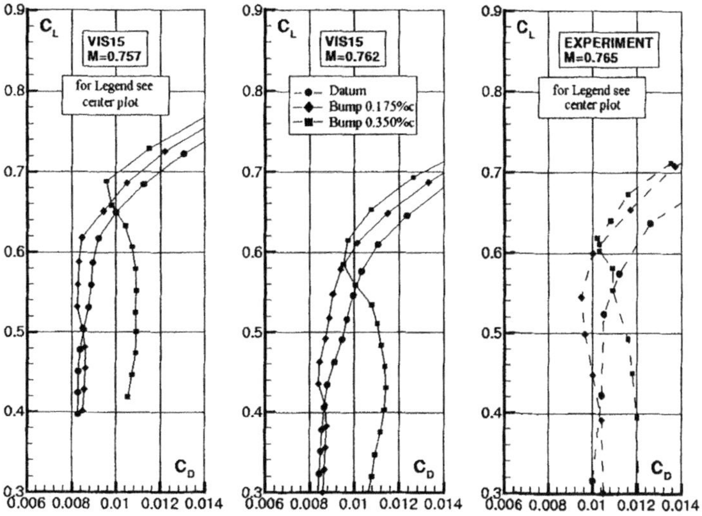

Figure 46 Drag polars for the ADIF airfoil with and w/o bumps at $M_\infty = 0.757$ and 0.762, ONERA computations [4.10], and $M_\infty = 0.765$, DLR experiments [4.15]

The reason for the reduced effectiveness of the higher bump at lower lift coefficients is clearly demonstrated in Figure 47 by the pressure distributions computed by EADS-Airbus D [4.5] and measured by DLR [4.15]: for the smaller bump, the pressure increase due to the shock is spread over a certain chordwise distance (a), while the higher bump causes an initial pressure rise followed by a strong expansion and, in turn, a second stronger shock (b). This constellation causes a considerable increase in wave drag as well as an increased load on the boundary layer which leads to the dramatic increase in total drag. It seems that the relative position of the bump (actually the bump crest, see Figure 49) with respect to the shock location and the bump height relative to the shock strength — with both shock location and strength dependent on angle of attack (lift) and Mach number, but also on the flow development initiated by the bump — are essential parameters for optimum bump effectiveness; the "correct" position and/or height with changing freestream conditions can, of course, be achieved with an adaptive bump. As an alternative, one may also consider applying a fixed bump of lower height at a reduced overall effectiveness.

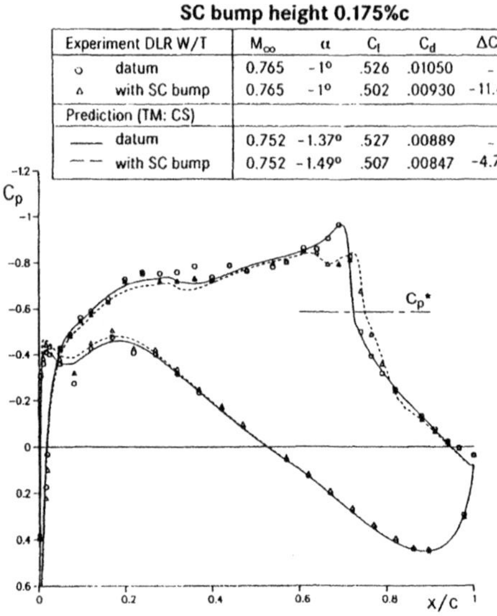

a. Bump height $h_{Bump} = 0.175\%$ chord

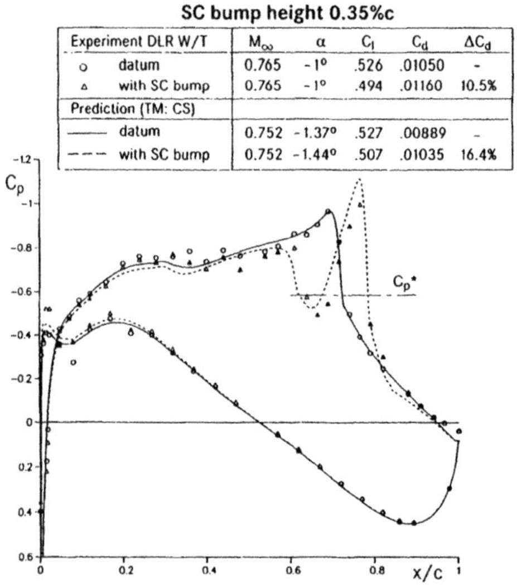

b. Bump height $h_{Bump} = 0.350\%$ chord

Figure 47 Comparison of numerical and experimental pressure distributions for two different bump heights, ADIF-airfoil, $M\infty_{nominal} = 0.765$, $Re = 8 \times 10^6$ [4.5, 4.15]

The strong double-shock system was also predicted in computations by CIRA [4.3] — and others — Figure 48. Also note that the agreement between numerical and experimental pressure distributions and in the corresponding (qualitative) drag changes due to control is quite good, while absolute levels again deviate.

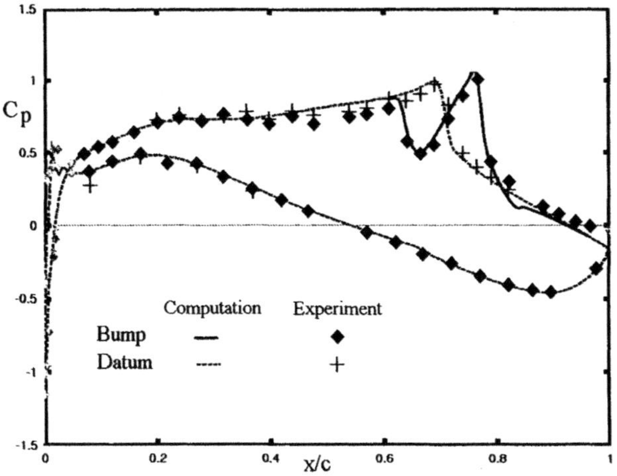

Figure 48 Comparison of computational and experimental pressure distributions w/o and with bump, ADIF-airfoil, $h_{Bump} = 0.35\%$-chord, $M\infty_{comp} = 0.757$, $M\infty_{exp} = 0.765$, $Re = 8 \times 10^6$, $C_L = 0.50$ [4.3, 4.15]

The preceding discussion indicates that a bump should actually be adjustable to changing Mach number and/or lift conditions, e.g., in height, so that at high lift coefficients a larger bump could be deployed, while, when forced to fly at reduced lift, the bump could be gradually retracted. A corresponding numerical optimization study, here for the laminar-type airfoil DA LVA-1A at $M_\infty = 0.76$, $C_L = 0.47$, $Re_c = 6 \times 10^6$, was carried out by EADS-Airbus D determining, for instance, the drag reduction potential dependent on geometric bump characteristics, Figure 49 [6.3]: for a symmetrical bump (left diagram), drag reduction generally increases with increasing bump height until, for the given lift coefficient, a height is reached where a further increase results in a rapid deterioration of the bump effectiveness, i.e., where the bump most likely causes an expansion and an additional strong shock; the bump effectiveness also increases with bump length. Keeping the length of the bump constant at 20% chord (right diagram), drag reduction again increases with increasing bump height up the "break point". It is, furthermore, indicated that the most efficient bump is an asymmetric bump with the crest location at 70% of the bump length. Drag reductions of up to 25% are obtained. (For a direct indication of the effect of the relative bump location, please consult Figure 76.)

Such an optimization process must, of course, be performed for all Mach number/lift combinations providing an adaptive-bump matrix for a given flight mission; this will be demonstrated in Chapter 6.2 when considering the application of a bump to an A340-type hybrid-laminar-flow-wing aircraft.

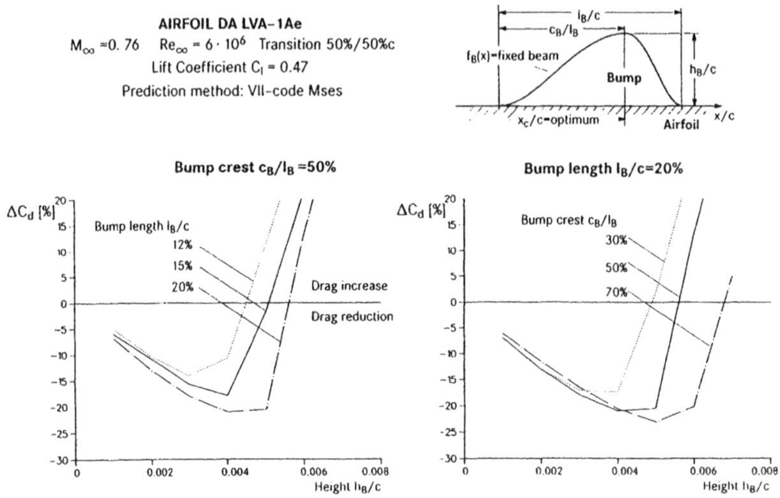

Figure 49 Effect of bump geometric characteristics on drag reduction, airfoil DA LVA-1A $M\infty = 0.760$, $Re = 6 \times 10^6$, $C_L = 0.47$ [6.3]

4.3.2.2 Steady flow conditions: sheared-wing studies

Computations for the ADIF-sheared-wing configuration with and w/o bumps of different height were initially, i.e., prior to the sheared-wing measurements, performed for the freestream conditions $M\infty = 0.840$, $Re_c = 9.9 \times 10^6$ and $C_L = 0.59$ and, correspondingly, for the 2D airfoil at $M\infty = 0.755$, $Re_c = 8 \times 10^6$ and $C_L = 0.710$. The results of the computations, here, e.g. by CIRA, indicated that the pressure distributions for the 2D and the sheared-wing configurations without and with control by the smaller bump ($h_{bump2D} = 0.175\%$-chord) are generally quite similar, Figure 50 [4.3]. This also holds for the drag reductions due to bump control with the drag reductions for the sheared wing, however, generally lower than the equivalent 2D reductions, here, for instance, 2.4% in the sheared wing case and 6% for the 2D airfoil. An equivalent bump height of $h_{bump2D} = 0.54\%$-chord caused a total drag reduction of 15.8% in the 3D case and a reduction of 17.6% for the equivalent 2D configuration. The reduced drag benefits are due to the lesser contribution of wave drag to total drag.

A further comparison between computational and experimental sheared-wing pressure distributions, here for the datum wing and bump heights of $h_{Bump} = 0.1573\%$-chord and 0.3146%-chord, is presented in Figure 51 [4.10]: at the higher lift coefficient, $C_L \approx 0.60$ (a), both bumps result in a spreading of the shock wave and a corresponding reduction in wave drag; at the lower lift coefficient, $C_L \approx 0.50$ (b), the small bump already generates a double shock system — experimentally as well as numerically — which again, as in the case of the airfoil, indicates an increase in wave drag and most likely in total drag as a result of the mismatch between shock location/strength and bump characteristics.

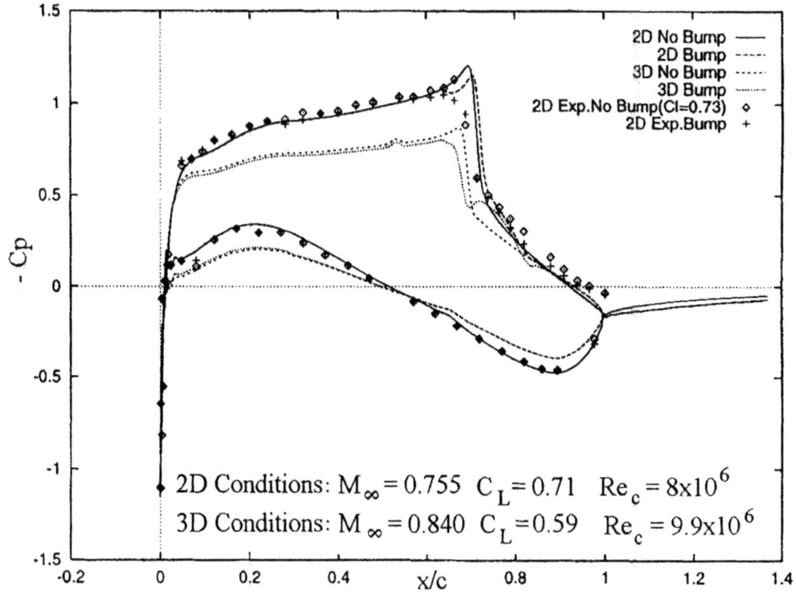

Figure 50 Computations for the ADIF sheared wing and airfoil with and without bump control, $h_{Bump2D} = 0.175$, CIRA computations, DLR KRG experiment [4.3, 4.15]

The latter is confirmed by the corresponding drag polars in Figure 52 which show the lift-dependent crossover in bump effectiveness similar to the one described above for the ADIF airfoil [4.5]. As a general conclusion one may, therefore, state that the flow development due to bump control is quite similar in the 2D and the 3D case and that *sweep effects* on bump-control effectiveness are minor. Note that we will return to bump control — also in conjunction with discrete suction — in Chapter 5 where we shall, more closely, examine the results of the experimental investigations on airfoils and the sheared wing.

4.3.2.3 Unsteady flow conditions – buffet

Here, the objective is again twofold: ① the assessment of unsteady or time-accurate codes in predicting buffet onset and the buffet process, and ② the evaluation of the effectiveness of a bump in suppressing or delaying buffet. To determine buffet onset, three criteria were suggested and defined by ALENIA [6.4], viz., C1 related to the magnitude of the shock-upstream Mach number, C2 related to the trailing-edge pressure divergence, and C3 related to the actual onset of flow unsteadiness in the computations or, similarly, in the experiments. The latter criterion was already employed when considering the effect of discrete suction on buffet in Chapter 4.3.1.1.

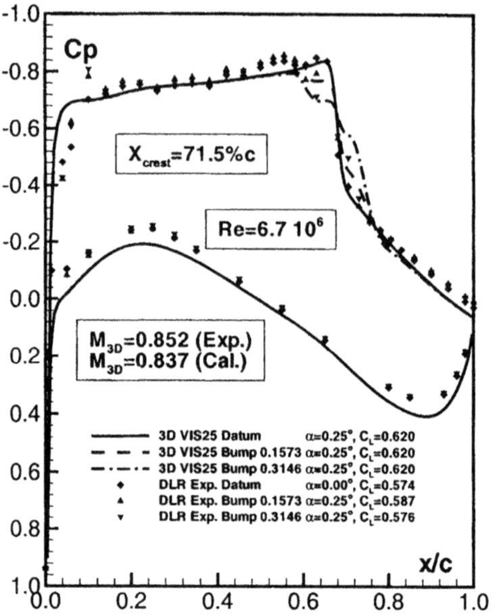

a. ADIF sheared-wing pressure distributions at $C_L \approx 0.60$

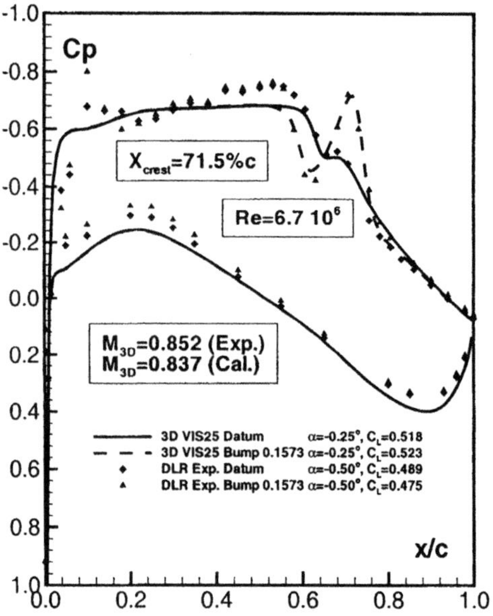

b. ADIF sheared-wing pressure distributions at $C_L \approx 0.50$

Figure 51 Experimental and numerical pressure distributions for the ADIF sheared wing with and without bump control, ONERA computations, DLR experiments [4.10, 4.16, 4.17]

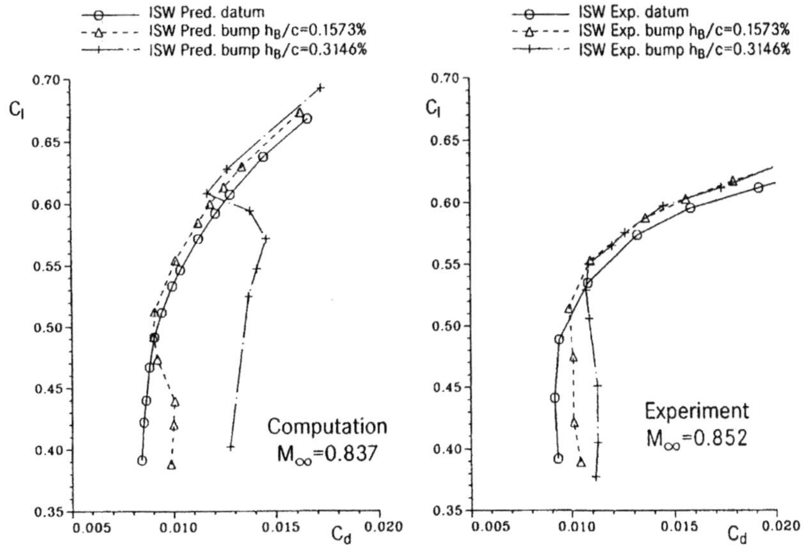

Figure 52 ADIF sheared-wing polars for bump control, EADS-Airbus-D computations, DLR experiments, $Re_c = 6.7 \times 10^6$, $(x/c)_{trans} = 0.10$ upper and 0.15 lower surface [4.5, 4.16, 4.17]

The use of the criterion C1 is demonstrated in Figure 53 by estimating the effect of a bump on buffet onset for the DRA-2303 airfoil; computations were performed by INTA employing the DLR time-accurate Navier-Stokes code [4.7]. Defined are three (Mach number) lines representative of conditions just prior to buffet onset (lower line), at buffet onset and at well-developed buffet. The lines are slanted to account for the condition of the boundary layer upstream of a shock, i.e., taking into account that a thin boundary layer may withstand a stronger shock before separation develops, while a thick boundary layer, as present closer to the trailing edge, tends to separate earlier, i.e., at a lower shock-upstream Mach number. Here, it will suffice to just consider the centerline. Buffet onset is reached when this line is tangent to the Mach number distribution. This is generally equivalent to a simpler criterion — frequently used and valid for shock positions at about 50% to 60% chord — namely, that buffet is imminent if the shock-upstream Mach number exceeds $M_L = 1.30$; this is the Mach number where shock-induced separation is incipient [4.18].

In the present representative example, i.e., the DRA-2303 airfoil at $M_\infty = 0.6829$, $Re_c = 19 \times 10^6$, Figure 53, the lift coefficient at buffet onset for the airfoil without control is $C_{LB} = 0.8619$ ($\alpha = 2.8°$), while for the airfoil with bump control this coefficient is raised to $C_{LB} = 0.9550$ ($\alpha = 3.5°$), corresponding to an increase in the buffet boundary of 10% due to the bump, which is considerable.

An example of the prediction of the onset of buffet using criterion C3, onset of unsteadiness, is given in Figure 54 by time-accurate Navier-Stokes computations of DLR, again for the DRA-2303 airfoil at $M_\infty = 0.68$, $Re_c = 19 \times 10^6$ [4.8]: flow oscillations, here represented by oscillations in the aerodynamic coefficients, commence in case of the datum airfoil at $\alpha = 5.5°$, while this angle is increased to

$\alpha = 6.0^{\circ}$ when bump control is applied. The corresponding lift coefficients are $C_{LB} = 1.130$ and $C_{LB} = 1.194$ for the no-control and the control case, respectively, representing an improvement of lift at the buffet onset of 5.7%.

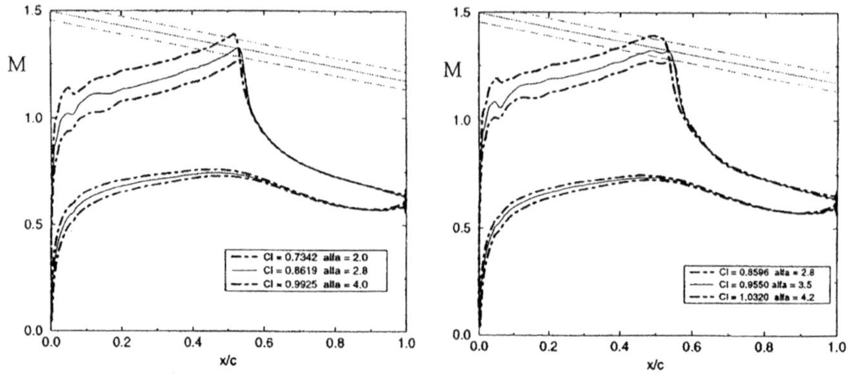

Figure 53 Buffet onset according to the Mach number criterion C1 as computed by INTA, airfoil DRA-2303, $M\infty = 0.6829$, $Re_c = 19 \times 10^6$ [4.7]

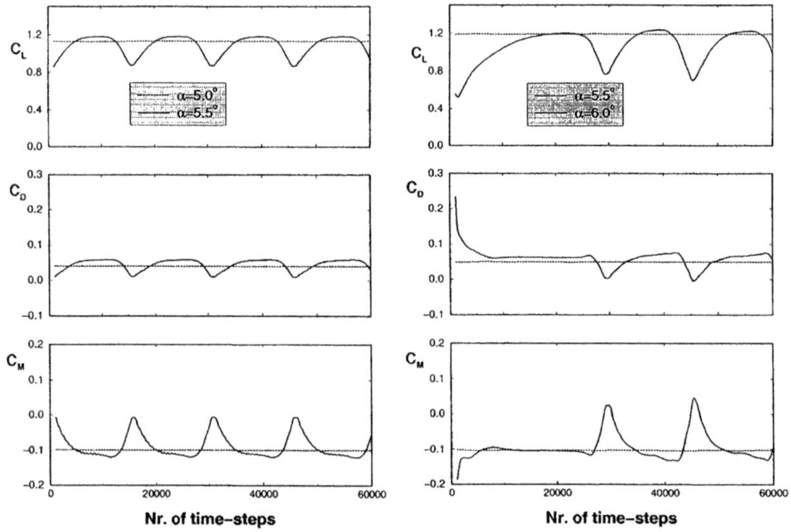

a. Datum airfoil b. Airfoil with contour bump

Figure 54 Effect of a bump on the onset of buffet for the DRA-2303 airfoil $M_\infty = 0.68$, $Re = 19 \times 10^6$, DLR Navier-Stokes computations [4.8]

Results obtained for buffet onset by the various codes, employing the three criteria, are summarized in Table 6: the buffet boundary is, independent of the criterion applied, generally increased due to a bump — although the bump is not explicitly designed for buffet control — which is probably the result of a fairly fixed shock position with increasing angle of incidence (also see Chapter 5.2.4.2). Increases in lift due to control are largest when applying criterion C1, while criteria C2 and C3 give about the same lower increases. The absolute lift coefficients at buffet onset are generally lower in case of C1 than they are for C2 and C3. The reason for this is that flow conditions for the criteria C2 and C3 to apply are much closer to total flow separation than conditions corresponding to C1 (centerline), where separation just starts to develop, and a further increase in angle of attack (lift) is needed for the actual buffet process to commence.

The computations show that bump control generally raises the actual buffet boundary by up to 5%, which is in close agreement with experiment; so it may be concluded that the time accurate codes considered seem well suited to predict steady and unsteady conditions and the boundary between these conditions.

Table 6 Effect of bump control on buffet, airfoil DRA-2303 at $M_\infty = 0.68$, $Re = 19 \times 10^6$

	C_{LB} w/o control			C_{LB} with control			$\Delta\ C_{LB}\%$		
Criterion:	C1 2)	C2 3)	C3	C1	C2 3)	C3	C1	C2 3)	C3
U. Naples (VII)	0.920	0.839	0.962	0.998	0.869	0.990	8.0	3.5	2.9
DLR (N.-S.)	0.939	1.108	1.130	0.985	1.156	1.194	4.9	4.3	5.7
ONERA 1) (VII)	0.820	0.976	0.976	0.937	1.011	1.007	14.3	3.4	3.2
INTA (N.-S.)	0.862	—	—	0.955	—	—	10.8	—	—

1) $M_\infty = 0.69$ 2) Centerline, see Figure 53 3) C2 refers to the deviation in trailing edge pressure from its value at $C_L = 0$ by an amount of $\Delta C_{Pte} = -0.04$ as C_L (or α) is increased.

4.3.2.4 Pneumatic bump / distributed blowing

CIRA has introduced the concept of distributed blowing from a multi-slot surface with individual separated plenum chambers in the shock region [4.3]. By controlling the mass flux from each slot, a *Pneumatic Bump* can be generated. Computations for the DRA-2303 airfoil with and without distributed blowing at $M_\infty = 0.68$, $Re = 19 \times 10^6$, $C_L = 0.60$ have shown that the drag coefficient can be reduced by 6.8%, quite similar to a geometric bump. By reversing the flow within the first slot, a combination of upstream suction and *bump* was simulated resulting, at the above conditions, in a drag reduction of 12.8%. The *pneumatic bump* was found to be similarly effective as the solid bump in raising the buffet boundary. For more details concerning this device refer to Chapter 14.

4.4 Evaluation of CFD Codes

One of the objectives of Task 2 was the evaluation of the participating codes not only with respect to their predictive capabilities but also with regard to their computational efficiency. The procedure for determining the latter was as follows: five test cases were selected from the DRA-2303 airfoil tests, viz., ① the datum airfoil, ② the airfoil with bump control, ③ the airfoil with discrete suction, and ④ the airfoil with passive and ⑤ active control via a perforated surface/cavity arrangement. The convergence histories for C_L and, where applicable, C_Q, were to be provided as function of the CPU-time divided by the number of grid points (NG) on the surface. The common convergence criterion was defined to be $\Delta C_L = 10^{-4}$. The CPU performance of the respective computers was compared in terms of MFLOPS required for the inversion of a 1000 x 1000 matrix using a standard LINPACK routine [4.3].

The result of the above exercise is presented in Table 7. It is clearly indicated that the lowest convergence time is associated with the ONERA VIS15 code, while the Navier-Stokes code employed by DLR and INTA had the highest normalized CPU time. The difference between the two Navier-Stokes calculations was attributed to the use of different grids. Further details of the comparison of computer code performances are given in Annex B of [1.5].

Table 7 Evaluation of computer code performance

Partner	Code	Computer	MFLOPS (Ratio)	Normalized CPU Time (sec) /NG				
		Test Case →		1	2	3	4	5
CIRA	EUBL2D (VII)	Power Challenge R10000	325 (1)	3.0	3.1	2.1	3.8	4.8
EADS-A.	VII	HP 715	85 (0.26)	4.8	4.3	7.8	—	—
DLR	NS-2D	NEC SX4	1944 (6)	37.6	37.6	—	—	—
INTA	NS-2D	CRAY Y-MP	107 (0.32)	9.0	10.0	12.4	17.4	19.5
U. Naples[1)]	EULSL (VII)	PENTIUM 200 264 MHz	N. A.	432	420	405	336	532
ONERA	VIS15 (VII)	DEC ALPHA 8400	764 (2.35)	1.5	1.9	1.6	—	—

1) CPU-time not scaled

4.5 Conclusions and Future Work

The overall objective of Task 2 was to extend and improve the participating numerical codes, including steady and time-accurate Viscous-Inviscid Interaction (VII) codes and Navier-Stokes solvers, to treat flows with shock and boundary layer control, and to employ these codes in subsequent parametric studies to evaluate potential control concepts, supplementing corresponding experiments.

In a first phase, several test cases were defined to assess the capabilities of the participating codes. These test cases included the following configurations: ① the laminar-type airfoil DRA-2303 with control by a contour bump, control by discrete suction, and control by a combination of a bump with discrete suction upstream of the bump, ② the turbulent airfoil RAE-5225 applying the same control concepts, and ③ the turbulent airfoil VA-2 with passive and active control via a perforated surface/cavity arrangement. During and as a result of the computations, codes were improved by grid refinements, especially in the control/shock region, and by the introduction of the new control law of Bohning and Doerffer. In a second phase, more detailed parametric studies were carried out, involving — besides the DRA-2303 airfoil with discrete slot suction and active and hybrid cavity ventilation — the turbulent A340-type ADIF airfoil, and the ADIF sheared wing without and with various control mechanisms.

With regard to CFD capabilities, the following conclusions may be drawn:

- For all configurations and control concepts investigated, the computer codes arrived — after improvements — at the same qualitative results concerning drag reductions/increases and the improvement of off-design conditions, i.e., essentially improvements in the buffet boundary due to control. However, absolute levels in drag as well as in the magnitude of drag changes due to control differed between codes and between codes and experiment. Here improvements, especially in turbulence modeling for flows with shock-induced and conventional (subsonic) separation w/o and with control, are still needed.
- Discrepancies between computations and experiments could be noticeably reduced by adjusting the freestream conditions, e.g., the Mach number, in the computations. Such a procedure is especially important in the case of shock control by contour bumps (surface modifications) since the effectiveness of such bumps is very sensitive to changes in the shock location and strength which depend on the freestream conditions, namely, Mach number and angle of attack.

For the control mechanisms investigated numerically, the effectiveness in improving design and off-design conditions was found to be — qualitatively in agreement with experiment — as follows:

- Discrete slot suction upstream of the shock boundary layer interaction region — or, at subcritical conditions, at the onset of the sustained rear adverse pressure gradients — reduces drag up to 6% with the experimentally determined maximum drag reduction being 7.5%. Slot locations downstream of the interaction region are less effective in reducing drag.

- Buffet onset and the buffet process could only be noticeably affected by discrete shock-upstream suction or suction at the foot of the shock when extreme suction rates, not realizable in actual aircraft installations, were applied; downstream slot locations were found to be superior even at realistic suction rates.
- Shock control by contour bumps control was found to be the most effective of the control mechanisms investigated with maximum drag reductions of up to 20% predicted for an optimized bump applied to a laminar-type airfoil. However, bump effectiveness is very sensitive to changes in the freestream conditions, suggesting an adaptive bump as the best control solution. Bump control also positively affects buffet onset. Bump control with upstream discrete slot suction has the potential to further increase control benefits.
- Bumps are somewhat less effective in case of sheared-wing flow with the flow development associated with control, however, being the same as in the 2D case.
- Control by a perforated plate/cavity arrangement with part-suction and by hybrid control, consisting of a passive cavity in the shock region and suction downstream, always resulted, at realistic suction rates, for the airfoils investigated here in an increase in drag due to the dominating effect of an increase in viscous drag caused by this type of control.

As in the case of the basic experiments, future work is envisaged to investigate a wider range of control mechanisms for aircraft performance improvements, such as sub-boundary layer devices, including vortex generators, mass-less air jets, reversed-flow flaps and other mini-flaps to control flow conditions at the trailing-edge; furthermore, double-bumps — one located close to the leading edge, one in the shock region of an airfoil or wing. Emphasis will also be placed on the combination of suitable control devices. The freestream conditions shall include high-lift, low speed regimes as well as the transonic speed range. Computational methods shall be extended, if needed, to treat these types of control but shall also be improved with regard to the incorporation of new turbulence models and computational procedures on the basis of the lessons learned during the present investigation.

5 Airfoil and Sheared-wing Experiments with and without Control (Task 3)

The objective of the present studies was to investigate, mainly experimentally, the effect of various shock and boundary layer control techniques on the design and off-design performance of airfoils and swept wings. The investigations concentrated on airfoils designed to have moderately to highly accelerating flow on the upper surface at transonic conditions — the latter typical of laminar designs — since these airfoils have a larger potential for improvements at design conditions, where shock waves are, generally, already present, and, in addition, exhibit smaller margins between the design point and the drag-rise and buffet boundaries compared to configurations with plateau-type pressure distributions characteristic of turbulent-airfoil designs.

The comprehensive wind tunnel test program performed was designed to provide an extensive data base for airfoil and wing configurations with and without control for 1) a large Reynolds number range, 2) different boundary layer characteristics ahead of the shock wave, and 3) different shock strengths. The data base thus established was used, as already discussed, extensively to assess the computer codes considered in Chapter 4 and to demonstrate the aerodynamic efficiency of potential control techniques up to conditions typical of flight. The incorporation of selected control concepts into actual wings and the assessment of the associated overall benefits and penalties shall be discussed in Chapter 6. The present work was carried out by **DERA** [5.1] and **DLR** [4.15, 4.16, 4.17]. Concerning details of the investigations, the reader is also referred to Chapters 19 and 20.

5.1 Experimental Program

The experimental program was divided into two parts: firstly, an investigation of the influence of Mach number and angle of attack (lift) on the effectiveness of various control systems was carried out by DLR on two airfoils, viz., the ADIF airfoil, also as a reference for the corresponding sheared-wing experiment, and the laminar-type DA LVA-1A airfoil already employed during the EUROSHOCK (I) project for the investigation of passive control by ventilation [4.14]. The focus of the present work was on the determination and analysis of airfoil performance gains due to control by contour bumps and its dependence on Mach number and angle of incidence [4.15, 4.17]. A similar program was carried out for the infinitely-swept sheared-wing model based on the ADIF airfoil [4.16, 4.17].

The second part of the investigation comprised detailed steady pressure measurements carried out by DERA [5.1]. The model of the DERA DRA-2303 airfoil was tested up to Reynolds numbers approaching of full scale at conditions up to and exceeding buffet onset. Control schemes included discrete suction, active control by a perforated plate / cavity arrangement with part-suction, and hybrid control by a passive cavity in the shock region in combination with suction downstream. The measurements allowed the flow development in the control region to be related to the overall performance of the airfoil.

5.2 DLR Airfoil and Sheared-wing Experiments

Experiments at various Mach numbers and associated angle of attack sweeps were carried out with the A340-type airfoil ADIF and the laminar-type airfoil DA LVA-1A, respectively, in the no-control datum configuration and with control by contour bumps of different heights. The measurements were performed in the Cryogenic Ludwieg-tube wind tunnel of DLR (KRG) [5.2]. In addition, tests with a 26-degree infinitely swept wing (sheared wing), based on the ADIF airfoil, were performed in the 1x1 Meter Transonic Wind Tunnel Göttingen (TWG) [5.3] in order to determine sweep effects on control effectiveness. Control mechanisms included here contour bumps, similar to the 2D configuration, a bump in conjunction with discrete suction upstream, and passive and active ventilation via a perforated plate / cavity arrangement, also in conjunction with discrete slot suction.

5.2.1 Wind Tunnel Characteristics

The airfoils were investigated in the Cryogenic Ludwieg-tube of DLR Göttingen (KRG): the tunnel is a short-duration facility with a test time of up to one second consisting of a 130-meter long tube, a contraction section with a contraction ratio of 3.6, an adaptive-wall test section, a combination of second throat for Mach number control and quick-opening valve and a dump tank, Figure 55 [5.2]. The test section has a cross section of $0.40 \times 0.35 m^2$ and a length of 2m which allows to obtain interference-free results for the model chords investigated here, i.e., $c \approx 0.18m$. The performance characteristics of the tunnel, which uses gaseous nitrogen as test gas, are summarized in Figure 55.

The sheared-wing experiments were carried out in the 1 x 1 Meter Transonic Wind Tunnel Göttingen (TWG) [5.3]. The TWG is a closed-circuit continuous tunnel with a cross-section area of $1 \times 1 m^2$. There are now three independent test sections available: a perforated test section, 6% open with 60°-slanted holes, for the Mach number range 0.40 to 1.3, an adaptive-wall transonic test section for the Mach number range 0.3 to 0.95, and a supersonic test section for Mach numbers of 1.4 to 2.2. The total pressure can be adjusted between 0.6 and 1.6 bar allowing for some Reynolds number variation. The present investigation was performed in the adaptive-wall test section to reduce wall-interference effects; the side walls of the test section were, for the present tests, contoured, as outlined below, in order to generate infinite swept-wing flow.

5.2.2 Airfoil Characteristics and Wind Tunnel Models

The **ADIF airfoil** is an A340-type modern airfoil with a thickness of about 12% chord. It has a moderately accelerating flow on the upper surface upstream of the shock at near-cruise conditions. The shock position varies slightly with changing freestream conditions, which is typical of airfoils with only moderate pressure gradients on the upper surface following the initial strong acceleration around the leading edge (see, e.g., Figure 58). The cryogenic wind tunnel model of the airfoil is depicted in Figure 56: the model has a span of 400 mm and a chord of 175 mm and is inserted between the wind tunnel side walls; the model instrumentation, mainly consisting of static and dynamic pressure orifices, is similar to the one for the sheared wing model, Figure 57. Various bumps, optimized in a parametric study using a 2D coupled Euler/boundary layer method [4.15, 5.4], and a discrete suction device upstream of the bump, which can be employed independently of and in conjunction with the bump, are available. Note that here only bump control has been investigated.

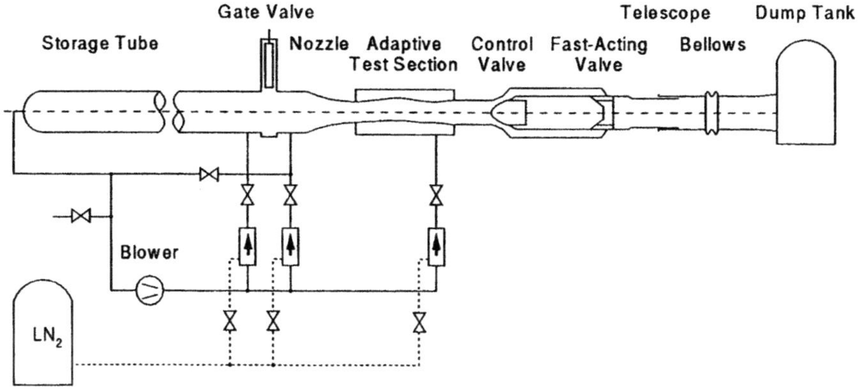

Tube	Diameter	0.8 m	Max. total pressure	10 bar
	Length	130 m	Temp. range	100 - 300 K
	Charge pressure	12.5 bar	Mach number	0.25 - 0.95
Test section	Cross section	0.40x0.35 m^2	Max. Reynolds No.	70x10^6
	Length	2.0 m	Run time	0.6 to 1.0 sec
	Model chord length	0.18 m		

Figure 55 Schematic and characteristics of the Cryogenic Ludwieg-tube [5.2]

The **LVA - 1A** airfoil is a transonic laminar-type airfoil with a thickness of 12% chord. It was designed by EADS-Airbus D to have natural laminar flow on the upper and lower surfaces up to 50% of the chord at a Mach number of M_∞ = 0.73, a lift coefficient of C_L= 0.4 and a Reynolds number of Re_c = 20x10^6 [5.5]. The pressure distribution is characterized by a moderate-strength shock wave occurring already at design conditions with strongly accelerating flow on the upper surface up to the shock to sustain laminar flow. An essential feature of this airfoil, especially important for shock control, is that with increasing lift coefficient, as the shock grows stronger, the position of the shock remains rather unchanged. The model design, similar to the ADIF models, provides an exchangeable insert to accommodate various bumps which have been optimized as described in Chapter 4.3.2.1 (see Figure 49). The model is equipped with two chordwise pressure plotting sections with a total of 64 pressure orifices and two dynamic pressure transducers positioned downstream of the shock control region to determine pressure fluctuations, hence buffet behavior. Thermocouples are located on the inside of the model wall to determine transition location.

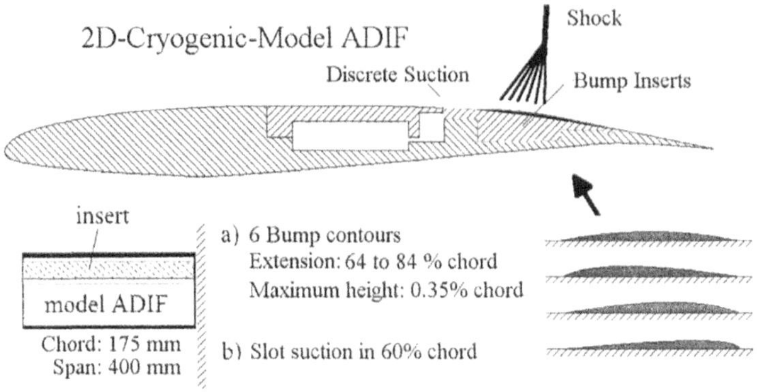

Figure 56 Wind tunnel model of the ADIF airfoil with bump inserts and suction slot [4.15]

The **Sheared-wing** model, corresponding to the ADIF airfoil, Figure 57, has been designed to study sweep effects on shock control [4.16, 4.17]. The wing has a chord length of 400mm and a sweep angle of 26°. On the upper surface of the wing there is a removable insert between 0.575c and 0.84c, similar to the ADIF airfoil model. It allows the contour of the datum wing to be exchanged with a bump contour or a perforated surface with a cavity underneath. Three different bump contours with the same asymmetric shape and the same length of 20% chord have been investigated with the location and the height of the bumps being as follows:

1) bump Ia: $x_{crest}/c = 0.715$, $h_{bump}/c = 0.1573\%$
2) bump Ib: $x_{crest}/c = 0.715$, $h_{bump}/c = 0.3146\%$
3) bump II: $x_{crest}/c = 0.760$, $h_{bump}/c = 0.1573\%$

The perforated plate / cavity arrangement was placed between $x/c = 0.60$ and $x/c = 0.75$; the porosity was 8% and the diameter of the holes 0.3mm. A suction slot was located at $x/c = 0.55$ and, in case of the ventilation insert, an additional perforated strip of 1% chord length was placed at $x/c = 0.80$.

The model was equipped with conventional pressure orifices and with dynamic pressure transducers for buffet detection, Figure 57. As was mentioned before, the effective wind tunnel side walls are contoured according to stream surfaces to achieve infinitely swept-wing flow conditions.

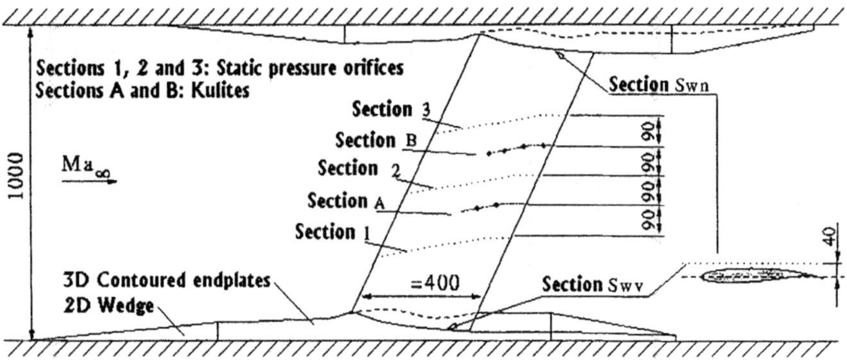

Figure 57 Test setup and instrumentation of the ADIF swept-wing configuration [4.16]

5.2.3 Experimental Results for the ADIF and DA LVA-1A Airfoils

Experiments with the ADIF reference airfoil model were carried out with and without bumps. The results served as a reference for the investigation of sweep effects and to verify the design and effectiveness of the surface bumps optimized as indicated above. Based on the 2D numerical study and the experimental results, a number of bump contours were chosen for the subsequent sheared-wing investigation.

In the tests with the **ADIF** airfoil model angle of incidence sweeps were performed at Mach numbers of M_∞ = 0.735, 0.755, 0.765 and 0.775 at a Reynolds number of $Re_c = 8x10^6$ with transition fixed at 30% chord on the upper surface and 7% chord on the lower surface. Tests were also performed with free transition, and the correspondingly thinner boundary layer, to investigate the influence of the boundary layer displacement thickness on the effectiveness of the contour bumps. The shape of the bumps is asymmetric with the initial ramp being 12% chord in length and the closure 8%. Two bump heights, viz., 0.175%c and 0.35%c, were investigated.

Typical pressure distributions and wake profiles for the datum airfoil and the airfoil with the respective bumps at a Mach number of M_∞ = 0.765 and a lift coefficient of C_L = 0.610 are presented in Figure 58: the bumps result in a considerable spreading of the shock, indicative of a reduction in wave drag; however, as is shown by the wake profiles, also viscous drag seems to be reduced resulting at the conditions considered in total drag reductions of ΔC_D = 16% and 20% for the higher and the lower bump, respectively. It should be noted that the higher bump exhibits here the lower drag reduction. Considering the corresponding shock spread, one notices that there is an incipient expansion following the initial pressure rise, and a second shock, causing the reduced bump effectiveness due to an increase in wave drag — compared to the smaller bump — also discernable in the wake profiles. When further decreasing the lift coefficient this shock pattern will resolve into a strong double-shock system due to a mismatch between shock and bump location, as already discussed in Chapter 4, leading to an increase in total drag relative to the datum airfoil as will be shown below.

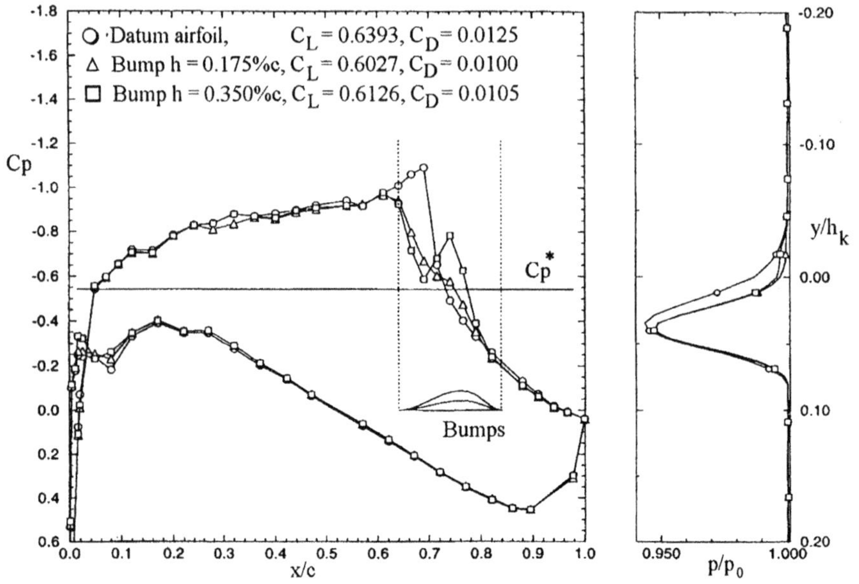

Figure 58 Pressure distributions and wake profiles for the ADIF airfoil without and with bump control, $M_\infty = 0.765$, $Re_c = 8x10^6$, $C_L \approx 0.610$ [4.15, 4.17]

The corresponding drag polars, Figure 59, confirm the development indicated by the single-point condition of the pressure distributions in Figure 58: drag reductions of up to 20 drag counts (ΔC_D=0.0020) at a lift coefficient of $C_L \leq 0.66$ and up to 35 counts at higher lift coefficients, amounting to drag reductions of 17% and 23%, respectively, can be achieved. Also indicated in Figure 59 is that the lower bump shows better performance at the lower lift coefficients, while the higher bump is more effective at higher lift, i.e., in the presence of stronger shock waves, but generates an increase in drag at lift coefficients below 0.55. Generally, the results indicate the need for an adaptive bump in order to achieve full control effectiveness.

In the test series with the **DA LVA-1A** airfoil in the KRG, two different bump contours with heights of h_{Bump}/c = 0.20% and 0.40%, respectively, located between x/c = 0.59 and x/c = 0.79, were investigated. An asymmetric shape was used, similar to the bump contour of the ADIF airfoil. Angle of attack sweeps have again been performed, here for three Mach numbers, viz., M_∞ = 0.76, 0.77, and 0.78, at Re = $6x10^6$ with transition fixed at 48% chord on both surfaces to allow long runs of laminar flow.

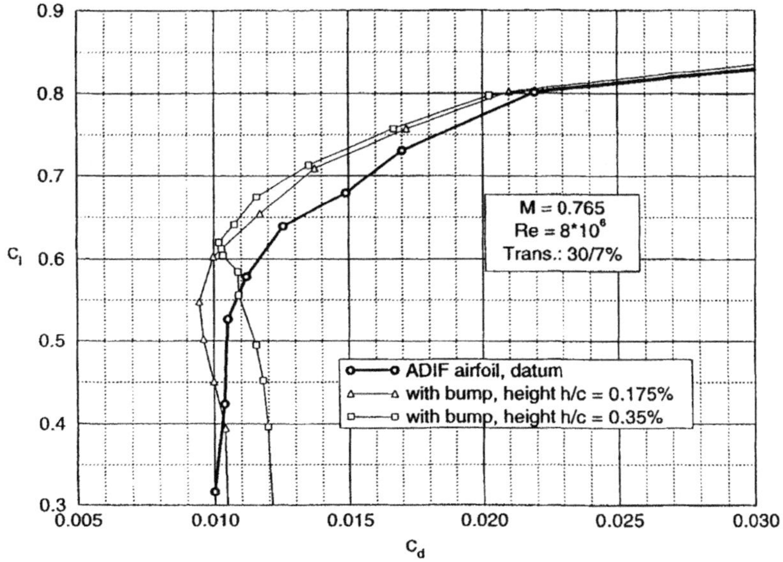

Figure 59 Drag polars for the ADIF airfoil without and with bump control $M_\infty = 0.765$, $Re_c = 8 \times 10^6$ [4.15, 4.17]

Representative results are presented in Figure 60 for a freestream Mach number of $M_\infty = 0.77$: the comparison of the drag polars with and without control indicates that for this laminar-type airfoil with strongly accelerating flow on the upper surface, hence stronger shocks, the effectiveness of the smaller bump starts already at lift coefficients between $C_L = 0.01$ and 0.02 with drag reductions reaching at, say, $C_L = 0.40$, already 19% [4.17]. The higher bump creates additional drag at low lift coefficients but somewhat higher benefits in the intermediate lift range, a result similar to the observations for the ADIF airfoil, Figure 59, but not as extreme.

Figure 61 shows a Mach number sweep at the constant angle of incidence of $\alpha = 1.0°$. The beneficial effect of the bump in increasing the drag-rise boundary is clearly demonstrated. Even more significant, lift is increased at the same time.

5.2.4 Experimental Results for the ADIF Infinitely Swept Wing

Tests with the **ADIF sheared-wing** model were performed in the 1x1 Meter Transonic Wind Tunnel two-dimensional adaptive-wall test section [4.16, 4.17]. The tests were mainly carried out at a Mach number of $M_\infty = 0.852$ and a Reynolds number of $Re_c = 6.7 \times 10^6$, i.e., the design condition for the contoured side walls. Due to a contamination of the leading edge, transition had to be fixed at 10% chord on the suction side; on the pressure side transition was fixed at 15% chord. Since a swept adaptation of the top and bottom walls — to achieve wall-interference-free flow — was not possible in the 2D adaptive-wall test section, the best approach for adaptation was investigated prior to testing with the 3-D Navier-

Stokes code FLOWer [4.16], simulating the flow about the model in the test section. It was found that the best agreement of the pressure distributions with the undisturbed swept-wing case was achieved with a wall contour corresponding to the center streamline of an unbounded flow around the model. Therefore, the pressure distribution at the centerline of the top and bottom wall was used for the adaptation process.

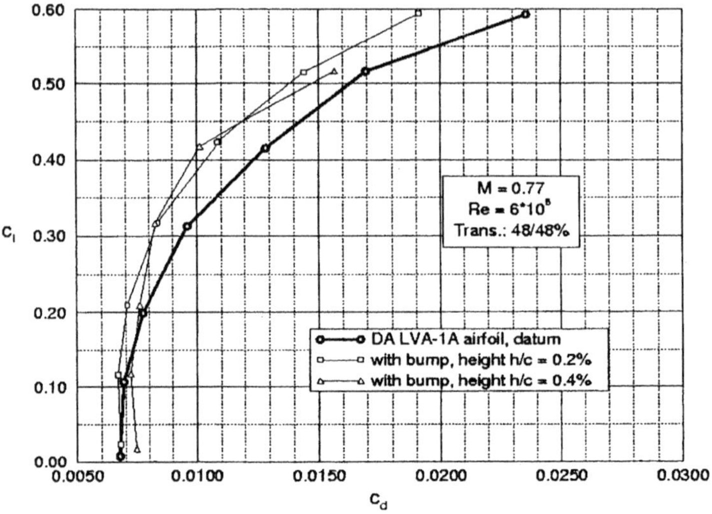

Figure 60 Drag polars for the DA LVA-1A airfoil without and with bump control $M\infty = 0.770$, $Re_c = 6x10^6$ [4.17]

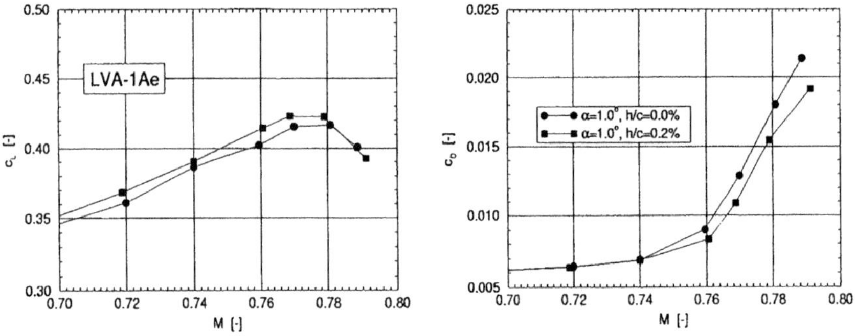

a. Lift coefficient b. Drag coefficient

Figure 61 Drag and lift dependence on Mach number, airfoil DA LVA-1A $\alpha = 1.0°$, $Re_c = 6x10^6$ [4.17]

In addition to the top and bottom walls, the side walls of the test section were, as mentioned above, contoured to achieve infinitely swept-wing flow. Since the wall contours depend on the wing geometry and the flow conditions, the contour is correct only for one (design) point, here $M_\infty = 0.852$, $Re_c = 6.7 \times 10^6$, $\alpha = 0.0°$. The pressure distributions for the three span-wise stations S1, S2 and S3 (see Figure 57) exhibit a rather small span-wise gradient indicative of the successful design of the wall contours, Figure 62. At off-design conditions, the flow differs slightly from the infinite swept-wing flow, mainly in that the sweep angle of the shock is not the same as the geometrical sweep angle and the trailing edge pressure is not constant in span-wise direction. The influence of the different means of shock control on drag reduction and buffet onset is, nevertheless, primarily judged at the design point of the contoured side walls.

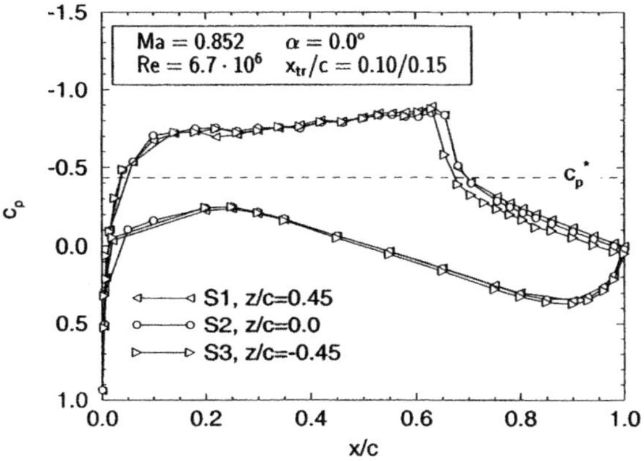

Figure 62 Effectiveness of contoured side walls at the design condition, ADIF sheared wing, $M_\infty = 0.852$, $Re_c = 6.7 \times 10^6$, $\alpha = 0.0°$ [4.16]

5.2.4.1 Hybrid Control by passive ventilation/suction

The effect of passive control by a perforated surface / cavity arrangement has been investigated extensively within the EUROSHOCK (I) project [1.5]. It was found that wave drag may be reduced substantially by this type of control due to the displacement effect of the ventilation in the shock region weakening the shock; however, primarily for laminar-type airfoils a dominating increase in viscous drag occurred, not only nullifying the reduction in wave drag but leading to an increase in total drag. For turbulent airfoils, the drag reduction was only marginal and limited to specific freestream conditions. Therefore, a combination of passive ventilation in the shock region with boundary layer suction downstream and upstream, respectively, of the passive control region has been studied during the present investigation [4.16, 4.17]. This type of hybrid control was expected to reduce the increase in viscous drag and hence reduce total drag.

Since hybrid control by discrete suction downstream of the passive cavity is also considered by DERA for the DRA-2303 airfoil and discussed below, we will here look at the effect of upstream discrete suction in conjunction with passive cavity ventilation in the shock region, Figure 63: pure passive ventilation leads to an increase in total drag of $\Delta C_D = 11\%$ at the design point of the side-wall contour — despite a decrease in wave drag — which can clearly be identified as resulting from the dominating increase in viscous drag. If the ventilation is coupled with boundary layer suction upstream of the interaction region, viscous drag is reduced — and wave drag possibly increased — resulting in a situation where any further decrease in viscous drag by suction is exactly compensated by a wave drag increase so that a total drag reduction is not achieved. These global results are, of course, in perfect consent with the local results obtained in the basic experiments (also see Chapter 3). A drag reduction is, however, achieved by suction at the lower lift coefficients where shock waves, hence wave drag, are of minor influence; furthermore, at high lift coefficients where the wave-drag increase is less severe. It will be seen in Chapter 5.3.3 that hybrid control employing downstream suction leads to similar results.

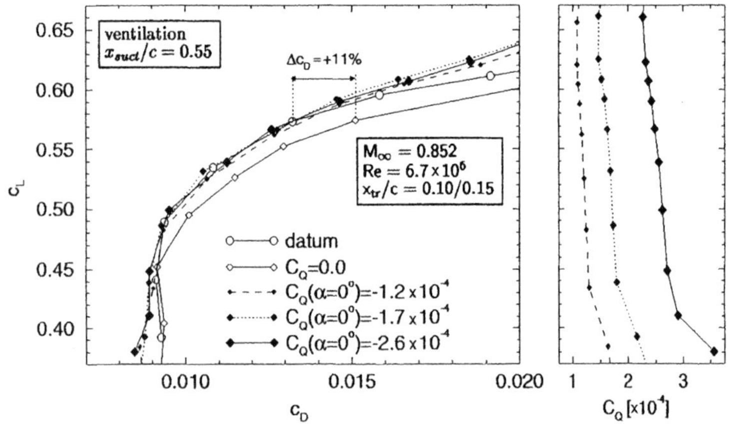

Figure 63 Hybrid control effects on the drag polars of the ADIF sheared-wing configuration, $M_\infty = 0.852$, $Re_c = 6.7 \times 10^6$ [4.16, 4.17]

5.2.4.2 Flow control by bump and by bump plus suction

The influence of contour bumps of different heights, here $h_{Bump} = 0.1573\%c$ and $0.3146\%c$, respectively, with crests located at 71.5% chord, is generally similar to the effect at 2D conditions: at the (side-wall contour) design Mach number $M_\infty = 0.852$, the higher bump leads at lower lift coefficients to a considerable increase in drag, while at lift coefficients of $C_L > 0.53$ a drag reduction occurs with the sensitivity of the drag reduction to bump height, however, negligible (see Figure 52). At the design point of the side walls, a drag reduction of about 8% is achieved for both bumps.

A more interesting test case, since not considered in the 2D experiments — but treated numerically in conjunction with the DRA-2303 airfoil in Chapter 4 — is the bump in combination with upstream discrete suction. The effect of this type of control on drag reduction is demonstrated in Figure 64: considering again the design point of the side walls, one observes that the bump by itself — denoted $C_Q = 0.0$ — reduces drag, as mentioned above, by 8%. Applying a low suction rate upstream reduces drag by 12%, while the somewhat higher suction rate causes a further drag reduction of up to 22%. The corresponding surface and wake pressure distributions in Figure 65 indicate that the bump reduces wave drag due to the spreading of the shock as well as viscous drag; adding suction does not necessarily increase wave drag — thanks to the now more effective bump — but reduces viscous drag which is indicated by the better pressure recovery over the rear of the wing but also by the narrowing of the wake [4.16, 4.17].

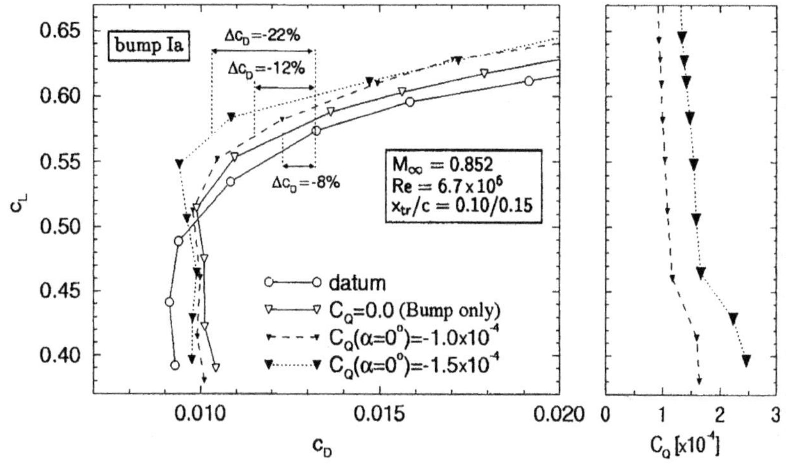

Figure 64 Drag polars for the ADIF swept wing with control by bump and bump plus upstream suction, $M_\infty = 0.852$, $Re_c = 6.7 \times 10^6$, $h_{Bump} = 0.1573\%c$ [4.16, 4.17]

Comparing now the 2D and 3D results with regard to the bump drag reduction potential, the following must be considered. In the case of the swept-wing experiments, transition had, as already mentioned, to be fixed at 10% chord on the upper surface. This lead to a larger boundary layer thickness, with some inherent spreading of the shock, than in the 2D case, where the effect of the high Reynolds number at flight conditions, i.e., a thin boundary layer, was simulated by duplicating the displacement thickness upstream of the shock by fixing transition at 30% chord. If one now regards the boundary layer suction upstream of the shock as a means of Reynolds number simulation, it is possible to determine the effect of sweep on drag reduction by the bump: at the design point of the side walls ($C_L \approx 0.57$), the drag reduction for the swept wing configuration is slightly lower than in the 2D case at similar conditions, i.e., 22% versus 23%, a result also predicted by the computations discussed in Chapter 4.

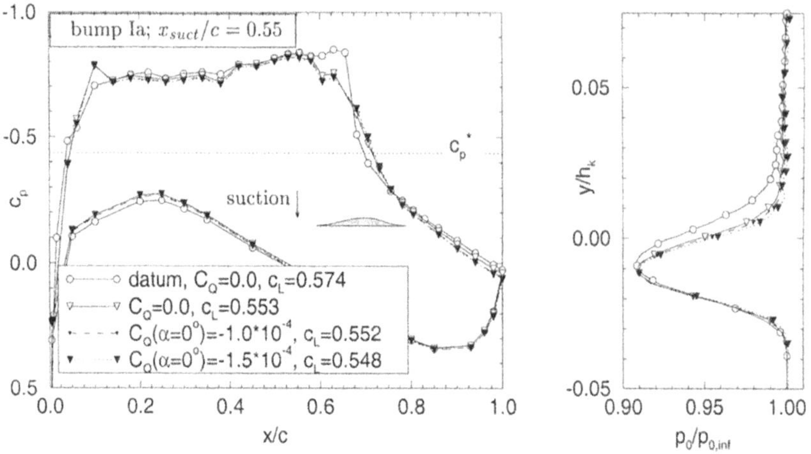

Figure 65 Surface and wake pressure distributions for the ADIF swept wing with control by bump and bump plus upstream suction, M_∞=0.852, Re_c=6.7 x 10^6, $C_L \approx 0.55$ [4.16, 4.17]

The bumps considered up to now were essentially designed for drag reduction. They were less effective in shifting the buffet boundary to higher lift coefficients. Therefore, Bump II, which is placed just downstream of the shock and does not reduce the shock strength at all, was introduced. As indicated in Figure 66, where the effect of various means of control on the buffet boundary are compared, this rear bump is very effective in shifting buffet onset to a higher lift coefficient by its positive influence on separation. The buffet boundary is here defined according to Criterion C3, onset of (major) unsteadiness [4.16].

5.3 DERA Large-scale Airfoil Experiments

5.3.1 Wind Tunnel Characteristics

The DRA-2303 airfoil was tested in the 8ft x 8ft Subsonic-Supersonic wind tunnel at DERA Bedford in the datum configuration and with various controls [5.1, 5.6]. The tunnel is a variable pressure, closed circuit, continuously running facility with a solid wall working section, Figure 67; therefore appropriate corrections have been applied to the data to account for tunnel-wall interference and blockage. Normal force and pitching moment coefficients were determined by appropriate integration of the surface static pressures around the airfoil and sectional drag was inferred from pitot- and static-pressure measurements obtained from the large wake rake depicted in Figure 67.

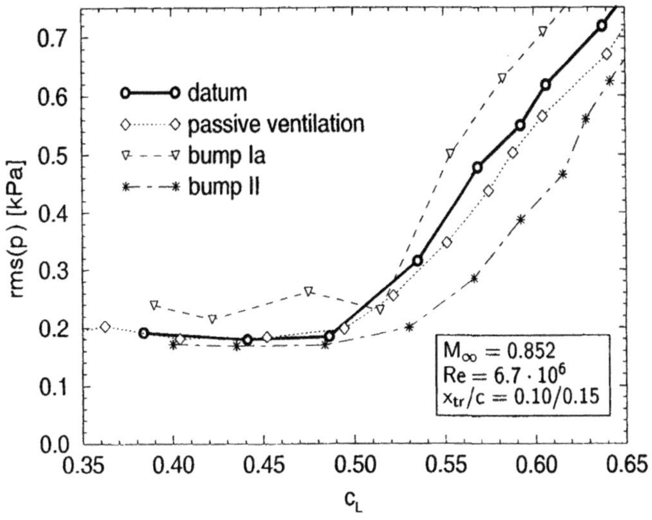

Figure 66 RMS-values of the pressures at 90%-chord, ADIF swept wing with control $M_\infty = 0.852$, $Re_c = 6.7 \times 10^6$ [4.16]

Figure 67 DRA 2303 airfoil in the working section of the DERA 8Ft x 8Ft tunnel [5.6]

5.3.2 Airfoil Characteristics and Wind Tunnel Model

The DRA-2303 airfoil section was designed to be representative of a laminar-flow section with long runs of favorable pressure gradients on both the upper and lower surfaces extending to 50% chord close to the design conditions, i.e., a freestream Mach number of $M_\infty = 0.68$ and a lift coefficient of $C_L = 0.50$ [5.6]. The 635mm-chord model consists of a main spar with detachable leading- (0 to 0.17c) and trailing-edge (0.7 to 1.0 c) sections, Figure 68. On the upper surface of the main spar there is a removable panel between x/c = 0.39 and 0.69 allowing various control systems to be inserted; an insert was also manufactured to form the original profile. The airfoil, including inserts, was equipped with static pressure holes at three spanwise plotting stations on the upper and lower surfaces. Various pressure measuring stations were located inside the cavity. The characteristics of the various inserts are as follows:

1) Porous surface between 0.475c and 0.625c (Active Suction)
2) Porous surface between 0.45c and 0.46c (Discrete Suction)
3) Porous surfaces between 0.474c and 0.625c and 0.65c and 0.66c (Hybrid Suction)

The open area ratio of the suction surface was in all cases 4% based on local area; the perforations were formed by laser-drilling with a nominal diameter of 0.076 mm. Calibrations of the surfaces showed that for both, flow into and out of the cavity similar characteristics existed which were consistent with the hole diameter specified.

5.3.3 Discussion of Results

The test conditions for the DRA-2303 airfoil investigations were: $M_\infty = 0.68$ at a nominal Reynolds number of $Re_c = 19 \times 10^6$ and transition fixed at 5% chord on the upper and lower surfaces. The angle of incidence was varied from approximately $\alpha = -2°$ to $4°$, the later being chosen to ensure that buffet onset conditions were reached [5.1].

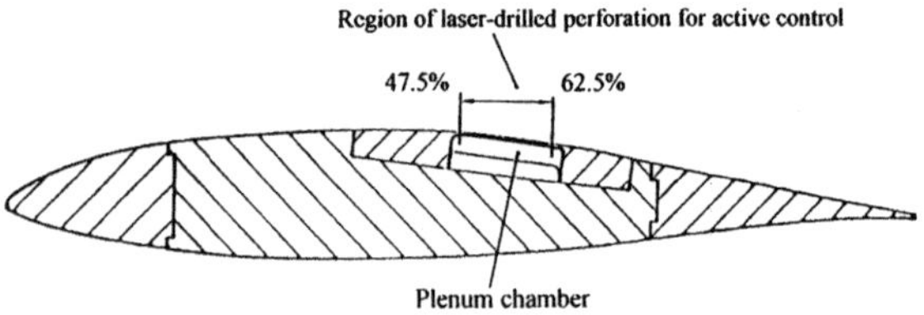

Figure 68 DRA-2303 airfoil model with insert for active control by part-suction [5.6]

5.3.3.1 Discrete suction

For the case of *Discrete Suction*, the variation of normal-force coefficient with angle of incidence is shown in Figure 69a. It can be seen that the effect of varying the rate of suction on lift is small. For the no-suction case, $C_Q = 0$, the effect is to reduce normal force by approximately $C_N = 0.04$ at a given angle of incidence at the higher normal-force coefficients. As the rate of suction is increased, the decrement in normal-force is restored; at the maximum suction coefficient of C_Q=0.00009, the normal force curve is almost indistinguishable from the datum (solid) airfoil case. This is somewhat in contrast to the computational results, Figure 37, where the effect of discrete suction is found to increase lift for a fixed angle of incidence, although only marginal. More pronounced lift increases were also predicted by DERA for a turbulent airfoil due to discrete suction as a direct result of increasing effective rear camber due to the reduced boundary-layer thickness and rearward movement of the shock wave. For the present laminar-type airfoil, there is, in the experiments, little evidence of camber increase or shock movement due to suction which is probably a feature of this type of (laminar) airfoil design where the favorable pressure gradient is ensuring that the boundary layer is kept thin so that the benefits of discrete suction on lift are less than for a turbulent design.

Figure 69b shows the variation of drag coefficient with normal-force coefficient. For the no-suction case drag is increased; however, as the suction coefficient is raised, drag is noticeably reduced. The increase in drag — as the reduction in lift — for the no-suction case is probably due to the aerodynamic roughness of the porous surface and some re-circulating flow in the porous suction-strip region. The latter is also indicated by the pressure distributions in Figure 70a where, at this condition, an initial pressure increase is seen.

The reduction in drag with suction coefficient appears to be monotonic, suggesting that for higher values of suction, larger drag reductions should be possible; furthermore, the use of a slot in place of the porous strip may be beneficial. Clearly, the effect of suction is to cause changes in the boundary-layer development downstream of the suction strip and, as already discussed in Chapter 4, a reduction in boundary layer thickness downstream of the interaction/control region down to the trailing edge, all leading to a dominant reduction in viscous drag. Ultimately — as the angle of incidence is increased — the tendency for trailing edge separation diminishes. This is confirmed by the fact that suction improves the normal-force coefficient for trailing-edge pressure divergence, Figure 70b.

It is worth noting that the present scheme is primarily a boundary-layer control device and therefore independent of flight condition. The true effectiveness of the suction systems can, of course, only be assessed once allowance is made for internal or 'pump' drag. The latter is depicted in Figure 71, where, for the maximum suction rate applied, the drag variation with normal force with and without an allowance for pump drag is plotted. It can be seen that pump drag approximately halves the drag reduction due to suction with the reduction, nevertheless, still being considerable [5.1].

5.3.3.2 Active and Hybrid Control

The case of *Active Suction* by a perforated plate/cavity arrangement shall not be considered here separately but referred to in discussing the case of hybrid control by means of a passive cavity in the shock region and active suction downstream (also see Chapter 19).

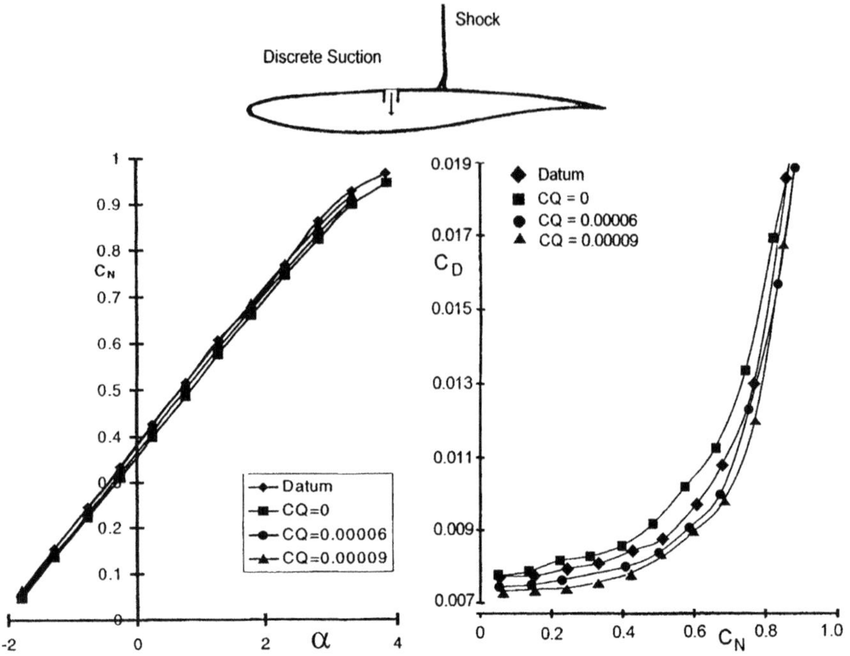

a. Normal force coefficient

b. Drag coefficient

Figure 69 Effect of discrete suction on lift and drag coefficient, airfoil DRA-2303, $M_\infty = 0.68$, $Re_c = 19x10^6$, $(x/c)_{trans} = 0.05$ [5.1]

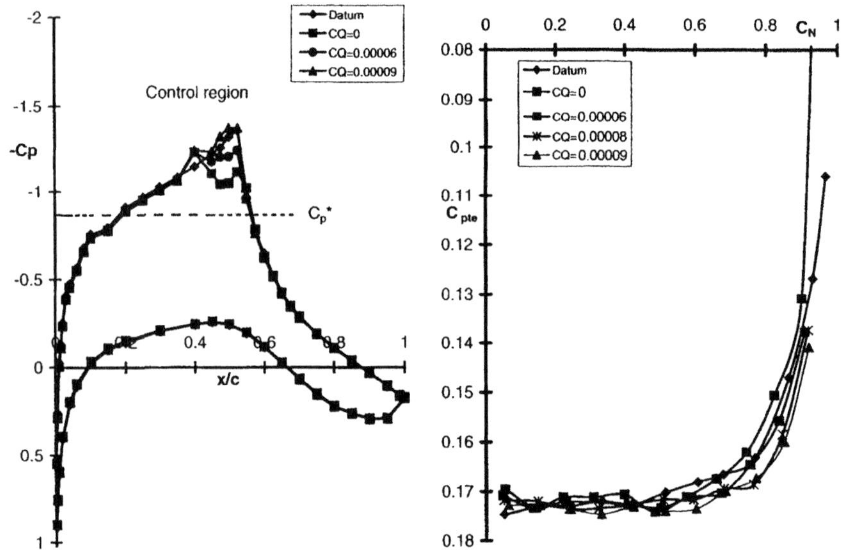

a. Surface pressure distributions

b. Trailing edge pressure coefficient

Figure 70 Effect of discrete suction on the pressure distribution, airfoil DRA-2303, $M_\infty = 0.68$, $Re_c = 19x10^6$, $C_L = 0.60$, $(x/c)_{trans} = 0.05$ [5.1]

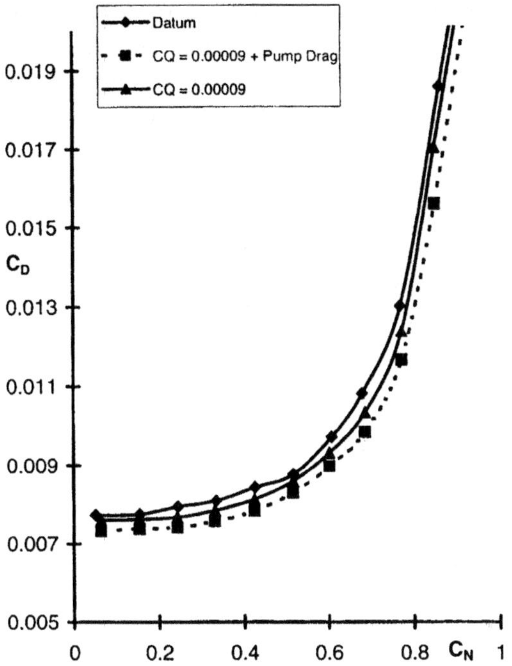

Figure 71 Effect of discrete suction on drag with and without accounting for pump drag, airfoil DRA-2303, $M_\infty = 0.68$, $Re_c = 19x10^6$, $(x/c)_{trans} = 0.05$ [5.1]

For the *Hybrid-Suction* case, the variation of normal-force coefficient with angle of incidence is shown in Figure 72a: for high values of normal-force coefficient ($C_N > 0.7$) there is a considerable loss of normal-force for this configuration, although it is not as large as that for the single-cavity active suction system. There appears, otherwise, to be no discernible effect of suction with the suction on and off cases essentially indistinguishable.

The variation in drag coefficient with normal-force coefficient is shown in Figure 72b for varying suction rates. Here, the secondary suction control region behind the 'passive' region is designed to reduce the additional viscous drag generated by the passive re-circulation. Indeed, increasing suction reduces the drag relative to the suction-off, passive case; however, it is clear that in order to obtain significant reductions, it would be necessary to resort to very high suction rates which, in turn, would result in a large 'pump' drag penalty. It is, therefore, debatable as to whether a nett drag reduction can be achieved with this arrangement. Another factor to be considered is that at high suction rates the flow will choke in the holes of the porous surface, rendering any further increases futile. This would suggest the possibility of either increasing the porosity or replacing the downstream porous surface with a slot, but at the high suction rates needed it would still appear unlikely that such a system would be viable. Generally, the present results confirm the results discussed in Chapter 3, namely, that unrealistically high suction rates are required to bring the boundary layer thickness parameters downstream of the interaction/control regions to levels approaching the ones for the no-control case. The dominating effect of the increase in viscous drag on total drag associated with hybrid control is here clearly demonstrated since wave-drag reduction is unimpaired by this control scheme.

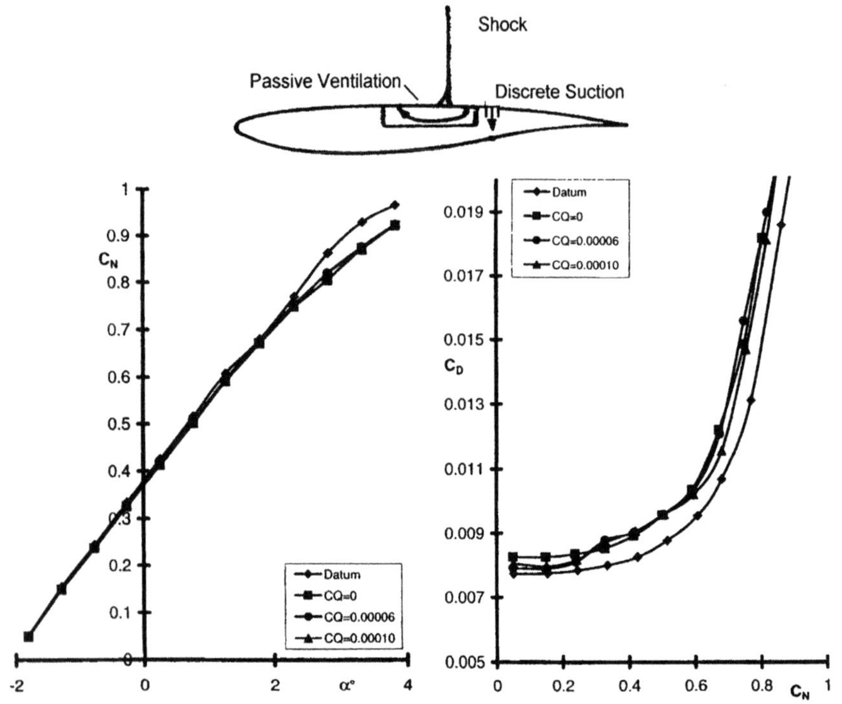

a. Normal force coefficient b. Drag coefficient

Figure 72 Effect of hybrid control on lift and drag coefficient, airfoil DRA-2303, $M_\infty = 0.68$, $Re_c = 19 \times 10^6$, $(x/c)_{trans} = 0.05$ [5.1]

The above statement is supported by the pressure distributions in Figure 73a at a lift coefficient of $C_L = 0.60$: clearly, the upstream passive cavity reduces the shock strength, hence wave drag; however, considering the trailing edge pressure as function of the normal force coefficient with and without hybrid control, Figure 73b, one sees that the pressure recovery in the case of control is much less than the one for the no-control configuration and that applying suction downstream of the passive cavity does not significantly improve the situation, a rather drastic demonstration of the large boundary layer displacement thickness present over the rear part of the airfoil upper surface causing the said increase in viscous drag.

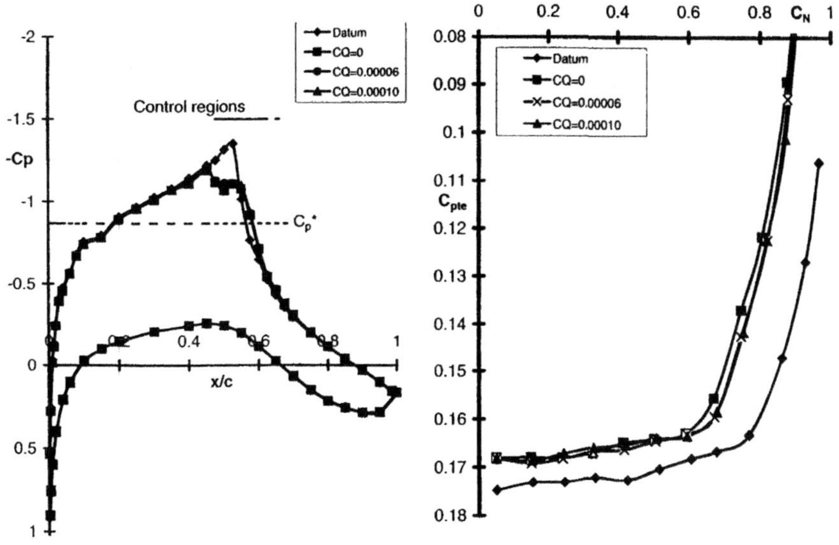

a. Surface pressure distributions b. Trailing edge pressure coefficient

Figure 73 Effect of hybrid control on the pressure distribution, airfoil DRA-2303, $M_\infty = 0.68$, $Re_c = 19\text{x}10^6$, $C_L = 0.60$, $(x/c)_{trans} = 0.05$ [5.1]

5.4 Conclusions and Future Work

The experimental results for airfoils and wings, designed for significant extents of laminar flow on the upper surface, show that the application of shock control in the form of 'bumps' significantly reduces total drag; analysis of the experimental data shows that this derives from a large reduction in wave drag whilst essentially having only a minor effect on viscous drag. For this class of airfoil or wing this is extremely beneficial since strong shock waves tend to form at a fixed chordwise position following the long favorable pressure gradient necessary to maintain natural laminar flow. Two potential implementations of the bump system therefore emerge: ① A fixed bump designed into the airfoil or wing from the outset; the results presented here suggest that there is no viscous drag penalty for having a bump present at all flight conditions, however, this type of bump must be limited in height. ② A bump of variable geometry which can be activated when necessary to reduce wave drag.

For 'conventional' airfoils and wings of 'turbulent' design the experimental work carried out suggests that similar benefits can be obtained, however, a complication for this class of wings is the potentially large chordwise movement of the shock wave. In this case a fixed bump would lead to large drag penalties at off-design conditions; however, a deployable bump could be utilized for certain parts of the flight envelope. The best solution for this class of wing would appear to be a 'smart bump' which could track the position of the shock wave and therefore always be at an optimum. Such a bump could also be placed, as shown in the DLR tests, at a chordwise position that would allow to effectively increase the buffet boundary. Both applications require a structure that can adapt to flight conditions through the use of smart structures or shape memory alloys. It is in the area of materials and structures that further work should be considered in order to arrive at a practical implementation of such a system.

Aerodynamically, the application of bumps to a wing should pose no significant problems. The studies of sweep effects carried out here suggest that the flows are predictable and realizable, which is perhaps not surprising since 'bumps' are merely a surface shape change which is equally applicable in three as well as two dimensions.

The use of suction, in general, does not appear to be a suitable solution to the problem of drag reduction. The exception to this seems to be the use of Discrete Suction by making the chord-wise extent of the suction region so small that a maximum reduction in boundary layer momentum thickness can be obtained whilst imparting the minimum increase in total skin friction. In the case of the DERA investigation into discrete suction the system has been realized by the use of a porous strip and it can be hypothesized that a slot may be better, since the magnification effect on skin friction of suction through a porous surface is eliminated. In addition, an inclined slot or one with a forward facing area, i.e., a scoop, would have the additional benefit of utilizing the "total pressure" within the boundary layer to help drive the system, as opposed to working with the static pressure as in the case of the normal slot. This may help to overcome the detrimental effect of pump drag on the overall drag reduction, although the aerodynamic effectiveness might stay the same. However, whatever system is adopted, the requirement for internal ducting which encroaches on the wing internal volume (fuel volume) and the additional weight may negate the advantages of such a system; these aspects will be further considered in Chapter 6.

The use of discrete suction has shown an additional benefit in delaying the onset of trailing edge separation and hence, by implication, buffet onset. This should be expected from the physical mechanism of the system: due to thinning the boundary layer and by taking advantage of the magnification effect of the local pressure gradients, the boundary layer at the trailing edge is thinner and more robust than it would have been without suction, therefore flow separation is delayed to higher angles of incidence or lift coefficients.

In summarizing, the wind tunnel tests on various airfoils and wings in several major European wind tunnels have generated the following conclusions:

- Shock control via Bumps is very effective at reducing wave drag by the ramp effect that it imparts on the flow in the shock region.
- For laminar flow-type airfoils and wings, the wave drag reduction is not accompanied by any increase in viscous drag, provided the placement of the Bump is correct, which generally suggests an Adaptive Bump, especially for turbulent airfoils.

- Bumps were also found to have a positive effect on the buffet boundary, especially when adapting the correct location for this purpose.
- The use of Discrete Suction leads to a nett reduction in drag for all values of suction coefficient considered. The effectiveness of the system is also independent of Mach number and lift coefficient, therefore making it of use at all flight conditions.
- The effects of Discrete Suction in combination with a Bump are additive.
- Active Suction by a perforated plate/cavity arrangement is ineffective at reducing drag for the configurations tested, the suppression of the passive re-circulation effect and the associated viscous drag being insufficient at feasible suction rates.
- The Hybrid Suction system is unable to overcome the development of the thickening boundary layer associated with the 'passive' control region. It is probable that in order to reduce the boundary layer thickness sufficiently to yield a reduction in drag, the rate of suction required would be such that the benefits in terms of wave drag reduction would be nullified by the pump drag.

Future work concerning the application of a bump should essentially consider the structural integration of adaptive bumps into existing and/or new wing designs, turbulent or laminar. Especially for the former, emphasis should be placed on how the incorporation of a bump may lead to the design of thicker wings with the benefit of reduced structural weight, hence reduced costs and time to market. Furthermore, other flow control mechanisms, aiming at drag reduction and the avoidance of separation, such as, for instance, sub-boundary layer vortex generators, mass-less air jets, and possibly trailing-edge mini flaps, should be considered in combination with bumps, especially to maintain the margin between the cruise regime — which may be extended by the bump — and the buffet boundary. Concerning control involving cavity suction, emphasis should be placed on applications where drag reduction is not the main driver, such as, for instance, in the case of supersonic air intakes where the avoidance of separation and shock oscillations as well as the establishment of the best possible flow uniformity are of prime interest.

6 Benefits of Control Application to Aircraft Wings / Aircraft (Task 4)

The objectives of this study were the introduction of various potential control methods, identified in the deliberations of all tasks, into existing or new wing designs and the assessment of the benefits and penalties associated with the aircraft implementation of control. In these investigations, **ALENIA** and **Dassault** have studied the application of shock and boundary layer control to regional-jet aircraft, while **BAE SYSTEMS-Airbus UK** and **EADS-Airbus D** considered the application of control to a large A340-type transport aircraft with a turbulent and a laminar wing design, respectively. Preceding the application studies were aircraft-specific investigations and optimization studies of control for the respective airfoils and wings.

6.1 Control Assessment Criteria

When considering penalties and benefits of control, certain criteria have to be established by which a design can easily be judged. Such criteria were suggested by BAE SYSTMS-Airbus UK [6.1] and by Dassault Aviation [6.2]. The former criterion is based on Cash Operating Costs (CoC) given by the formula

$$\Delta\, \mathrm{CoC/CoC} = 0.49\ \Delta\, C_D/C_D + 1.9\mathrm{x}10^{-3}\ \Delta\, W + 0.113\ \Delta\, \mathrm{mc/mc}.$$

Here, $\Delta\, C_D$ is the change in total aircraft drag due to control, Δ W (tons) is the weight increase due to control installation and Δ mc/mc is the increase in costs caused by the additional maintenance needed due to the control system. Dassault Aviation has defined for a medium-size jet for the purpose of the evaluation of control concepts a criterion based on the maximization of the range in the form

$$C = -10^5\ \Delta\, C_D - 0.068\ \Delta\, W - 9.27\ \Delta\, V - 241.5\, \dot{m}\,,$$

with $\Delta\, C_D$ being the change in drag coefficient, Δ W (lb) the change in weight and Δ V (ft^3) the loss in fuel capacity due to a system installation inside the wing box; $\dot{m}$ (lbs./s) is the total suction mass-flow rate removed through a suction strip. The factor "C" must be positive for a control device to be considered for application; C=0 just covers the penalties without any net gain.

6.2 Control Application to an A340-type HLF-Wing Aircraft

Work, carried out by EADS-Airbus D [6.3], is focussed here on the application of the most effective shock and boundary layer control concept to a Hybrid-Laminar-Flow (HLF) wing of a long-range aircraft. The aircraft treated approximates in wing and fuselage size and planform and in the wing structural and control surface layout the A340 aircraft. The main driver for control is drag reduction at design and off-design conditions. Note that this chapter treats the control implementation in somewhat more detail since the work is characteristic for the approach to be taken.

6.2.1 Assessment of Control Concepts

A laminar wing design as envisaged here requires — in addition to suction around the leading edge — a continuously accelerating flow on the suction side and, to maximize the benefit from laminarization, a shock location as far downstream on the wing as possible. The latter is, of course, associated with an increase in shock strength, hence wave drag, possibly nullifying the benefits of laminarization, thus providing an ideal scenario for shock and boundary layer control. Consequently, in a first step and in unison with the work within the other tasks, various control concepts were investigated with regard to their suitability for application to a laminar wing design [6.3]. Among these concepts were active control by means of a perforated plate/cavity arrangement, by discrete slot suction and by contour bumps, Figure 74.

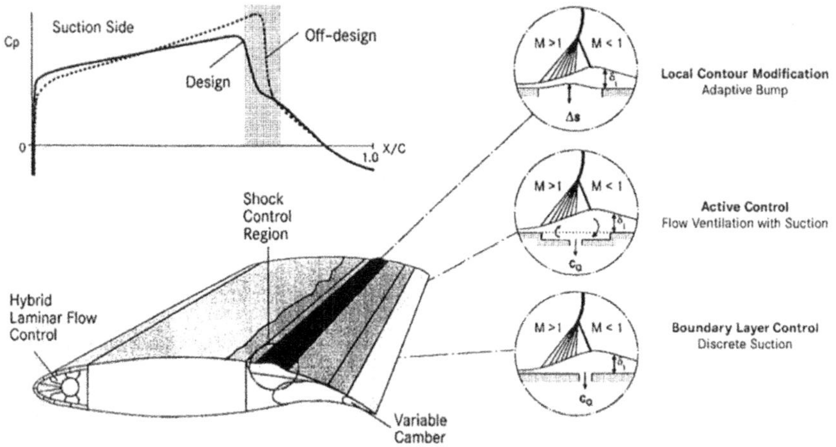

Figure 74 Control concepts considered in conjunction with the HLF wing [6.3]

The results of applying these control devices to the laminar-type airfoils considered within the present project, i.e., the DRA-2303 and the DA LVA-1A airfoil, respectively, can be summarized as follows, especially reflecting the present EADS-Airbus D numerical studies [4.5, 6.3]:

- Active shock control by part-suction reduces the detrimental viscous drag present for passive control, but as the suction rate is increased, wave drag also increases, and, accounting for pump drag, an insufficient nett-gain is realized.
- Discrete suction mainly reduces viscous drag while wave drag increases, Figure 75. Nevertheless, a reduction in total drag of up to 10% can be achieved. The suction rates are, however, such that pump drag will nearly nullify the aerodynamic drag reduction. Even if pump drag accounts for only half the gain, as found by DERA and discussed in the Chapter 5.3.3.1, a drag reduction of 5% seems not sufficient to render the installation of a suction system worthwhile as will be demonstrated in Chapter 6.5.
- The contour bump device seems to be able to produce a substantial drag reduction, as shown in the preceding discussions. Its effectiveness is, however, sensitive to changes in the freestream conditions so that only an adaptive bump, associated with increased installation penalties, might be suitable.

Nevertheless, due to its superior performance, the bump was selected for implementation on the HLF wing.

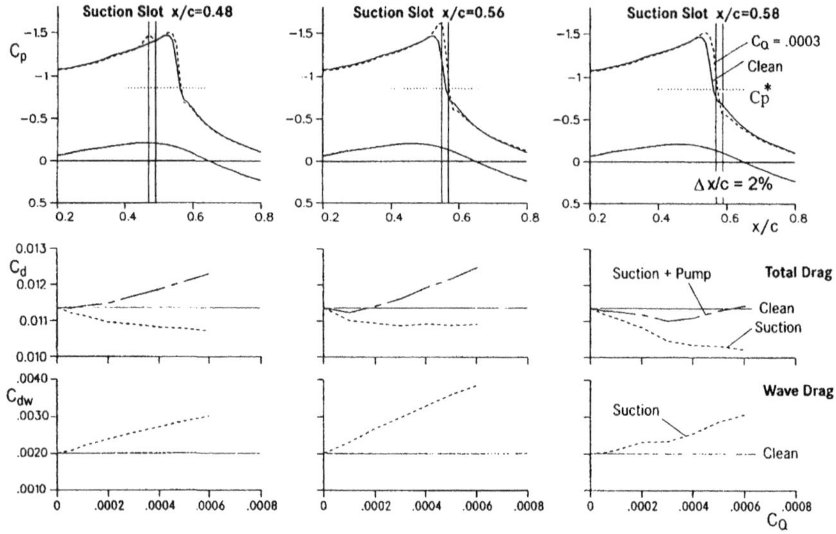

Figure 75 Drag balance for discrete suction at different slot locations, airfoil DRA-2303, $M_\infty = 0.68$, $Re_c = 19x10^6$, $C_L = 0.74$, $(x/c)_{trans} = 0.05$ [6.3]

6.2.2 Bump-Control Optimization Studies

As already indicated in the conclusions to Chapter 5, there exist two possible bump arrangements for a laminar wing with a relatively fixed shock position: ① a fixed bump of small height which is moderately effective over a larger range of lift coefficients and does not lead to significant drag increases in the basic design range of the airfoil, and ② a bump, with fixed location but variable height to realize the large potential in drag reduction already demonstrated (see, for instance, Figures 59 and 60). Such an "adaptive" bump may, however, require strong modifications in the wing design.

A bump optimization study was initially carried out for the laminar-type airfoil DA LVA-1A, already addressed in Chapter 4, Figure 49, at a Mach number of $M_\infty = 0.76$ and a lift coefficient of $C_L = 0.47$. This study lead to the following results [6.3]:

- *Bump location* The bump is in an optimum position when the pressure rise due to the shock is converted into an isentropic compression and a subsequent weak shock. When the shock is too far forward, an initial compression occurs followed by a rapid expansion and a strong shock considerably increasing wave drag, hence total drag, Figure 76, upper diagram.
- *Bump shape* The shape of the bump, for instance, a ramp, a fixed beam, a 3rd order polynomial, or a sinus function, has only a minor effect on drag ($\Delta C_D \approx 3\%$), with the fixed beam, nevertheless, giving the largest drag reduction.

- *Bump length and crest location* For a fixed crest location, l_B/c_B, an increase in bump length results in an increase in drag reduction due to the spreading of the pressure rise associated with the shock over a larger chordwise region. For a fixed bump length, here at its structurally feasible optimum of l_B/c = 20%, the maximum drag reduction is a function of the crest location with a maximum drag reduction of about 18% occurring for c_B/l_B = 50%, i.e., for a symmetrical bump; the crest is located about 2% chord downstream of the shock, Figure 76, lower diagram. It should, however, be noted that a more rearward (asymmetrical) crest location is somewhat less sensitive to shock movement due to its wider minimum and might, therefore, be preferred.
- ***Bump height*** The effect was discussed in Chapter 4, Figure 49. The bump effectiveness generally increases with bump height up to a certain height where, when exceeded, a rapid deterioration of effectiveness occurs. For optimized conditions drag reductions of up to 25% may be realized for a bump height of h_B/c = 0.5%.

The present optimization, although generally valid, was carried out at a fixed freestream Mach number and lift coefficient indicating that an optimized bump has the following characteristics: bump length 20%c, crest location at 70% of the bump length to achieve a somewhat reduced sensitivity to shock movement, fixed-beam contour, bump height at the lift coefficient considered (C_L = 0.47) 0.5%c, but otherwise variable to adjust to shock strength (lift), with the bump crest located about 2% downstream of the inviscid-outer-flow shock position.

6.2.3 Hybrid Laminar Flow Wing Section and Bump Design

For the evaluation of the benefits and penalties associated with introducing bump control into a hybrid laminar flow wing of a long-range A340-type aircraft, a wing section was employed originally designed for a wing glove to cover the basic A340 wing at a spanwise station of η = 0.50 [6.3] (see Figure 80). The pressure distributions for this wing section are shown in Figure 77 for the chosen design Mach number M_∞ = 0.82 and the off-design Mach number of M_∞ = 0.84 with the lift coefficient as parameter. The distributions are based on a sweep angle of 26° and infinite swept-wing conditions assuming a flight Reynolds number of $Re_c = 35x10^6$. Strong shocks but limited shock movements with changing Mach number and lift are characteristic features of this design.

The off-design Mach number of M_∞ = 0.84 and the local lift coefficient of C_L = 0.48 were selected for the bump design. Based on the optimization described above, a bump length of 20%c, a 70%-asymmetric shape and a bump height of 0.5%c were selected; the bump location for maximum drag reduction at these conditions was found to be between x/c = 0.63 and 0.83 with the crest located at x/c = 0.77, i.e., about 5% downstream of the shocks, the latter centered around 72% chord.

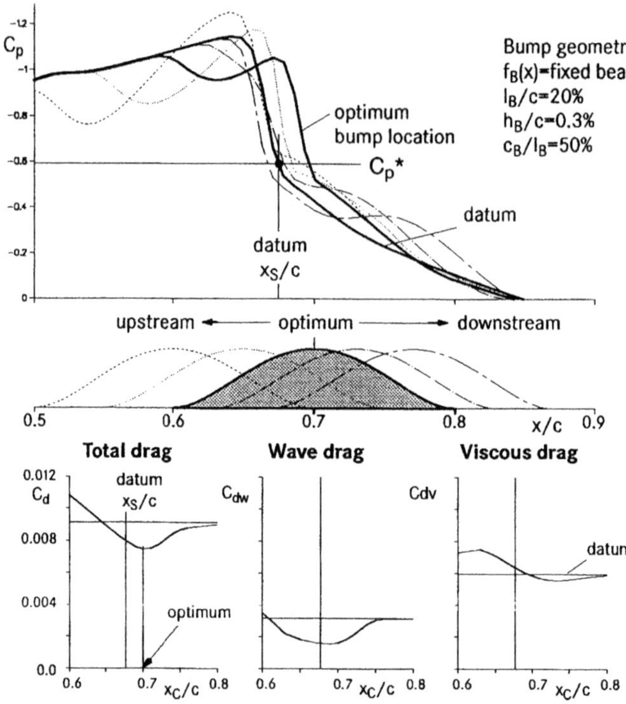

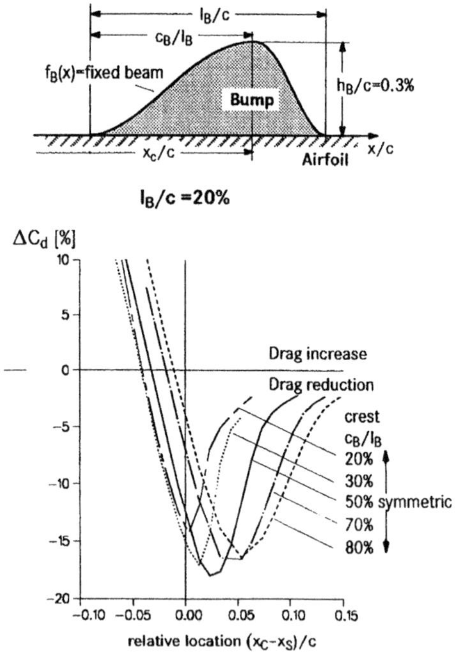

Figure 76 Effect of bump location and relative crest position on drag, airfoil DA LVA-1A, $M_\infty = 0.76$, $Re_c = 6x10^6$, $C_L = 0.47$, $(x/c)_{trans} = 0.50$ [6.3]

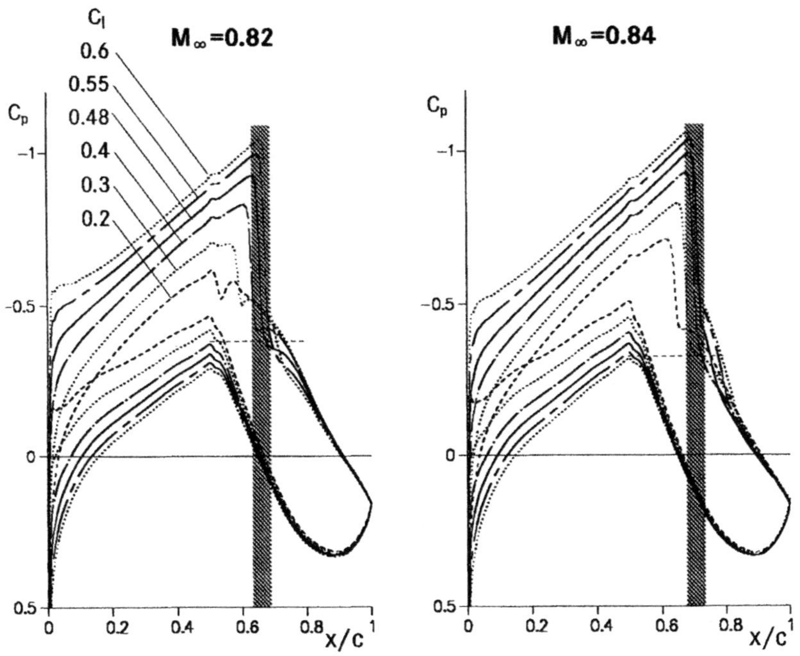

Figure 77 Pressure distributions for the HLF wing section at flight conditions, wing section PHLF1, $\phi = 26°$, $Re_c = 35x10^6$, $(x/c)_{trans} = 0.50$ [6.3]

The drag polars for the design and off-design Mach number, respectively, without and with the optimized bump, are shown in Figure 78: it is indicate that at the design point of the wing section, $M_\infty = 0.82$, $C_L = 0.48$, the bump is not effective, although the shock is relatively strong, which is due to the bump being located too far downstream of the shock; furthermore, a variable-height bump is required — as expected — to cover the range of lift coefficients of interest.

Structurally, the optimized wing/bump design, as specified above, interferes with the rear spar of the wing box, located at x/c = 0.64, and the spoiler, located between x/c = 0.74 and 0.86, of the original A340 wing so that wing modifications are needed. The wing box must be shortened and, assuming the integration of the bump into the spoiler, one option considered here, the spoiler has to be enlarged from 12% chord up to 23% chord, Figure 79. Furthermore, since strong shocks mainly appear at the outer wing, the spoiler has to be extended to the wing tip, Figure 80; for roll-control capability, a variable-camber (VC) flap is introduced over the whole span replacing the aileron. This flap can possibly also be used to fine-tune the shock location relative to the bump. The modification of the wing structure for shock control is estimated to carry a *weight increase of 0.25 tons.*

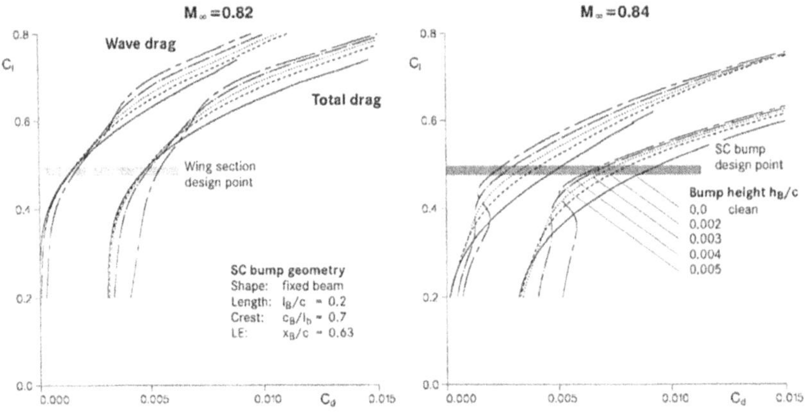

Figure 78 Drag polars of the A340 HLF wing section with variable-height adaptive-bump device at design and off-design conditions, $Re_c = 35x10^6$, $(x/c)_{trans} = 0.50$ [6.3]

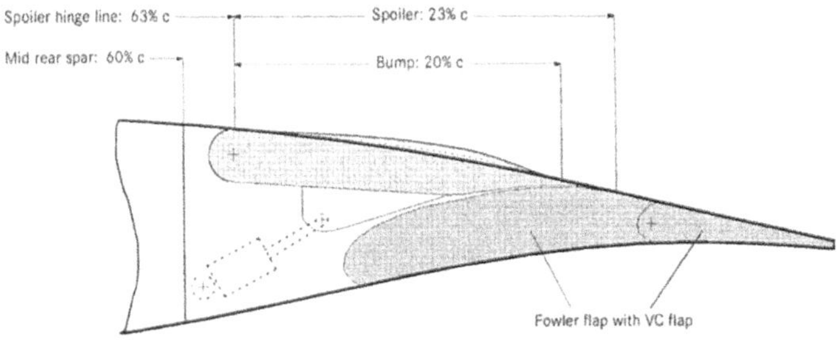

Figure 79 Integration of the adaptive variable-height bump control device into the modified A340-wing spoiler [6.3]

6.2.4 Flight Mission Benefits

Based on the wing section drag characteristics at 50% span, the laminar wing drag was corrected by the drag reduction due to the adaptive variable-height bump device. The corresponding flight polars and the drag balance for the complete aircraft at the two Mach numbers considered are shown in Figure 81. The highest drag reduction of about 4% is obtained for $C_L = 0.38$ at the off-design Mach number $M_\infty = 0.84$ and for $C_L = 0.58$ at the design Mach number $M_\infty = 0.82$.

With the given flight polars and the weight penalty, a standard North Atlantic flight mission with a range of 3500 nm was investigated assuming 600 trips per year at the cruise Mach number $M_\infty = 0.82$ and 620 trips per year at the off-design Mach number $M_\infty = 0.84$, Figure 82. The flight profile has been predicted by the standard procedure with steps of 4000ft for optimal flight altitude. At these conditions, a reduction in fuel consumption per year due to shock control of about 353 tons (1.23%) at $M_\infty = 0.82$ and of about 792 tons (2.11%) at $M_\infty = 0.84$ is obtained from flight-mission computations. Using the formula for Cash Operating

Costs (CoC) given above, a decrease of Δ CoC/CoC = 1.3% is achieved assuming an average aircraft drag reduction of $\Delta C_D/C_D = 3\%$, the weight penalty of ΔW = 0.25 tons and an increase in maintenance costs of Δ mc/mc = 0.5%. A final assessment of the bump control benefits/penalties can, of course, only be made after a much more detailed device-integration study. For possible solutions to the realization of an adaptive bump, which is actually beyond the scope of the present investigation, the reader is referred to Reference 6.3 and Chapter 21.

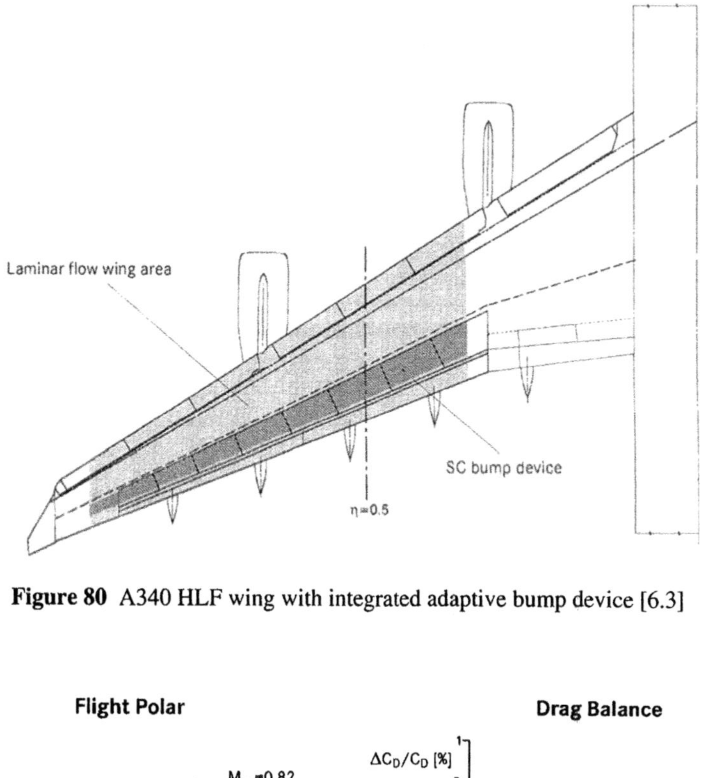

Figure 80 A340 HLF wing with integrated adaptive bump device [6.3]

Figure 81 Predicted flight polar and drag balance for the A340 HLF-wing aircraft with bump control [6.3]

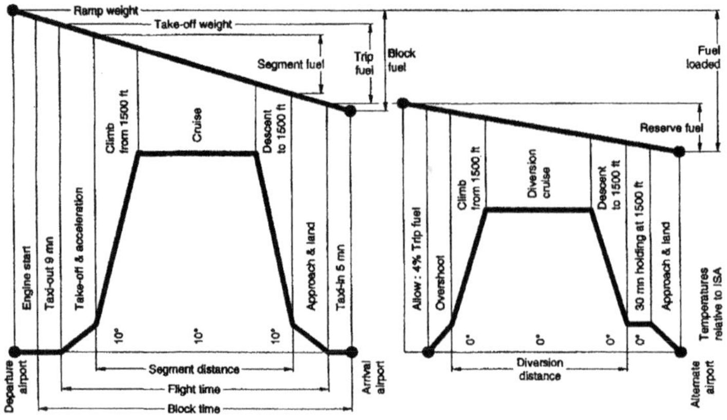

Range 3500 NM

Mach No 0,82	Utilization	Blocktime	Blockfuel	Fuel/Year	Gain in Fuel	Gain in Fuel
	Trips/Year	hr	kg	to	to	%
HLF	600	7,9	47998	28799		
HLFSC	600	7,9	47410	28446	353	1,23
Mach No 0,84						
HLF	620	7,6	55797	34594		
HLFSC	620	7,6	54622	33866	729	2,11

Figure 82 Flight mission profile for the A340 HLF-wing aircraft with bump control [6.3]

6.3 Control Application to an A340-type Aircraft with a Turbulent Wing

The objective of the present work, carried out by BAE SYSTEMS-Airbus UK, is, similar to the study of EADS-Airbus D, to apply shock and boundary layer control to an A340-type aircraft wing of existing turbulent design and a to new turbulent wing, designed under consideration of the potential of control, and to determine benefits and penalties of control implementation into the operational aircraft [6.1]. The control mechanism considered is foremost the bump device.

6.3.1 Bump Control Optimization

Initially two-dimensional and swept-wing computations were performed for a characteristic A340-type wing section using a full-potential VII code (BVGK) and the BAE SYSTEMS Euler VII code, respectively, the latter in the infinite swept-wing mode. Since the predicted shock location at the near-design Mach number $M_\infty = 0.84$, i.e., $(x/c)_{Shock} = 0.70$, did not match the shock location of the three-dimensional wing, i.e., $(x/c)_{Shock} = 0.63$ — which was considered important for subsequent structural considerations — the computations were repeated and further computations carried out with the BAE SYSTEMS-Airbus 3D Euler VII code for the complete wing which will be discussed here [6.1].

In optimizing bump control for a turbulent wing, characterized by a rather flat, slowly increasing upper surface pressure distribution, difficulties arise from the rapid movement of the shock with changing C_L and/or Mach number. This is demonstrated in Figure 83 (for the wing plan-form see Figure 80): the shock moves in the C_L-range of interest, i.e., $C_L = 0.5$ to 0.6, over nearly 15% of the chord, and over about 10% chord for a change in Mach number of $\Delta M=0.01$. The problem is exacerbated by the effect of the bump on shock movement as a result of the sensitivity of the shock location to small changes in the trailing edge flow conditions, here caused by the bump, which is characteristic for the "turbulent-type" pressure distribution considered.

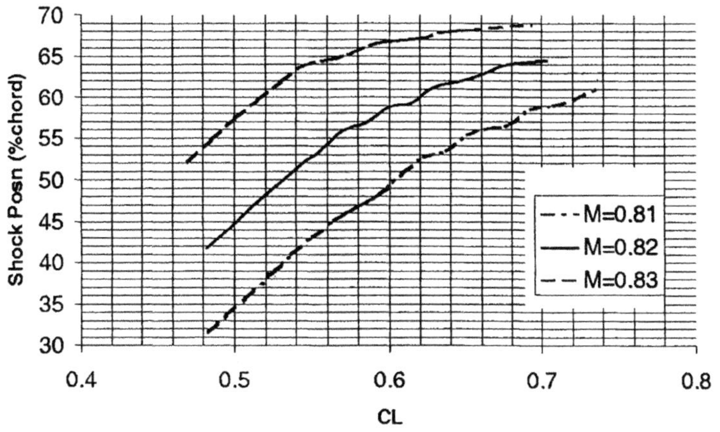

Figure 83 Effect of Mach number and lift coefficient on shock position, A340 wing [6.1]

This shock behavior suggests that a fixed bump will give an effective benefit only over a narrow lift and Mach number range centered around the design point. This conclusion is supported by the results in Figure 84 for a bump of 0.2%c height and a length of 12.5%c located 2% downstream of the shock at the design condition $M_\infty = 0.82$, $C_L = 0.59$. Here, the drag increments are plotted versus M^2C_L/M^2_{design}, i.e., effectively versus aircraft weight, at a fixed altitude. The C_L-range over which the bump shows a benefit at $M_\infty = 0.82$, the design condition, is approximately $C_L = 0.565$ to 0.655, with the maximum drag reduction being $\Delta C_D = 0.00021$ (2.1 counts); this level is unacceptably low. The bump is most effective at off-design conditions, i.e., $M_\infty = 0.81$, $C_L = 0.723$, with $\Delta C_D = 0.0007$. The aircraft operating C_L-range is about 0.5 to 0.6 at $M_\infty = 0.82$.

In an attempt to improve the effectiveness of the bump in general and, in particular, over the required operating range, a study was performed, similar to the EADS-Airbus D study for the laminar-type wing, varying the bump shape parameters and its position. The largest effect was achieved by increasing the bump height to $h_B = 0.30\%c$ with the maximum drag reduction in the lift range of interest being about $\Delta C_D = 0.0005$ (5 counts); however, as already found for the laminar-type wings, the lift range of effectiveness also narrowed down as the bump height was raised, Figure 85. Other parameters, such as the detailed bump shape, including the crest position with respect to the bump chord, had little effect on drag in the C_L range of interest, any positive changes tending to occur at high values of C_L. Moving the bump location on the wing shifted, as expected, the point of maximum drag reduction to a different lift coefficient.

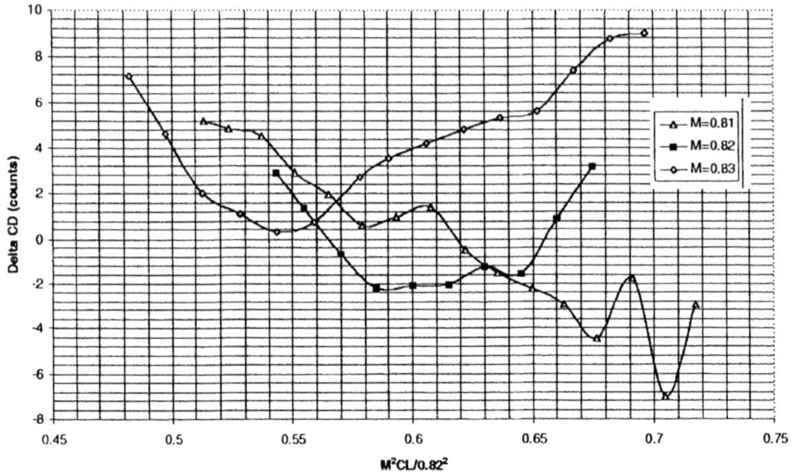

Figure 84 Bump effectiveness dependent on Mach number and lift coefficient A340 wing, $h_B = 0.2\%c$, $l_B = 0.125c$ [6.1]

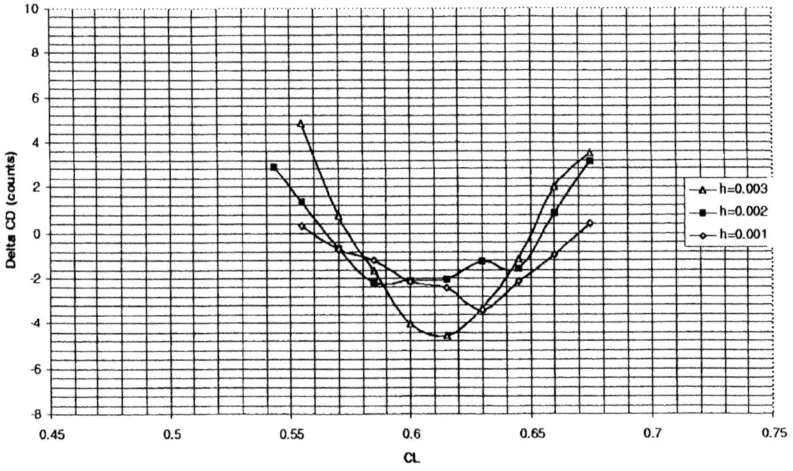

Figure 85 Effect of bump height on drag reduction A340 wing, $M_\infty = 0.82$, $l_B = 0.125c$ [6.1]

Concerning the implementation of bump control, the following must be considered: Over the operating C_L-range at the design Mach number $M_\infty = 0.82$, the wave drag of the basic wing varies from about $\Delta C_{DW} = 0.0005$ (5 counts) at $C_L = 0.5$ to about $\Delta C_D = 0.0019$ (19 counts) at $C_L = 0.60$. Assuming a drag reduction of 4.5 counts possible at the bump design lift coefficient of $C_L = 0.59$ gives a maximum overall drag benefit of about 25% of the wave drag. Assuming further, for the benefit of argument, the total aircraft drag to be $C_D = 0.030$ (300 counts), the drag reduction for the complete aircraft would be about 1.5%. An average drag reduction of 3% was achieved and utilized for the laminar-type aircraft mission calculations considered above. There, the wave drag at off-design ($M_\infty = 0.84$), indicative of the potential for wave drag reduction, amounted to about $C_{DW} = 0.0040$ (40 counts). The lesser drag reduction will, of course, be reflected in the mission benefits considered below.

6.3.2 Assessment of Bump Control Effects on Aircraft Performance

Assessments have been made of the benefits attainable over typical mission profiles, taking into account the aerodynamic benefits over the operating C_L-range as well as the weight penalties associated with the bump installation, here assumed to be $\Delta W = 0.50$ tons for a variable-geometry bump and $\Delta W = 0.20$ tons for a fixed bump. The drag reduction characteristics for the "variable" bump were taken to be the same as for the fixed bump, where that offered a drag benefit, with zero penalty at off-design, assuming that the bump could be retracted. No variation in the chord-wise position of the bump was assumed. The results are tabulated below for both bump types. The maximum reduction in ΔCoC is 0.4% compared with the 1.3% for the mission of the laminar-type-wing aircraft.

Table 8 Performance benefits due to bump control for typical missions

AIRCRAFT	ΔBLOCK FUEL			ΔSPP* RANGE	ΔCOC
	3000nm	5000nm	7000nm		
With Fixed Bump	-0.3%	-0.4%	-0.4%	20nm	-0.1%
With Variable Bump	-1.1%	-1.5%	-1.5%	95nm	-0.4%

* SPP – Standard Passenger Payload

The CoC could, of course, be further reduced by also making the bump adaptive in the chord-wise direction, especially in conjunction with a further increase in height. Furthermore, operating the aircraft at off-design conditions, e.g., within the drag rise, which can often not be avoided, would definitely lead to larger benefits as shown by the study of the laminar A340-type aircraft. Unfortunately, the study of the application of the bump control concept to a new wing design, incorporating the bump into the design, as intended, could not be performed due to time limitations.

6.4 Control Application to a Regional-Jet Aircraft

The objective of the work undertaken here was to select suitable airfoil/wing configurations for a regional-jet aircraft and to incorporate shock and boundary layer control mechanisms into these wings. For the wing/control configurations selected, benefits and installation penalties were to be determined. Concerning the control mechanisms considered, emphasis was, following the results of the other tasks, essentially placed on control by a contour bump and discrete suction, respectively. The present work was performed by **ALENIA** [6.4] and **Dassault Aviation** [6.2].

6.4.1 Bump Evaluation and Optimization

For the application to a regional-jet aircraft two airfoils were considered: the laminar-type airfoil DRA-2303 and a Falcon-wing-equivalent turbulent airfoil. For the former, computations were carried out by ALENIA, employing their steady/ unsteady Navier-Stokes code, for the datum airfoil and the airfoil with contour bump obtaining results similar to the ones described in Chapter 4. The DRA-2303 airfoil was then incorporated by ALENIA into a regional-jet wing including bump control; performance benefits for the aircraft are described below [6.4].

For the second — Falcon-equivalent — turbulent airfoil, two Mach numbers were defined, viz., $M_\infty = 0.72$ equivalent to a nominal cruise Mach number, and $M_\infty = 0.765$ corresponding to an off-design condition. The original airfoil was slightly modified to reproduce the 3D-wing shock locations at these conditions.

Initially, a *symmetrical bump* device was optimized by ALENIA for the design Mach number but a higher-than-cruise angle of incidence, $M_\infty = 0.72$, $\alpha = 2.6°$; the optimization resulted in a bump length of $l_B = 20\%c$, a height of $h_B = 0.20\%c$ and a crest location at 60%c. Possible drag reductions are shown in Figure 86 with a representative airfoil pressure distribution depicted in Figure 87 [6.4]. Drag reductions of up to 7% are achieved at the design point of the bump. Dassault Aviation has carried out similar computations for the design Mach number — with similar results — but also for the off-design Mach number of $M_\infty = 0.765$; here, the bump showed only negligible benefits due to a mismatch between shock and bump location [6.2].

As a consequence of these somewhat unsatisfactory results, Dassault optimized an *asymmetric bump* with the relative crest location at $l_c/l_B = 80\%$, a bump length of $l_B/c = 20\%$, and bump heights of $h_B = 0.20\%c$ and $0.40\%c$, respectively [6.2]. At the design Mach number $M_\infty = 0.72$, the bump crest was located at 58% of the airfoil chord, at $M_\infty = 0.765$ the crest was located at 66%c, i.e., about 3% downstream of the shock position. The results, Figure 88, show drag reductions for the higher bump at $M_\infty = 0.72$ of about 11% throughout the range of lift coefficients investigated, while at the higher Mach number generally drag reductions of 20% are obtained due to the higher bump. One may conclude that, although drag reductions for the regional-jet airfoil investigated are small at cruise conditions, a worthwhile enlargement of the flight domain may at least be achieved. We will return to bump control in conjunction with the incorporation of a bump (and of a suction slot, respectively) into the Falcon wing in Chapter 6.4.3.

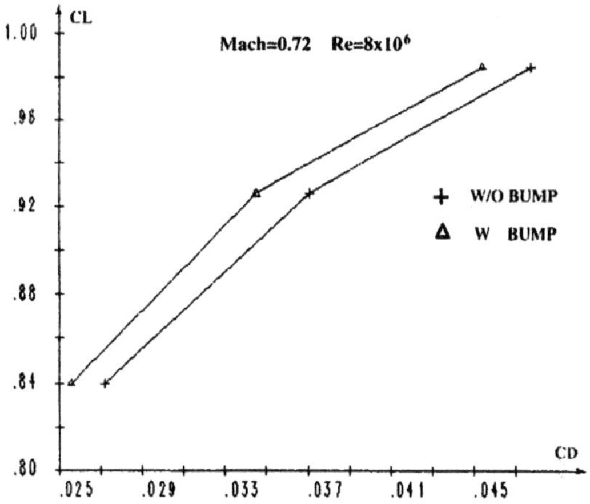

Figure 86 Effect of a bump on drag, Falcon-type airfoil, $M_\infty = 0.72$, $Re_c = 8x10^6$ [6.4]

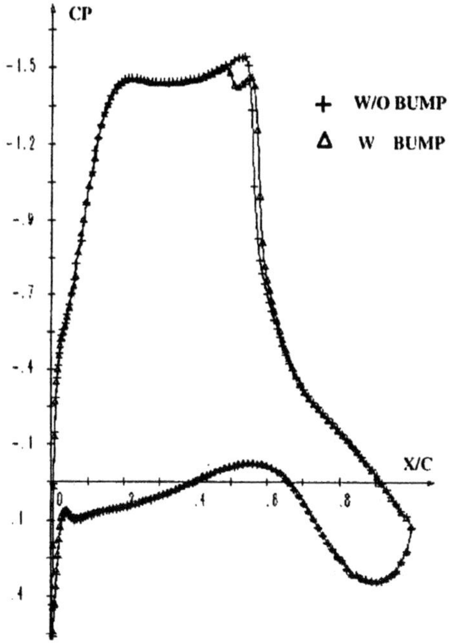

Figure 87 Effect of bump control on the pressure distribution, Falcon-type airfoil, $M_\infty = 0.72$, $\alpha = 2.6°$, $C_L = 0.95$, $Re_c = 8x10^6$ [6.4]

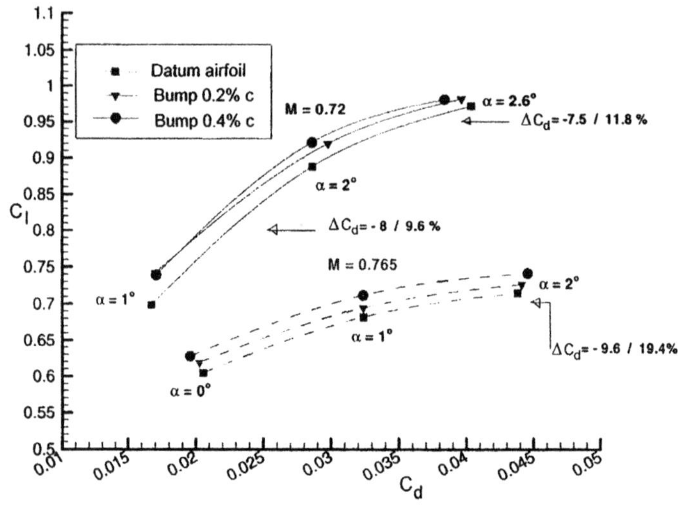

Figure 88 Effect of control by an asymmetric bump on drag, Falcon-type airfoil $M_\infty = 0.72$, $Re_c = 8x10^6$ [6.2]

6.4.2 Performance Improvement for a Laminar-Wing Regional-Jet Aircraft

In order to assess performance improvements possible for a regional-jet aircraft due to bump control, a corresponding (simplified) aircraft was generated deriving its characteristics from data for existing jets: the wing has a 26°-swept leading edge without twist and a constant airfoil section in span-wise direction, Figure 89 [6.4]. The airfoil corresponds, as mentioned above, to the DRA-2303 airfoil without and with bump. The bump, when present, is fitted over the entire span. The horizontal and vertical tails have constant symmetrical NACA-0100 airfoil sections and the nacelles are axis-symmetric.

The code used for polar computations is a 3D Full Potential/Boundary Layer code; computations were performed for the design Mach number $M_\infty = 0.73$ at lift coefficients between $C_L = 0.15$ and 0.68. Results of the computations without and with bump at a constant lift coefficient of $C_L = 0.54$ are depicted in Figure 89 in form of surface pressure distributions: quite obvious is the reduction in shock strength due to the bump. It should be noted that with the bump present, the same lift is obtained at a lower angle of attack.

The reduction in shock strength indicates that the bump reduces wave drag — and, as an analysis of drag components has indicated, also viscous drag — which leads, as shown in Figure 90, to the expected reduction in total drag. It is interesting to note that, at the Mach number considered, the bump is effective over the whole lift-range investigated: the drag reduction for the complete business-jet aircraft ranges from 2% to 6%, i.e., it is higher than the minimum given by Dassault for making the introduction of control worthwhile, as will be discussed below.

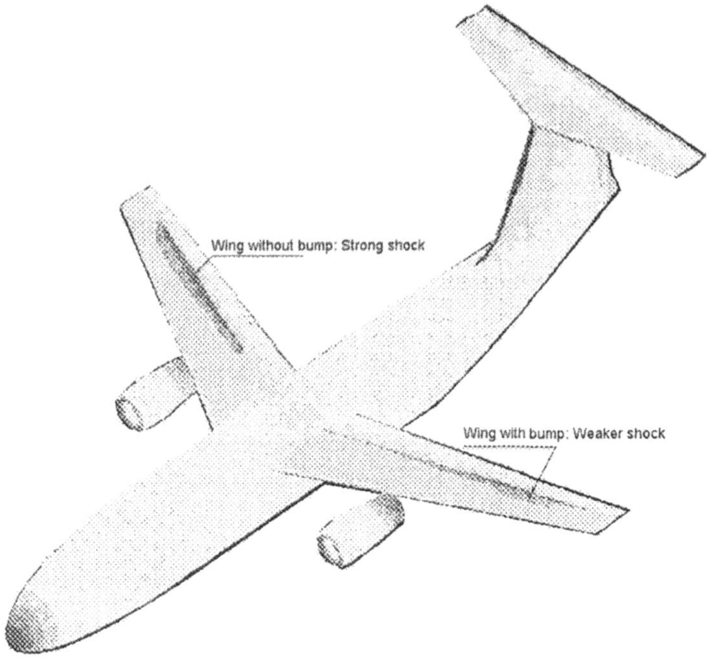

Figure 89 Effect of bump control for the regional-jet aircraft with laminar wing based on the DRA-2303-airfoil, $M_\infty = 0.73$, $C_L = 0.54$, $Re_c = 6.7 \times 10^6/m$ [6.4]

Computations at the off-design Mach numbers $M_\infty = 0.80$, $C_L = 0.45$, and $M_\infty = 0.50$, $C_L = 0.60$, have indicated a negligible drag reduction of 1% for the former and essentially no effect for the latter, again stressing the need for an adaptive bump, preferably in conjunction with a wing design incorporating a bump from the start.

6.4.3 Minimum Drag Reduction Requirements

Dassault Aviation concentrated strongly on the assessment of the penalties associated with the introduction of control into a real aircraft and on the minimum drag reduction required for making an introduction of control into a wing design worthwhile. Considered here are the two concepts of control that have shown the largest potential for performance improvement, namely, the adaptive contour bump and discrete suction.

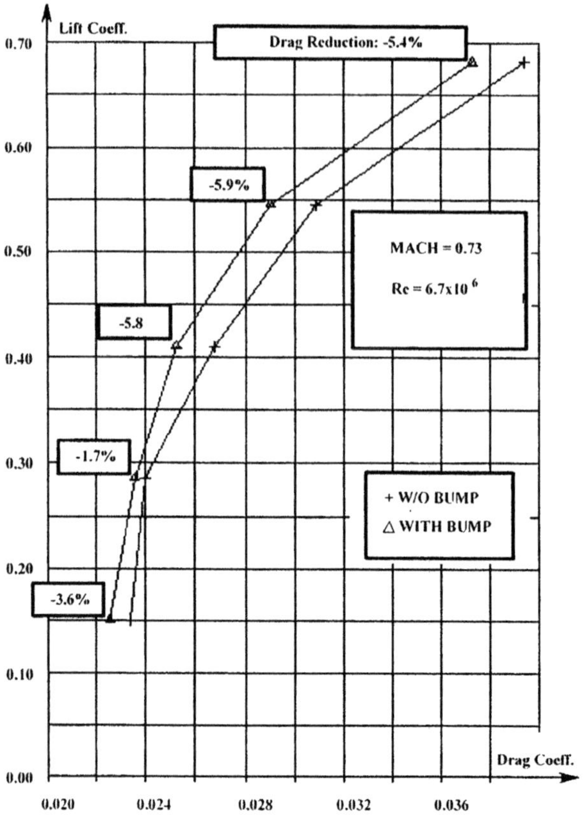

Figure 90 Effect of bump control on drag reduction for the laminar-wing regional-jet aircraft, $M_\infty = 0.73$, $Re_c = 6.7x10^6/m$ [6.4]

Following the evaluation of the asymmetric bump for the "Falcon-equivalent" airfoil, technical implementation and evaluation studies for integrating a bump system into the Falcon aircraft were performed [6.2]. Firstly, transferring the 2D-results to wing conditions indicates that at the aircraft cruise Mach number M_∞ = 0.80 the bump should extend from 50% to 70% chord with the crest at x/c = 65%; for a buffeting condition at M_∞ = 0.85, it should be located between 60% and 80% chord with the crest at 72% chord, Figure 91. This means, of course, that a fixed bump either degrades cruise or buffet performance. Furthermore, the position optimized for buffet control interferes with the wing box, the air-brakes and the aileron. Thus, a fixed or an adaptive bump is not a viable solution for retrofitting the existing aircraft. (Of course, the buffet boundary can also be improved, if needed, by other means, such as, for instance, retractable flaps, as indicated in the conclusions to Chapter 5.)

A new wing with an **adaptive bump** system — for details see Reference 6.2 and Chapter 24 — which is partly located in the fuel wing box and partly downstream of the rear spar, thus affecting fuel capacity and the height of the spar, would result, in comparison to a conventional wing design, in a weight penalty of 264 lbs. and a fuel-capacity penalty of 4.6 ft^3. Applying the second criterion proposed for the classification of concepts, viz.,

$$C = 10^5\,\Delta C_D - 0.068\,\Delta W - 9.27\,\Delta V - 241.5\,\dot{m},$$

where ΔC_D is the change in drag, ΔW the weight penalty, ΔV the change in fuel capacity due to the wing-box system installation and $\dot{m}$ the total suction mass-flow rate, where applicable, results, for $C = 0$, in

$$10^5\,\Delta C_D = -60.5$$

or a minimum drag reduction of 3% for the complete aircraft to balance the penalties, and in a drag reduction of at least 5% to make the installation worthwhile. It will take a very careful and dedicated design to accomplish this but it is, following the discussions in the preceding chapters, quite possible. Of course, these considerations are very crude and very preliminary and restricted to the aircraft and complexity of the bump control design considered. Nevertheless, these estimates are in consent with the EADS-Airbus D evaluations discussed in Chapter 6.2.

The second concept investigated is discrete suction with the suction slot, having a width of 10mm and a span of 3m, consisting of 6 individual slots divided by the wing ribs; the slots are located at 50% chord, Figure 92. Taking the maximum allowable mass-flow rate for the aircraft to be 0.35 lbs./s, given by engine restrictions, leads to a suction rate of $C_Q = 0.00015$ which is compatible with the rates considered within the other tasks. The wing integration study leads to a weight penalty of 140 lbs. and a total loss of fuel capacity of 4.7 ft^3 resulting, according to the formula given above, in

$$10^5\,\Delta C_D = -137.6$$

or a minimum "zero-benefit" drag reduction for the complete aircraft of 6.8% which is, following the investigations within the other tasks, hard to achieve by pure discrete suction.

6.5 Conclusions and Future Work

The objectives of the present study were the introduction of various potential control methods, identified in the deliberations of all tasks, into existing or new wing designs, and the demonstration of the benefits and penalties associated with the implementation of control. In these investigations, ALENIA and Dassault have studied the application of boundary layer and shock control to regional-jet aircraft, while BAE SYSTEMS-Airbus UK and EADS-Airbus D considered the application of control to a large A340-type transport aircraft with a turbulent and a laminar wing design, respectively. The main driver for control application was considered to be drag reduction. Preceding the application studies were aircraft-specific investigations and optimization studies of control for the corresponding airfoils and wings.

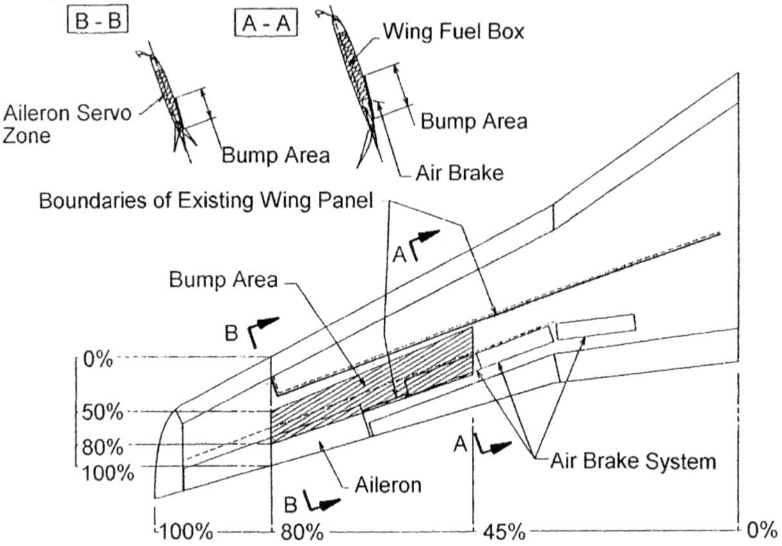

Figure 91 Wing-bump location for a Falcon-wing configuration [6.2]

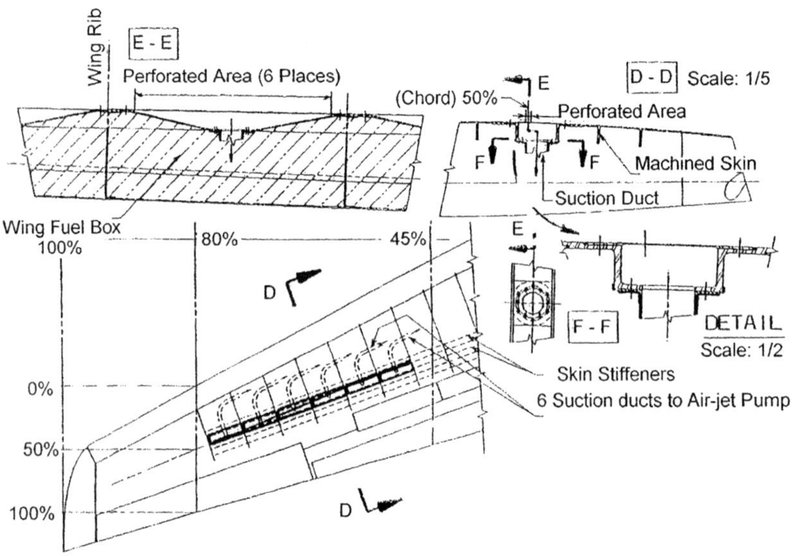

Figure 92 Boundary layer suction concept for a Falcon-wing configuration [6.4]

Generally, studies concerned with the efficiency of control devices in reducing drag have led to the conclusion that the most potential device is the contour bump, followed by discrete slot suction. In the present control assessment study, emphasis was, therefore, placed foremost on the bump control device with some consideration, especially with respect to installation penalties, given to discrete suction.

Concerning the **Regional-jet Aircraft**, the introduction of a laminar wing, based on the DRA-2303 airfoil, has shown that drag reductions of up to 6% are attainable near cruise conditions due to bump control, rendering this type of wing/ control combination worthwhile pursuing for future aircraft. Applying control to existing turbulent wings is only effective in enlarging the operating range of the aircraft. The potential of bump control can, however, be exploited for turbulent-wing aircraft if the bump is adaptive and fully integrated into the design. Concerning installation penalties, it was derived that for an adaptive bump system a drag reduction of 3% is required to offset the installation penalties and that a total drag reduction of about 5% is needed to render an installation attractive. For a suction system installation, a "zero-benefit" drag reduction of 6.8% was estimated to be required which is, following the discussions in Chapters 4 and 5, hard to achieve.

For a large **A340-type Transport Aircraft with a Hybrid Laminar Flow wing** it was found that variable-height bump control results in fuel reductions of up to 2.11% on typical North-Atlantic missions, corresponding to savings in the Cash Operating Costs (CoC) of 1.3%, rendering such an aircraft design, as in the case of the corresponding regional-jet aircraft, worthwhile pursuing. For a similar aircraft with an **existing turbulent wing,** bump effectiveness is rather limited due to the large shock movements generally associated with a turbulent wing. Here, reductions in CoC due to a retractable bump, considering typical missions, are about 0.4% with higher gains to be achieved at off-design flight conditions which are generally unavoidable but were not considered here.

For 'conventional' airfoils and wings of 'turbulent' design, the studies carried out here have indicated that benefits similar to the ones for laminar-wing designs can be achieved; however, the complication for this class of wings is the potentially large chord-wise movement of the shock wave and the, by design, rather weak shock at cruise conditions. Future work should, therefore, firstly consider new wing designs with a bump integrated into the design. This way thicker wings could be built with less structural weight and reduced time for manufacturing — in addition to possible gains in performance. Furthermore, adaptive bumps should be utilized which could track the position of the shock wave and therefore always be at an optimum. Realizing such a bump, and especially providing suitable materials and (smart) structures for such a bump, is where further work should be considered in order to arrive at a practical implementation of such a system. Of course, as already mentioned in conjunction with the other tasks, other control devices, such as for instance vortex generators and unconventional deploying devices, among them Gurney flaps, divergent trailing edges and reversed-flow flaps, must be considered, individually and in combination with the devices already investigated, for future improvements in air vehicle performance.

7 Assessment of Shock and Boundary Layer Control – A Summary

During the EUROSHOCK (I) project, passive control of shock boundary layer interaction by means of a perforated plate/cavity arrangement was thoroughly investigated in order to determine the improvements in the design and off-design performance of transonic airfoils possible due to this type of control. It was found that passive control was very successful in reducing wave drag associated with the shock but that total drag was, for laminar-type airfoils, always increased and that the gains for turbulent airfoils were rather marginal. The drag increase could be traced to the dominating effect of an increase in viscous drag caused by the perforation and the re-circulating flow in the cavity region and the amplification effect due to the sustained rear adverse pressure gradients on the airfoil upper surface. The buffet boundary could, however, be noticeably improved by passive shock control [1.5].

The negative effect of passive shock control on drag lead to the numerical and experimental investigation of the influence of active shock and boundary layer control on the design and off-design performance of airfoils and wings. The control mechanisms considered here were, based on the EUROSHOCK (I) results:

- Active control by means of a perforated plate/cavity arrangement with part suction to overcome the detrimental effect of passive control on viscous drag.
- Hybrid control, consisting of a passive cavity in the shock region with active suction downstream of the passive cavity, with the same objective of reducing viscous drag.
- Discrete slot suction upstream and downstream of the shock and at the foot of the shock, respectively, to reduce viscous drag and to delay the onset of separation.
- A contour bump in the shock region designed to reduce wave drag without creating additional viscous drag; the bump was also investigated in conjunction with discrete suction upstream of the bump, to possibly reduce both, wave drag and viscous drag.

The effectiveness of control originates, of course, in the region where control is applied. It is therefore appropriate to first consider the influence of control on the (local) flow development in the shock boundary layer interaction region, and here especially the effect on the boundary layer development, which was determined in ***basic experiments*** with the following results:

- Active single-cavity control with part-suction can reduce the boundary layer thickness parameters downstream of the interaction/control region considerably and, dependent on the suction rate, to the level of the no-control boundary layer. Since some shock spreading still remains, wave drag is being reduced while viscous drag (on airfoils and wings) is likely to stay the same. The suction rates are, however, at these conditions prohibitive corresponding to almost 8 times the rate considered reasonable for aircraft installation.

- Hybrid control has a similar effect on the boundary layer development with the shock spreading, however, contrary to the case of single-cavity control, not affected by the amount of suction applied.
- Discrete slot suction reduces the boundary layer thickness downstream of the interaction region to any level desired; however, as the suction rate is increased — applying suction upstream of or at the foot of the shock — wave drag also increases; downstream suction is less effective due to the thickening of the boundary layer by the shock, has, however, the advantage that wave drag is not or positively affected.
- A contour bump in the shock region, if placed correctly with respect to the shock, reduces wave drag by shock-smearing, possibly without increasing viscous drag, and, of course, without needing additional energy input as in the case of suction.

The local flow development and the conditions downstream of the interaction/control region as affected by control are, of course, reflected in the global flow development on airfoils and wings and, correspondingly, in the aerodynamic parameters of interest here, namely drag and the buffet boundary. The following results were obtained for the airfoils and the sheared-wing configuration investigated numerically and experimentally:

Active Suction by a perforated plate/single-cavity arrangement and ***Hybrid Control*** always resulted — at suction rates considered realizable in aircraft installations — in an increase in total drag due to the dominating influence of the increase in viscous drag caused by the cavity and the amplification effect of the sustained rear adverse pressure gradients. It is believed that in order to reduce the boundary layer thickness (parameters) downstream of the interaction/control region sufficiently to yield a reduction in total drag — which is, following the basic experiments, quite possible — the rate of suction required would be such that the benefits in terms of wave drag reduction would be nullified by pump drag. The buffet boundary is, as in the case of passive control, positively affected.

Discrete Suction leads, if applied upstream of the shock, even at feasible suction rates (here $C_Q \approx -0.00009$), to drag reductions of up to 7.5%. Pump drag may account for about half that gain with a net drag reduction of 4% realistically remaining. Applying suction at the foot of the shock or further downstream is less effective in reducing drag, but more efficient in delaying buffet onset with maximum gains in lift at buffet onset of 5%.

Bump Control in the shock region was found to be the most effective control mechanism investigated with airfoil drag reductions of up to 23% at near-design conditions for laminar-type airfoils and, similarly, at off-design for turbulent ones. In case of the infinitely-swept sheared wing, drag reductions were slightly less due to the lesser contribution of wave drag to total drag. The effectiveness of bumps is, however, dependent on freestream conditions, i.e., on the position of the shock relative to the bump and on shock strength, suggesting an adaptive bump as the most effective device, especially for turbulent airfoils where large shock movements may prevail. Contour bumps were also found to have a positive effect on the buffet boundary with increases of up to 10% in lift at buffet onset for bumps located downstream of the shock. The effect of a bump in combination with upstream ***Discrete Suction*** was found to be additive since both, wave drag and viscous drag are being reduced.

The final step concerned the assessment of benefits and penalties associated with the implementation of control into actual aircraft (wings). Here, mainly the contour bump and discrete suction — the latter only with respect to installation penalties — were considered. For an ***A340-type HLF-wing Aircraft*** a variable-height bump resulted in savings in fuel consumption on typical long-range missions of up to 2.11%, corresponding to reductions in Cash Operating Costs (CoC) of 1.3%. For the aircraft with an ***existing turbulent wing***, the range of bump effectiveness is rather limited due to the large shock movements associated with turbulent wings, and the inherent low wave drag at design. The benefits are, therefore, also limited with reductions in CoC amounting to 0.4% for a retractable bump. Higher gains can, of course, be expected if off-design flight conditions, which are, in praxis, unavoidable, are included in the mission profile.

For a ***Regional-jet Aircraft*** with a ***laminar wing***, based on the DRA-2303 airfoil, aircraft drag reductions of up to 6% were achieved near design due to bump control, while with an ***existing turbulent wing***, only the operating range of the aircraft could be enlarged by bump control. Concerning installation penalties, it was determined that for an adaptive bump a drag reduction of 3% is required to offset the penalties, while for a suction system installation, a "zero-benefit" drag reduction of 6.8% is required; the former is quite achievable, while the latter, following the results obtained within the present project, seems hard to realize. Of course, these results must be considered very preliminary and applicable foremost to the aircraft and the specific systems installed.

In summarizing one may conclude that adaptive bump control, possibly in conjunction with upstream suction, seems the most promising control approach when drag reduction is the main driver. A laminar wing is inherently well suited for bump control — actually without bump control hardly realizable — while for a turbulent wing the benefits are rather seen in designing thicker, hence lighter wings, reducing the associated higher wave drag by adaptive-bump control.

8 Overall Conclusions and Future Work

The specific objective of the research described here was to study, experimentally and numerically, the effect of various means of active shock and boundary layer control on cruise performance in terms of cruise drag and/or speed and on the off-design performance, i.e., essentially the drag-rise and buffet boundaries, of transonic airfoils and wings. It was, furthermore, aimed at clearly defining the benefits and penalties associated with incorporating potential control methods into existing and new wing designs for typical transonic long-range transport aircraft and, correspondingly, for regional-jet aircraft, considering characteristic aircraft missions. Also to be provided were the tools needed for the design of transonic wings with control, i.e., essentially the numerical codes able to treat control, which requires a clear understanding of the flow phenomena associated with control and the establishment of corresponding control laws and boundary conditions.

The project was divided into four major, interrelated tasks, discussed in Chapters 3 to 6, namely:

- ***Task 1*** - Modeling of Active Control Phenomena, with contributions by ONERA and the Universities of Cambridge and Karlsruhe.
- ***Task 2*** - Prediction of Transonic Airfoil/Wing Flow with Control, with contributions by CIRA, DLR, ONERA, INTA, the University of Naples and EADS-Airbus D and ALENIA.
- ***Task 3*** - Wind Tunnel Experiments on Airfoils and Wings with Control, with contributions by DERA and DLR.
- ***Task 4*** - Control Application Aspects, with contributions by BAE SYSTEMS-Airbus UK, EADS-Airbus D, ALENIA and Dassault Aviation.

Task 1

Basic channel-flow experiments of shock boundary layer interaction and its control at typical transonic shock-upstream Mach numbers were performed for two-dimensional and three-dimensional swept-shock interactions, supplemented by Navier-Stokes computations. The area of investigation was essentially from upstream of the influence sphere of the shock to downstream of the interaction/ control region, i.e., studied was the local flow development as affected by shock and boundary layer control which, of course, strongly shapes the global developments associated with airfoil and wing flow.

The results, already discussed in detail in the preceding chapter, were, in brief: by active single-cavity control with part suction the boundary layer thickness parameters downstream of the interaction/control region can be considerably reduced – compared to the passive control case – and, at extreme suction rates, the no-control boundary layer can be reestablished with some shock spreading still remaining. Hybrid control has a similar effect. Slot suction allows to reduce the boundary layer thickness parameters downstream of the interaction/control region to almost any level desired. The slot geometry and the location of the slot seem to be of secondary influence on the effectiveness of the mechanism except that the rear slot location is somewhat less efficient due to the thickening of the boundary layer by the shock. The type of control and the suction rate applied determine, of course, the magnitude of the contributing drag components, i.e., wave drag, viscous drag and pump drag, which must be carefully balanced to achieve a net gain in the case of airfoil and wing flow. On the other hand, a contour bump was found to reduce wave drag (without energy input) without increasing the boundary layer displacement thickness downstream of the interaction/control region, which is indicative of only minor effects on viscous drag.

Concerning the numerical simulation, the computations by Navier-Stokes codes, employing various turbulence models, generally predicted the trends with respect to the effect of control quite well. However, the levels in the pressure distributions as well as in the boundary layer parameters were not always satisfactorily determined. This was considered to be due to an insufficient grid resolution in the control region and due to deficiencies in the turbulence models employed. Concerning the latter, the Baldwin-Lomax turbulence model was generally found to predict the flow development closer to the experimental results than the other models investigated. Furthermore, the new control law of Bohning and Doerffer, taking into account an outer transonic tangential stream, was introduced; it gives, in the case of suction through perforated surfaces, more accurate normal-velocity distributions in the control region than the previously employed linear transpiration law.

Task 2

Computational methods, including steady and time-accurate Viscous-Inviscid Interaction (VII) and Navier-Stokes codes, were extended and improved to treat flows with shock and boundary layer control and subsequently employed to evaluate control concepts in parametric studies. In a first phase, test cases were defined to assess CFD capabilities as well as control concepts, then computations for various airfoils and control mechanisms were performed. During and as a result of the computations, codes were improved by grid refinements, especially in the shock/control region, and by the introduction of the new control law of Bohning and Doerffer.

Concerning the performance of the computational codes, the following can be stated: for all configurations and control concepts investigated, the computer codes arrived at the same qualitative results concerning drag and the improvements in the buffet boundary. However, absolute levels in drag as well as in drag changes due to control differed between codes and between codes and experiment. Discrepancies between computations and experiments could, however, be noticeably reduced by small changes in the freestream Mach number in the computations, i.e., by matching effective freestream conditions. The latter is especially important if the results are sensitive to the freestream conditions as is the case for bump control.

For the control mechanisms investigated numerically the effectiveness in improving design and off-design conditions has already been reported in the preceding chapter, but briefly: shock and boundary layer control by discrete suction upstream of the shock reduces drag by up to 6% — as compared to 7.5% in the corresponding experiments — with slot locations downstream of the interaction region being less effective. Buffet could only be positively affected when applying unrealistically high suction rates. Active single-cavity and hybrid control resulted, at realistic suction rates, always in an increase in total drag. Bump control was found to be the most effective control mechanism with drag reductions of up to 23% attainable; the introduction of additional upstream discrete suction resulted in a further decrease in drag. However, bump effectiveness was determined to be very sensitive to changes in the freestream conditions, suggesting an adaptive bump as the best control solution.

Task 3

Experiments with the ADIF airfoil and the DA LVA-1A laminar-type airfoil, respectively, and the sheared wing, based on the ADIF airfoil, were carried out by DLR without and with control to assess the effect of control and to determine the influence of sweep on control effectiveness. Mainly considered were bump control in the case of the airfoils, and bump and single-cavity control, respectively, both also in conjunction with discrete suction, in case of the sheared wing. Discrete suction, single-cavity active control and hybrid control on the large-scale model of the laminar-type airfoil DRA-2303 were investigated by DERA.

The experimental results for airfoils and wings designed for significant extents of laminar flow on the upper surface, show that the application of shock control by bumps significantly reduces total drag which is essentially due to a large reduction in wave drag with only little effect on viscous drag. For this class of airfoil or wing this is extremely beneficial since strong shock waves tend to form at a fixed chordwise position following the long favorable pressure gradient necessary to maintain natural laminar flow. For "conventional" airfoils and wings of "turbulent" design, the experimental work suggests that similar benefits can be obtained, however, a complication for this class of wings is the potentially large chordwise movement of the shock wave. In this case a fixed bump would lead to large drag penalties at off-design conditions; however, a deployable bump could be utilized for certain parts of the flight envelope with the best solution being, however, a fully adaptive bump.

The use of suction, in general, does not appear to be a suitable solution to the problem of drag reduction. The exception to this seems to be the use of discrete suction where drag reductions of up to 7.5% were obtained by DERA. The results could also be further improved by optimizing the suction slot geometry, replacing the presently employed porous surface by a slot or a scoop taking advantage of the total pressure in the boundary layer.

Task 4

The objective of the present study was the introduction of potential control methods into existing or new wing designs and the assessment of the benefits and penalties associated with the implementation of control. In these investigations, ALENIA and Dassault studied the application of boundary layer and shock control to regional-jet aircraft, while BAE SYSTEMS-Airbus UK and EADS-Airbus D considered the application of control to a large A340-type transport aircraft with a turbulent and a laminar wing design, respectively. The main driver for control application was considered to be drag reduction. As a consequence of the aerodynamic control efficiency studies within all tasks, emphasis was here placed foremost on the bump control device with some consideration, especially with respect to installation penalties, given to discrete suction.

The results of the control application studies are briefly: for a Regional-jet aircraft, equipped with a laminar wing, aircraft total drag reductions of up to 6% near cruise could be realized, while for the aircraft with an existing turbulent wing only the operating range, i.e., off-design conditions, could be improved by bump control. Similar results were obtained for an A340-type aircraft where, on typical long-range missions, for the aircraft with an HLF-wing savings in Cash Operating Costs of up to 1.3% were achieved, while for an existing turbulent wing only 0.4% were realized due to bump control. For the latter, it must, however, be reemphasized that when considering flight mission profiles with and without control, off-design conditions must be included into the mission profile since these conditions can generally not be avoided. Concerning installation penalties, studies for a Regional-jet Aircraft have indicated that for an adaptive bump installation drag reductions of at least 5% are needed to render such an installation attractive, while for a suction-system implementation at least 6.8% drag reduction are needed; the former is quite feasible while the latter seems hard to realize.

Future Work: The "bump" has been identified to be the most effective device when drag reduction is the main driver — with additional benefits related to buffet. Aerodynamically, the application of bumps to a wing should pose no significant problems. The studies of sweep effects carried out here suggest that the flows are predictable and realizable. However, bumps should be adaptive to changing flight conditions for maximum gain. This requires smart structures and materials and it is here that further work should be considered in order to arrive at a practical application of such a system. Furthermore, the potential of bumps in allowing the design of thicker (turbulent) wings should be explored.

Future work should also consider a wider range of boundary layer and flow control mechanisms, aiming at drag reduction and the avoidance of separation, comprising, for instance, sub-boundary layer devices, including vortex generators and mass-less air jets, reversed-flow flaps and other mini-flaps to control flow conditions at the trailing-edge. Especially a combination of control mechanisms is worth considering, a candidate being, for instance, a bump in conjunction with a trailing-edge device to provide the required margins between the cruise regime — enlarged by the bump — and the buffet boundary. The freestream conditions should include high-lift, low speed regimes as well as the transonic speed range. In these control studies it is important to consider the aerodynamic performance improvements but also the improvements in structural design possible due to control.

Computational methods must be extended to treat these types of control; they must also be improved by incorporating new turbulence models, this being the key to any successful numerical treatment. Generally, physical models and boundary conditions related to control must be studied further and improved models and conditions incorporated into the computational codes. Finally, the efficiency of the control mechanisms in improving aircraft performance must be demonstrated in large-scale tests and the benefits and penalties associated with the implementation of new control methods assessed.

9 References

Chapters 1 and 2

[1.1] Stanewsky, E. and Krogmann, P. "Transonic Drag Rise and Drag Reduction by Active/Passive Boundary Layer Control", AGARD Report No. 723, Lecture Series Aircraft Drag Prediction and Reduction, July 1985

[1.2] Délery, J., "Shock Wave/Boundary Layer Interaction and its Control", Progress in Aerospace Science, Vol. 22, 1985

[1.3] Thiede, P., Krogmann, P. and Stanewsky, E., "Active and passive shock/boundary layer interaction control on supercritical airfoils", AGARD-CP-365, Brussels, 1984

[1.4] Stanewsky, E., Délery, J., Fulker, J. and Geissler, W., " Synopsis of the Project EUROSHOCK", in EUROSHOCK – Drag Reduction by Passive Shock Control, Notes on Numerical Fluid Mechanics, Volume 56, Friedr. Vieweg & Sohn Verlagsgesellschaft mbH, Braunschweig/Wiesbaden, 1997, pp. 1 - 81

[1.5] Stanewsky, E., Délery, J., Fulker, J. and Geissler, W. (Ed.), "EUROSHOCK – Drag Reduction by Passive Shock Control", Notes on Numerical Fluid Mechanics, Volume 56, Friedr. Vieweg & Sohn Verlagsgesellschaft mbH, Braunschweig/ Wiesbaden, 1997

[1.6] Bur, R., Délery, J. and Corbel, B., "Basic Study of Passive Control Applied to a Two-dimensional Transonic Interaction", in EUROSHOCK – Drag Reduction by Passive Shock Control, Notes on Numerical Fluid Mechanics, Volume 56, Friedr. Vieweg & Sohn Verlagsgesellschaft mbH, Braunschweig/Wiesbaden, 1997, pp. 89 - 111

[1.7] Bohning, R. and Doerffer, P., "Passive Control of Shock Wave Boundary-layer Interaction and Porous Plate Transpiration Flow", in EUROSHOCK – Drag Reduction by Passive Shock Control, Notes on Numerical Fluid Mechanics, Volume 56, Friedr. Vieweg & Sohn Verlagsgesellschaft mbH, Braunschweig/ Wiesbaden, 1997, pp. 113 - 133

[1.8] Fulker, J.L. and Simmons, M.J., "An Experimental Investigation of Passive Shock/Boundary-layer Control on an Aerofoil", in EUROSHOCK – Drag Reduction by Passive Shock Control, Notes on Numerical Fluid Mechanics, Volume 56, Friedr. Vieweg & Sohn Verlagsgesellschaft mbH, Braunschweig/ Wiesbaden, 1997, pp. 379 - 400

[1.9] de Matteis, P. and Dima, C., "Numerical Investigation of the Passive Shock Control on Transonic Airfoils through an Euler/Boundary-layer Coupling Technique", in EUROSHOCK – Drag Reduction by Passive Shock Control, Notes on Numerical

Fluid Mechanics, Volume 56, Friedr. Vieweg & Sohn Verlagsgesellschaft mbH, Braunschweig/Wiesbaden, 1997, pp. 151 - 170

[1.10] Wagner, W., "An Experimental Investigation of Passive Shock/Boundary Layer Interaction Control on an Airfoil: Unsteady Measurements", in EUROSHOCK – Drag Reduction by Passive Shock Control, Notes on Numerical Fluid Mechanics, Volume 56, Friedr. Vieweg & Sohn Verlagsgesellschaft mbH, Braunschweig/ Wiesbaden, 1997, pp. 401 - 414

Chapter 3

[3.1] Bur, R., Benay, R., Soares-Morgadinho, R. and Soulevant, D., "Study of control devices applied to transonic shock wave/boundary layer interaction", EUROSHOCK II TR BRPR-95-76/1.1 and ONERA RT 126/7078 DAFE/Y, 1999 (also see Chapter 11)

[3.2] Bohning, R. and Doerffer, P., "Hybrid and Active Control of the Shock Wave – Turbulent Boundary Layer Interaction and Porous Plate Transpiration Flow", EUROSHOCK II TR BRPR-95-76/1.2, 1999, University of Karlsruhe Final Report, (also see Chapter 12)

[3.3] Babinsky, H., "Active Control of Swept Shock Wave/Boundary Layer Interactions", EUROSHOCK II TR BRPR-95-76/1.3 and Cambridge University Engineering Department Report CUED/A-AEROTR.25, 1999 (also see Chapter 13)

[3.4] Poll, D.I.A., Danks, M. and Humphreys, B.E., "The Aerodynamic Performance of Laser Drilled Sheets", First European Forum on Laminar Flow Technology, Hamburg, March 1992, Paper 92-02-02, 1992

[3.5] Bur, R., Corbel, B., Délery, J. and Soulevant, D., " Transonic shock wave/boundary layer interaction control. Complementary experiments on suction slot and bump control", ONERA RT 131/7078 DAFE/Y, 2000 (also see Chapter 11)

[3.6] Ashill, P.R., Fulker, J.L. and Simmons, M.J., "Simulated Active Control of Shock Waves in Experiments on Airfoil Models", Proceedings ICEFM, Turin, July 1994

Chapter 4

[4.1] Arina, R., Ceresola, N. and Pianta, P.G., "Application of Unsteady Turbulence Models to the Numerical Prediction of Transonic Buffet", Proceedings of the ICAS-Conference, Sorrento, September 1996 (also see [6.4] and Chapter 23)

[4.2] de Matteis, P. and Dima, C., "Numerical Investigation of Passive Shock Control on Transonic Airfoils Through an Euler/Boundary-Layer Coupling Technique", in EUROSHOCK – Drag Reduction by Passive Shock Control, Notes on Numerical Fluid Mechanics, Volume 56, Friedr. Vieweg & Sohn Verlagsgesellschaft mbH, Braunschweig/Wiesbaden, 1997, pp. 151 - 170

[4.3] Dima, C., "Numerical Investigation of the Active Control Technique of Shock/Boundary-layer Interaction", EUROSHOCK II TR BRPR-95-76/2.1 and CIRA-TR-AET-99-108, 1999 (also see Chapter 14)

[4.4] Dargel, G., "Extension, Validation and Application of the DA VII Transonic Airfoil Code with Passive Shock Control", in EUROSHOCK – Drag Reduction by Passive Shock Control, Notes on Numerical Fluid Mechanics, Volume 56, Friedr. Vieweg & Sohn Verlagsgesellschaft mbH, Braunschweig/Wiesbaden, 1997, pp. 195 - 220

[4.5] Dargel, G., "Validation of DA VII Airfoil/Swept-wing Code for Shock and Boundary Layer Control", EUROSHOCK II TR BRPR-95-76/2.2 and DA-Report No. EF-068/99, 1999 (also see Chapter 21)

[4.6] Geissler, W., "Transonic Airfoil Flow Prediction with Shock Boundary Layer Interaction Control (SBLIC) by a Time-accurate Navier-Stokes Code", in EUROSHOCK – Drag Reduction by Passive Shock Control, Notes on Numerical Fluid Mechanics, Volume 56, Friedr. Vieweg & Sohn Verlagsgesellschaft mbH, Braunschweig/Wiesbaden, 1997, pp. 245 - 263

[4.7] Caballero, N., "Numerical Investigation of Shock and Boundary Layer Control Techniques", EUROSHOCK II TR BRPR-95-76/2.3, 1999, INTA Final Report, (also see Chapter 16)

[4.8] Geissler, W., "Steady and Unsteady Numerical Navier-Stokes Calculations on the DRA-2303 Airfoil with and without Bump", EUROSHOCK II TR BRPR-95-76/2.4, 1999, DLR Task 2 Final Report, (also see Chapter 18)

[4.9] Le Balleur, J.C., Girodroux-Lavigne, P. and Gassot, H., "Development of Viscous-Inviscid Interaction Codes for Prediction of Shock Boundary-Layer Interaction Control (SBLIC) and Buffet over Airfoils", in EUROSHOCK – Drag Reduction by Passive Shock Control, Notes on Numerical Fluid Mechanics, Volume 56, Friedr. Vieweg & Sohn Verlagsgesellschaft mbH, Braunschweig/Wiesbaden, 1997, pp. 221 - 244

[4.10] Le Balleur, J.C., Girodroux-Lavigne, P. and Gassot, H., "Prediction of ransonic airfoil/wing flow and buffet with control using a time-accurate strong viscous-inviscid interaction approach ", EUROSHOCK II TR BRPR-95-76/2.5, 1999, and ONERA Technical Report RT 130/7078 DAFE/DSNA, Oct. 1999 (also see Chapter 17)

[4.11] de Nicola, C., "Introduction of Passive Shock Control into an Interactive Boundary Layer Method", in EUROSHOCK – Drag Reduction by Passive Shock Control, Notes on Numerical Fluid Mechanics, Volume 56, Friedr. Vieweg & Sohn Verlagsgesellschaft mbH, Braunschweig/Wiesbaden, 1997, pp. 171 - 194

[4.12] de Nicola, C. and Cirino, V.; "Euler/Boundary Layer Coupling to Predict Steady and Unsteady Transonic Flow past an Airfoil with and without Shock Control", EUROSHOCK II TR BRPR-95-76/2.6, 1999, University of Naples Final Report (also see Chapter 15)

[4.13] Archambaud, J.P. and Rodde, A.M., "Qualification by Laser Measurement of the Passive Control on the LVA-1A Airfoil in the T2 Wind Tunnel", in EUROSHOCK – Drag Reduction by Passive Shock Control, Notes on Numerical Fluid Mechanics, Volume 56, Friedr. Vieweg & Sohn Verlagsgesellschaft mbH, Braunschweig/ Wiesbaden, 1997, pp. 331 -353

[4.14] Rosemann, H., Knauer, A. and Stanewsky, E., "Experimental Investigation of the Transonic Airfoils DA LVA-1A and VA-2 with Shock Control", in EUROSHOCK – Drag Reduction by Passive Shock Control, Notes on Numerical Fluid Mechanics, Volume 56, Friedr. Vieweg & Sohn Verlagsgesellschaft mbH, Braunschweig/ Wiesbaden, 1997, pp. 355 -378

[4.15] Knauer, A., "Performance Improvement of Transonic Airfoils Through Contour Modifications in the Shock Region", DLR Research Report 98-03, January 1998, and Ph.D. Thesis University Hannover (also see Chapter 20)

[4.16] Birkemeyer, J., "Drag Minimization on a Transonic Wing by Ventilation and Adaptive Contour Bump", DLR Research Report 1999-28, July 1999, and Ph.D. Thesis University Hannover

[4.17] Rosemann, H. and Birkemeyer, J., "Wind Tunnel Experiments on Airfoils and Sheared Wings with Control", EUROSHOCK II TR BRPR-95-76/3.2, 1999, DLR Task 3 Final Report (also see Chapter 20)

[4.18] Stanewsky, E., "Shock Boundary Layer Interaction", in: Boundary Layer Simulation and Control in Wind Tunnels, AGARD-AR-224, 1988, pp. 271 - 305

Chapter 5

[5.1] Fulker, J.L. and Simmons, M.J., "An investigation of active, suction, shock and boundary layer control techniques", EUROSHOCK II TR BRPR-95-76/3.1 and DERA/MSS4X/CR980817/1.0, 1999 (also see Chapter 19)

[5.2] Rosemann, H., "The Cryogenic Ludwieg-tube Tunnel at Göttingen", in AGARD-R-812, Special Course on Advances in Cryogenic Wind Tunnel Technology, 1997

[5.3] Binder, B., Riethmüller, L., Tusche, S. and Wulf, R., "Upgrading the Transonic Wind Tunnel Göttingen", in Yearbook of the DGLR, Volume 1, 1992

[5.4] Drela, M., "A User's Guide to MSES 2.95", MIT Computational Aerospace Sciences Laboratory, 1996

[5.5] Dargel, G. and Thiede, P., "Design and Manufacture of the Cryogenic Laminar-type Airfoil Model DA LVA-1A with Passive Shock Control", in EUROSHOCK – Drag Reduction by Passive Shock Control, Notes on Numerical Fluid Mechanics, Volume 56, Friedr. Vieweg & Sohn Verlagsgesellschaft mbH, Braunschweig/ Wiesbaden, 1997, pp. 319 - 329

[5.6] Fulker, J. L. and Simmons, M. J., "An experimental investigation of passive shock/boundary-layer control on an aerofoil", in EUROSHOCK – Drag Reduction by Passive Shock Control, Notes on Numerical Fluid Mechanics, Volume 56, Friedr. Vieweg & Sohn Verlagsgesellschaft mbH, Braunschweig/Wiesbaden, 1997, pp. 379 - 400

Chapter 6

[6.1] Doe, R., "Control Application Aspects", EUROSHOCK II, Task 4 Final Report, BAE SYSTEMS Airbus UK Report B61/R&T/000926, Dec. 2000

[6.2] Vallee, J.J., "Application of Shock and Boundary Layer Control to a Business-jet Aircraft", EUROSHOCK II TR BRPR-95-76/4.2, 1999, Dassault Aviation Final Report, Reference DGT 78414 (also see Chapter 24)

[6.3] Thiede, P. and Dargel, G., "Assessment of Shock and Boundary Layer Control Concepts for Hybrid Laminar Flow Wing Design", EUROSHOCK II TR BRPR-95-76/4.3 and DA-Report No. EF-069/99, 1999 (also see Chapter 21)

[6.4] Catino, N. and Ceresola, N., "Drag Reduction and Buffet Damping by a Contour Bump Control Device and Regional-jet Application", EUROSHOCK II TR BRPR-95-76/4.4, 1999, ALENIA Final Report (also see Chapter 23)

B. INDIVIDUAL CONTRIBUTIONS

10 Introduction to the Individual Contributions

E. Stanewsky

DLR, Institute of Aerodynamics and Flow Technology
Bunsenstrasse 10, D-37073 Göttingen

In the preceding chapters an extended summary and résumé of the results obtained during the EUROSHOCK II research program was presented. EUROSHOCK II was, however, a team effort with the individual work adding up to the final product. The Synopsis is therefore followed by a more detailed description of the work performed by the various members of EUROSHOCK II, written by these members. The contributions, Chapters 11 through 24, are assembled in the order of the four tasks which comprise the research project. The sequence of the individual contributions and the main contents of the research performed are briefly outlined below.

Chapter 11 by R. Bur, R. Benay, B. Corbel and J. Délery, ONERA: Experimental work was focused on a local analysis of various controlled interactions in order to establish the flow field, including both mean and turbulent properties, and to study control effectiveness. The channel-flow experiments were supplemented by Navier-Stokes computations to evaluate physical models and boundary conditions prescribed, also utilizing several turbulence models. The control mechanisms considered were: a combination of several cavities through which hybrid control was applied, discrete slot suction, and a contour bump in the shock region.

Chapter 12 by R. Bohning and P. Doerffer, University of Karlsruhe: Investigated were active control by part-suction through a perforated plate/cavity arrangement and hybrid control, consisting of a passive cavity in the shock region and active suction downstream, placing emphasis on a careful examination of the influence of the size (length) and location of the downstream active cavity. In further work, the important problem of the mass-flow determination through a porous plate in the presence of an outer transonic stream was examined. Supplementing the experiments, 2D and 3D Navier-Stokes computations, simulating the channel-flow experiments, were performed.

Chapter 13 by H. Babinsky, University of Cambridge: Considered was active shock and boundary layer control in three-dimensional flows; control mechanisms investigated were a perforated plate/cavity arrangement with part-suction and discrete slot suction with emphasis on the effect of the slot location with respect to the shock. Again, Navier-Stokes computations, using different turbulence models, supplemented the swept-shock channel-flow experiments.

The work described in these chapters was performed within Task 1 of EUROSHOCK II "Modeling of Active Control Phenomena". The objective was to study the influence of fundamental parameters involved in shock and boundary layer control and to optimize control devices; furthermore, to improve and to validate physical models for the control region based on experiments and computations.

The contributions in Chapters 14 through 18 relate to Task 2 "Prediction of Transonic Airfoil/Wing Flow with Control". Objectives were the extension of available steady and unsteady two-dimensional and steady three-dimensional computational methods to predict flows with shock and boundary layer control and to perform first parametric studies to assess control concepts; furthermore, to validate the computational methods involved, employing, among others, results obtained within Task 3. Note that all participants in this task carried out a control-law simulation, based on a prescribed pressure distribution and Poll's control law, to determine the transpiration velocity. Also note that ALENIA and EADS-Airbus D participated in Task 2 for code validation; their contribution will, however, be discussed under Task 4. In detail, the contributions are:

Chapter 14 by P.P. de Matteis and C. Dima, CIRA: A 2D Viscous-Inviscid-Interaction (VII)-method and its quasi-3D extension, validated through comparisons with experiments, was employed to assess the effectiveness of various shock and boundary layer control concepts. A new concept, the "Pneumatic Bump" was introduced.

Chapter 15 by C. de Nicola and V. Cirino, University of Naples: A numerical method based on an Euler/boundary layer coupling has been extended to unsteady flows and employed to investigate control effects on drag reduction and, especially, buffet behavior.

Chapter 16 by Nieves Caballero, INTA: In a first step, steady calculations have been performed to study the effect of various control devices on drag reduction. In a second phase, unsteady calculations were performed to investigated the effect of control on buffet onset. For the computations a time-accurate Navier-Stokes code was employed.

Chapter 17 by J.C. Le Balleur, P. Girodroux-Lavigne and H. Gassot, ONERA: Different concepts of shock wave and/or boundary layer control have been numerically investigated with the 2D and 3D time-accurate strong viscous-inviscid interactions codes VIS15 and VIS25. Drag reduction as well as buffet behavior as dependent on control have been considered.

Chapter 18 by W. Geissler, DLR: New turbulence models have been introduced into the DLR 2D time-accurate Navier-Stokes code. Computations were then carried out for the DRA-2303 airfoil with and without contour bump to study its effect on steady as well as unsteady flow behavior, also considering the influence of different turbulence models. Hysteresis effects related to the buffet process were investigated for a circular-arc airfoil.

The objectives of Task 3 "Wind Tunnel Experiments on Airfoils and Wings with Control" were to investigate, mainly experimentally, the effect of various shock and boundary layer control techniques on the design and off-design performance of airfoils and swept wings. A comprehensive wind tunnel test program was, correspondingly, executed to provide an extensive data base for cases with and without control to demonstrate the efficiency of various control techniques up to conditions typical of flight, to assess computer codes considered in Task 2, and to assess the practicalities of applying the control techniques considered to actual aircraft wings, this topic being treated in Task 4. Related to Task 3, the following contributions are being presented:

Chapter 19 by J.L. Fulker and M.J. Simmons, DERA: Detailed steady pressure measurements were carried out with the DRA-2303 laminar-type airfoil, at Reynolds numbers up to those approaching full scale, at conditions up to and including buffet onset. Control schemes included discrete suction, active control by perforated plate/cavity arrangement with part suction and hybrid control. The experiments were supplemented by VII computations.

Chapter 20 by H. Rosemann and J. Birkemeyer, DLR: Investigated were the effect of Mach number and angle of attack (lift) on the effectiveness of various control systems for two airfoils, viz., the A340-type ADIF airfoil and the laminar-type airfoil DA LVA-1A; furthermore investigated, especially in conjunction with bump control and control by a bump with upstream slot suction, was an infinitely-swept sheared wing based on the ADIF airfoil.

The objectives of Task 4 "Control Application Aspects" were the introduction of various potential control methods, identified in the deliberations of all tasks, into existing or new wing designs and the assessment of the benefits and penalties associated with the aircraft implementation of control. Aircraft configurations considered included regional-jet and A340-type transport aircraft with laminar and turbulent wings, respectively, mainly considering control by contour bumps. Minimum drag reductions needed to balance installation, operation, and maintenance penalties were identified.

Chapter 21 by G. Dargel and P. Thiede, EADS-Airbus D: Work was focused on the assessment of different control devices for the application to a hybrid laminar flow wing of a long-range transport aircraft, following the improvement and extension of the VII airfoil/swept-wing code. A bump parametric optimization study was then carried out and the optimized variable-height bump integrated into the HLF wing. Assessed were benefits and penalties associated with employing bump control based on typical long-range aircraft missions.

Chapter 22 by R. Doe, BAE SYSTEMS-Airbus UK: The work performed involved the optimization of a bump for an existing turbulent airfoil/wing and the incorporation of bump control into the turbulent wing of a long-range transport aircraft. Benefits and penalties associated with a typical long-range mission were assessed. *The present contribution was, unfortunately, not available for inclusion into the present volume. The reader is requested to accept our apologies for this, but also referred to Chapter 6 for a summary of this work..*

Chapter 23 by N. Catino and N. Ceresola, ALENIA Aerospazio: An evaluation and optimization of contour bumps at steady and unsteady conditions for two airfoils has been performed to explore their capability to reduce drag in transonic flow and to dampen aerodynamic buffet. A simplified configuration of a regional-jet aircraft with a laminar-type wing has then been generated and the wing fitted with a contour bump; drag reductions possible for the complete aircraft were estimated.

Chapter 24 by J.J. Vallee, DASSAULT Aviation: An aerodynamic assessment of the effect of bumps on the flow development about a Falcon-type (turbulent) airfoil was performed looking at cruise as well as at off-design conditions. A technical solution for the integration of an adaptive bump and of slot suction, respectively, into a Falcon-type wing was then derived and minimum drag reductions needed to cover installation penalties determined.

Concerning the individual contributions, the reader is advised that the interpretation of certain flow developments given in these contributions does not necessarily agree with the interpretations given in the Synopsis, as is not uncommon when discussing the physical background of specific developments. Here, the reader must, of course, use his own judgement when studying the results. It is, furthermore, suggested to consult Chapter 4.1.3 when in doubt about the definition of the various drag components discussed within these contributions.

The test cases considered in Task 2 and the experiments performed within Task 3 are summarized in Table 1. Note that the test conditions given in the table are approximate with the exact conditions provided in the corresponding chapters of the Synopsis and the individual contributions.

Table 1 Essential computational test cases, parametric study cases, and experiments

Airfoil/ wing	Type	M_∞	α °/C_L	$Re_c x10^{-6}$	Control mechanism	Wind tunnel data from:
Computational test cases						
DRA-2303	laminar	0.681	2.1/0.747	19	Bump	DERA 8 Ft 1)
RAE-5225	turbulent	0.730	2.95/0.756	19	Bump	DERA 8 Ft 1)
RAE-5225	turbulent	0.725	2.90/0.756	18.7	Slot suction C_Q=-0.0002	
DRA-2303	laminar	0.68	2.0/0.740	19	Bump + Slot suction, C_Q=-0.0001	
RAE-5225	turbulent	0.725	2.90/0.756	18.7	Bump + Slot suction, C_Q=-0.0002	
VA-2	turbulent	0.740	2.0/0.80	2.5	Passive and active cavity C_Q=-0.00088	TWG [5.3]
Parametric studies						
DRA-2303	laminar	0.68	Variable	19	Slot suction C_Q=-0.00007	DERA 8 Ft 1)
DRA-2303	laminar	0.68	Variable C_L = 0.608 2)	19	Active cavity C_Q=-0.00009 and higher	DERA 8 Ft 1)
DRA-2303	laminar	0.68	Variable C_L = 0.608 2)	19	Hybrid contr. C_Q=-0.00009 and higher	DERA 8 Ft 1)
ADIF airfoil	turbulent	Variab. 0.765 2)	Variable	8	Bump	KRG [5.2]
ADIF wing	turbulent	Variab. 0.84 2)	Variable C_L = 0.59 2)	9.9	Bump	TWG [5.3]
Further experimental test cases						
ADIF wing	turbulent	Variab.	$C_D = f(C_L)$	9.9	Bump + Slot suction, C_Q=-0.0001	TWG [5.3]
ADIF wing	turbulent	Variab.	$C_D = f(C_L)$	9.9	Active/Hybr. cavity C_Q=-0.0002	TWG [5.3]
DA LVA-1A	laminar	Variab.	$C_D = f(C_L)$	6	Bump	KRG [5.2]

1) See Chapter 19   2) Computational test case

11 Study of Control Devices Applied to a Transonic Shock Wave/Boundary Layer Interaction

R. Bur, R. Benay, B. Corbel and J. Délery

ONERA, Fundamental / Experimental Aerodynamics Department
8, rue des Vertugadins, 92190 Meudon, France

Summary

The aim of this study was to investigate in detail the flow field resulting from transonic shock wave / boundary layer interactions under control conditions. Several control techniques have been investigated: 1) active control consisting in sucking a part of the boundary layer flow through a slot, 2) hybrid control, which is a combination of a passive control cavity placed underneath the shock region and a suction slot (or cavity) located downstream, and 3) control by a local deformation of the surface, the so-called bump concept. In the experimental part of the study, flow surveys have been executed with a two-component LDV system, including mean velocity and Reynolds tensor component measurements. Results show that active control by slot suction produces an increase in wave drag, but a significant decrease in friction drag, especially when the slot is located just downstream of the shock. The tendencies are inverted when bump or hybrid control is applied, the bump control having nevertheless a moderate effect on the friction drag. Moreover, active control by a slot efficiently diminishes the turbulence level downstream of the interaction domain. The aim of the theoretical study was to discuss the problems involved in the modeling of the vertical velocity component at the wall in the control region and of the effect of control on the turbulent field. The RANS research solver Nasca has been used for numerical simulations. Both, the algebraic Baldwin-Lomax model and the transport equation [k-ε] Chien model were investigated, the latter having shown deficiencies in the prediction of the channel flow. On the other hand, Poll's law and the Bohning-Doerffer law give a similar and fair reproduction of the vertical velocity at the wall in the control region.

11.1 Introduction

The work reported in this chapter is the Onera contribution to Task 1 of the EUROSHOCK II project devoted to drag reduction by shock and boundary layer control. This work has been started within the EUROSHOCK (I) project where the technique used to control the interaction was a passive control device [1]. It is recalled that the principle of passive control consists in establishing a natural circulation between the downstream high pressure face of a shock and its upstream low pressure face. This circulation is achieved through a closed cavity, placed underneath the shock region, the contact with the outer flow being made by a perforated plate. It has been shown that, in some instances (e.g., for a turbulent airfoil), passive control may produce a reduction in airfoil drag, while postponing buffet onset to higher incidences.

Within the framework of the EUROSHOCK II project, it has been decided to investigate other control methods which could lead to a larger improvement of airfoil performance essentially by a more significant drag reduction. Several control techniques have been used to achieve this goal, namely : 1) active control consisting in sucking a part of the boundary layer flow through a slot or a cavity covered by a perforated plate, 2) hybrid control which is a combination of passive and active control, and 3) control by a local deformation of the surface, the so-called bump concept. The bump action can be combined with active control (suction) to improve its efficiency. The location of these control devices relative to the shock wave is an important parameter that has to be carefully examined.

The objective of Task 1 of the EUROSHOCK II project is to contribute to the understanding and modeling of the physical phenomena involved in shock wave / boundary layer interaction under control conditions. Corresponding work is based on: 1) the execution of basic experiments aiming at a detailed description of the interacting flow field, including both, its mean and turbulent properties, and 2) the exploitation of the results obtained to discuss both, the definition of the boundary conditions to be prescribed at the wall in the control region and the repercussion of control techniques on turbulence behavior and its modeling.

The test set-up and the measurement techniques used to investigate the flow are first described. A presentation of experimental results is then given, the precise objective being to define, mainly by means of LDV explorations, the properties of the interaction domain, firstly in the absence of control (solid wall reference case), secondly under control conditions. The numerical approach used to compute the transonic shock wave / boundary layer interaction is presented, particularly the modifications accounting for the presence of passive or hybrid control in the interaction region. Comparisons of experimental and numerical results are shown for the reference case and one case with passive control (one specific perforated plate has been selected). The two laws, i.e., Poll's law and the law of Bohning and Doerffer, employed to obtain the surface velocity distribution along the perforated plate (control region) are compared. One hybrid control case has also been treated numerically and compared to the experiments.

11.2 Experimental Conditions

11.2.1 Test Set-up Arrangement and Techniques of Investigation

The experiments have been executed in the S8Ch basic research transonic-supersonic wind tunnel of the Onera Fluid Mechanics Laboratory at Chalais-Meudon. This facility is a continuous wind tunnel supplied with desiccated atmospheric air. The stagnation conditions, nearly constant throughout the tests, are : $p_{st0} = 96\,000 \pm 800$ Pa, $T_{st0} = 300 \pm 4$ K.

The test set-ups used for the present study are shown in Figs. 1a, 1b and 1c, respectively, for the active, hybrid and bump control devices. The basic set-up consists of a transonic channel having a height of 100mm and a span of 120mm in the test section itself. The lower wall is rectilinear, the upper wall being made of a contoured profile designed to produce a uniform supersonic flow of a nominal Mach number equal to 1.4. The two side walls are equipped with high quality glass windows to allow visualizations and LDV measurements. A second throat, of adjustable cross section, is placed at the test section outlet (at X=510mm, the streamwise distance X being measured from the throat section of the half-nozzle) allowing: 1) to produce, by choking, a shock wave whose position, and hence intensity, can be adjusted in a continuous and precise manner, and 2) to isolate the flow field under study from pressure perturbations emanating from downstream ducts. Such a device notably reduces unwanted shock oscillations.

Several control configurations, installed on the rectilinear lower wall, have been tested, Figure 1:

— An active control system which consists in sucking a part of the boundary layer through a slot (a).
— A hybrid control device which consists of a combination of passive and active control (b). The passive control cavity and the perforated plate are those used for the EUROSHOCK (I) experiments [2].
— Control by a local deformation of the surface, the so-called bump concept (c).

The flows under study were quantified by means of Schlieren visualizations, measurements of wall pressure distributions and probings with a multi-component LDV system. For the present study, where the flows were nominally two-dimensional, the two-component version of the Onera three-component LDV system was used. The light source is a 15W - Argon laser, used in the present tests at a power of 5W, whose beam is separated into two beams of wave length 0.488μm (blue color) and 0.5145μm (green color) by means of a semi-transparent mirror. The two original beams are first split by classical beam splitters and then traverse Bragg cells to enable the system to detect the velocity direction. The four beams are focused by the emission lens to generate two fringe patterns inside the measuring volume whose diameter was equal to 200μm. The two fringe patterns were rotated at $\pm$ 45° with respect to the X-direction, allowing the simultaneous measurements of two velocity components in a vertical plane. The blue and green fringe spacings are 13.14μm and 14μm, respectively. In order to obtain a correct signal-to-noise ratio, the LDV system was operated in the forward scattering mode. The collecting part of the LDV system is equipped with a Cassegrain telescope, having a diameter of 200mm, which collects the light scattered by a particle when it crosses the probe volume. The light from the telescope passes through interference filters that separate the green and blue components which are then applied to two photo-multipliers whose signals are processed by an IFA 755 counter connected to the acquisition system. The flow is seeded with sub-micron

(0.5μm diameter) droplets of olive oil injected in the wind tunnel settling chamber.

The flow fields have been explored along lines normal to the wall (Y-direction) extending from the surface (Y=0) to a height of Y=22mm and contained in the test section center plane. This extent was chosen to be sure to cover the entire dissipative layer and a part of the outer inviscid flow. Each exploration contained 44 measurement points unevenly distributed in order to refine the probing in the vicinity of the wall. The interaction domain has been explored along 26 (29) vertical lines whose streamwise locations were in the range 115mm $\leq X \leq$ 260mm (270mm) for the reference case, the active control device, and the hybrid control devices with the 50mm-long (and 70mm long) passive control cavity. Reliable measurements with the LDV system, in the two-component version, were limited to a minimum distance of 0.3mm from the wall.

At each measurement point, a sample of couples of the instantaneous values of the velocity components u and v was acquired for further processing. The size of the sample, equal to 5000, gave an acceptable statistical uncertainty for the first and second order statistical moments. The determination of the mean velocity components allows the local Mach number to be computed and the boundary layer global properties (δ^*, θ and H_i) to be defined. The density and the velocity of sound were computed by assuming constant stagnation temperature throughout the flow. The two-component version of the LDV system did not allow the determination of the spanwise velocity component w. Thus, the turbulent kinetic energy k has been estimated by the following formula:

$$k = \frac{1}{2}\left[\overline{u'^2} + \overline{v'^2} + \frac{1}{2}\left(\overline{u'^2} + \overline{v'^2}\right)\right]$$

The LDV components give the field quantities with an accuracy depending on uncertainties affecting the LDV system calibration (uncertainties in the fringe distance, in the Bragg frequency), the determination by the counters of the frequency of the light scattered by the particles, and the statistical treatment of the sample. For the present experiments, the field properties have been determined with a precision of: 1) 1 % of the maximum velocity modulus for the mean velocity components, 2) less than or equal to 8 % for the normal stress components of the Reynolds tensor, and 3) less than or equal to 10 % for the turbulent shear stress component.

11.2.2 Configurations Tested

The solid wall reference case has been tested for two different shock locations (X_s=155mm and 165mm) in order to allow comparisons with all the control configurations. This location was monitored by considering the pressure distribution on the channel upper wall, the shock position being accurately defined by adjustment of the second throat section.

The first type of control is an active control device which consists in sucking a part of the boundary layer flow through a slot. Three locations of the slot relative to the shock, which was fixed in the outer inviscid flow at X_s=165mm, were tested: 1) about one incoming boundary layer thickness upstream of the shock, between X=157mm and X=162mm, 2) centered, and 3) downstream of the shock between X=168mm and X=173mm. The streamwise width of the slot was one and a half incoming boundary layer thicknesses (5mm). The shape of its leading edge (blunt or sharp) and its inclination (90° and 60° relative to the main flow direction) have been tested to optimize the control. Three different suction mass flow rates were considered, their values being measured by sonic throats located within the suction duct.

The second type of control is a hybrid control device consisting of a combination of passive and active control. One hybrid control device has been selected in common with the University of Karlsruhe: the original (EUROSHOCK I) 70mm-long cavity has been divided into two parts by a 5mm-thick solid wall. The upstream part is a 50mm-long passive control cavity and the downstream part a 15mm-long active control cavity. Both cavities are covered by a perforated plate. The shock location was fixed at mid-distance between the leading and trailing edges of the passive control cavity, i.e., at X_s=155mm. A second hybrid control device has been selected in order to keep the shock location the same as for the other control devices investigated. Here, the 70mm-long passive control cavity was used, the shock being centered at X_s=165mm. The active control system, located downstream of the passive cavity, is here a 15mm-long open cavity similar to like a suction slot. Several suction mass flow rates have been tested for both hybrid control devices. The experiments were carried out with a plate whose nominal characteristics are : porosity 5.67 % and 0.3mm-diameter holes, inclined at 45° with respect to the surface (in flow direction) in the upstream half of the plate and normal to it in its downstream half.

The third type of control is the control by a local deformation of the surface, the so-called bump concept. The shape of the bump is similar to the ones tested by DASA-Airbus on the A340 Hybrid-Laminar-Flow wing (see Chapter 21). Its (chord) length is 80mm, extending from X=130mm to X=210mm. Three locations of the shock have been investigated: 1) on the rising part of the bump (X_s=165mm), 2) at the bump crest (X_s=186mm), and 3) at the end of the bump (X_s=210mm).

Concerning the control involving suction, a definition of the suction mass flow rate coefficient, C'_Q, has been proposed to have here the same order of magnitude as the C_Q-coefficients used in external (airfoil-like) flows. In fact, it is not possible in channel-flow experiments to compute the C_Q coefficient, since an upstream uniform flow does not exist (the streamwise Mach number varies continuously in the channel) and there is no appropriate chord length. For this reason, the reference length L has been chosen equal to 1 meter, which is of the order of magnitude of the airfoil chord length used in the definition of the C_Q coefficient. Then, the C'_Q-coefficient was defined as:

$$C'_Q = \frac{q}{\rho^* U^* L}$$

where ρ^*, U^* are taken at the sonic state in the channel; q is the control mass flow rate per unit span.

This definition is inspired by the following considerations: the (ρU) product is not very sensitive to the Mach number within the range of interest here. In airfoil flows, the freestream Mach number is around M=0.8 and the shock upstream Mach number in the channel experiments is around M=1.3. The corresponding difference between the (ρU) values is small because (ρU) passes through a maximum in this Mach number range. Thus, the values of C'_Q are comparable to C_Q values, which makes it more meaningful for airfoil designers, in particular for the contributors to the other Tasks.

11.3 Presentation of the Experimental Results

11.3.1 Flow Visualizations

The Schlieren photographs in Fig. 2 reveal the flow structure for the reference case (solid wall, Fig. 2a) and for three different (active, hybrid and bump) control devices. For the active control by a suction slot (see Fig. 2b), the lambda shape structure of the shock (the λ-foot) has almost disappeared. The oblique line visible upstream of the shock is a weak perturbation wave caused by the junction between two parts of the lower wall. The single shock crossing the entire channel is strengthened and impinges on the wall slightly downstream compared to the reference case, i.e., in the vicinity of the slot location. As far as an airfoil is concerned, this strengthening of the shock leads to an increase of the wave drag. However, the thickening of the boundary layer downstream of the slot is strongly reduced. For the hybrid control case (see Fig. 2c), the structure of the shock takes a well defined lambda shape due to the boundary layer starting to thicken suddenly at the passive cavity origin. This lambda shape structure also appears when bump control is applied (see Fig. 2d), the location of the triple point being in this case at a lower height compared to the hybrid control case. Thus, hybrid control and bump control produce a decrease in wave drag, a single strong shock being replaced by two weaker shocks over a great part of the channel flow. The total entropy production by the shock waves will be less compared to the reference case or the active control case where the crossing shock occupies nearly the entire channel. To prevent the important thickening of the boundary layer in the hybrid control case, suction is applied downstream of the interaction region. However, the suction mass flow rates must be high to have a positive effect on the boundary layer behavior.

11.3.2 LDV Measurements

Mean Flow Field Properties. The Mach number contour lines are traced in Fig. 3 by adopting the same scale for the X and Y distances in order to have an "objective" view of the flow field structure. The traces for the reference case (Fig. 3a) show the system of compression waves generated by the initial thickening of the boundary layer inner region. The apparent thickness of the shock is due to an insufficient refinement of the measurement mesh (X-wise spacing of 5mm) which did not permit the correct "capture" of discontinuities oriented along a direction normal to the wall. When active control is applied (Fig. 3b), the foot of the shock moves slightly downstream compared to the reference case and fixes itself in the vicinity of the slot location. The strength of the shock has increased, especially for the high value of the suction mass flow rate. The boundary layer thickness downstream of the control region is considerably reduced by the suction effect, the low Mach number region having almost disappeared. For the hybrid control case (Fig. 3c), the tracing shows that, in the outer inviscid flow, the leading separation shock provokes a first supersonic compression of the flow from an upstream Mach number of 1.3 to a value around 1.2. In the triangular supersonic region between the leading and the trailing shocks, the Mach number varies from 1.2 to 1.1, the flow undergoing an isentropic compression. A region of supersonic flow exists downstream of the trailing shock. The boundary layer downstream of the interaction region is thickened compared to the reference case. The (moderate) suction mass flow rate is too small to re-stabilize the boundary layer. Only a very high mass flow rate has a significant effect, a large part of the boundary layer being then captured by the slot. When bump control is applied (Fig. 3d), the shock structure is similar to that of the hybrid control case. A continuous compression takes place over the forward rising part of the bump. The boundary layer starts to thicken over the rear part of the bump and reaches a thickness comparable to that of the reference case far downstream of the interaction region.

The development of the boundary layer displacement and momentum thicknesses δ^* and θ when active control is applied is plotted in Figs. 4 and 5, respectively, to show suction mass flow rate and slot location effects. In the reference case, δ^* starts to increase rapidly reaching a maximum value δ^*_{max} such that $\delta^*_{max}/\delta^*_0 \approx 6.3$. This rapid rise is felt by the outer still supersonic flow as a ramp effect of angle $\tan^{-1}(d\delta^*/dX) \approx 4°$. When slot suction is applied, this ramp effect almost disappears, the maximum value δ^*_{max} being such that $\delta^*_{max}/\delta^*_0 \approx 2.2$ for the higher suction mass flow rates (Fig. 4a). The downstream values of δ^* are lower when the slot is in its downstream location (Fig. 5a). In the reference case, θ steadily rises during the interaction to a final value θ_f such that $\theta_f/\theta_0 \approx 6.8$. When slot suction is applied, the rise of θ is much slower: for example, for a moderate mass flow rate (C'_Q=6.1x10^{-4}), the amplitude of the momentum thickness rise is such that $\theta_f/\theta_0 \approx 2.5$ (Fig. 4b). Thus, when active slot suction control is applied, the behavior of θ should lead to a significant reduction of the friction drag, this reduction being larger when the slot is located just downstream of the shock (Fig. 5b).

The variations of δ^* and θ for one hybrid control case (with the 50mm-long passive control cavity) are plotted in Fig. 6. For the flow with control, the maximum of δ^* has increased compared to the reference value (Fig. 6a). This increase is due to passive control which provokes a greater destabilization of the boundary layer which is not compensated by the downstream suction, even for the highest suction rate (C'_Q=3x10^{-4}). The final value θ_f of the momentum thickness is greater than for the reference case when small suction mass flow rates are applied (Fig. 6b). A very slight decrease is obtained when strong suction is utilized (only for the hybrid control case with the 70mm-long passive control cavity). At this stage, it can be concluded that friction drag cannot be decreased by hybrid control with downstream suction at feasible suction rates.

The variations of δ^* and θ when bump control is applied are plotted in Fig. 7. Over the front part of the bump, the rise in δ^* and θ is similar to the one of the reference case. In this first part of the interaction, the compression in the reference case is close to an isentropic ramp-like compression as the one caused by the bump. Further downstream, the rise in δ^* and θ is far less rapid than in the reference case since there is no separation in the case of the bump. After the bump crest, δ^* and θ rise strongly because of the rapid compression occurring over this part of the bump. The downstream values of δ^* are almost the same as those of the reference case (Fig. 7a). The final value θ_f of the momentum thickness is slightly reduced when bump control is applied (Fig. 7b), but this reduction is not significant compared to the one obtained with active control by slot suction. It can be concluded that the bump control has nearly no effect on friction drag.

Turbulent Field Properties. For active, hybrid and bump control, the streamwise variation of the local maximums of the (pseudo) turbulent kinetic energy is plotted in Fig. 8. The turbulence level downstream of the interaction is strongly reduced by active control, this level being the lowest when the highest suction mass flow rate is applied (Fig. 8a). To obtain a similar result with a hybrid control device, a much higher suction mass flow rate has to be applied (Fig. 8b). Bump control does not change the downstream turbulence level (Fig. 8c). Moreover, in the bump and hybrid control cases, the location of the highest value of k_{max} is shifted downstream (to the end of the control region) compared to the reference and the active control case. This is correlated to the modification of the shock structure (a large λ-foot) by the former two control devices.

11.4 Numerical Simulations

11.4.1 Flow Modeling

The numerical simulations were performed with the Nasca code which solves the classical Reynolds Averaged Navier-Stokes (RANS) equations. The code, using a finite volume technique, is totally implicit, including its modules treating the turbulence transport equations. The numerical scheme is an extension of the Osher and Chakravarthy scheme [3] to the case of a mesh which can be locally non-uniform. Turbulence modeling is carried out by means of the Baldwin-Lomax algebraic model [4] or the [k-ε] transport equation model of Chien [5].

The calculation domain is a part of the experimental channel extending from a well chosen section of the divergent expansion zone, where experimental velocity profiles and cross correlations are imposed, to the end of the channel, where the experimental pressure is imposed. The upstream section, where the profiles are prescribed, has been chosen at X=60mm to give well defined supersonic inflow conditions. At the downstream boundary, coinciding with the end of the channel at X=380mm, a numerical procedure is developed for fixing the experimental pressure, which is an adaptation of the characteristic extrapolation method due to Yee [6].

Passive control is simulated by prescribing the unit mass flow ρv at the wall, the other variables remaining unchanged. The ρv value at the wall is obtained by means of relations expressing a direct dependence of the wall vertical velocity on the pressure difference between the cavity and the external flow. The cavity pressure is taken as the experimental one. The relations used for the computations are the calibration law of Poll [7] and the Bohning-Doerffer law [1].

In the case of the hybrid control calculation, the slot has been approximated by imposing a negative value of ρv as boundary condition at the mesh points located in the slot area. The ρv value at these points has been set equal to the measured suction mass flow rate divided by the number of points. This very rough approximation shows how the phenomena could be simulated without changing the calculation domain.

11.4.2 Comparisons with Experiment

Calculations of the reference case flow have been performed with the [k-ε] transport equation model of Chien. These calculations led to an overprediction of the thickening of the boundary layer during the interaction process and downstream of it. There resulted two important viscous effects in the channel, a "viscous" choking and a breakdown of the supersonic flow (the effective second throat becomes smaller than the nozzle throat). This phenomenon points out the difficulties for two-equation [k-ε] models to correctly reproduce viscous flows in long channels with nearly constant cross section. Thus, the forthcoming computational results have been obtained by means of the algebraic Baldwin-Lomax turbulence model.

Calculations of the interaction with passive control are first examined. Both vertical velocity laws were implemented in the code and their respective results compared to the experimental results obtained in the EUROSHOCK (I) experiments [2]. The Mach number contour lines plotted in Fig. 9, here obtained using Poll's law, show a good representation of the large spreading of the shock structure, which begins at the origin of the perforated plate. The difference in the location of the crossing (quasi-normal) shock is small. The great extent of the viscous zone due to the control device is well reproduced by the calculation.

Figure 10 shows the Mach number contour lines when hybrid control is applied. The predicted leading shock originating at the beginning of the passive control cavity is too weak. A second, oblique shock, emanating approximately in the center of the cavity, is predicted. The calculated lambda shock pattern has a smaller size compared to the experimental one visible in the Schlieren photograph (Fig. 2c). The calculation gives a considerable overprediction of the viscous zone in the interaction domain. These deficiencies in the prediction of the interaction region lead to suspect important defects in the modeling of hybrid control.

The wall pressure distributions for the reference case and the passive and hybrid control cases, respectively, are plotted in Fig. 11. For the reference case, the calculated wall pressures are in good agreement with the measured ones. The interesting fact is that the strong pressure gradient due to the interaction occurs at the correct location, with no need to modify (even slightly) the downstream pressure imposed as boundary condition. For the passive control case, there is a rather good agreement between the calculated and measured wall pressure distributions. The steep rise in the calculated pressures at X=130mm corresponds to the beginning of the cavity. This first rise is followed by a drop in the computed pressure distributions which does not exist in reality. This drop is due to an insufficient mesh refinement in the longitudinal direction. These results lead to the conclusion that both, Poll's law and the Bohning-Doerffer law predict a similar behavior in passive control modeling. When hybrid control is applied, the prediction of the wall pressure distribution is less satisfactory than for passive control. For this kind of control where suction is imposed just after the passive control cavity, the mesh size is of fundamental importance. Indeed, discrepancies occur in the calculated pressure at the beginning of the perforated plate (as in the passive control case) and the beginning of the slot. Moreover, for modeling of the slot region, the ρv value at the corresponding mesh points has, as mentioned above, been set equal to the measured suction mass flow rate divided by the number of mesh points. The mesh used for the computations has a lack of points in the slot region (only four points). In further calculations, the slot should be represented by a more refined mesh to improve the results.

11.5 Conclusion

A detailed experimental investigation of transonic shock wave / boundary layer interaction under control conditions has been performed in channel-type flow. The solid wall reference case (with two different shock locations), active slot-control devices (with several suction slot geometries and locations), hybrid control devices (with two different passive control-cavity lengths) and control by a local deformation of the surface, the bump concept (with three different shock locations) have been studied, mainly using a two-component LDV system to determine the mean and turbulent flow field properties.

The evaluation of the results shows the following facts :

— When active control by slot suction is applied, the shock is strengthened since the spreading caused by the interaction is reduced. Consequently, the entropy production through the shock is increased and the wave drag higher. On the other hand, the momentum loss in the boundary layer is considerably reduced downstream of the control region, which leads to a reduction in friction drag. Active control has a significant effect on the turbulence level downstream of the interaction region, this level being reduced even for a moderate suction mass flow rate. The most favorable location of the slot is slightly downstream of the shock location, the upstream and centered locations leading to a greater rise in the boundary layer displacement and momentum thicknesses.

— When hybrid control is applied, the above tendencies in terms of drag are inverted. The reduction of the wave drag due to the smearing of the compression is impaired by the high friction drag along the perforated plate. The role of the suction slot (or cavity) placed downstream of the passive control cavity is to reduce friction losses. For the hybrid control devices tested and for suction mass flow rates similar to those used for active control, the friction drag is always increased compared to the reference case without control. A possible way to improve the efficiency of such devices could be to reduce the distance between the interaction region and the suction slot placed downstream.

— Bump control is very effective in spreading the compression in the near wall region, thus reducing wave drag. At the same time, the pressure gradients generated by the bump have only a modest effect on the boundary layer properties. Thus the bump combines the advantage of a natural shock induced interaction, i.e., reducing wave drag, without the strong destabilization of the boundary layer, the friction drag staying the same as for the reference case without control. A closer investigation of the shock location relative to the bump is likely to lead to an optimization of the benefit of such a control device.

Interactions under passive and hybrid control conditions have been computed by using a code solving the RANS equations and employing the Baldwin-Lomax algebraic turbulence model, the [k-ε] Chien model having shown deficiencies in the prediction of the boundary layer development in the channel flow. Modifications of the boundary conditions were introduced to take into account an imposed downstream pressure and the injection/suction velocities at the wall in the control region. The transpiration velocities at the wall in the passive control region have been computed by using the calibration law of Poll and the Bohning-Doerffer law. In the present calculations, both laws gave close results and a satisfactory prediction of the transpiration velocities.

In the passive control case, the results of the computations were generally satisfactory, the scatter observed being explained by discrepancies in the prediction of the shape of the viscous zones in the interaction region. In the case of hybrid control, future computations must represent the suction slot by a more refined mesh, or include the entire slot in the computation domain, to correctly reproduce its effect.

References

[1] E. Stanewsky, J. Délery, J. Fulker and W. Geissler (Eds), "Euroshock - Drag Reduction by Passive Shock Control", Notes on Numerical Fluid Mechanics, Vol. 56, Friedr.Vieweg & Sohn, 1997.

[2] R. Bur, B. Corbel and J. Délery, "Study of Passive Control in a Transonic Shock Wave / Boundary Layer Interaction", AIAA Journal, Vol. 36, No. 3, 1998, pp. 394-400.

[3] S. Osher and S. Chakravarthy, "Very High Order Accurate TVD Schemes", ICASE Report 84-44, 1984.

[4] B.S. Baldwin and H. Lomax, "Thin Layer Approximations and Algebraic Model for Separated Turbulent Flows", AIAA Paper 78-0257, 1978.

[5] K.Y. Chien, "Prediction of Channel and Boundary-Layer Flows with a Low-Reynolds-Number Turbulence Model", AIAA Journal, Vol. 20, No. 1, 1982, pp. 33-38.

[6] H.C. Yee, R.M. Beam and R.F. Warming, "Boundary Approximations for Implicit Schemes for One-dimensional Inviscid Equations of Gas Dynamics", AIAA Journal, Vol. 20, No. 9, 1982, pp. 1203-1211.

[7] D.I.A. Poll, M. Danks and B.E. Humphreys, "The Aerodynamic Performance of Laser Drilled Sheets", Proceedings of the First European Forum on Laminar Flow Technology, Hamburg, Germany, 1992, pp. 274-277.

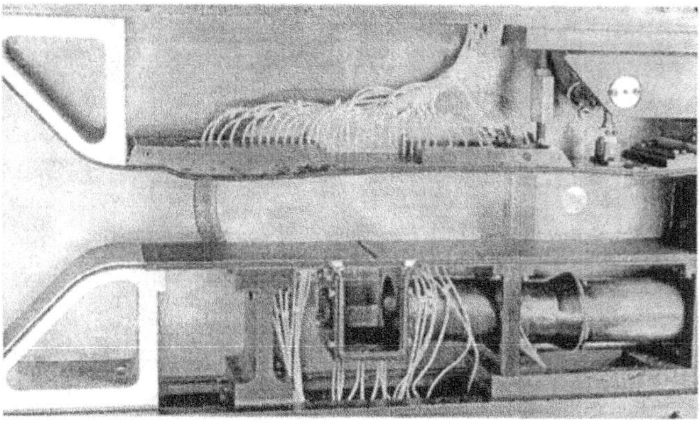

a - active control device

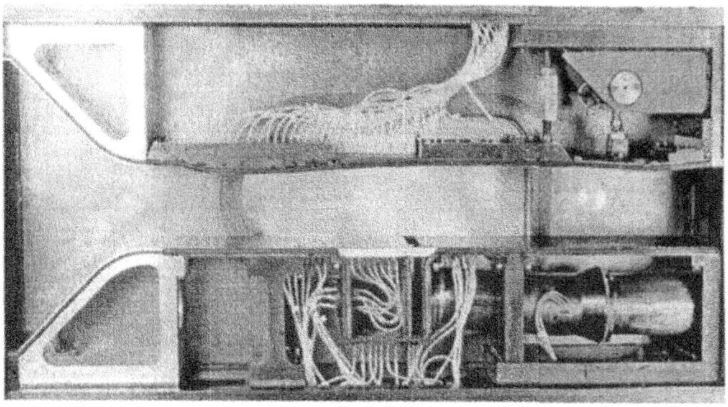

b - hybrid control device

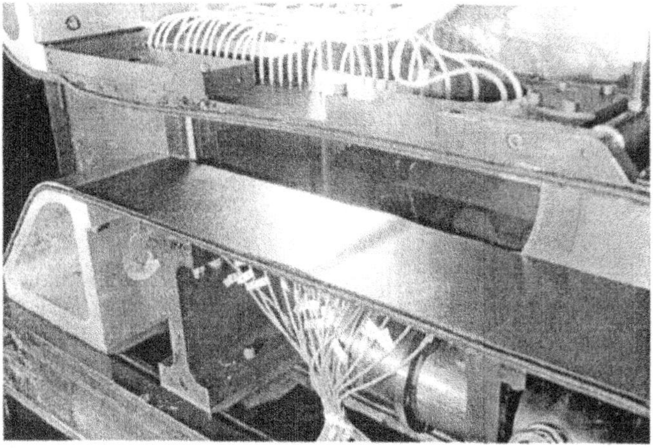

c - bump control device

Figure 1 Photographs of the test set-up

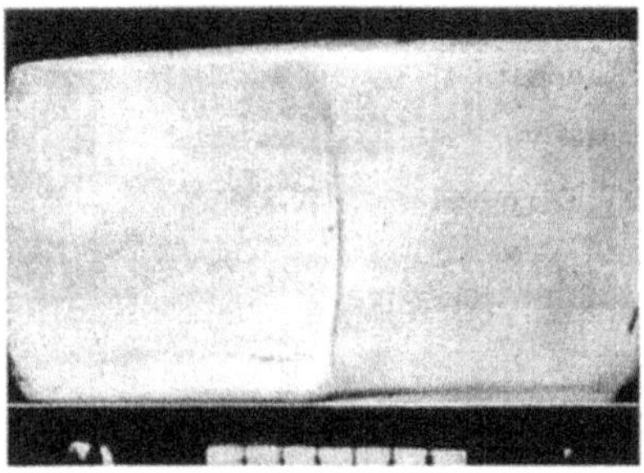

a – reference case (X_s = 165mm)

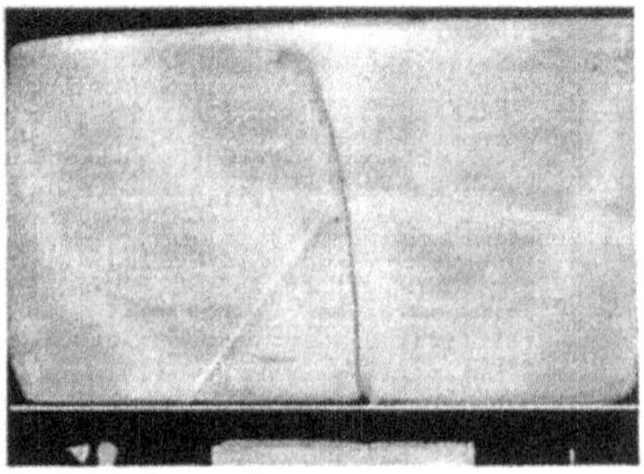

b – active control case
(with the downstream slot)

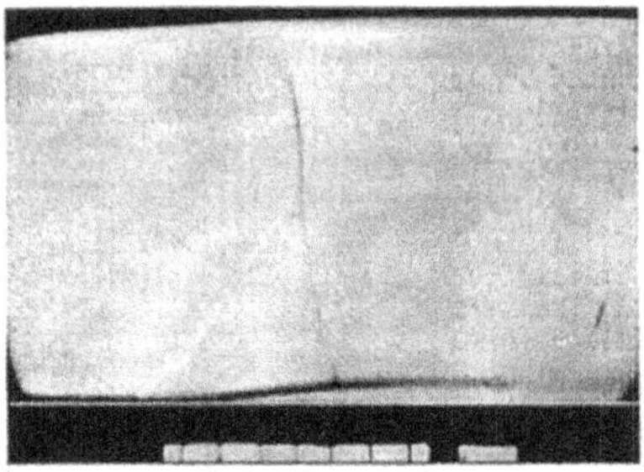

c – hybrid control case
(with the 70mm-long passive cavity)

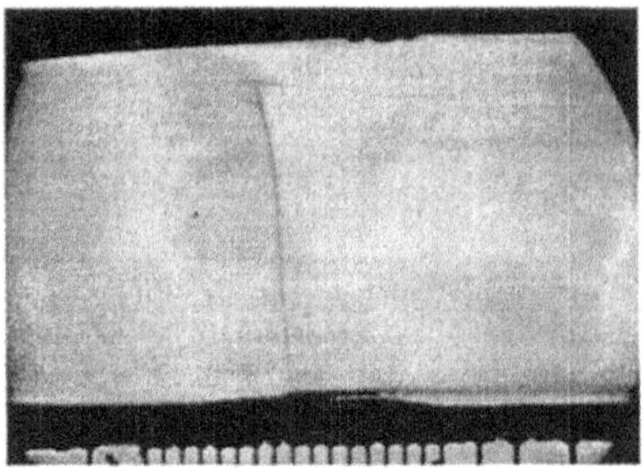

d – bump control case

Figure 2 Schlieren photographs of the flow field

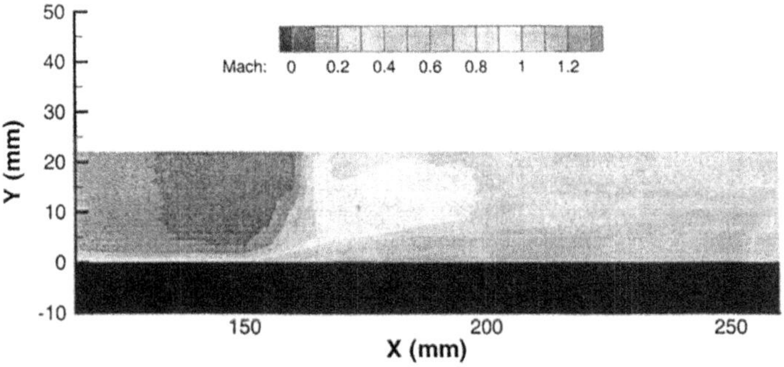

a – reference case (X_S = 165mm)

Figure 3 Mach number contour lines deduced from LDV measurements

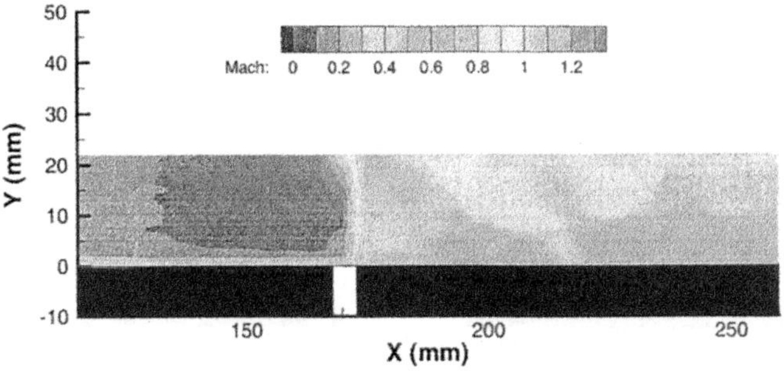

b – active control case (with the downstream slot)

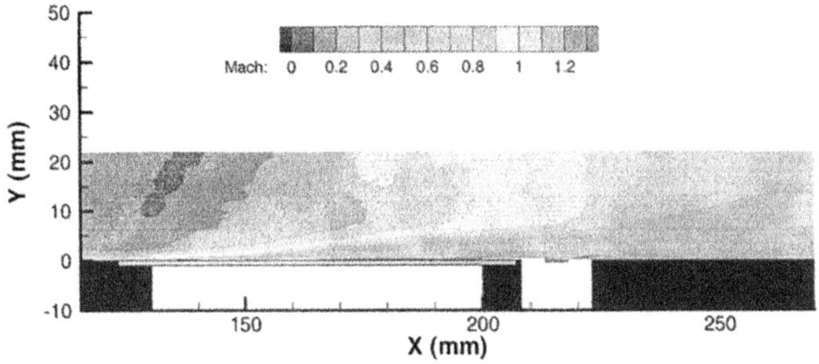

c – hybrid control case (with the 70mm-long passive control cavity)

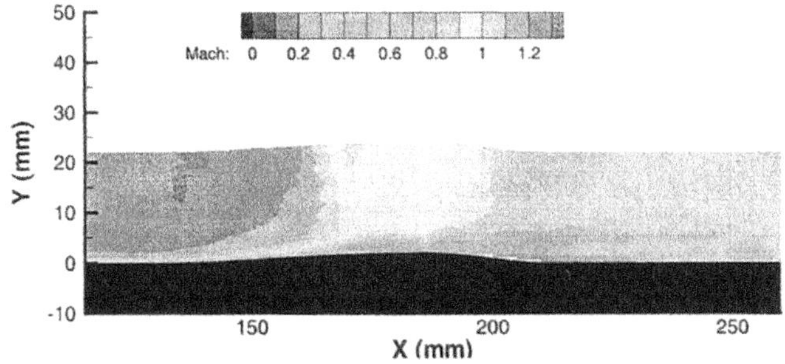

Figure 3 Mach number contour lines deduced from LDV measurements (Concluded)

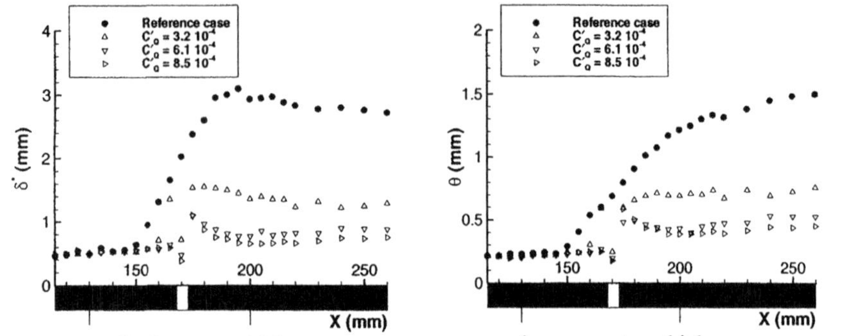

a – displacement thickness

b –momentum thickness

Figure 4 Boundary layer global characteristics – Active control cases (downstream slot) Effect of the suction mass flow rate

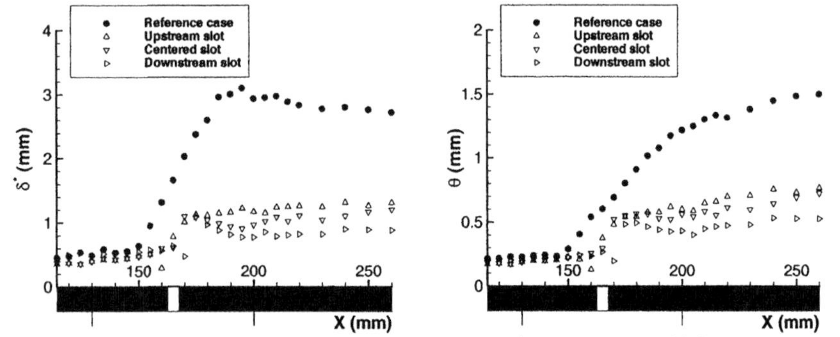

a – displacement thickness

b – momentum thickness

Figure 5 Boundary layer global characteristics – Active control cases Effect of slot location

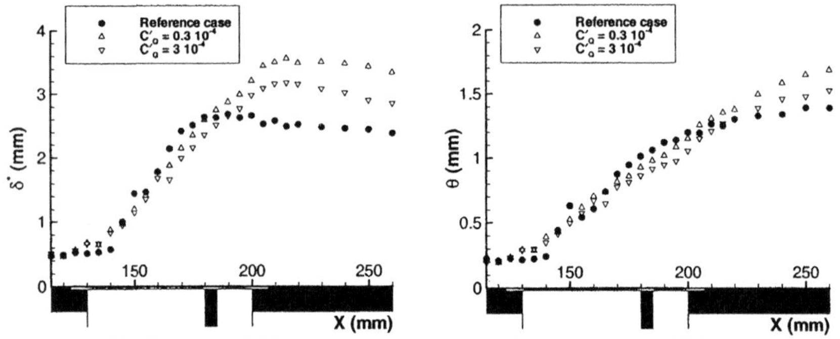

a – displacement thickness

b – momentum thickness

Figure 6 Boundary layer global characteristics – Hybrid control cases (with the 50mm-long passive control cavity)

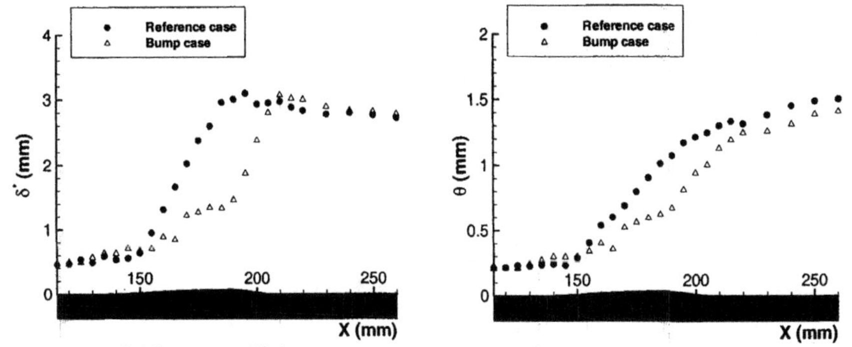

a – displacement thickness

b – momentum thickness

Figure 7 Boundary layer global characteristics – Bump control case

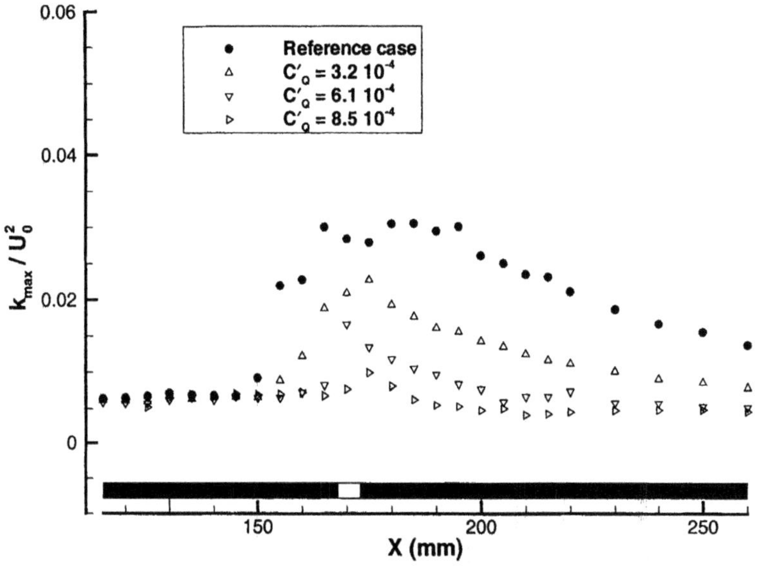

a – active control cases (with the downstream slot)

Figure 8 Streamwise variations of the maximum turbulent kinetic energy

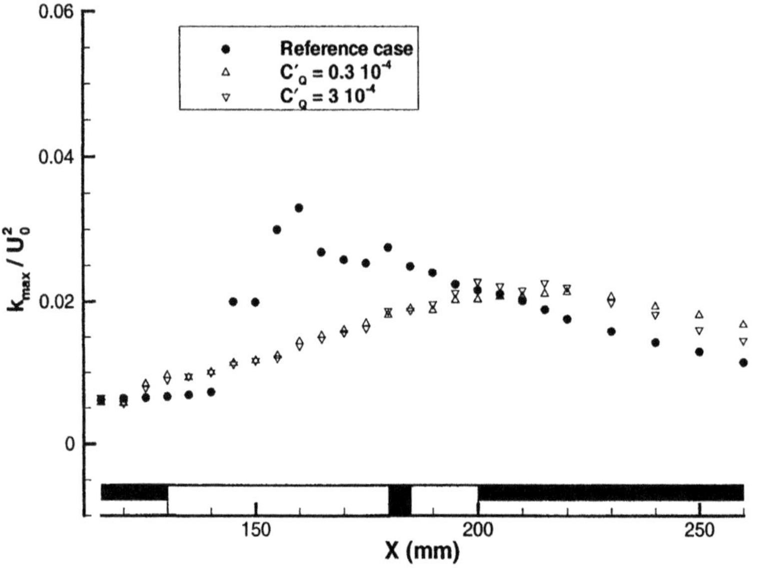

b – hybrid control cases (with the 50mm-long passive control cavity)

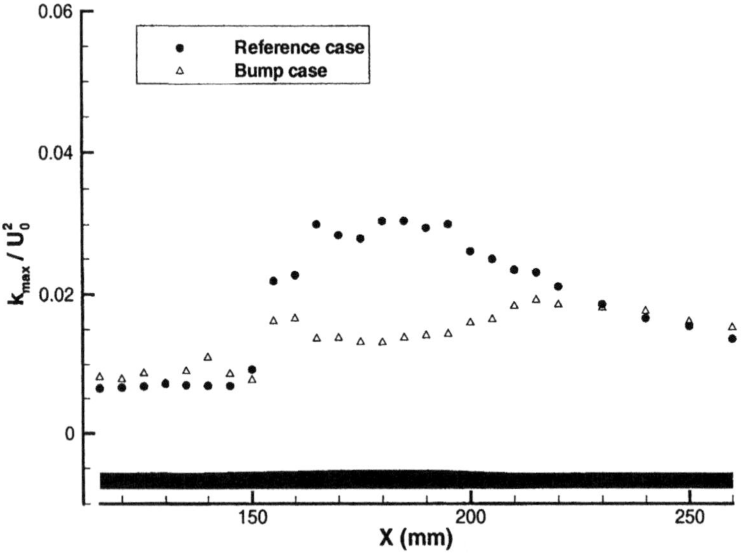

c – bump control case

Figure 8 Streamwise variations of the maximum turbulent kinetic energy (Concluded)

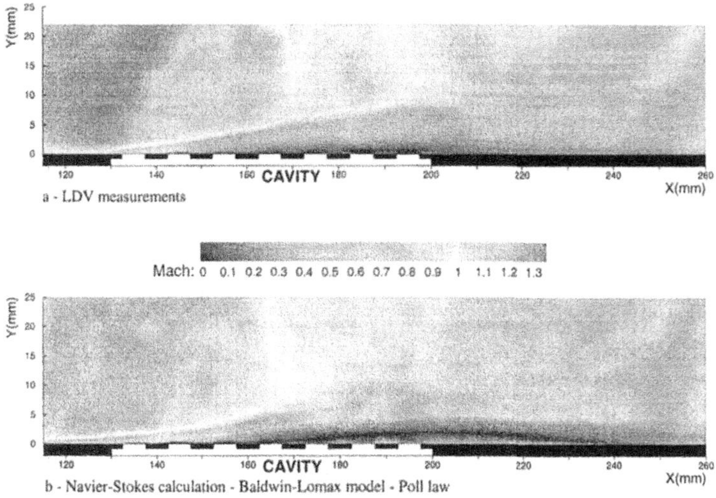

Figure 9 Mach number contour lines – Passive control case

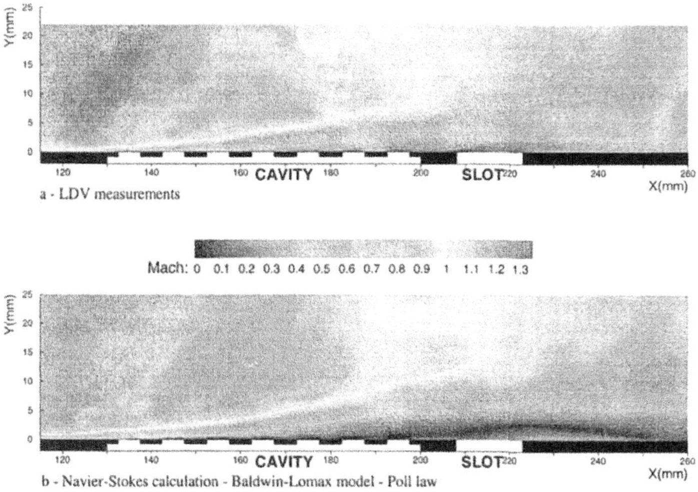

Figure 10 Mach number contour lines – Hybrid control case

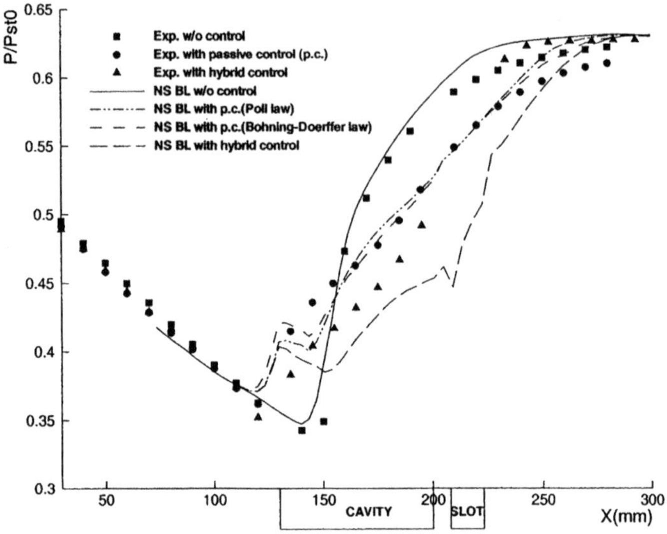

Figure 11 Comparison of calculated and measured wall pressure distributions

12 Hybrid and Active Control of Shock Wave - Turbulent Boundary Layer Interaction and Perforated Plate Transpiration Flow

R. Bohning and P. Doerffer

ISL, University of Karlsruhe
Kaiserstrasse 12, D- 76 128 Karlsruhe, Germany

Summary

The present investigation concerned hybrid and active control of shock wave - turbulent boundary layer interaction. It has been shown that additional suction downstream of the passive control cavity is able not only to suppress separation but even to restore the boundary layer to the velocity profile shape present without control. Active control by suction through a single cavity is even more effective. The boundary layer resulting from active control is nearly as thin as the one upstream of the shock wave.

Further work on the physical modelling of transpiration flow has been undertaken. An extension of the transpiration law has been derived which takes into account the effect of tangential stream along a porous plate.

Numerical simulations have been extended to a 3-D analysis of the flow in a channel. Here, it was found that 3-D effects are responsible for the discrepancy between experiment and 2-D calculations performed earlier.

12.1 Introduction

The work reported in the present paper concerns the contribution of the University of Karlsruhe to the EUROSHOCK II project. The subjects addressed constitute a continuation of work carried out in the previous (EUROSHOCK) project. Therefore, the results and discussions presented here are based on six years of research experience.

The starting point of the present project was the lack of positive effects of passive control in terms of drag reduction determined in the EUROSHOCK (I) investigation. In order to improve the state of the boundary layer, which was disturbed by passive control, an additional active suction cavity, located just downstream of the passive cavity, has been used. Such an arrangement was termed "hybrid control". In the present investigation, various suction cavity lengths and suction intensities have been studied. The results confirmed that the state of the boundary layer may be considerably improved by hybrid control.

Besides hybrid control, so called "active control" has been investigated. Here, suction has been applied to the single cavity used as passive cavity in earlier experiments.

An important issue was the continuation of research concerning the transpiration flow model. Since some of control methods involved suction upstream of the shock wave, hence in the presence of a fast tangential stream over the porous plate, it became important to be able to include the effect of a tangential stream in the transpiration law.

In EUROSHOCK (I) it has been observed that 2-D numerical simulations of the channel flow showed significant deviations from the experimental results downstream the shock wave even in the wall static pressure distribution. In the present project an effort has been undertaken to clarify the reasons for such discrepancy by undertaking a 3D numerical simulations.

12.2 Hybrid Control

12.2.1 Measurements

The measurements were carried out using the configuration presented in Fig.1. In the reference case an undisturbed shock, with a shock-upstream Mach number of M=1.3, is located at the center of the 50mm long passive cavity. A suction cavity, 15mm long in the presented sketch, is located 5mm downstream of the passive cavity. In the present research program its length has been varied between 15 and 45mm.

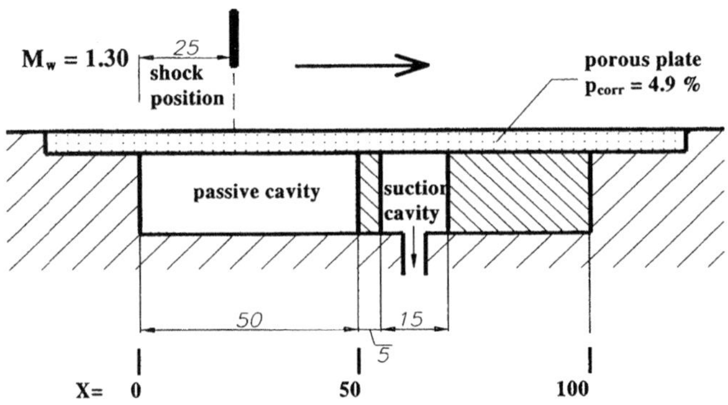

Figure 1 Experimental configuration

Boundary layer profiles have been measured by means of a pressure probe at the following locations:

— at the upstream edge of the passive cavity X = 0 mm - common for all test cases;
— at the location X = 100 mm for each test case.

The measurements also incorporated static pressure distributions within the interaction area; the structure of the shock system has been visualized by a Schlieren system and interferograms.

The experimental program presented here shows the effect of the suction cavity size and the suction intensity on the boundary layer profile at X=100.

12.2.2 Mass Flow Rate in the Case of Suction

The mass flow rate through the porous wall is given by the mass flow coefficient C_Q defined as:

$$C_Q = q/(\rho_\infty U_\infty L), \tag{1}$$

where q is the mass flow per unit span. In our measurements the span is 0.05m. This definition may cause some misunderstanding between external and internal aerodynamics. In the former, (ρU) is taken from the far field condition and the length L is usually a airfoil chord. In case of internal aerodynamics, the local shock-upstream condition is used for (ρU) and the length scale is the boundary layer displacement thickness upstream of the interaction. These two approaches lead to values that are several orders of magnitude different.

It should be pointed out that the (ρ·U) value is not very sensitive to the Mach number within the range of interest. In external flow, the freestream Mach number is about M=0.8 and the shock upstream Mach number is about M=1.3. The difference between the (ρ·U) values of interest is very small because (ρ·U) reaches a maximum at M=1.

For the present (channel-flow) application the coefficient would take the form:

$$C_\mu = \frac{q}{(\rho U)_{M=1.3}\delta_1} \tag{2}$$

In order to unify the suction mass flow rate description, a new definition has been proposed. The (ρU) value is to be taken at M=1.0 and the unit length as L=1m. In that case the mass flow rate would be universal for all applications, viz.,

$$C'_q = \frac{q}{(\rho U)_{M=1} L} \tag{3}$$

12.2.3 Effect of Hybrid Ccontrol

Different suction intensities and suction cavity lengths have been investigated. Examples of boundary layer profiles measured at X=100 are presented in Fig. 2 which includes two reference profiles, one being the profile obtained for the interaction without control. It is a strongly disturbed profile, however, the flow is not yet separated. The other one is the profile obtained with passive control. This profile indicates a strong separation.

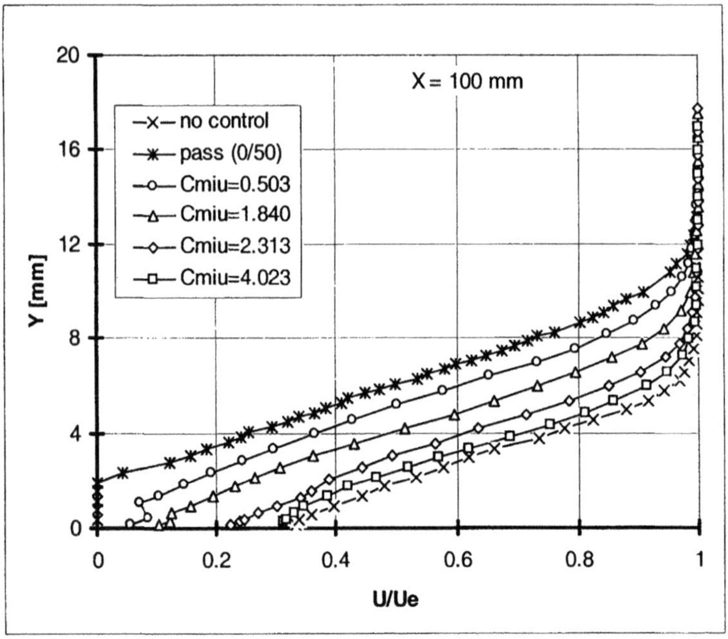

Figure 2 Hybrid control effect on the boundary layer profile

Even quite weak suction applied has a noticeable effect on the whole boundary layer profile. For the smallest C_μ value a very short suction cavity has been used - 15 mm. This suction length is only about 20% longer than the boundary layer thickness. For the highest C_μ value the maximum suction intensity has been used. It corresponds to an effective hole Mach number of M_h=0.57 at which choking in the porosity holes takes place. The suction cavity was 45mm long in this case.

Fig.2 indicates that additional suction downstream of the passive control cavity significantly influences the boundary layer. It is not only possible to eliminate separation but also to restore the boundary layer to a similar velocity profile shape as in the case without control.

Fig.3 shows the boundary layer thickness, displacement thickness and shape parameter dependence on the suction mass flow rate. All cases investigated have been included in these plots. The results suggest that the most important quantity is the total amount of suction. The way how it is realized (cavity length, intensity) plays a minor role.

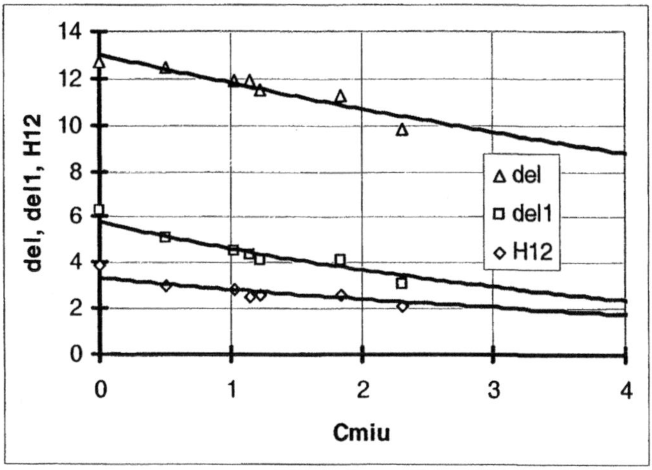

Figure 3 Hybrid control effect on the integral parameters of the boundary layer

12.2.4 Flow Visualization

The illustration of the effect of the suction rate on the boundary layer profile is presented in the flow visualization pictures in Fig.4, showing a sequence of six pictures. The picture on top presents a full view of the channel. Because the interesting part of the flow is the boundary layer close to the wall, all other pictures are showing only a strip with the boundary layer in view. The four center pictures relate to hybrid control; the amount of suction is increasing from top to the bottom. The bottom picture shows the boundary layer development in the case without control. In all pictures the flow direction is from left to right.

The fringes in Fig 4, produced by the Mach-Zehnder interferometer set to the infinite fringe mode, correspond to the iso-lines of density. Due to this setting, the shock wave is well visible as a few lines located very close to each other because the density changes over a very short distance in a shock wave. The general character of channel flow is that it accelerates upstream the shock and decelerates in the part of the channel downstream of the shock. Therefore the fringes mostly run vertically perpendicular to the channel axis and the main flow direction. Close to the upper and lower walls a significant change in the fringe direction takes place. Due to the changes of density in a (supersonic) boundary layer, fringes become parallel to the walls. The area where fringes change their direction from vertical to horizontal corresponds to the edge of the boundary layer. On the basis of the visualization pictures, it is easy to compare the thickness of the boundary layer for the various cases.

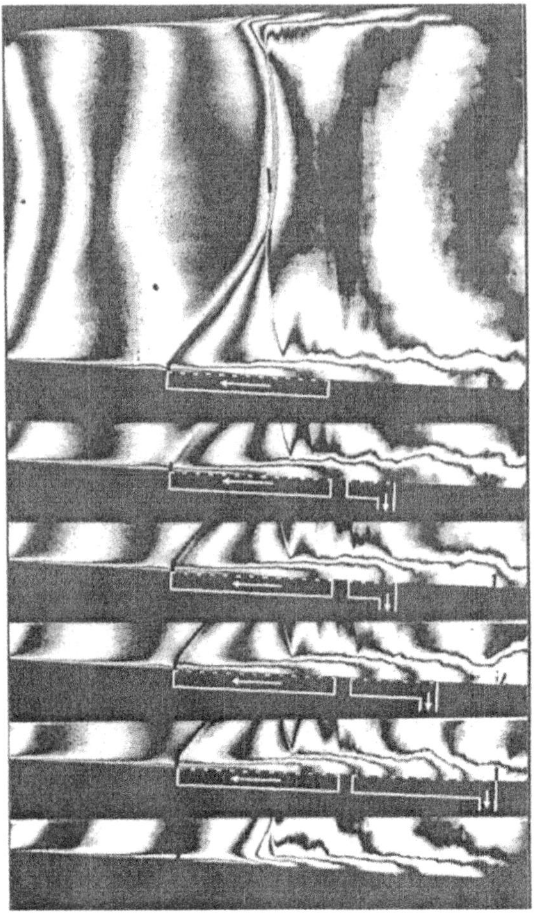

Figure 4 Interferometric visualization of the boundary layer development at different suction rates: -passive control, *top*, four hybrid control cases with increasing suction towards the bottom, no control, *bottom picture*

The most important part of the flow is the lower right corner of the flow pictures. Here, the boundary layer leaving the interaction region is to be studied: in case of passive control (top), the boundary layer is very thick, much thicker than in the no-control case (bottom). In the center pictures, the gradual change of the boundary layer thickness from the thick layer at low suction rates to the much thinner layer at maximum suction is very well illustrated by the behavior of the fringes. At maximum suction the boundary layer thickness is similar to the no-control case shown in the bottom picture, confirming the conclusions based on the boundary layer profile measurements.

12.3 Contribution of Boundary Layer and Shock Wave to Pressure Losses and Drag

The losses induced by the interaction are, in terms of drag, comprised of two components:
— shock (wave) drag which is reduced by passive control due to the spreading of the compression and
— viscous drag induced due to the boundary layer thickness being significantly increased by passive control but reduced by additional suction.

The measurement of the stagnation pressure has previously been carried out only sufficiently far from the wall to allow a boundary layer profile determination. There is, however, the possibility to continue the stagnation pressure measurement further into the flow field to measure the losses induced by the shock waves. This will allow to distinguish between the contribution of the boundary layer and the shock wave, respectively, and to compare passive, hybrid, and active control under these aspects. In order to capture all effects, the measurement traverses have been carried far enough into the field to include the complete λ-foot of the shock in the case of passive control into the surveys.

12.3.1 Measurements

Five flow cases have been considered:

1- regular shock wave boundary layer interaction without control at M_w=1.3;
2- passive control, 50 mm cavity centered on the shock;
3- hybrid control, i.e., passive control as above plus active suction via the 45mm long cavity at maximum mass flow rate;
4- active control, i.e., suction through the passive cavity, at maximum mass flow rate.

In all cases was the undisturbed shock wave above the λ-foot kept on the same position.

The effect of active control is shown in the interferograms in Fig.5 together with the regular no-control interaction and passive control. In an active control case the suction takes essentially place upstream the shock wave. Therefore, not only the λ-foot is not formed but also the small compressions which precede the shock wave in the case without control disappear. The shock remains very strong down to the wall and the boundary layer is very thin. Upstream of the shock small disturbances are observed which are generated by the holes of the porous plate. It is evident that the shock losses are very high in this case but the boundary layer downstream of the interaction is very thin. Suction is also present downstream of the shock and nearly no effect of the shock on the boundary layer is observed.

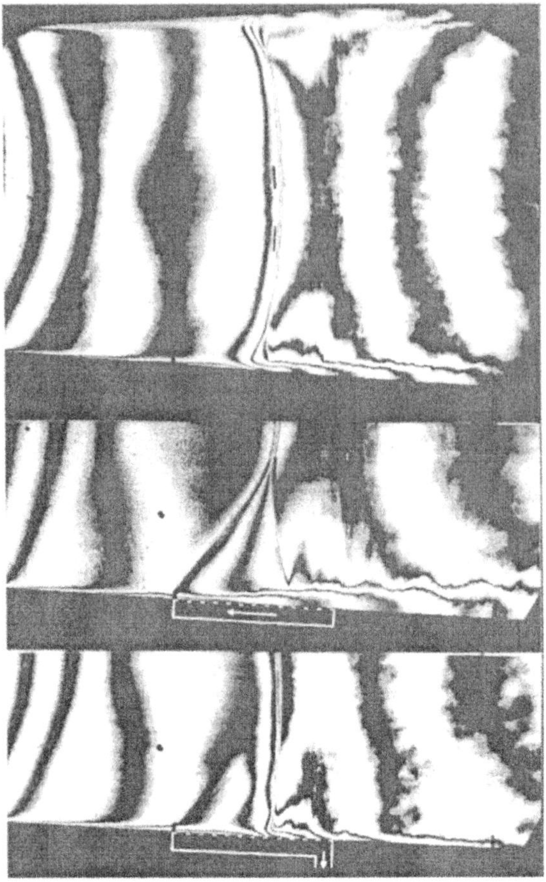

Figure 5 Interferograms of no control, *upper*, passive control, center, active control, *lower* picture

Measured stagnation pressure profiles are presented in Fig.6. They include a profile denoted "upstream", which has been measured at the upstream edge of the passive cavity, and should be understood as a reference (X = 0). Stagnation pressures measured are normalized with the stagnation pressure (P00) in the settling chamber of the wind tunnel. The "upstream" profile shows small losses occurring between the settling chamber and the traverse location for obvious reasons.

One should be reminded that when a compression is realized by a sequence of shocks, the losses are smaller than the losses occurring if the same compression is performed by a single normal shock. One may, therefore, expect that, when the measurement probe is located downstream of a normal shock (Y>45mm), it should indicate lower stagnation pressures than downstream of the λ-foot

(Y<45mm) due to the smaller losses, hence higher stagnation pressures in case of the latter.

All these features are displayed correctly by the results shown in Fig.6. Close to the wall all profiles display a typical boundary-layer-like behavior. First of all, the upstream profile is very thin compared to all others. The "no-control" profile is much thicker but still attached. Implementation of passive control induces a strong separation. When suction is added downstream with maximum intensity and length, the boundary layer profile is significantly improved. It actually coincides with the profile of the uncontrolled interaction. In case of active control when the passive cavity pressure is decreased to induce maximum suction, the boundary layer remains very thin and the compression upstream of the shock disappears. The shock wave remains very strong down to the wall. Downstream of the shock wave the boundary layer remains very similar to the upstream profile. This is because suction also takes place downstream of the shock, preventing separation or an excessive thickening of the boundary layer.

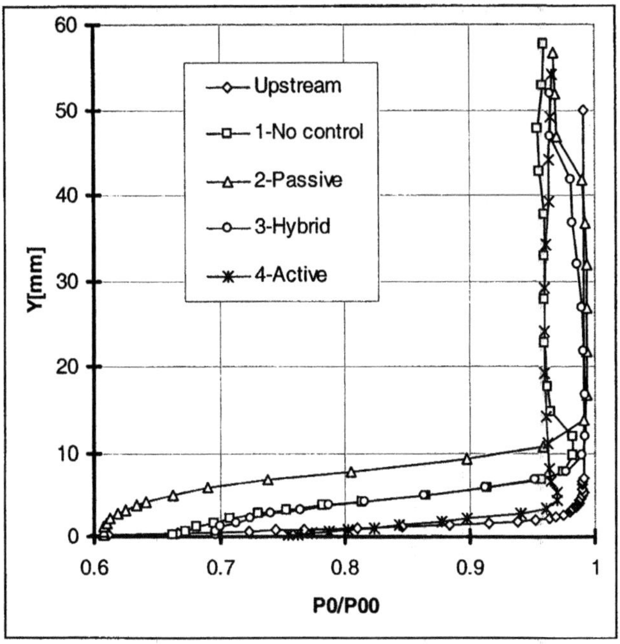

Figure 6 Stagnation pressure profiles for no-control and different control methods

Away from the wall, outside the boundary layer, one can analyze the direct contribution of the shock waves to the pressure losses. First of all, one should point out that at the end of the measurement range (Y ≈ 50 mm) all downstream profiles exhibit the same stagnation pressure. This confirms that the shock at this distance is no longer affected by control which is in agreement with flow visualizations indicating that the λ-foot height is about 45 mm. Closer to the wall, the profiles for passive and hybrid control show almost no stagnation pressure loss due to the shock. These two control methods produce the λ-foot structure causing the reduction in shock losses. The other two profiles (no-control and active control) indicate significant losses which correspond to normal shock loss at M=1.34. This Mach number is somewhat higher than the shock-upstream Mach number at the wall due to the compression at the wall preceding the main shock. All these effects are well displayed by the presented measurements.

The no-control case indicates a local maximum in the total pressure distribution at the edge of the boundary layer. This is due to the existence of a compression preceding the main shock which reduces the shock losses in this region.

12.3.2 Stagnation Pressure Defect

There are many ways to present the integral effect of the distribution profiles; important is to apply the same method to all cases. Considering a boundary layer stagnation pressure profile (as in Fig.6), its defect is represented by an area between the line P0/P00 = 1.0, corresponding to an isentropic process, and the measured profile. In order to analyze the effects of the boundary layer and shock wave contribution to the overall losses, let us use the following definition of the profile defect:

$$\zeta_Y = \int_0^Y \frac{\Delta P_0}{P_{00}} dY \tag{4}$$

This definition provides an area between the isentropic line and the respective profile from the wall to a certain height "Y". It has no physical meaning but it displays the intensity of loss production at different distances from the wall.

Results are presented in Fig.7 allowing to observe the development of losses with changing distance from the wall. Close to the wall, the curves show high gradients due to the increased losses within the boundary layer. Further away, the increase of losses is due to the shock contribution.

The gradient of defect increase within the boundary layer is close to the wall similar for all cases but extends over different lengths. Further away from the wall it decreases quickly at the edge of the boundary layer. Here, all curves display a significant change from a rather flat distribution to a nearly vertical one. Due to different boundary layer thicknesses, the losses obtained at this point are very different. For passive control, the losses at the edge of the boundary layer are significantly higher than in all other cases considered.

Outside the boundary layer, the "upstream" profile displays an increase in defect due to the losses in the channel inlet. The other profiles, downstream of the shock system, display two distinct slopes in the distribution: the steep one, indicating small shock losses, is displayed by flow cases with a λ-foot. The others indicate much higher shock losses since only a single strong shock is present.

Further into the field (Y>50), all profiles indicate the same gradient because in all cases the same shock intensity exists.

On the basis of Fig.7, one may conclude that:

— passive control does not have a positive effect on the interaction since the total defect is the same as without control,
— hybrid control and active control produce nearly the same total positive effect.

The approach presented above lacks generality but allows to distinguish areas of loss contributions due to different effects. It also allows to judge the relative quality of the control methods considered. The most illustrative effect is the difference in curve slopes for cases with and without a λ-foot structure.

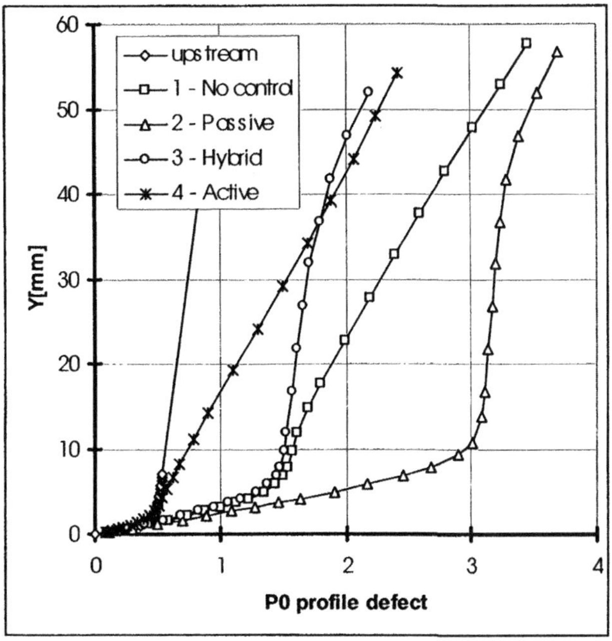

Figure 7 Increase in stagnation pressure defect with distance from the wall

12.3.3 Flow Efficiency

In order to discuss the boundary layer and shock wave contribution to drag or losses it is useful to show the analysis process in an h-s plane (enthalpy-entropy diagram) as illustrated in Fig.8. The stagnation parameters P_0 and T_0 determine the corresponding starting point in h-s diagram.

An expansion from the stagnation parameters to some static pressure P_2 may take place in an isentropic way at s = constant. In such a case the enthalpy difference between points "0" and "is" is $\Delta h_{0\text{-}is} = U_{is}^2/2$. However, it is also possible that the expansion takes place with losses, hence the entropy increases and the enthalpy difference between points "0" and "2" is $\Delta h_{0\text{-}2} = U^2/2$. The termination of the process is at the static pressure P_2 line. The stagnation conditions for the air at condition "2" correspond to an isentropic "return" to the total enthalpy of the flow, h_0. It means that the stagnation pressure is lower for flow with losses, $P_{02}<P_0$. Regarding the boundary layer, this stagnation pressure is measured by the pitot probe.

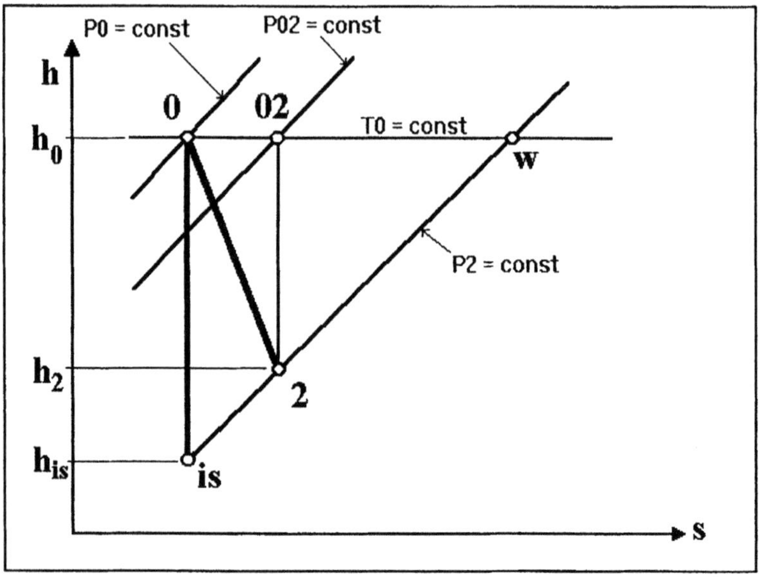

Figure 8 Flow parameters in the enthalpy - entropy diagram

Points of a boundary layer profile are located along the constant pressure line P2 between points "is" and "w". The point "is" is located outside the boundary layer where the flow may be considered as isentropic. The last point, "w", is located at the wall where the static pressure is equal to the stagnation pressure and the velocity is U = 0.

For each expansion, hence for each point within the boundary layer, one may calculate the increase in entropy or the decrease in stagnation pressure. In order to characterize the irreversibility of this process, the typical approach, commonly used, is the determination of efficiency. This is the ratio of the enthalpy drop in reality and in an isentropic expansion:

$$\eta = \frac{\Delta h}{\Delta h_{is}} \tag{5}$$

with: $\Delta h = h_0 - h_2$ and $\Delta h_{is} = h_0 - h_{is}$.

Knowing that $h_0 = h + U^2/2$, the efficiency may be expressed by the velocity ratio squared, viz.,

$$\eta = \frac{U^2}{U_{is}^2} \tag{6}$$

For each point of the traverses, the static pressure P_w and the stagnation pressure P_{02} are known allowing to calculate the Mach number

$$M = \sqrt{\frac{2}{\kappa - 1}\left[\left(\frac{P_0}{P_w}\right)^{\frac{\kappa-1}{\kappa}} - 1\right]}.$$

The velocity is expressed by

$$U = \frac{M}{\sqrt{1 + \frac{\kappa - 1}{2}M^2}}\sqrt{\kappa R T_0}$$

and hence the efficiency at a given point of the measured traverse by:

$$\eta = \frac{M^2}{M_{is}^2}\frac{1 + \frac{\kappa - 1}{2}M_{is}^2}{1 + \frac{\kappa - 1}{2}M^2} \tag{7}$$

The efficiency is plotted in Fig.9 for all measured profiles. The figure is similar to Fig.6 but this presentation is more general thus allowing comparisons with other results. All distributions start at the wall at $\eta=0$ and reach nearly $\eta=1$ away from the wall. The similarity with the stagnation pressure distributions in Fig.6 shows that the stagnation pressure may adequately represent the flow efficiency. All conclusions are, therefore, the same.

In order to present the efficiency development with the distance from the wall, the integral efficiency of a stream between the wall and a given "Y" has been defined as:

$$\eta_Y = \frac{1}{Y}\int_0^Y \eta \, dY.$$

The corresponding profiles are presented in Fig.10. The effects of the boundary layer, the shock wave and the λ-foot are not as distinct as in the previous presentation, Fig.7; however, all previous conclusions are, as mentioned, confirmed here:

— passive control has a particularly low efficiency in the boundary layer area,

— the efficiency in the boundary layer area for the "no control" case and the hybrid control case are nearly the same,

— the total efficiency of passive control (obtained at Y=50mm) and the one for the case without control are the same,

— the total efficiency in the case of hybrid control and in the case of active control are the same and the highest of all cases considered.

The higher efficiency for hybrid and active control is caused by the energy supplied by suction. One must, therefore, when considering overall benefits, also account for "pump" drag.

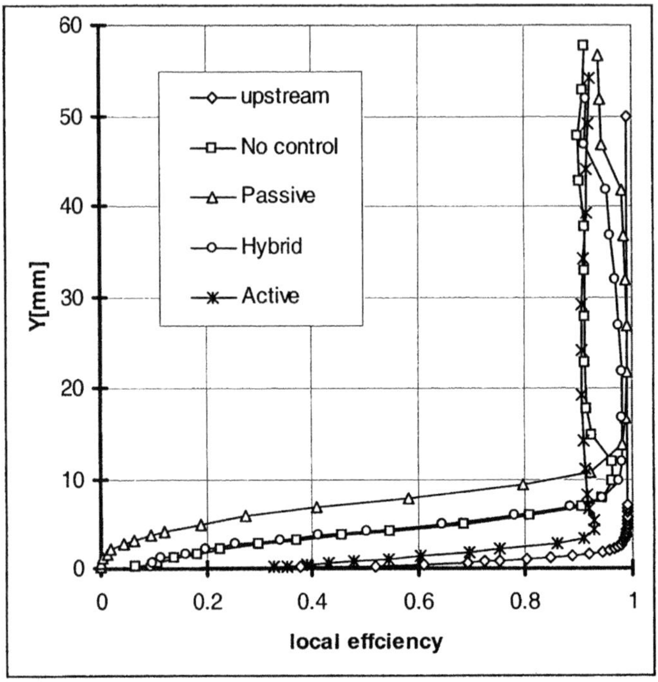

Figure 9 Distribution of the efficiency for different control methods

The present results show the effectiveness of various control schemes at a single streamwise location where pressure profiles have been measured. The approach accounts for all effects related to shock losses. However, the boundary layer has influence further downstream of the present traverse. Therefore, it is not sufficient to state that passive control has the same efficiency as the flow without control. It must be added that passive control may have a negative effect in all flow applications where a thicker boundary layer is detrimental, such as in airfoil and wing flow so that the final statements should read:

- passive control may increase drag and losses since it leaves a thicker boundary layer downstream,
- active control is more favorable than hybrid control since it leaves a thinner boundary layer downstream.

- active control is more favorable than hybrid control since it leaves a thinner boundary layer downstream.

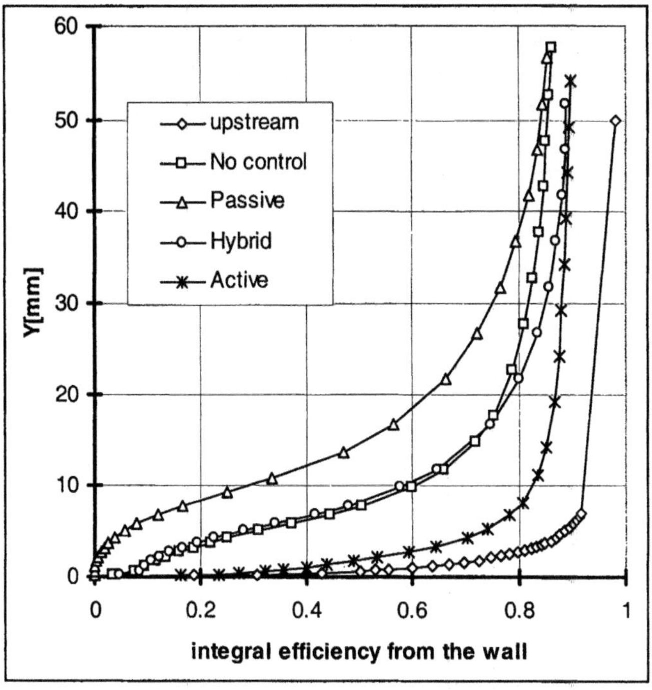

Figure 10 Efficiency of the stream between a given Y and the wall

12.4 Transpiration Flow with Outer Tangential Stream

A special insert to the test section has been built in order to investigate the effect of an outer tangential stream along a porous wall on the transpiration flow. The mass-flow-rate measurement system, consisting of interchangeable sonic nozzles, has been built and tested. A series of eight nozzles has been prepared and carefully calibrated.

It turned out that in case of the existence of a tangential stream, the transpiration flow encounters a blocking effect, i.e., that for the same pressure drop across the porous plate the mass flow rate is lower than in the case without a tangential stream. The effect is present only for suction. In case of blowing, no effect has been observed.

The stagnation pressure for the transpiration flow corresponds to the static pressure of the tangential stream since in a boundary layer the stagnation pressure drops to the value of the static pressure at the wall. In the wind tunnel different absolute static pressures are reached for different Mach numbers of the flow which implies different stagnation pressures for the transpiration flow dependent on the channel Mach number.

In order to confirm this effect, transpiration measurements without tangential stream have been carried out for different pressures in the test section. The results coincided with the Bohning/Doerffer law derived in the EUROSHOCK (I) project [1]. This indicates that not the static pressure but only the existence of a tangential stream or some other property of this stream affects the transpiration flow.

12.4.1 Measurements

In order to study transpiration flow in the case of suction and dependent on the outer-stream Mach number, a plate with normal holes of diameter D=0.185mm (T-3) has been chosen. The plate porosity has been checked after installation. The value obtained for suction was 4.9%. The transpiration flow was measured for various Mach numbers of the tangential stream, Table 1. The Mach numbers were kept constant in the experiments to within 1 to 3%.

Table 1 Description of flow cases by the main stream Mach number

Case No.	1	2	3	4	5	6
Mach No.	0.0	0.11	0.285	0.494	0.705	0.865

The results are plotted in Fig.11. They confirm the significant effect of a tangential stream on the wall characteristics in the case of suction. The measurements provided a systematic set of data well suited for the modeling of the phenomenon observed.

For M=0.0 there is no tangential stream along the porous plate. The measured points should, therefore, fit the Bohning/Doerffer law developed during EUROSHOCK (I). This law is represented by a thick line in the plot, Fig. 11. All measured points of the M = 0 case are well represented by the earlier law.

In the experiments five tangential stream Mach numbers have been applied with each suction (metering) nozzle. For a given nozzle (and the porous plate considered) nearly the same M_h values are reached independent of the stagnation conditions. In the nozzle throat, M=1 is obtained and, therefore, the Mach number in the holes of the perforated plate is only dependent on the cross section ratio. In Fig.11 this dependence is indicated by nearly the same M_h-value being obtained for all tangential stream velocities. The values differ somewhat because the flow losses depend on the Reynolds number which changes here with the Mach number of the tangential stream.

The application of the tangential stream causes a shift of the characteristic transpiration curve towards an increased pressure drop or to a lower hole Mach number M_h. It means that the tangential stream imposes a blocking effect on the transpiration flow. In case of blowing, no such effect has been observed. All measurement points coincided with the line denoted "func" in Fig.11.

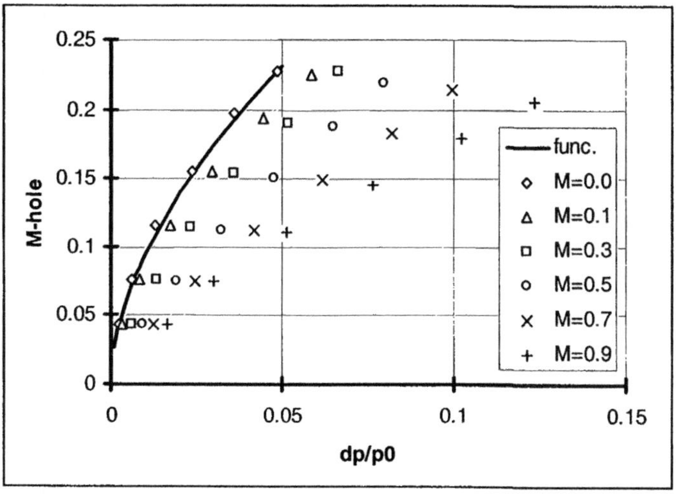

Figure 11 Effect of tangential stream on the transpiration flow characteristic

The results indicate that the shear stress at the inlet to the holes of the porosity is responsible for the effect observed. In the experiments, the suction has been applied to an undisturbed boundary layer assumed to be turbulent. For a typical turbulent profile it is possible to couple shear stress at the wall to the main stream parameters and the boundary layer thickness. The boundary layer thickness may be approximated from Schlieren visualizations.

For a flat plate flow, the fully turbulent boundary layer profile may be assumed to be exponential with the exponent of 1/7:

$$\left(\frac{u}{U_\infty}\right)=\left(\frac{y}{\delta}\right)^{\frac{1}{7}}. \tag{8}$$

For such a profile the wall shear stress, constant within the laminar sub-layer adjacent to the wall, is given by:

$$\tau_0=\rho u_\tau^2=0.0225\rho U_\infty^2\left(\frac{\nu}{U_\infty\delta}\right)^{\frac{1}{4}}. \tag{9}$$

The wall shear stress normalized with the kinetic energy of the main stream provides the skin friction coefficient:

$$c_f=\frac{\tau_0}{\frac{1}{2}\rho U_\infty^2}=0.045\left(\frac{\nu}{U_\infty\delta}\right)^{\frac{1}{4}} \tag{10}$$

The kinematic viscosity "ν" is defined by $\nu=\rho\,\mu$ where μ is determined from Sutherlan's law:

$$\mu = \mu_0 \left(\frac{T}{T_0} \right)^{\frac{1}{5}} \frac{T_0 + 110}{T + 110} \quad \text{with } \mu_0 = 17.1 \text{x} 10^{-6} \text{ Pa and } T_0 = 273.16 \text{ K}$$

The above expressions show that the skin friction at the wall is dependent on three main factors:

— the velocity at the edge of the boundary layer or the Mach number of the main stream,
— the stagnation parameters, hence the Reynolds number, and
— the boundary layer thickness.

Employing the above method, one can determine the wall shear stress τ_0 for each main stream Mach number. For the main stream Mach numbers considered, the wall shear stress values are as follows:

Table 2 Skin friction values for the flow cases considered

M_{flow}	0.0	0.11	0.285	0.494	0.705	0.865
τ_0 [N/m^2]	0	4	20	49	80	103

In the experimental set up the suction cavity is connected to the vacuum tank via the nozzle measuring the mass flow rate. The pressure ratio between suction cavity and vacuum tank must be above the critical ratio in order to secure sonic conditions in the nozzle throat. When the tangential stream is generated, the pressure in the wind tunnel decreases with a corresponding decrease in cavity pressure taking place. During a blow-down, the pressure in the vacuum tank is very low at the start and increases with time. At a certain pressure level in the tank, the sonic condition in the metering nozzle brakes down ending the measurement. As a consequence the blow-down time is shorter for a higher Mach number in the test section. Due to this limitation in the run time, the tests carried out were restricted to subsonic velocities of the tangential stream.

12.4.2 Formulation of the Transpiration Law

Detailed considerations and testing different approaches have lead to the transpiration law in the following form:

$$\frac{dP}{P_0} = M_{hole}^{\frac{1}{0.55}} \left[\left(\frac{1}{1.2} \right)^{\frac{1}{0.55}} + b \left(B M_{hole} \right)^{\frac{1}{a}} \right]; B = \frac{|\tau_W|}{\frac{\rho_{hole} u_{hole}^2}{2}}; \; a = 1.52; b = 25. \qquad (11)$$

This is an extension of the formula proposed in EUROSHOCK (I). When the wall shear stress is $\tau_W = 0$ it takes the identical form as the original law. The "dP" used here denotes the pressure difference across the porous plate. P_0 is the stagnation pressure on the inlet side of the plate. If a tangential stream is present its static pressure becomes the stagnation condition for the hole. (For a summary of the nomenclature used, refer to Table 3.)

The shear force at the wall is, as mentioned above, the driving mechanism for blocking the flow through the porosity holes. The influence of this shear force is dependent on the intensity of the transpiration flow. Therefore, a factor formed by the ratio of τ_w and the kinetic energy of the flow in a hole has been introduced corresponding to the normalization of τ_w with $(\rho u^2/2)_h$ in a hole.

In considering control by cavity ventilation, one has to distinguish between two cases, blowing and suction; blowing corresponds to the case where $p_w < p_c$, suction takes place when the $p_w > p_c$. Since the static pressure in the boundary layer is constant, $p_w = p_e$. In case of suction, one has to provide an appropriate value for the shear stress τ_w, which depends on the boundary layer thickness, the Mach number of the outer stream, and on the disturbance to the boundary layer, since for suction, the tangential flow, or more precisely the wall shear stress, plays a very important role. Its increase means an increase of the pressure drop across the porous plate at the same transpiration mass flow rate. In other words, at a constant pressure drop, the presence of a tangential flow will cause a lower suction mass flow rate. In the case of blowing, no such effect was observed. This is because the inlet to the plate is located on the cavity side where $\tau_w = 0$.

Table 3 Description of the nomenclature used

INDEX	DESCRIPTION OF THE MEANING
0	Stagnation parameters of the flow field considered
∞	Free stream parameters or stresses normalization conditions (conditions at which $\rho u^2/2$ is determined to calculate c_p or c_f)
w	Parameters at the wall (inside the boundary layer) with one exception that M_w means an isentropic Mach number, obtained from p_w and p_0
e	Parameters outside the boundary layer
c	Parameters in a cavity
h	Parameters in a porosity hole

Consequently, the transpiration laws may, using the above nomenclature, be written in the following form:

Suction

$$\frac{P_w - P_c}{P_w} = M_h^{\frac{1}{0.55}}\left[\left(\frac{1}{1.2}\right)^{\frac{1}{0.55}} + 25\left(BM_h\right)^{\frac{1}{1.52}}\right]; B = \frac{|\tau_w|}{\frac{\rho_h u_h^2}{2}} \qquad (12)$$

Blowing ($\tau_w = 0$)

$$\frac{P_c - P_w}{P_c} = \left(\frac{M_h}{1.2}\right)^{\frac{1}{0.55}} . \tag{13}$$

Actually, it is not necessary to distinguish between these two cases because in the case of blowing the shear stress is equal to zero, hence B = 0, and the formula for suction reduces automatically to the formula for blowing. As mentioned above, the simple formula for blowing is identical to the one proposed in EUROSHOCK (I) [1].

The proposed formula is plotted in Fig.12 together with the experimental points. The agreement of the Bohning/Doerffer (B/D) law with experiment is very good and the proposed law reflects well the physical behavior of transpiration flow.

It should be mentioned once again that the values of τ_w were not measured but were calculated assuming the exponential boundary layer velocity profile. It is difficult to say to what degree this holds for the corresponding measurements. It would be helpful to make the final tuning of the law by comparison with measurements that included the determination of the wall shear stress. These are, however, not available.

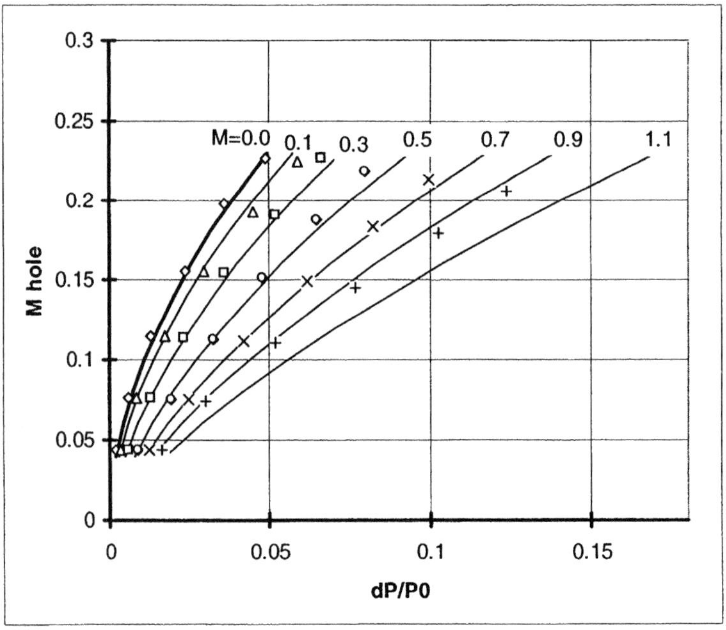

Figure 12 Bohning/Doerffer law for transpiration flow

12.4.3 Calculation of the Parameter "B"

For given τ_w values

In order to calculate the parameter "B", it is necessary to know the shear stress at the wall, τ_w, and the quantity $(\rho_h u_h^2/2)$. The shear stress must either be taken from measurements or calculations; to determine $(\rho_h u_h^2/2)$, the following procedure may be used.

First, in order to simplify the equations, it is proposed to make the substitution

$$O=\left(1+\frac{\kappa-1}{2}M^2\right).$$

Then the main relations for the stagnation parameters become:

$$\frac{T_0}{T}=O;\frac{p_0}{p}=O^{\frac{\kappa}{\kappa-1}};\frac{\rho_0}{\rho}=O^{\frac{1}{\kappa-1}}.$$

The velocity in the hole may be expressed as

$$u_h=M_h a=M_h\sqrt{\kappa RT_h}.$$

For the flow in a hole, the stagnation temperature is equal to the temperature at the porous wall, $T_{0\text{-}h}=T_w$. The temperature at the wall is equal to the stagnation temperature assuming, with a certain loss of accuracy, a temperature recovery factor of 1, so that $T_{0\text{-}h}=T_0$. It follows that

$$u_h=\frac{M_h}{\sqrt{O_h}}\sqrt{\kappa RT_0}.$$

The stagnation density for the flow in a hole is equal to the tangential flow density at the wall, i.e., $\rho_{0\text{-}h}=\rho_w$. The density at the wall may be calculated from the main stream parameters using the assumption made above about the temperature:

$$\rho_w=\frac{p_w}{RT_w}=\frac{p_w}{RT_0}=\frac{p_e}{RT_0}=\frac{p_0}{RT_0}\frac{1}{O_e^{\frac{\kappa}{\kappa-1}}}.$$

Knowing that ρ_w is the stagnation parameter for the flow through the hole, one may write

$$\rho_h=\frac{p_0}{RT_0}\frac{1}{O_h^{\frac{1}{\kappa-1}}O_e^{\frac{\kappa}{\kappa-1}}},$$

and, consequently, for the dynamic pressure

$$\frac{\rho_h u_h^2}{2} = \frac{1}{2} \frac{\kappa p_0 M_h^2}{(O_h O_e)^{\frac{\kappa}{\kappa-1}}} .$$

Finally, one obtains:

$$B = \frac{2\ \tau_w}{\kappa p_0 M_h^2} (O_h O_e)^{\frac{\kappa}{\kappa-1}} . \tag{14}$$

For given c_f values

In the case of a flow where the skin friction distribution is provided, one may write B in a somewhat different way: the skin friction coefficient can be expressed as

$$c_f = \frac{\tau_w}{\frac{1}{2} \rho_e u_e^2} .$$

It follows that:

$$B = c_f \frac{\rho_e u_e^2}{\rho_h u_h^2} .$$

If the freestream conditions are given by the Mach number M_e, one may easily calculate the reference density by

$$\rho_e = \frac{\rho_0}{O_e^{\frac{1}{\kappa-1}}} .$$

From previous consideration we have

$$u_e = \frac{M_e}{\sqrt{O_e}} \sqrt{\kappa R T_0}$$

Therefore, the dynamic pressure (skin friction normalization factor) has the following form:

$$\rho_e u_e^2 = \frac{\rho_0}{O_e^{\frac{1}{\kappa-1}}} \frac{M_e^2}{O_e} (\kappa R T_0) = \kappa p_0 \frac{M_e^2}{O_e^{\frac{\kappa}{\kappa-1}}} ,$$

and, as was shown above,

$$\frac{\rho_h u_h^2}{2} = \frac{1}{2} \frac{\kappa p_0 M_h^2}{(O_h O_e)^{\frac{\kappa}{\kappa-1}}} .$$

Finally, one obtains $$B = c_f \frac{M_e^2}{M_h^2} (O_h)^{\frac{\kappa}{\kappa-1}}. \tag{15}$$

If the skin friction normalization factor corresponds to the far field conditions, the index "e" in above the formula must be substituted by "∞".

12.4.4 Maximum mass flow rate through a perforated plate

Measurements with active suction downstream of the passive control cavity has inspired to also use the maximum possible suction mass flow rate. It is obtained by a direct connection of the suction cavity to the vacuum tank. In this case the flow is choked at the holes of the porous plate and there is no possibility of the mass flow rate measurement by means of a sonic nozzle. It became essential to carry out special measurements to determine the effective Mach number in the porosity holes corresponding to the case of choking.

Mass flow rate measurements are based on the following pressure data:

-static pressure on the main-flow side of the porous plate,

-pressure in the suction cavity,

-pressure in the throat of the sonic nozzle.

The pressure drop across the porous plate is thereby determined and, assuming that the stagnation temperature is constant in the measurement system, the mass flow rate obtained from the throat pressure.

Measurements have been carried out for three sonic nozzles corresponding to three different mass flow rates. For each nozzle the length of the suction plate has been changed from its maximum of 45 mm (corresponding to 44 rows of holes) down to the minimum, i.e., a single row of porosity holes.

At a certain length of the suction plate, a transfer of the choking location should take place. At the large plate sizes the choking takes place in the sonic (metering) nozzle, for short plates it should take place at the porous plate. The change in the choking location was expected to be noticeably displayed by the following effects:

- a sudden drop of the nozzle pressure and the suction cavity pressure,
- a decrease in the pressure difference between the suction cavity and the nozzle,
- if choking takes place in the sonic nozzle, the nozzle and suction cavity pressures are independent of the downstream condition, i.e., essentially the vacuum tank pressure,
- if choking takes place at the porous plate, the suction cavity and nozzle pressures should depend on the tank pressure.

The main aim was to determine the M_{hole} at which the flow is choked at the plate. This is an effective Mach number and should, of course, be lower than unity. In the analysis of the present test results, this Mach number was determined to be $\mathbf{M_{hole} = 0.57}$

The knowledge of this value allows the determination of the maximum mass flow rate through a porous plate at different flow parameters. It is very useful for the processing of experimental data in cases where choking at the plate takes place. It is also a guideline for numerical simulations providing a limit to the mass flow rate.

12.5 Numerical simulations

Numerical simulations, carried out earlier, have shown that two dimensional calculations displayed significant deviations from the experimental results, especially downstream of the shock wave. It was, therefore, decided to carry out three-dimensional calculation to determine whether 3-D effects were possibly responsible for the observed discrepancies or whether other effects, like turbulence modeling, were to blame.

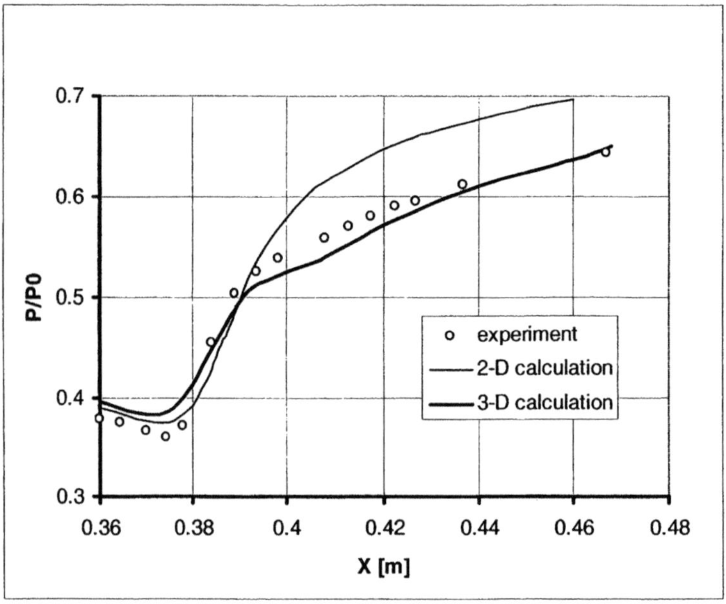

Figure 13 Wall static pressure in the interaction area

The 3-D simulations have been performed by the Navier-Stokes code KAPPA used at the University of Karlsruhe. Flows with passive control as well as regular shock boundary layer interactions have been calculated. In Fig.13 the results concerning the no-control case presented. The comparison between the experimental wall static pressure distribution and the corresponding 2-D and 3-D calculations indicates that the main discrepancy between experiment and 2-D calculation has been eliminated by the 3-D approach. This confirms the existence of major three-dimensional effects in the channel flow which have, it is judged, however, no influence on the conclusions with regard to control efficiency.

References

[1] E. Stanewsky, J. Délery, J. Fulker and W. Geissler (Eds), "Euroshock - Drag Reduction by Passive Shock Control", Notes on Numerical Fluid Mechanics, Vol. 56, Friedr.Vieweg & Sohn, 1997.

[2] Bur, R., Corbel, B., Délery, J., "Study of Passive Control in a Transonic Shock Wave/Boundary-Layer Interaction", AIAA Journal Vol. 36. No. 3, March 1998

[3] P. Doerffer, R. Bohning "Aerodynamic performance modeling of porous plates", Journal of Aerospace Science & Technology, Vol.4, No 8, 2000

13 Active Control of Swept Shock Wave / Boundary Layer Interactions

H. Babinsky

University of Cambridge, Department of Engineering
Trumpington Street, Cambridge, CB2 1PZ, England

Summary

The control of a swept shock wave / turbulent boundary layer interaction was studied experimentally and numerically. Suction was applied to a passive control cavity and through a slot at various positions relative to the inviscid shock location. It was observed that the application of suction from a passive control cavity lowered the plateau pressures and that strong suction effectively removed any beneficial shock smearing. Slot suction applied upstream and underneath the interaction was found to steepen the shock induced pressure rise whereas downstream suction had the opposite effect. In all cases, suction was found to be successful in counteracting the increase of boundary layer parameters (displacement and momentum thickness and shape factor) through the interaction. Data obtained from the numerical simulation of active control generally confirmed these trends. Of the three turbulence models investigated (Baldwin-Lomax, Johnson-King, Cebeci-Smith) none was found to be superior.

13.1 Introduction

While the other investigations of Task 1 focus on two-dimensional configurations, the work presented here was performed on an oblique shock wave / boundary-layer interaction in order to study the influence of sweep in addition to the other significant physical parameters. This is of relevance to transport aircraft which invariably feature swept wings. Previous work performed as a contribution to the Euroshock (I) programme [1] investigated three different angles of shock sweep while keeping the shock-normal Mach number approximately constant at 1.15. During these investigations no fundamentally different results were observed when passive control was applied to these three configurations and for this reason the present study was limited to the investigation of only one shock sweep angle as described below. The angle chosen was the largest of the previous investigation. This was thought to maximise any potential sweep effects while also providing the largest spanwise extent of the interaction zone, thus reducing end effects.

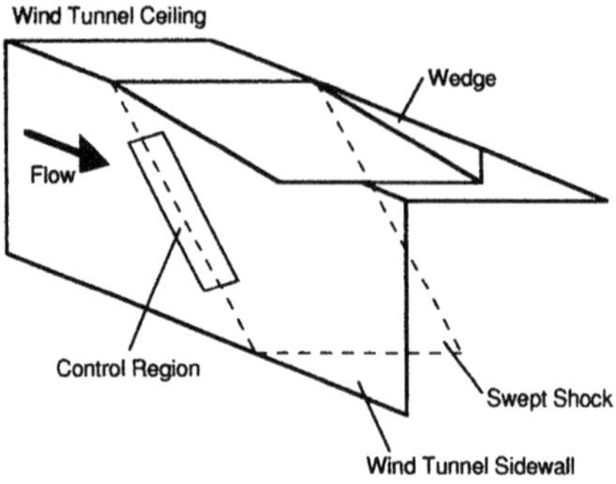

Figure 1 Sketch of experimental configuration

Table 1 Experimental conditions

M_∞ Free-stream M	δ deflection angle	β shock angle	p_0 reservoir pressure	M_n shock-normal Mach number	p_2/p_1 pressure jump through shock	Re (m^{-1}) free-stream Re
1.80	6°	39.5°	0.24 MN/m^2	1.14	1.36	3.3×10^{-7} m^{-1}

13.2 Experimental conditions

The work was carried out in the Department of Engineering's supersonic wind tunnels, which are capable of simulating a naturally grown turbulent boundary-layer at a Reynolds number equivalent to cruise conditions on an aircraft wing at 3m chord. The wind tunnel (working section: 0.15m x 0.11m) is operated at a Mach number of 1.85 and a swept shock wave is formed by a 6° wedge placed on the wind tunnel ceiling as seen in Fig. 1. The shock-normal Mach number is 1.15, which is comparable to that observed on transonic aircraft wings. A number of important flow parameters are summarised in Table 1. A plenum chamber (0.2m x 0.05m) is placed underneath the swept shock in the side-wall of the working section at an angle of 40° to the free stream direction. This can be covered with either a porous surface or various solid plates featuring a single suction slot (see Fig. 2). The porous surface is covered by a large number of laser drilled holes to give an open-to-closed ratio of 8%. As seen in Fig. 2b) three suction slot positions were tested, namely, upstream, underneath and downstream of the inviscid shock surface. The slot width in each case was 5mm.

Suction was applied by connecting the plenum chamber to a high-pressure driven ejector and the suction mass flux was controlled by inserting orifice plates in the suction line and ensuring choked conditions across them. Consequently, the suction mass flux is a function of the orifice diameter and the total pressure in the plenum chamber (see Table 2). With the porous surface fitted and no suction applied (suction line blocked off), the configuration is equivalent to passive control.

For comparison with other research it is useful to define the following parameter:

$$C_\mu = \frac{q}{\rho_\infty \, U_\infty \, \delta^*} , \qquad (1)$$

where q is the suction mass flux per span, ρ_∞ and U_∞ are the density and flow velocity at the boundary layer edge and δ^* is the displacement thickness of the incoming boundary layer. Using the corresponding values (for δ^* refer to sub-section 13.4.3) and the spanwise extent of the suction plate (0.2 m × sin 40°), the relationship between suction mass flux for each orifice plate and C_μ is given in Table 2. Note that the actual mass flux for each orifice plate depends on the pressure in the suction cavity and therefore varies slightly between experiments depending on the surface pressure distribution. The values given in Table 2 are averages for each orifice plate.

Table 2 Suction mass flux for various orifice plates

Orifice Ø [mm]	3.3	4.6	5.6	6.5	7.2	10.2	12.5
$\dot{m}$ [g/s]	1.0	1.8	2.5	3.5	4.2	8.0	11.0
C_μ [-]	0.013	0.023	0.033	0.046	0.055	0.104	0.143

Flow visualisation was carried out using a standard z-type single pass schlieren system and surface oil flow visualisation was performed in some cases. Due to the small size of the laser drilled holes in the porous plate, oil flow visualisation was not possible for active and passive control configurations as the oil would have affected the porosity.

Surface pressures were recorded along three approximately streamwise rows (X_1, X_2, X_3), as seen in Figs. 2. For these measurements, Druck pressure transducers were connected to the pressure taps via a short length of tubing. Boundary-layer Pitot traverses were performed upstream and downstream of the interaction at the locations shown in Fig. 2. The Pitot pressures were measured with a flat-head boundary layer probe measuring 0.13 × 3 mm. The internal dimensions of the opening were 0.05 × 2.7 mm. The surface pressure at the traverse location was measured simultaneously with the traverse signal. To calculate velocities, constant pressure across the boundary layer as well as Crocco's velocity-temperature relationship were assumed.

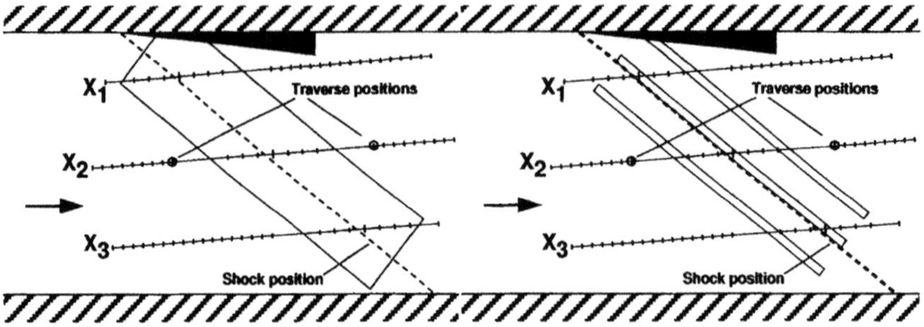

a) porous surface b) slot suction (3 positions)

Figure 2 Control positions and measurement co-ordinate systems

13.3 Numerical simulation

All computations were performed with a finite-volume implicit algorithm which solves the full three-dimensional Reynolds-averaged Navier-Stokes equations as described by Dawes [2]. The code was originally developed to investigate turbomachinery flows and was later modified by Leung and Squire [3] for the current configuration. Note that in the present calculations the energy equation is not solved and that total enthalpy is assumed constant. This results in considerable savings of computing time with little effect on the overall accuracy, as discussed by Holmes and Squire [4]. During Euroshock (I), the same code was used to study passive control. Since computing power has improved in recent years it was possible to significantly increase the number of grid points to generate a higher mesh resolution. The mesh itself is similar to that employed in Euroshock (I), featuring refinement at the walls and throughout the interaction region. The number of grid points and equivalent physical dimensions are given in Table 3. For comparison, the relevant figures for Euroshock (I) are also included.

Table 3 Computational domain and grid size

	Physical Dimensions ($x \times y \times z$)	Number of Grid Points
Euroshock I [1]	$235 \times 76.2 \times 240$	$64 \times 46 \times 80$
Present Investigation	$235 \times 76.2 \times 240$	$90 \times 64 \times 106$

The porous surface was simulated by removing the no-flux condition from the wall-facing side of some of the cells adjacent to the side-wall. Since the shock aligned grid did not overlap the porous plate exactly, all cells cutting the edge of the porous region were treated as part of the porous plate. This increased the effective area of the porous region by less than 0.5% of the plate size. The flow inside the plenum chamber was not simulated, instead, the static pressure inside the plenum cavity was evaluated to estimate the flow velocity through the surface. This was achieved by using an analytical relationship between the mass-flow through the porous surface and the pressures on either side and iteratively varying the cavity pressure until mass conservation was satisfied, i.e. the net mass-flux was equivalent to the suction mass flux applied to the control cavity (within 5×10^{-3} g/s). In order to save computing time, this iteration was performed only once every five time steps, during which the cavity pressure was kept constant.

The functional relationship between mass-flux and static pressures was determined from a number of experiments with a calibration rig, similar to the apparatus described in Chokani and Squire [5]. It was found that suction and blowing were not strictly symmetrical and the best fit to the experimental data was achieved from the following relationship:

$$\frac{\rho_w v_w}{p_{mean}} = A\left(\frac{p_a - p_b}{p_{mean}}\right)^B, \tag{2}$$

where $p_{mean} = (p_a + p_b)/2$ and A and B were set to the values given in Table 4. A comparison between the experimentally measured values of mass flux and the predictions using equation (2) can be seen in Fig. 3.

Table 4 Parameters used in Equation 2

	A (sm^{-1})	B
suction	3.15×10^{-4}	0.6052
blowing	2.98×10^{-4}	0.6298

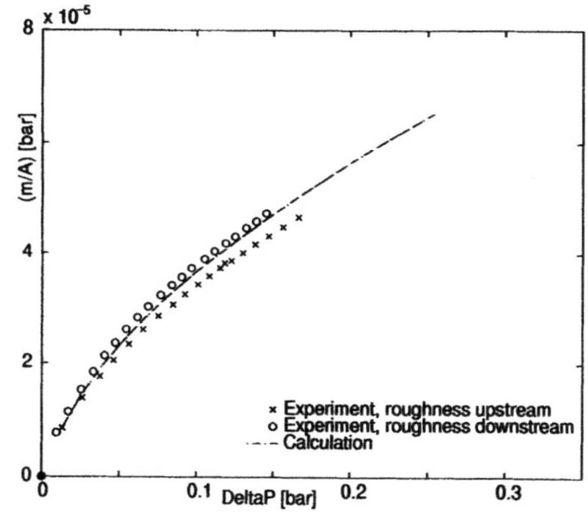

Figure 3 Mass flux through porous surface, comparison of exp. and polynomial curve-fit

In the general case it appears reasonable that the temperature as well as the tangential wall velocity influence the mass-flux. In the present experiments the temperature of the calibration tests was similar to the wind tunnel test and a temperature dependence is therefore neglected. The effects of tangential velocity, however, have not been investigated and are subject to future studies.

The preliminary calculations used three turbulence models, namely Cebeci-Smith, Baldwin-Lomax and Johnson-King. The effect of wall transpiration on turbulence is known to be confined to the inner region of a boundary-layer [6]. Modifications to the Van Driest damping term to account for wall transpiration effects were implemented into all three turbulence models by Chokani and Squire [5]. To account for the crossflow effects, the T-model proposed by Rotta [7] for the anisotropic modification was used with a modelling constant of 0.7.

13.4 The Uncontrolled Interaction

Before discussing both active control and slot suction, the following section describes measurements and simulations performed on the interaction without control. The understanding of the uncontrolled flowfield is crucial for the evaluation of various means of control and therefore more detail is provided here.

13.4.1 Flow Visualisation

In the case of an uncontrolled interaction, it is possible to replace the side walls of the working section with optical windows. A schlieren photograph taken with this arrangement is given in Fig. 4. The wedge mounted at the top of the working section as well as the main swept shock wave can be seen clearly. Originating from the trailing edge of the wedge an expansion fan can also be observed penetrating into the flow domain. Furthermore, there is evidence of a number of disturbance waves originating form the floor of the tunnel and penetrating into the centre of the flowfield.

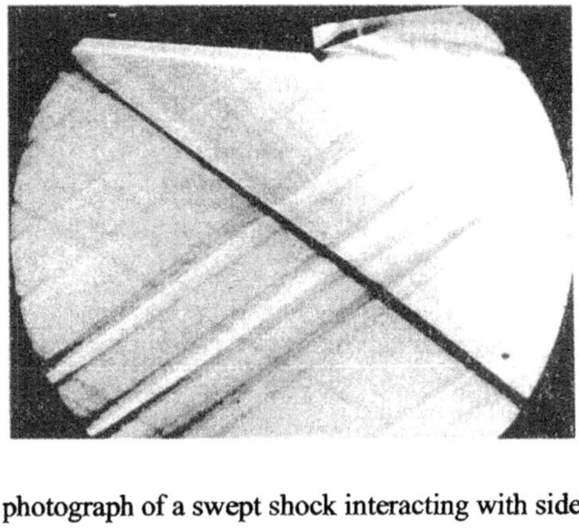

Figure 4 Schlieren photograph of a swept shock interacting with side-wall boundary layer

Figure 5 shows a photograph of the surface oil-flow pattern observed on the side-wall. It can be seen that the surface flow is deflected downwards by the wedge and the influence of the expansion fan downstream can also be seen clearly. Closer examination reveals that quasi-2-dimensional conditions can be achieved in the centre of the flowfield. As expected for the relatively small shock-normal Mach number, there is no indication of any flow separation in the interaction.

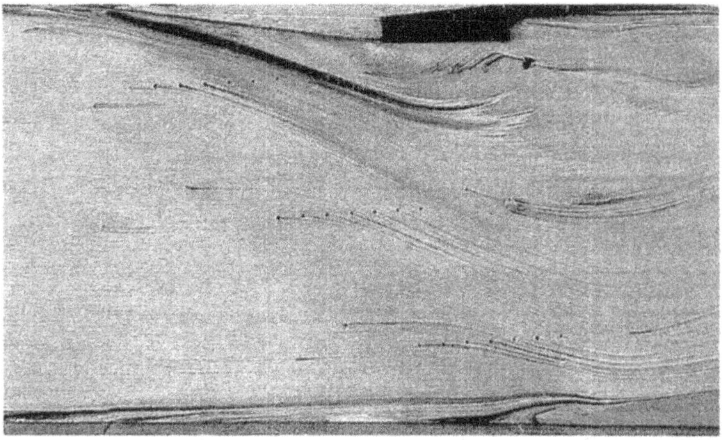

Figure 5 Surface oil-flow visualisation of uncontrolled interaction

13.4.2 Surface pressures

Figure 6 gives the surface pressure distributions observed along the lines X_1, X_2 and X_3. In each case, X=0 is the location of the inviscid wedge shock. It can be seen that the flow is not entirely uniform across the working section and that the strength of the shock induced pressure rise as well as its shape vary depending on the spanwise position. The pressure distributions along X_1 and X_2 also show the influence of the expansion fan downstream of the shock (see also Fig. 4).

Unfortunately the pressure rise along X_2 is distinctly less than that expected and measured elsewhere and it is believed that this is due in part to the disturbances seen in the schlieren photograph (Fig. 4). However, for the purposes of this study the data recorded along X_2 is believed to be the most useful since it suffers least from end-effects. Furthermore, all configurations will be compared to each other and any disturbances originating upstream are likely to be present in all cases.

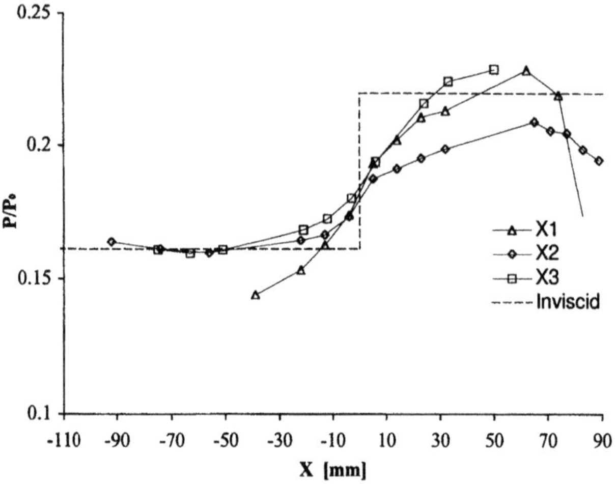

Figure 6 Surface pressure distributions along X_1, X_2 and X_3

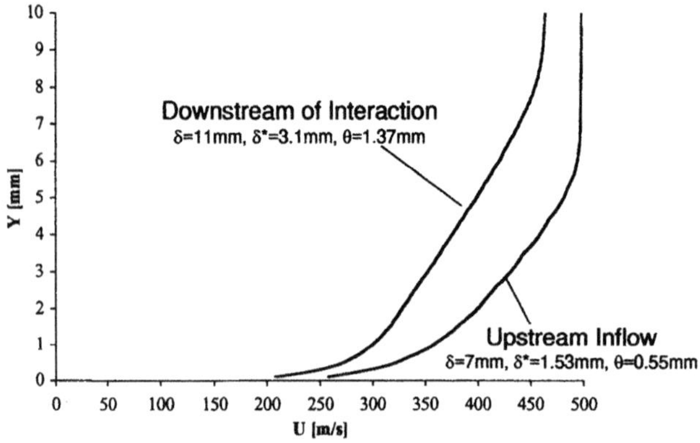

Figure 7 Boundary layer profiles upstream and downstream of uncontrolled interaction

13.4.3 Boundary layer profiles

Figure 7 compares the boundary layer velocity profiles upstream and downstream of the interaction (for the exact location refer to Fig. 2). It can be seen that the interaction increases boundary layer displacement and momentum thicknesses by factors of 2 and 2.5, respectively.

13.4.4 CFD results

Surface pressures:

Figure 8 shows the surface pressure contours obtained for the uncontrolled interaction. This simulation was performed using the Baldwin-Lomax turbulence model. Similar to the oil-flow visualisation obtained experimentally it can be seen that a quasi-two-dimensional region exists in the centre of the flowfield, where the interaction width is more or less independent of spanwise position. A detailed comparison of CFD prediction and experiment for the uncontrolled interaction has already been performed as part of Euroshock (I) and this is not repeated here. However, for flow fields involving suction a detailed comparison will follow below.

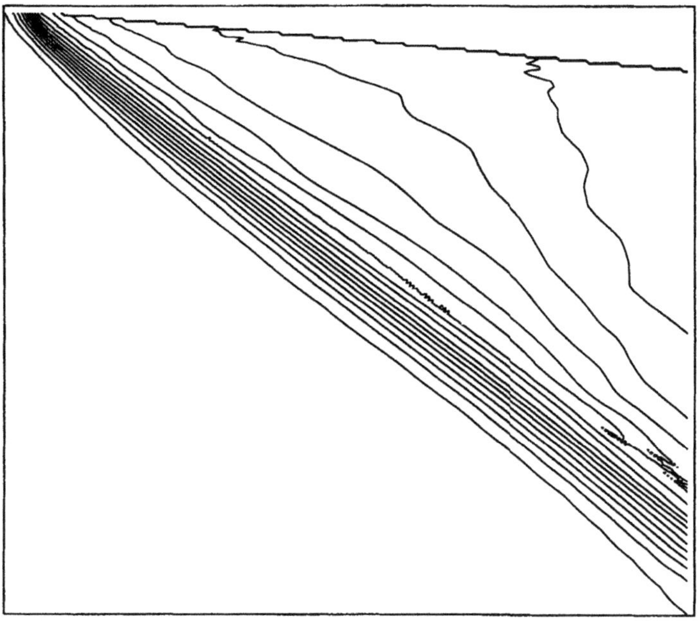

Figure 8 Surface pressure contours predicted by CFD

Velocity profiles:

Figure 9 shows the velocity profiles predicted upstream of the interaction for all three turbulence models compared to the experimentally measured values. It can be seen that the choice of turbulence model has only little effect on the prediction and that all numerical profiles are slightly fuller than the experimentally observed ones. Nevertheless, the agreement is reasonable and it is felt that all three simulations captured the inflow satisfactory. By comparison, Fig. 10 shows the profiles obtained downstream of the interaction and again it can be seen that the choice of turbulence model has only little influence on the shape of the profile. To some extent, the Johnson-King and Cebeci-Smith models appear to capture the flow slightly better, however, the differences are small. Overall, the agreement between CFD and experiment is good.

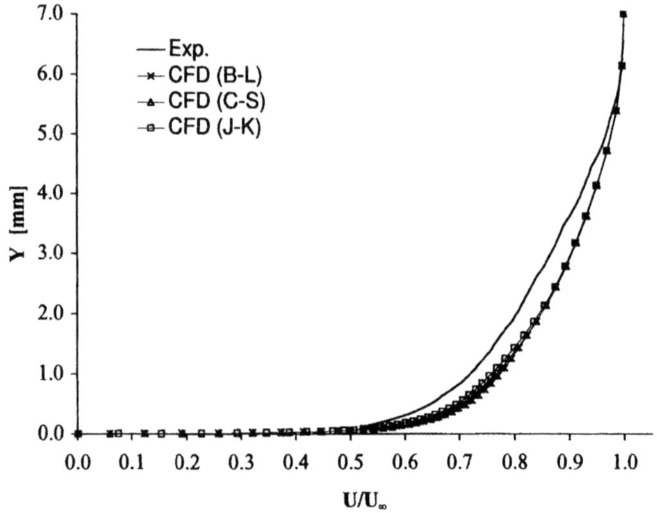

Figure 9 Velocity profiles upstream of the interaction (CFD and exp.)

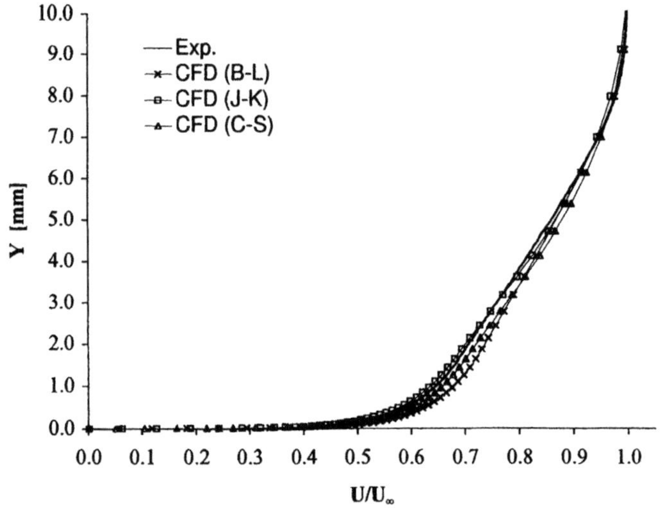

Figure 10 Velocity profiles downstream of the uncontrolled interaction (CFD and exp.)

13.5 Active Control-Suction Applied to Passive Control Cavity

13.5.1 Surface Pressures

Figure 11 shows the surface pressures recorded for an uncontrolled interaction as well as for active control at various suction levels along all three streamwise co-ordinate systems. It can be seen that the effect of suction is similar at all three spanwise positions, although along X_1 the influence of control is only small. This is due to the fact that this location is close to the edge of the control cavity (see Fig. 2) and that the interaction itself is not yet fully developed. Along X_2 the maximum pressure reached behind the interaction falls some way short of the expected inviscid pressure rise, which is due to local non-uniformities in the oncoming free-stream (as discussed earlier).

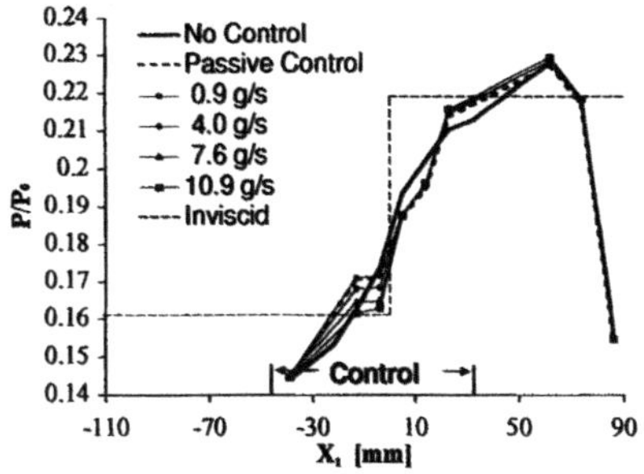

a) along X_1

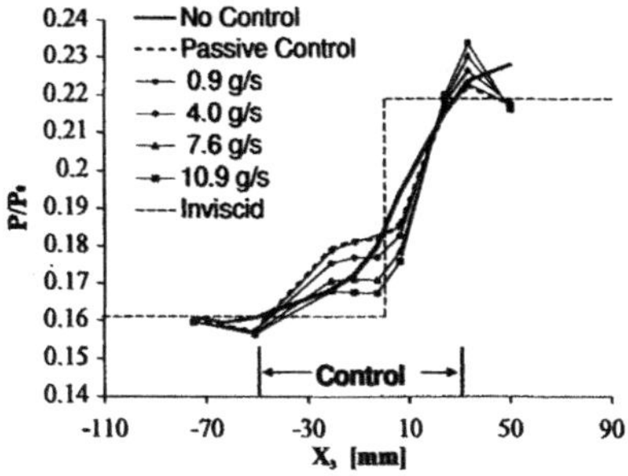

b) along X_3

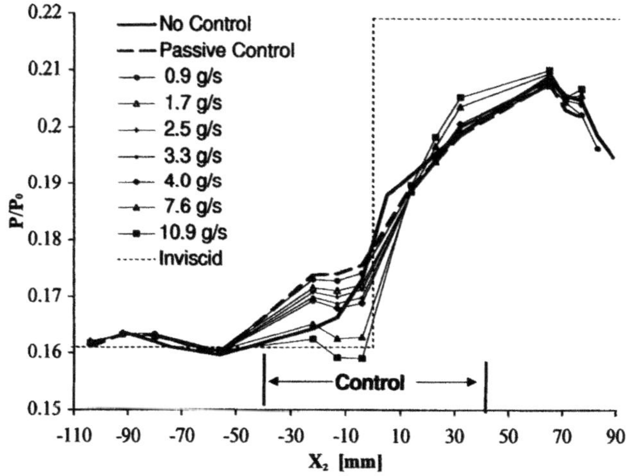

c) along X_2

Figure 11 Surface pressure distributions for active control at various suction levels

It can be seen that for the case without suction (i.e. equivalent to passive control), a distinct plateau is formed in the pressure distribution. This indicates that the shock wave has been split into a lambda structure and, if the control were applied to a transonic wing, a saving of wave drag could be expected. The application of suction from the plenum chamber has the effect of reducing the pressure level at the plateau. This is due to the fact that suction shifts the balance between upstream blowing and downstream suction across the porous plate (see also Fig. 15). It appears that low to moderate suction levels maintain some of the beneficial upstream blowing, although large values of mass removal eradicate the plateau and remove any beneficial shock smearing. For the maximum suction level, there is also an indication of a reduction in the surface pressure immediately upstream of the shock. In this case, the mass removal out of the cavity is large enough to generate suction throughout the porous surface and hence cause a local thinning of the oncoming boundary layer and an associated expansion fan. This in turn strengthens the shock and reduces the interaction length, thus steepening the pressure rise through the interaction.

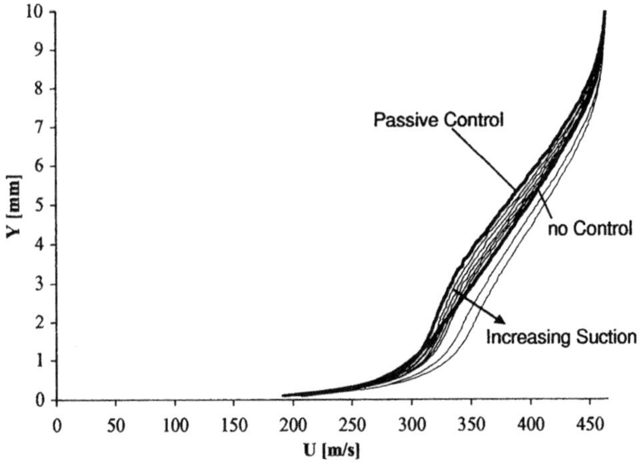

Figure 12 Velocity profiles behind the interaction for active control at various suction levels

13.5.2 Boundary Layer Profiles

Figure 12 confirms the well known effect that passive control thickens the boundary layer and reduces the shape factor; the velocity profile for the flow without suction (passive control) is visibly less full than the solid wall (no control) case. The presence of suction causes the profiles to be fuller as well as giving a reduction in boundary layer displacement and momentum thickness. The two cases featuring the strongest suction show a distinctly improved profile over the uncontrolled case. However, the same two cases were found to eradicate any beneficial effect of shock smearing in the surface pressure traces of Fig. 11c).

13.5.3 CFD Results

One of the objectives of this study was to investigate the performance of three different turbulence models. Before this is discussed in detail, a comparison is made between the numerical simulations and the experimental observations as well as the effects of control. All numerical results shown for this purpose have been obtained using the Bladwin-Lomax turbulence model. It will be shown later that the choice of turbulence model does not significantly alter the general conclusions drawn here.

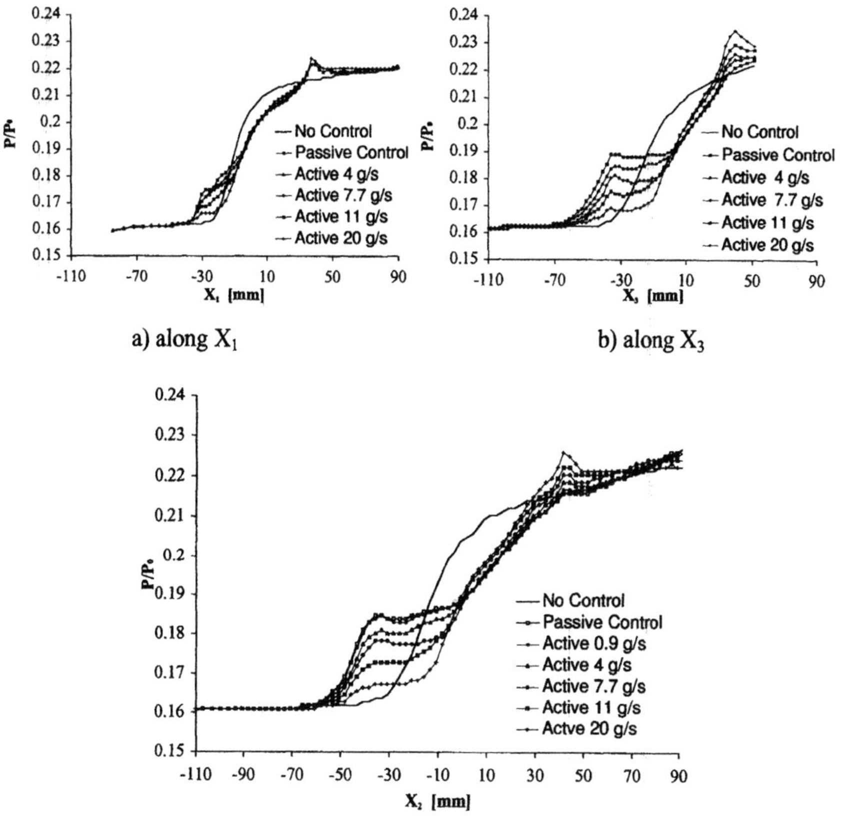

a) along X_1 b) along X_3

c) along X_2

Figure 13 Surface pressure distributions for active control at various suction levels (Baldwin-Lomax turbulence model)

Surface pressures

Figure 13 shows the pressure distributions obtained from a numerical simulation of the active control configuration. It can be seen that the trends in the data are very similar to those seen in the experimental data of Fig. 11. In particular it can bee seen that the interaction along X_1 is shorter in length and less affected by suction. This is due to geometrical effects as discussed earlier. As in the experimental data, the pressure distributions are similar along X_2 and X_3, and suction is seen to reduce the level of the pressure plateau observed. However, the level of the plateau pressure is not accurately predicted and, even for a suction mass flux well in excess of the experimental maximum, the plateau itself remains. This is seen more clearly in Fig. 14, which compares the average level of plateau pressure for experiment and simulation.

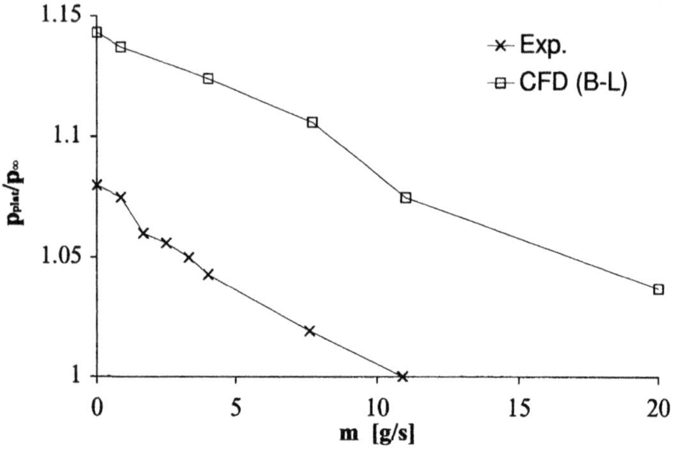

Figure 14 Average plateau pressure obtained from experiment and simulation

While the numerical simulation can be seen to significantly overpredict the plateau pressure itself, the effect of suction mass flux is remarkably similar for experiment and simulation showing an almost linear decrease with increasing suction levels. Due to the overprediction of the plateau pressures by the simulation a significantly larger mass-flux is required to eradicate the pressure completely.

Transpiration velocities

Figure 15 shows the distribution of the transpiration velocity along X_2 for a variety of suction levels. It can be seen that suction reduces the blowing in the upstream part of the cavity. However, even suction levels in excess of the largest experimentally tested case, do not completely eradicate any blowing. This is somewhat in contradiction to the experimental observation and is likely to be connected to the difficulties in predicting the correct plateau pressure.

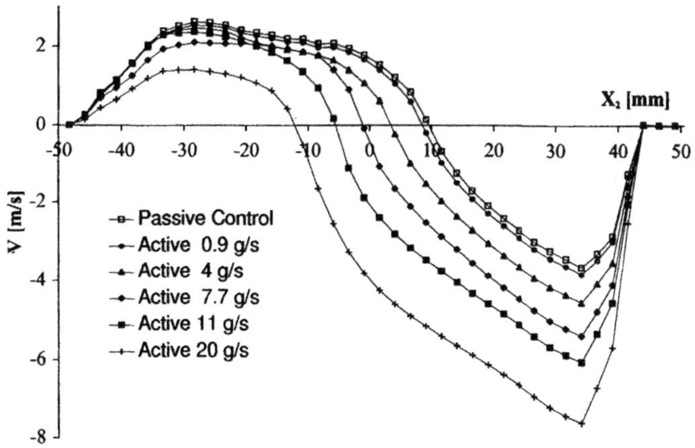

Figure 15 Transpiration velocity distribution along X_2 for various suction levels

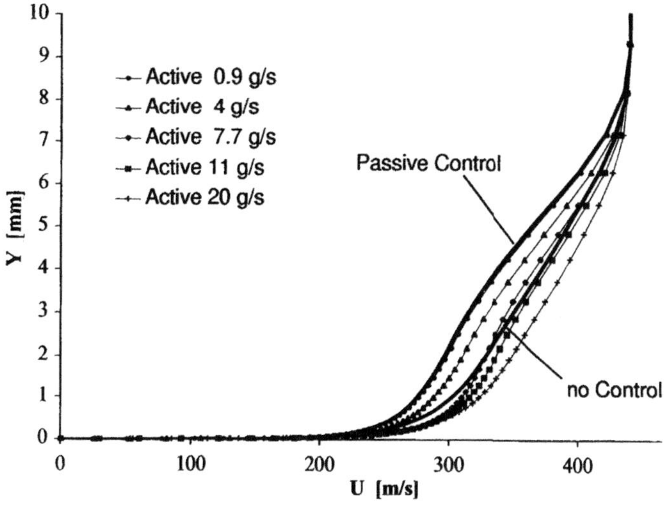

Figure 16 Velocity profiles downstream of the interaction (active control, CFD)

<u>Velocity profiles and integral boundary layer parameters</u>

Figure 16 shows the velocity profiles obtained from CFD for the actively controlled interaction. The trends are very similar to those observed in the experimental data presented in Fig. 12. Again it can be seen that large suction mass fluxes can reduce boundary layer thickness and shape factor to values below those observed for the solid wall (uncontrolled) interaction. This is seen more clearly in Fig. 17, which compares the displacement and momentum thicknesses downstream of the interaction for all cases. It appears that in each configurations there is an almost linear reduction of displacement thickness with mass flux. While the experimental and CFD data exhibit the same trend of decreasing thicknesses with increasing suction, the actual levels can be seen to be quite different and CFD is seen to consistently underpredict the displacement thickness downstream of the interaction.

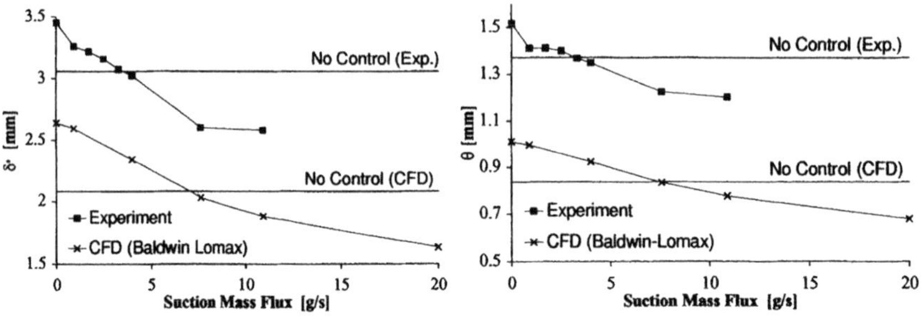

a) displacement thickness b) momentum thickness

Figure 17 Boundary layer thickness downstream of the interaction for various levels of suction

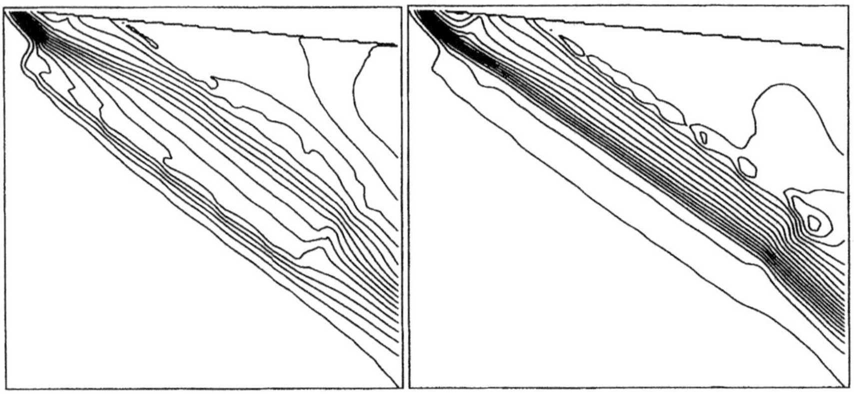

a) no suction (passive control) b) max. suction (20 g/s)

Figure 18 Contours of surface pressure from CFD (Baldwin-Lomax)

Three-dimensional effects

Figure 18 shows surface contours for the controlled interaction without suction and with maximum suction (20 g/s). It can be seen that in both cases there is a significant region of quasi-two-dimensional flow. Near the upstream edge of the control cavity there is a short region of adjustment, where the shock induced pressure rises changes from the uncontrolled behaviour near the wedge corner to that observed throughout most of the control region. After this transition there is no sign of any unexpected three-dimensional effects and in general the pressure contours are well aligned with the control cavity. A similar conclusion can be drawn from the experimental and numerical pressure distributions shown earlier in Figures 11 and 13. This confirms the observation from Euroshock (I) [1] that the swept case shows a very similar behaviour to a comparable two-dimensional case.

Influence of turbulence modelling

Figure 19 shows the surface pressure distributions along X_2 for all turbulence models and various control configurations. It can be seen that the choice of turbulence model has very little effect on the pressure distribution, although it was noted that the Johnson-King model was prone to spurious oscillations in the case of passive control. It was generally found that the computation of passive control proved to be more difficult than the simulation of cases with suction. Overall, however, it is not felt that any turbulence model provides significantly superior predictions of surface pressure distributions.

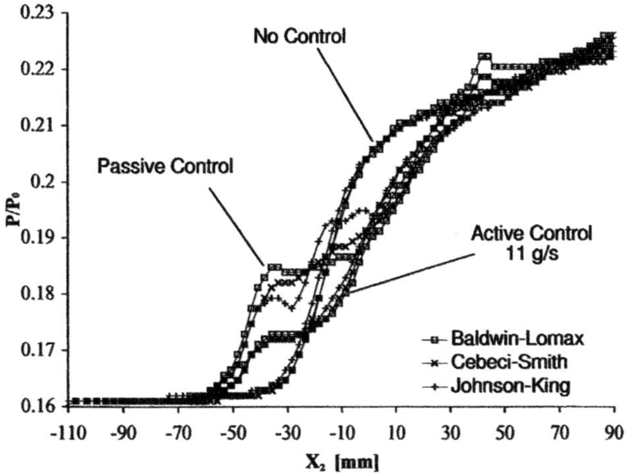

Figure 19 Surface pressure distributions along X_2 for different turbulence models

Figure 20 shows the velocity profiles obtained from the CFD predictions using all three turbulence models. In the case of passive control (Fig. 20a), there is little difference between each prediction, although the Baldwin-Lomax model is slightly closer to the shape of the experimental data. For an actively controlled interaction (Fig.20b), there is a pronounced difference between the Baldwin-Lomax model, which fits the experimental data very well, and the other two simulations. Overall it appears that no model can give consistently better results and it is felt that the Baldwin-Lomax turbulence model may be preferable since it is relatively simple and well established.

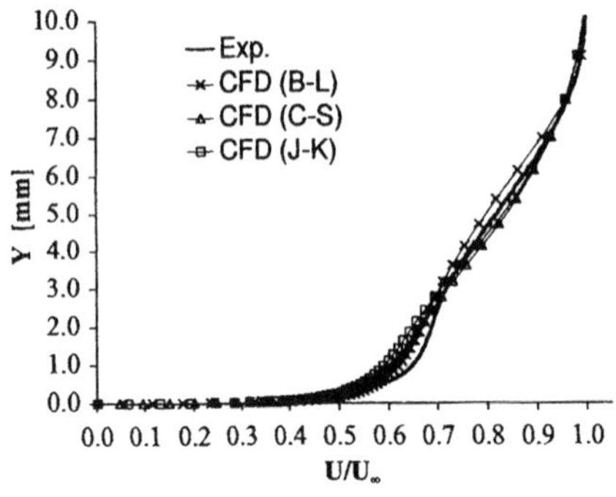

a) No suction (passive control)

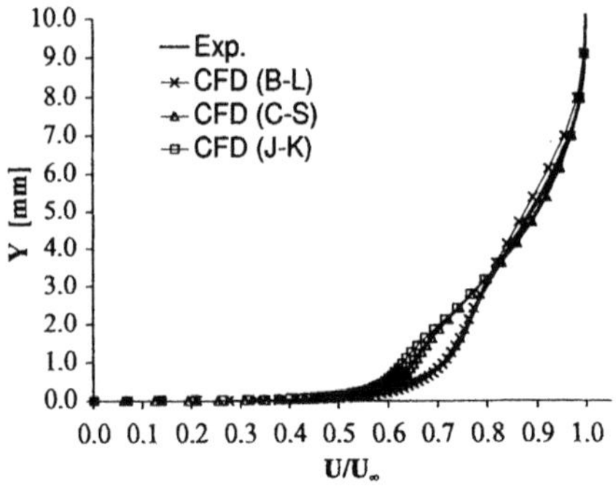

b) suction mass flux: 11 g/s

Figure 20 Velocity profiles downstream of the interaction for different turbulence models

13.6 Slot suction

As part of the current investigation, a second control mechanism was tested. This consisted of discrete suction slots of approximately one boundary layer thickness in width, located in three locations relative to the inviscid shock. While this device is not strictly speaking a shock control, but rather a boundary layer control device, it promises to reduce the viscous drag without increasing the wave drag. It may even be possible to locate a suction slot in a position were it has some beneficial effect on the shock induced pressure rise. No numerical simulation was attempted for this control mechanism.

13.6.1 Flow Visualisation

Figure 21 shows the surface oil flow pattern obtained with the suction slot upstream of the inviscid shock position (see also Fig. 2) without and with maximum suction. It can clearly be seen that the application of suction changes the flow downstream of the slot. There is also a small change in the flow upstream of the suction slot: without suction there is a small amount of flow turning in this region, whereas with suction the streamlines enter straight into the slot. For the other two slot positions, the effects observed were relatively small and no fundamental conclusions can be drawn.

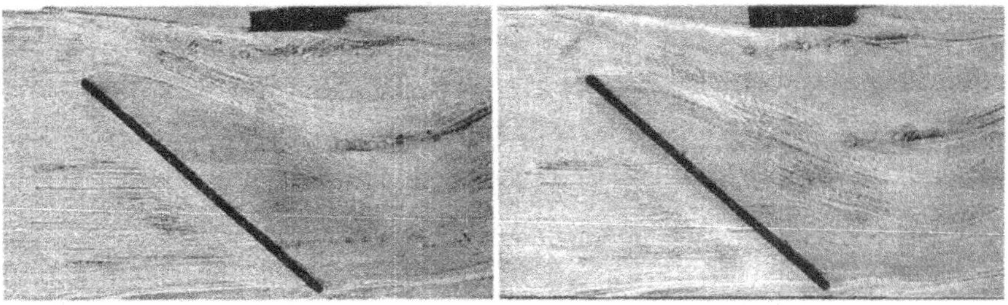

a) no suction b) max. suction (10g/s)

Figure 21 Oil flow visualisation with suction slot upstream of inviscid shock location

13.6.2 Surface Pressures

Figure 22, a – c, show the pressures recorded for slot suction, with the slot being at three different locations. Comparing the pressure distributions for a solid wall ('No Control') with the traces for 'No Suction' highlights the effects the slot itself has on the flowfield. It can be seen that slot positions underneath and downstream of the inviscid shock location cause severe disruption of the downstream pressure distribution. Behind the shock the velocity relative to the slot-normal direction is subsonic (or transonic) and the flow is therefore very sensitive to surface variations. In contrast, a slot upstream of the swept shock does not cause significant flow disturbance.

When comparing the data gathered for the three different slot positions it should be noted that the suction levels vary depending on the configuration. This is due to the fact that the pressure in the plenum chamber is approximately equivalent to the static pressure at the slot location. The mass flux through the same orifice plate is therefore larger when the slot is downstream of the inviscid shock where the wall pressures are higher. In practice it would be easier to achieve high suction rates in positions where the wall pressures are high and this, to some extent, illustrates this effect.

The introduction of suction shows that the effects are very different, depending on the slot location. As before, the upstream slot position causes the smallest effects. Oil flow visualisation indicated that suction changes the flow immediately upstream of the slot, however, no pressure taps were present in this region and elsewhere the effects of suction are relatively small. However, suction does slightly reduce the pressures upstream of the shock and steepen the subsequent pressure rise. This effect is most likely due to the thinning of the boundary layer as it enters the interaction.

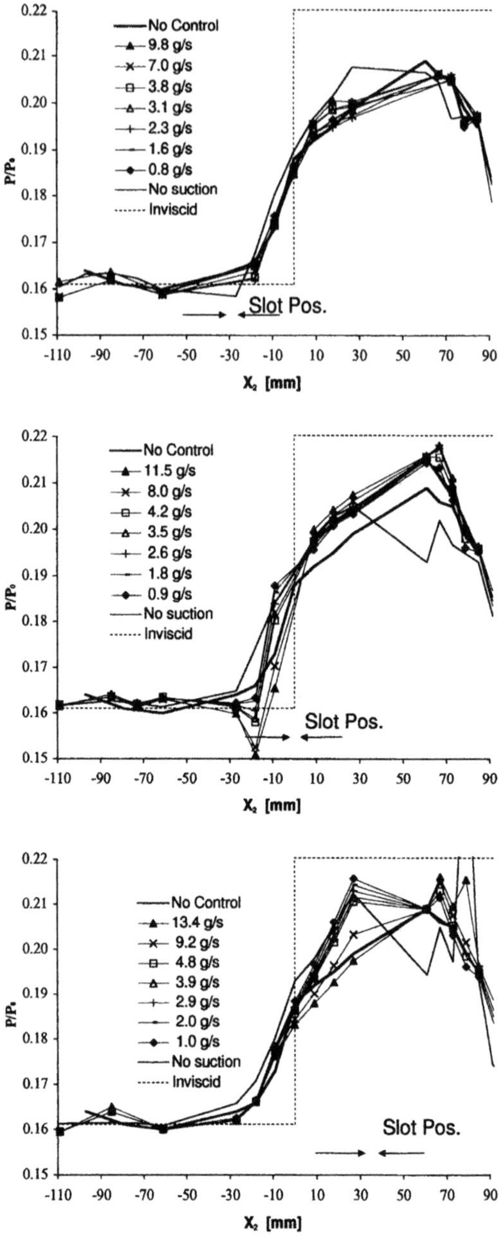

a) upstream slot position

b) central slot position

c) downstream slot position

Figure 22 Surface pressure distributions for slot suction at various suction levels

In the case of the slot being positioned behind the shock, suction does reduce the pressure rise through the interaction, which had been steepened significantly by the presence of the slot. With maximum suction applied, the pressure rise is restored to values comparable to the solid wall interaction. Compared to the effects caused by the slot upstream, the influence of suction is much more pronounced here.

The case of the slot directly underneath the inviscid shock (Fig. 22b) is the most complex. While the presence of the slot alone appears to have widened the interaction, even small levels of suction strongly affect the upstream influence and large suction levels generate an expansion just upstream of the shock and considerable steepening of the subsequent pressure rise. This case exhibits the largest upstream effect of slot suction.

13.6.3 Boundary Layer Profiles

Figure 23, a – c, show the velocity profiles measured downstream of the interaction for all slot positions at various suction levels. Also included are measurements taken with the suction slot in place but without any suction applied. It can be seen that the presence of a suction slot has different effects on the boundary layer profiles: for the slot in the upstream position (Fig. 23a) there is relatively little change, while the slot underneath and downstream of the inviscid shock position (Figs. 23 b and c) cause an increased free-stream velocity. Additional experiments confirmed that for the two latter cases there is a region of increased Pitot pressure which extends for several boundary layer thicknesses into the free-stream flow. Comparison with the surface pressure measurements (Fig. 22) show that the boundary layer traverse is in a region where the flow is considerably disturbed. Therefore it is likely that the slot underneath or downstream of the shock causes a region of enhanced flow velocities behind the shock. However, it is also possible that the assumptions of constant static pressure throughout the boundary layer are not valid in this region and that the velocity data is therefore flawed.

The application of suction has very similar effects in all cases. The profiles measured in the presence of suction are visibly fuller and thinner and in all cases it was possible to generate profiles which were fuller and thinner than that observed for the uncontrolled interaction. However, it can also be seen that this was hardest for the case of the suction slot upstream of the interaction. The maximum mass flux in this case was less than for the other cases due to the lower pressures at this point (reducing the density of the air being sucked away). There is no evidence of any 'magnification effect' through the interaction and it appears that downstream slot positions are more effective in changing the boundary layer parameters through suction.

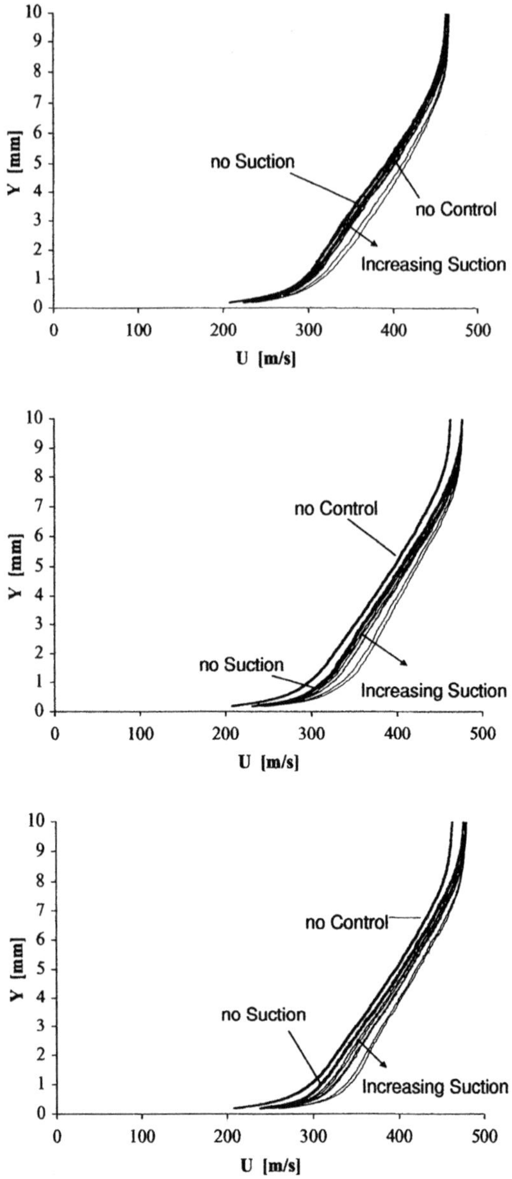

a) upstream slot position

b) central slot position

c) downstream slot position

Figure 23 Velocity profiles behind the interaction for various slot positions and suction levels

In order to compare the ability of suction slots to reduce the boundary layer thickness, Fig. 24a shows the displacement thicknesses measured for all slot locations and suction levels. For comparison the data obtained for active control is also included. Similar to active control, there is a steady drop of displacement thickness with increasing suction mass flux. However, the boundary layer thicknesses for zero mass flux are all different. This is caused by the differences in flow disturbance caused by the presence of the control device. For active control this disturbance was greatest since the zero suction case is equivalent to passive control which is well known to significantly disrupt the boundary layer. To gain a relative comparison of the ability of suction to reduce the boundary layer thickness, Fig. 24b shows the same data but normalised using the displacement thickness obtained for each configuration when the suction mass flux was zero. Plotted in this fashion, the CFD and experimental data for active control agree very well with each other and the slot results also coincide. However, the two groups of results show different trends and it can be seen that, for the same mass flux, suction through a porous surface has a larger effect on the boundary layer development.

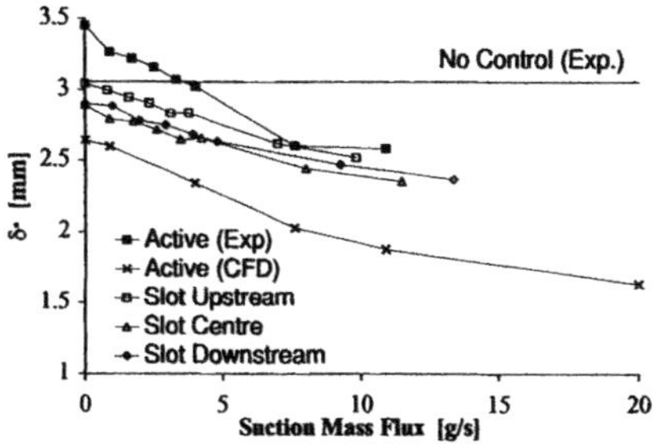

a) actual data

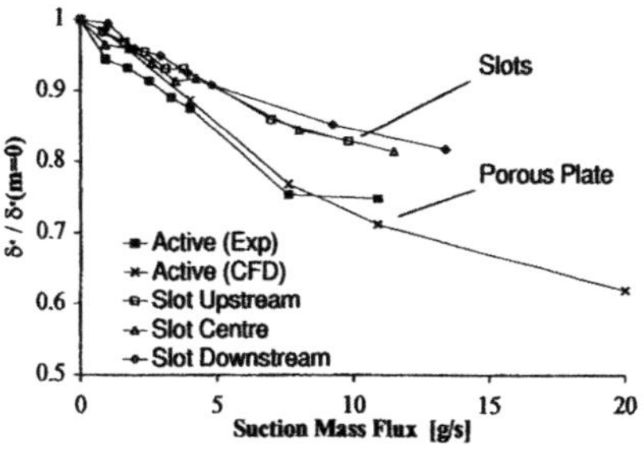

b) normalised by δ*

Figure 24 Displacement thickness downstream of the controlled interaction

This is somewhat surprising and may be due to the fact that suction through the porous surface not only improves the boundary layer directly (similar to slot suction) but also reduces the detrimental effects of blowing as observed for passive control. In this case suction has therefore two simultaneous effects which both help to reduce the boundary layer thickness downstream of the interaction. However, considering a practical application, suction through a porous surface requires significantly more pump power to overcome the losses encountered in the small diameter holes. This can be seen in the present data by the differences in maximum mass flux reached for the different configurations. The same orifice plate achieved approximately 25% more mass flux for slot suction (downstream slot) compared to the porous surface.

13.7 Conclusions

Suction applied to the control cavity of a typical passive control arrangement reduces the beneficial effects of shock smearing while also reducing boundary layer thickness and shape factor. Relatively small amounts of suction are capable of restoring boundary layer parameters to levels similar to those observed behind an uncontrolled interaction while still retaining some of the beneficial effects of passive control. Large amounts of mass flux, however, remove and even reverse any shock-smearing due to control.

A numerical simulation has been found to model the effects of suction from a passive control cavity relatively well, but it overpredicts the plateau pressure observed in the controlled interaction and therefore underpredicts the effects of suction on the removal of this plateau region. While the increase in boundary layer thickness through the interaction is not well predicted, the influence of suction on the boundary layer displacement thickness is predicted very well.

Three turbulence models have been investigated, namely, Baldwin-Lomax, Cebeci-Smith and Johnson-King. The influence of these models on the simulations has been limited and no model was found to provide the best result in all cases. However, throughout, the Baldwin-Lomax model performed relatively well and since it is well established in the field and relatively easy to implement it may be the best choice.

From both, experimental and numerical data of active control it can be concluded that sweep effects are secondary. The control of swept interactions behaves fundamentally similar to comparable two-dimensional flows.

Suction applied through a narrow slot upstream of an interaction has relatively little effect on the surface pressure distributions. Large suction levels reduce upstream influence and steepen the shock, but compared to other slot positions the effect is small. Slots underneath and immediately downstream of the interaction have a noticeable effect on the pressure distribution. Suction underneath the shock reduces upstream influence and steepens the shock markedly, whereas suction just after an interaction has the opposite effect.

If the aim of suction is to improve boundary layer parameters while also preventing the shock induced pressure increase from getting too steep, the best configuration appears to be either the application of small amounts of suction from a passive control cavity or slot suction immediately downstream of the shock / boundary layer interaction.

Acknowledgements

The author would like to thank H A Holden for her invaluable help in the investigation of slot suction.

References

[1] E. Stanewsky, J. Délery, J. Fulker, and W. Geissler (Ed.), "EUROSHOCK-Drag Reduction by Passive Shock Control", Results of the Project EUROSHOCK, AER 2-CT92-0049, Notes on Numerical Fluid Mechanics, Vol. 56, Vieweg & Sohn, Braunschweig/Wiesbaden, 1997.

[2] Dawes, W.N., "A Numerical Analysis of the Three-Dimensional Viscous Flow in a Transonic Compressor Rotor and Comparison with Experiment", *Trans. ASME, J. Turbomachinery*, Vol. 109, 1997, pp. 83-90.

[3] Leung, A.W.C. and Squire, L.C., "Reynolds Number Effects in Swept-Shock-Wave / Turbulent-Boundary-Layer Interaction", *AIAA Journal*, Vol. 33, 1995, pp. 798-804.

[4] Holmes, S.C. and Squire, L.C, "Numerical Studies of the Supersonic Flow over a Compression Corner", *Aeronautical Journal*, Vol. 96, 1992, pp. 87-95.

[5] Chokani, N. and Squire, L. C., "Transonic Shockwave / Turbulent Boundary Layer Interactions on a Porous Surface", *Aeronautical Journal*, May 1993, pp. 163-170.

[6] Squire, L.C., "Turbulent Boundary-Layers with Suction and Injection", *The 1980-81 AROSR-HTIM-Stanford Conference on Complex Turbulent Flows*, Proc. Vol. 1, 1981, pp. 112-129.

[7] Rotta, J.C., "A Family of Turbulence Models for Three-Dimensional Boundary Layers", *Turbulent Shear Flows I*, Springer-Verlag, 1979, pp. 267-278.

14 Numerical Investigation of Active Shock Wave and Boundary-Layer Control on Airfoils and Sheared Wings

C. DIMA and P. DE MATTEIS
CIRA, Centro Italiano Ricerche Aerospaziali
Via Maiorise, 81043 Capua(CE), Italy

Summary

The present work reports the overall research activity conducted by CIRA within the EUROSHOCK II project (Task 2). The aim was to improve aerodynamic airfoil performance (drag reduction and buffet onset) using shock and boundary-layer active control techniques. The numerical flow simulation has been performed using the computational capabilities developed within EUROSHOCK (I) and the results have been validated through comparison with the experiments performed in Task 3. The method is based on a viscous-inviscid coupling technique and on different models for the calculation of the transpiration velocity on the airfoil surface which have been implemented and tested during the first phase of the activity. The control techniques which have been applied are: bumps, discrete slot suction, passive and active ventilation through perforated plates, and their combinations. A new concept of "pneumatic" bump, based on distributed blowing, has been tested and compared with the classical contour bump. Tests performed on both, laminar and turbulent airfoils have confirmed the results of the present numerical analyses.

14.1 Introduction

The interaction of a shock wave with the boundary layer has received large attention in the last decades, and it still represents a topic of high interest for modern transport aircraft. Drag rise, flow separation and buffet, adverse aerodynamic loading, poor engine inlet performance are some examples of its effects, with negative consequences on the aircraft overall performance. The study of the viscous effects, present in all these phenomena, has led to the development of many concepts for the control of both, shock wave and boundary layer. These can be divided in active (energy consuming) control techniques, such as intelligent bumps, suction/blowing, and passive control concepts which make use of devices such as a passive plenum, LEBUs and riblets.

The study of SaBLC (Shock and Boundary Layer Control) techniques was the subject of the European project EUROSHOCK II (European Shock Control Investigation II). It was aimed at assessing active shock control techniques in order to improve the aerodynamic characteristics of transonic aircraft wings by reducing total drag and by raising the buffet limits at higher speeds [1]. The necessity of studying active control was the consequence of the limitations of passive control treated in EUROSHOCK (I) especially regarding drag reduction [2]. In fact, reduction of drag was obtained for turbulent type airfoils [3] at very limited conditions, while the efficiency of passive control in alleviating buffet has always been demonstrated.

The present work describes the overall activity conducted by CIRA in the framework of the above project. Calculations have been performed for two types of airfoils, viz., turbulent (VA-2, RAE-5225) and laminar (DRA-2303, DA-LVA-1A). The aim of this numerical test campaign was to validate computational methods modified for SaBLC modelling and to assess the effectiveness of the active control. 2D calculations have been performed using the EUBL2D code, based on the viscous/inviscid coupling between an Euler solver and an integral boundary-layer code, previously modified to include the effect of transpiration [4]. A study of different transpiration laws, including Poll's law [5] and the law proposed by the University of Karlsruhe [6] within Task 1, has been performed. Some studies on the grid spacing, as well as on Mach number correction, to be applied in the calculations, have been performed. The quasi-3D flow analysis on infinite swept wings has been performed using the EUBLSW code based on the sheared wing theory. Special attention has been payed to the evaluation of the effect of control on drag and on buffet alleviation and to the comparison with the experiments performed in Task 3[7].

The present paper has been structurated as follows: after a brief introduction, the second section gives a general description of the control concepts investigated here and of the corresponding numerical tests. The third section includes a brief description of the computational capabilities, of the tests performed to assess the control law and of the modifications made in order to improve the agreement with the experiments, such as the optimization of the grid step, or the calculation of the Mach number correction. The following sections are entirely dedicated to the results of the computations performed for different types of airfoils and wings, using different control techniques. An important feature is the comparison with the resuts of the experimental measurements.

14.2 Description of the Control Concepts

Different active control techniques have been tested in Task 2 of EUROSHOCK II. As illustrated in Figure 1, they are based on bumps or on localized or distributed ventilation. Their combinations (hybrid techniques) have also been considered. The study also included the optimization of the control parameters (geometry and position of the bump, intensity of ventilation and location of the holes, etc.).

The control of the interaction region can be performed either by applying a passive device, like a perforated surface with a plenum cavity underneath generating a natural ventilation, or applying active control devices, based on local surface modifications (bumps) or forced ventilation through a single slot or through the perforated plate, mentioned above, with a given pressure in the cavity. In the cases in which the passive control effect is not effective in reducing the overall drag, the use of active control techniques or their combinations represents a promising alternative. In Figure 1, a sketch illustrating different control concepts is depicted.

The active control technique through a cavity (Figure 1.1) is aimed at controlling the flow in the plenum by fixing the total mass flow or the cavity pressure p_{cav}. For the optimum value of these parameters, the maximum overall drag reduction might be obtained by balancing wave drag reduction due to a decrease in pressure difference between the regions upstream and downstream of the shock, and friction drag increase due to the flow through the holes of the perforation. The "bump" configuration (Figure 1.2) produces the effect of the boundary-layer growth in the shock region upstream of the shock wave. In case of the latter, the upstream diffusion of the high pressure gradient due to the shock wave, which takes place in the subsonic layer, determines a δ^* growth over the interaction length L^*. This generates compression waves in the adjacent supersonic region along L^*, which reduce the shock wave intensity. The growth effect is more visible if the shock wave/boundary-layer interaction generates a separation bubble.The discrete and distributed suction techniques (Figure 1.3) are mainly aimed at diminishing the tendency of the boundary layer to separate. The location and the intensity of suction are to be optimized for each particular case. Localized suction is more attractive than the distributed suction since a lower friction drag increase is expected.

Since the individual techniques composing the hybrid methods (Figure 1.4) are aimed, separately, at reducing the shock wave intensity and the tendency of the boundary-layer to separate, the methods are refered to as SaBLC ('Shock and Boundary-Layer Control') concepts.

An interesting extension of the porous wall control concept is the "multi-box" plenum device, proposed by CIRA. As illustrated in Figure 2, it consists of a divided plenum chamber underneath the airfoil surface which communicates with the external flow by means of a "multi-slot" surface, allowing a separate control of the internal pressure in each box, and thus of the intensity of the ventilation (suction/blowing) through each slot. The device, based only on ventilation, has the advantage of easily simulating discrete (distributed) suction, blowing ("pneumatic", flexible bumps) and their combinations at locations, intensities and type of ventilation (constant, parabolic, etc.) optimized for each flight regime. As demonstrated by the computations performed within the EUROSHOCK project, the improvement of the airfoil performance due to the device, such as drag reduction and buffet alleviation, warmly recommends it as a valid alternative to the adaptive bump.

Of the control techniques enumerated above, bumps, including "pneumatic" bumps, slots and perforated plates have been tested for both, turbulent (RAE-5225) and laminar airfoils (DRA-2303, DA-LVA-1A) and the results compared with the experiments performed in Task 3 and others.

An important task of the EUROSHOCK II project was the investigation of the effectiveness of the above control concepts on swept wings of transport aircraft. To this end, a numerical investigation of the ADIF sheared wing has been performed, using a quasi-3D (2.5D) flow approximation, focusing on the bump. The results have then been compared with the measurements performed in the TWG wind tunnel of DLR Göttingen.

14.3 Improvement of Computational Capabilities

14.3.1 Theoretical Background

A viscous/inviscid method, EUBL2D [8], originally developed for transonic flow calculation has been extended to include flow control through ventilation [4]. The method is based on the iterative coupling of 2D Euler and integral boundary-layer equations. The unsteady Euler equations in the integral form are solved using a finite volume technique for the spatial discretization and a 4th order Runge-Kutta scheme for the time integration [9]. Artificial viscosity is introduced by means of a blend of second and fourth differences with adaptive coefficients. Multigrid technique [10], enthalpy dumping, residual averaging and the maximum allowed time step in each cell are used as convergence acceleration methods. The boundary-layer equations are solved using Cohen-Reshotko's method [11] in the laminar case and Green's compressible integral lag-entrainment method in the turbulent case. Using the procedure indicated in [12] for the case with control, Kays and Moffat's law for the C_f correlation allows to take ventilation velocity into account by means of the correction term B_f, which depends on the transpiration.

$$C_{f0} = C_{f0s}\left[\frac{\ln(1+B_f)}{B_f}\right]^{1.25}(1+B_f)^{0.25} \tag{1}$$

$$B_f = 2\frac{m_w}{C_{fo}} \tag{2}$$

$$\frac{C_f}{C_{f0}} = 0.9\left(\frac{\overline{H}}{\overline{H}_0} - 0.4\right)^{-1} - 0.5 \tag{3}$$

$$1-\frac{1}{\overline{H}_0}=6.55\left[0.5C_{f0}\left(1+0.04M^2\right)\right]^{0.5} \tag{4}$$

In the above notations, referring to the boundary-layer variables, the subscripts 0 and s indicate the flat plate and the solid surface, respectively. A semi-inverse scheme consisting of solving the direct Euler equations and the inverse boundary-layer equations is used to couple the viscous and inviscid solutions. The displacement thickness has been chosen as input parameter for the inverse formulation; its updating is based on the difference between viscous and inviscid velocities at the edge of the boundary layer in the attached flow region.

14.3.2 Assessment of the Transpiration Law

The simulation of the flow through the slot has been performed either by considering a constant mean transpiration velocity at the grid points covering the slot region or by using more precise laws such as Poll's law [5] or Bohning and Doerffer's law [6]. The results of the calculations have demonstrated that the first approach leads to a good approximation of the physical phenomenon.

Poll's law is the calibration law for laser-drilled plates; it has the following quadratic form:

$$Y=\frac{1}{K}\left[40.7X+1.95X^2\right] \tag{5}$$

with

$$Y=\frac{\Delta p}{\rho}\left(\frac{d}{\nu}\right)^2\left(\frac{d}{t}\right)^2 \tag{6}$$

$$X=\frac{\dot{m}}{\mu t} \tag{7}$$

where d is the hole diameter, t the thickness of the plate, Δp the pressure difference across the surface, ρ the density, ν the kinematic viscosity, $\dot{m}$ the mass flow rate per hole, μ the dynamic viscosity and K a calibration coefficient.

A new transpiration law implemented in the EUBL2D code aimed at taking into consideration not only the pressure difference across the perforated plate, but also the effect of the tangential component of the outer flow. It was developed within Task 1 of the EUROSHOCK II project at the University of Karlsruhe by Prof. Bohning and Dr. Doerffer. The transpiration law takes the form:

$$\frac{\Delta p}{p_0} = \left(\left(\frac{1}{1.2} \right)^{\frac{1}{0.55}} + b(BM_{hole})^{\frac{1}{a}} \right) * M_{hole}^{\frac{1}{0.55}} \tag{8}$$

$$B = \frac{\tau_w}{0.5 \rho_{hole} u_{hole}^2} \tag{9}$$

$$b = 25. \tag{10}$$

$$a = 1.52 \tag{11}$$

The first term in (8) corresponds to the linear form of the law proposed by the same authors in EUROSHOCK (I). It can be used in the case of blowing or when the Mach number in the hole due to suction is very low. The presence of the shear force at the inlet side of the wall, τ_w, causes a blocking of the flow through the holes. The shear effect (τ_w) is dependent on the intensity of the transpiration flow. It has been quantified by introducing, in the second phase of the EUROSHOCK project, a supplementary term depending on B, which is a function of τ_w, normalised by the dynamic pressure in the hole, $0.5(\rho u^2)_{hole}$, as indicated in (9).

The advantage of using Bohning and Doerffer's law instead of Poll's law may be the use of dimensionless quantities as well as the use of the effective porosity, measured in the calibration of the perforated plate, instead of the geometric porosity used by Poll.

In order to check the calculation accuracy when computing the flow through the holes, three common test cases have been considered by all Task 2 EUROSHOCK II partners [13]. These computations have been performed for the DA-LVA-1A airfoil with passive (C_q=0.0) and active control (C_q=-0.0007 and C_{pcav}=-0.72). The corresponding non-dimensional mass flow distributions, calculated by CIRA, are depicted in Figure 3.

The corresponding global variables of the flow, namely the non-dimensional total mass flow coefficient C_Q, the non-dimensional average ventilation velocity $\frac{V_p}{U_\infty}$, the non-dimensional pump coefficient C_{pp}, and the suction drag coefficient C_{dQ}, are presented in Table 1.

An important aspect which has been considered during EUROSHOCK II, concerned the agreement between the results when cavity pressure, C_{pcav}, or the total mass flow coefficient, C_Q, have been imposed on the transpiration flow through the perforated surface of the airfoil. Differences have been found between numerical results and experiments in the case of the DRA-2303 airfoil, especially when prescribing the mass-flow coefficient. Since pressure measurement in the cavity has been assumed to be more precise than the mass flow measurement, the following numerical tests have been performed by imposing C_{pcav}.

Table 1 DA-LVA-1A, M=0.77, Re=4.64*10[6]

	C_Q	Vp/U_∞	C_{pp}	C_{dQ}
C_{pcav}=-0.72	0.00037	0.00529	1.87702	0.00069
C_Q=0.0	0.00000	0.00113	1.93027	0.00000
C_Q=-0.0007	-0.00070	-0.00091	2.03055	0.00142

Other tests refer to the agreement between Poll's law and the two laws proposed by Bohning and Doerffer of the University of Karlsruhe (i.e., the linear and the non-linear variant) [6]. Inspite of the fact that the non-linear law takes into account the tangential component of the external flow, it was found here to be in good agreement with Poll's law for both, an experimental porosity of 4% and for a corrected value of 2.48%, the latter proposed by the University of Karlsruhe, Figure 4.

A preliminary study of the linear and the non-linear variant of the new law has shown that, especially for transonic Mach numbers, the velocity of the flow through the holes becomes high enough (Mach number about 0.2-0.3) to make the non-linear term quite important (about half of the total value).

14.3.3 Grid Refinement

A realistic simulation of the flow in the presence of the slot requires a sufficiently high density of the grid points in order obtain a local C_q close to the theoretical one (C_q=C_Q/Ls). If the grid spacing is larger than the slot width (Ls), the total suction has to be distributed over a larger length and the suction level is reduced correspondingly. Since the typical width of the slot is about 1.0% of the chord, and in order to use more points in the slot region, the grid step size along the surface has to be reduced to at least 0.5%c. This allows the presence of at least two points in the suction slot region, which represents a compromise between accuracy and stability. The influence of the grid refinement was initially investigated for the NLF DRA-2303 datum airfoil, using grids of different dimensions. As indicated in Figure 5 and Table 2, tests have been performed at both, constant incidence $\alpha = 1.978^{\circ}$ and constant C_N=0.747.

Along the airfoil surface 96, 144, 176 and 264 points, respectively, have been used in the grid, while in the direction normal the airfoil surface the number of points was of 48 and 72, respectively. As illustrated in Figure 5, the increased number of points in the normal direction did not improve the agreement with the experiments regarding the shock position and the total drag value, Table 2. Concerning the surface point distribution, the use of an increased number of grid points, keeping the number of points in normal direction constant (48 points), indicates an improvement with regard to the shock position and the Cp-curve slope in the shock region determined experimentally. The pressure distribution near the trailing edge also improves. The total drag value does not improve by increasing the number of points. On the contrary, the best agreement with the experiments is shown by the case with the lowest number of nodes of the grid. This encourages one to look for other potential corrections to be performed, like, for instance, a Mach number correction, to improve agreement between experiment and computation.

As described in the latter sections, the use of a refined mesh is required in the case of suction through a slot. In fact, because of the width of the slot of 1% c, and in order to insert at least two points in the control zone, the use of a grid step of 0.5% c is necessary. This requirement was fulfilled using the 264x48 grid, which, together with the Mach number correction, led to the best results. This grid has been selected as the best of all the grids considered.

Table 2 DRA-2303 Datum airfoil, M=0.6809, Re=19x10^6

Grid	α	C_N	C_m	C_d	C_{dpress}
96x48	1.978000	0.757640	-0.104073	0.012438	0.008438
176x48	1.978000	0.737400	-0.100366	0.011423	0.007353
176x72	1.978000	0.749637	-0.101474	0.010453	0.006424
144x48	2.238000	0.747067	-0.093173	0.011927	0.008082
264x48	2.420000	0.746918	-0.089997	0.011042	0.008323
Exp. DERA	2.0960	0.7469	-0.1001	0.01293	-

14.3.4 Mach Number Correction

As shown in the preceding section, the grid refinement alone was not sufficient to improve the value of the calculated total drag. Thus, a study of the influence of a Mach number correction was performed among other, for the DRA-2303 airfoil. As illustrated in Table 3 and Figure 6, a small Mach number correction of about 0.6% is sufficient to improve the total drag value and the shock wave position in comparison with the experiments performed with the DRA-2303 airfoil (and, similarly, for the ADIF-3D sheared wing). The best agreement has been shown by the results obtained using the corrected Mach number M_{cor}=0.685 for DRA-2303 airfoil (andM_{cor}=0.839 for ADIF-3D wing), instead of the experimental values of 0.6807 (and 0.852).

Table 3 DRA-2303 Datum airfoil, M=0.6807, Mcor=0.685, Re=18.89x10^6, CN=0.6075

	M	α	C_N	C_m	C_d	C_{dp}/C_{dw}
Datum cor.	0.685	1.605	0.607184	-0.086767	0.009849	0.005778
Datum	0.6807	1.638	0.607448	-0.085266	0.009251	0.005151
Exp. DERA	0.6807	1.285	0.6075	-0.2463	0.00973	-

14.4 Numerical Results of 2D-Tests on Airfoils

A theoretical analysis of the application of active control has been performed for the turbulent airfoil RAE-5225 as well as for the laminar airfoil DRA-2303. Optimisation of the control parameters has also been performed. Results are presented in the next sections.

14.4.1 Tests with the Turbulent Airfoil RAE-5225

Different control techniques like the bump, the suction slot placed upstream of the shock, and a combination of both (hybrid method) were tested on the RAE-5225 airfoil; the results are shown below.

- Bump

  The RAE-5225 geometry was investigated with a bump in accordance with DERA experiments previously performed. A circular shape of the bump, positioned at x/c=[0.4:0.6], was chosen. The maximum thickness (0.25% chord) was located at half-chord. Table 4 illustrates the favourable effect of the bump on total drag, which is reduced by 8.61% (computation) and by 13.67% (experiment), respectively. Figure 7 depicts the control effect on the C_p distribution. A good agreement with the experiments is to be noted. The conclusion is that the bump mainly reduces the shock wave intensity, thus contributing to the pressure drag reduction. This result is explained by the ramp effect that transforms the strong, concentrated shock into a compression wave system spread over a wider region. Compared to the techniques based on ventilation, the bump has the advantage of a smooth surface; thus the excrescence drag due to the roughness caused by the perforations and the additional drag due to the flow through the holes, and pump drag are avoided.

- Suction Slot

  The suction slot has been placed at x/c= [0.385:0.395]. Suction influences mainly the boundary-layer by reducing its thickness and the tendency to separation, even if skin friction C_{df} is increasing. Table 4 indicates a much higher drag reduction in the case with bump (8.61%) than in the case when suction has been applied (2.89%).

- Hybrid Control Technique (Bump plus Suction)

  As illustrated in Table 4 by the computations, the combined effect of bump + suction is superior to each of the individual control techniques: in fact, since the presence of the bump does not influence the C_f values, the higher drag reduction (9.40%) obtained in the hybrid case, due mainly to the bump, is also associated with the reduced tendency of the boundary layer to separate, which is an effect of the suction. An optimisation of the intensity of suction has been performed in both control cases, i.e., with suction and bump + suction. For instance, for the suction test case, an intensity of C_q=-0.06, results in an increased total drag reduction ($\Delta C_d = -3.15\%$) compared to $\Delta C_d = -2.89\%$ calculated with the experimental value of the suction intensity, i.e., C_q=-0.02.

Table 4 RAE-5225, M=0.73, Re=19x10^6, CN=0.756, Cq=-0.02, x/c=[0.385:0.395

	α	C_N	C_m	C_d	C_{dpres}	C_{df}	ΔCd (%)
Datum	2.925	0.755112	-0.099131	0.014577	0.010855	0.003722	-
Exp. DERA Datum	2.959	0.755392	-0.0918	0.01192	-	-	-
Bump	2.911	0.756817	-0.099308	0.013322	0.009592	0.00373	-8.61
Exp. DERA Bump	3.016	0.757207	-0.0903	0.01029	-	-	-13.67
Suct.	2.596	0.756109	-0.110194	0.014155	0.008902	0.005253	-2.89
Bump+ Suct.	2.609	0.755959	-0.109369	0.013207	0.008007	0.005200	-9.40

14.4.2 Tests with the DRA-2303 NLF Airfoil

The study of a laminar airfoil is significant regarding the efficiency of the control techniques because in this case the shock wave is stronger. The control effect on drag reduction has been studied at the transonic cruise conditions M=0.685, Re=19x 10^6, C_N=0.6, while the buffet study has been performed at higher lift (C_N=1.05) and M=0.67, Re=19x10^6. Different control techniques, like the contour bump, the pneumatic bump, distributed/concentrated suction/blowing and their combination (hybrid method), have been tested for the NLF airfoil; results will be shown in the next sections. Special emphasis has been placed on overall drag reduction and on buffet onset.

- Bump

  In accordance with the DERA experiments [7] for the DRA-2303 airfoil with bump control, an asymmetric bump with a relative height of 0.25% chord was placed at x/c = [0.485:0.685] on the airfoil contour. The bump consists of two arcs; at their junction, situated at 57.5% of the bump length, the slope of the bump contour is equal to zero.

  The pressure distributions are depicted in Figure 8. As shown in Table 5 for the drag reduction, an excellent agreement can be noted between the experiments (11.98% reduction) and the calculations performed (11.86% reduction).

Table 5 NLF Airfoil with and w/o Bump, M=0.6801, Re=19x10^6, CN=0.74

	α	C_N	C_m	C_d	C_{dp}/C_{dw}	$\Delta Cd(\%)$
Datum	2.200	0.7393	-0.09249	0.011548	0.007621(p)	-
Exp. DERA Datum	2.096	0.7469	-0.1001	0.01293	-	-
Bump	1.898	0.7399	-0.10243	0.010178	0.006172(p)	-11.86
Exp. DERA Bump	2.098	0.7399	-0.0989	0.01138	-	-11.98

- Suction Slot

  In the experiments performed by DERA in Task 3, suction has been applied over a length of 1% of the chord, the control zone being situated upstream of the shock. For optimisation purposes, numerical test cases also took into consideration other slot positions, namely, downstream of and centred at the shock wave.

The pressure distributions are illustrated in the case of the shock-upstream suction slot in Figure 9. The latter also contains a refined grid test case (0.5%c) compared to the coarse grid step of 1.0%c. The porosity of the plate covering the slot was 4%.

The calculations performed at a lower porosity (2.48%) led, at least for the University of Karlsruhe law, to the elimination of the problem of inconsistency which was noted, using a porosity of 4%, between the results obtained imposing the cavity pressure and the total mass flow, respectively. For the effective porosity of 2.48%, the total drag reduction (about 4.8% for both laws) has been much closer to the experimental value (7.5%) than the values obtained for 4% porosity (about 1% specifying the cavity pressure) as indicated in Table 6.

Table 6 NLF Airfoil with and w/o Suction Slot, M=0.6807, Re=18.89x10^6, CN=0.6075

	α	C_N	C_m	C_d	C_{dp}/C_{dw}	$\Delta Cd(\%)$
Datum	1.638	0.607448	-0.085266	0.009251	0.005151(p)	-
Exp. DERA Datum	1.285	0.6075	-0.2463	0.00973	-	-
Suct. (CQ=-0.00007)-Poll	1.580	0.595787	-0.084239	0.00906	0.004570(p)	-2.1
Suct. (CQ=-0.00007)-U. Karl.	1.963	0.595222	-0.073376	0.009079	0.004816(p)	-1.9
Suct.(Cpcav=-1.2975)-Poll	1.600	0.595320	-0.083708	0.009195	0.004778(p)	-0.6
Suct.(Cpcav=-1.2975)-U. Karl.	1.700	0.587726	-0.079680	0.009181	0.004913(p)	-0.8
Suct.(Cpcav=-1.2975)-U.Karl.(por. 2.48%)	1.578	0.595024	-0.084181	0.008812	0.004583(p)	-4.75
Suct. (Cp=-1.2975)-Poll(por. 2.48%)	1.600	0.598879	-0.084272	0.008855	0.004631(p)	-4.8
Exp. DERA Suct.(M_{exp}=0.6793, por. 4.%)	1.285	0.5953	-0.2428	0.00900	-	-7.5

- "Multi-box" plenum device

  The new control device ("multi-box" plenum) which has been tested is described in Chapter 14.2. Placed successively at x/c=[0.4:0.6] and x/c=[0.55:0.75], the device is composed of 10 boxes, each one corresponding to an open slot. Thus, the airfoil contour with the device underneath becomes a "multi-slot" surface. By activating/deactivating different slots at different locations, the "multi-box" plenum device can be used for the simulation of distributed blowing, Table 7, or discrete suction/blowing, Table 8. In order to reduce pump drag, a perforated plate of variable porosity covering the "multi-box" plenum could be a viable solution.

Table 7 NLF Airfoil with Distributed Blowing (CQ=+0.00007, M=0.685, Re=18.56x10^6)

	α	C_N	C_d	C_{dp}	$\Delta Cd(\%)$
Cq=0.00100,x/c=[0.4:0.6]	2.000	0.587972	0.010211	0.0071023	+3.68
Cq=0.00050,x/c=[0.4:0.6]	2.000	0.604082	0.009892	0.0067403	+0.44
Cq=0.00005,x/c=[0.4:0.6]	1.850	0.594570	0.009253	0.0060673	-6.05
Cq=0.00010,x/c=[0.4:0.6]	1.865	0.595333	0.009319	0.0061383	-5.38
Cq=0.00010,x/c=[0.55:0.75]	1.827	0.595375	0.009180	0.0059913	-6.79
Const.bl.Cq=.0001, x/c=[0.57:0.75] + + Discr.suct.Cq=-.01 x/c=[0.55:0.56]	1.683	0.595290	0.008589	0.0052963	-12.80
Parabolic bl.Cq=.0001, x/c=[0.57:0.75]+ +Discr.suct.Cq=-.01, x/c=[0.55:0.56]	1.680	0.595317	0.008584	0.0052897	-12.84

An optimum of the intensity has been calculated in the case of distributed blowing since a very strong blowing can lead to an increase in drag, indicated with the "+" sign in Table 7, viz., +3.68% and +0.44%. The Mach number correction (M=0.685 instead of M=0.6807, Table 3) has improved drag reduction from 2.11% to 7.66% which is much closer to the experimental value of 7.5%, Table 8. Table 8 also shows that total drag reduction is not very sensitive to the slot position, and that discrete blowing is less efficient than localised suction.

Table 8 NLF Airfoil with Discrete Blowing/Suction
(CQ=+/-0.00007, M=0.685, Re=18.56E+6)

	α	Cl	Cd	Cdp	$\Delta Cd(\%)$
Upstr.suct.x/c=[.45:.46]	1.835	0.596466	0.009095	0.005799	-7.66
Centr.suct.x/c=[.52:.53]	1.810	0.596984	0.009071	0.005797	-7.90
Downstr.suct.x/c=[.55:.56]	1.825	0.600005	0.009068	0.005846	-7.93
Upstr.blow.x/c=[.45:.46]	1.957	0.595372	0.009676	0.006525	-1.76
Centr.blow.x/c=[.52:.53]	1.924	0.595421	0.009591	0.006427	-2.62
Downstr.blow.x/c=[.55:.56]	1.910	0.595960	0.009506	0.006335	-3.48
Exp.DERA Suct.x/c=[.45:.46] M=0.6793	1.285	0.5953	0.00900	-	-7.5
Datum Mcor=0.685	1.605	0.607184	0.009849	0.005778	-
Exp. DERA Datum M=0.6807	1.285	0.6075	0.00973	-	-

14.4.3 Comparison of Active Control Techniques in 2D

The effects of different active control concepts have been tested in the case of the RAE-5225 turbulent airfoil: the bump aimed at reducing wave drag by transforming the concentrated shock wave into a compression shock wave system, and the suction slot which alleviates buffet by reducing the boundary-layer separation tendency. The hybrid control concept, suction + bump, applied to the RAE-5225 airfoil, combines the advantages of both techniques leading to the highest drag reduction (ΔC_d=9.4%, Table 4).

The control devices which have been tested on the DRA-2303 NLF airfoil are the asymmetric contour bump, placed at x/c = [0.485:0.685], with a relative height of 0.25%c, and the "multi-box" plenum device, placed at x/c = [0.55:0.75]. The "multi-box" plenum device, described in Chapter 14.2, was used as a "multi-technique" device allowing the simulation of different control techniques based on ventilation (concentrated/distributed suction/blowing). An optimisation study, concerning characteristic parameters of these control techniques, and a comparison with both, classical contour bump and the "pneumatic" bump have been performed.

Results for the bump test case and the optimised ventilation test cases, realised by using the "multi-box" plenum device and described in detail in [14], are illustrated in Table 9 and Figure 10.

Table 9 NLF Airfoil, Re=19x10^6 - Comparison of Different Active Control Techniques

	M	α	C_N	C_d	C_{dp}	ΔCd (%)
Datum	0.6850	1.605	0.6072	0.009849	0.005778	-
Exp. Datum	0.6807	1.285	0.6075	0.00973	-	-
Contour Bump	0.6850	1.295	0.5954	0.008960	0.004951	-9.03
Distr.bl. Cq=.0001, x/c=[0.55:0.75]	0.6850	1.827	0.5954	0.009180	0.005991	-6.79
Discr.Suct.Cq=-.007, x/c=[.45:.46]	0.6850	1.835	0.5965	0.009095	0.005799	-7.66
Exp. Discr.Suct., Cq=-.007, x/c=[.45:.46]	0.6793	1.285	0.5953	0.00900	-	-7.50
Distr.bl.Cq=.0001, x/c=[0.57:0.75]+ +Discr.suct.Cq=-.01, x/c=[0.55:0.56]	0.6850	1.680	0.5953	0.008584	0.0052897	-12.84

As shown in Table 9, the bump is an efficient shock/boundary-layer control concept for reducing drag. The drag reduction of over 9% is explained mainly by the important reduction of the pressure drag, C_{dp}, due to the weakening of the shock wave (Figure 10).

Regarding the ventilation techniques, the suction slot located upstream of the shock has led to a drag reduction of 7.66%, in good agreement with the experiments. Distributed suction is less effective than discrete suction since a friction drag increase has been noted. Discrete suction is also efficient in diminishing the tendency of the boundary layer to separate. Discrete blowing results in a lesser total drag reduction than discrete suction, wherever the slot is located. Distributed blowing through a "multi-slot" surface produces a "pneumatic" bump effect. Two types of blowing distribution have been tested: the parabolic distribution has shown to be more efficient in reducing drag than constant blowing. The intensity of blowing is very important: a very strong blowing can lead to an increase of the drag. An optimum for the location and length of the "multi-box" region, the type and the intensity of the blowing distribution has resulted in a total drag reduction of 6.79%.

The hybrid suction/blowing concept is much more efficient than discrete suction and distributed blowing applied separately; in the optimised test case, the total drag reduction was 12.84%. This gain can be maintained over a range of flight conditions due to the possibility of regulating the parameters of ventilation (location, intensity, type of distribution) correspondingly.

Buffet has been studied using a semi-empirical criterion, defined in a statistical manner, which defines buffet as the condition where the Mach number curve in front of the shock is tangent to a "Mach number limit" linear function [15]. The three linear functions, represented by the three straight lines (lower, middle, upper) in Figure 11, indicate, in this order, the intensity of buffet also taking into account, by their slope, where, on the airfoil, the probability of buffet onset is higher.

Regarding buffet alleviation, the contour bump and the "pneumatic" bump, generated by distributed blowing (Cq=0.0001 at x/c = [0.55:0.75] via the "multi-box" plenum device, are the most efficient concepts. Thus, as depicted in Figure 11, at the same lift coefficient (C_N=1.05) where on the datum airfoil an extended buffet is noted, the airfoil with the (pneumatic) bump only indicates the onset of buffet. The influence of localised suction is minor, at least for the level of suction (C_Q=-0.00007) used. Further studies [16] have shown that a ten times higher level of suction can lead to the suppression of buffet.

As a conclusion, the bump and hybrid localised suction/distributed blowing seem to be the most efficient concepts for drag reduction and buffet alleviation. Since for the bump, varying bump profiles leads to different aerodynamic characteristics, an optimisation of its geometry has to be performed at each flight regime of interest. For example, for a fixed bump position on the airfoil, a particular crest location may be associated with efficient buffet alleviation, whilst another crest position corresponds to the highest drag reduction. But since manufacturing of such an "intelligent" bump contour is still a challenge, the use of the adaptive "multi-box" plenum device can be a valid alternative. In fact, the device, based only on ventilation, has the advantage of easily simulating discrete or distributed suction, blowing ("pneumatic", "flexible" bumps), and their combinations at locations, intensities and type of ventilation (constant, parabolic, etc.) optimised for each flight regime.

14.5 2D/3D Tests for the ADIF Wing with Bump

14.5.1 Extension of the EUBL2D Method to Infinite Swept-Wing Calculations

The CIRA EUBLSW method, which is an extension of the EUBL2D code to sheared wings, has been used to perform 3D calculations for the ADIF swept wing. The use of this strip theory approach has been permitted by the characteristics of the present wing: an infinite swept wing whose sweep angle

($\Lambda = 26^\circ$) corresponds to that of a contemporary civil transport aircraft; the incidence of the flow represents the usual cruise angle of attack ($\alpha = 1^\circ - 2^\circ$). The sheared wing approximation, which allows de-coupling of streamwise and cross flow, has been applied to perform aerodynamic calculations for the flow about infinite swept wings. The method, which is rigorously exact for laminar flow, introduces an approximation due to cross-coupling for turbulent flow. As shown in Figure 12, the CIRA method (EUBLSW Code) consists of 2D Euler calculations in the direction normal to the leading edge using the effective asymptotic flow conditions

$$M_{eff} = M_\infty \sqrt{1 - \cos^2 \alpha_\infty sin^2 \Lambda} \tag{12}$$

$$\alpha_{eff} = \tan^{-1}(\alpha_\infty / \cos \Lambda) \tag{13}$$

and boundary-layer calculations along 3D streamlines using M_∞ and α_∞.

14.5.2 Numerical Results

Both 2D and 3D numerical tests have been performed for the ADIF airfoil and the ADIF sheared wing. Control has been applied using 2D bumps of three different heights, viz., H_{bump2D} = 0.60%c, 0.35%c, and 0.175%c, respectively, corresponding to 3D bumps of, respectively, H_{bump3D} = 0.54%c, 0.3146%c, and 0.1573%c, placed in the region of the shock wave. The bump has been defined as the deformation of an elastic straight beam with both ends fixed under a load acting at 75% of its length. The maximum deformation is therefore at 60% of the bump length. The bump length is 20% of the airfoil chord.

- Test Cases with the Higher Bump ($H_{bump2D} = 0.60\%$, $H_{bump3D} = 0.54\%$)

The first test case (H_{bump2D} = 0.60%c) showed convergence problems, which may be due to the excessive height of the bump which causes separation. To get a converged solution, the use of the smoothing of the boundary-layer variables has been necessary. Tables 10 and 11 illustrate the variation of the total drag due to bump control: the C_D reduction is 17.3% in the 2D case and 15.8% for the 3D calculations which are mainly due to pressure drag reductions of 25% and 22%, respectively. The present bump has been successfully used to reduce the separation tendency of the boundary layer in the 3D case but not in 2D. In fact, for $\alpha = 1.0^\circ$, the 2D bump shifts separation from x/c=0.9 (datum), to x/c=0.8.

Table 10 ADIF-2D, M=0.755, Re=8x10^6, Cl=0.755

	α	C_l	C_m	C_d	C_{dp}	ΔCd (%)
w/o bump	0.0020	0.754821	-0.196423	0.016997	0.0130516(p)	-
with bump	-0.1880	0.755466	-0.204579	0.014060	0.009779(p)	-17.28

Table 11 ADIF-3D, M=0.84, Re=9.9x10^6, Cl=0.61

	α	C_l	C_m	C_d	C_{dp}	$\Delta Cd(\%)$
w/o bump	-0.3810	0.609438	-0.154564	0.014124	0.009735(p)	-
with bump	0.0370	0.610827	-0.137904	0.011889	0.007567(p)	-15.82

- Tests Cases with the Lower Bump
  ($H_{bump2D} = 0.175\%$, $H_{bump3D} = 0.1573\%$)

Numerical tests have been performed for the 3D test case when experimental results from tests performed by DLR with the ADIF infinite swept wing became available. Since the experimental 3D data for C_N and the Mach and Reynolds numbers (C_N=0.574, M=0.852, Re=6.7x10^6) as well as the transition location have been modified with respect to the values of the initial (numerical) test case, the numerical test cases have been re-calculated for the updated freestream flow conditions. The new value of Re=6.7x10^6 represents the physical limit of the TWG tunnel at DLR-Göttingen. Due to leading-edge contamination from dust particles in the wind tunnel, transition has been changed from x/c = 30% on the upper surface to x/c = 10%; on the lower surface transition fixing stayed at x/c = 15%. For the new tests, the bump geometry has also been corrected by shifting the crest position upstream, from 76% c to 71.5% c, in order to make the bump more efficient by adjusting its position with respect to the real position of the shock wave. Thus, the shock was positioned on the compression ramp formed by the bump. The new geometric characteristics of the bumps are correspondingly

$$x_{crest}/c=0.715,\ x_{bumpcrest}/l_{bump} = 0.6,\ H_{bump} / c = 0.1573\% .$$

New updated Mach number corrections were necessary for both, datum (Table 12 and Figure 13) and bump (Table 13 and Figure 13) test cases. Incidence and Mach number corrections have been performed by fitting the shock location in both, datum and bump test cases at the same C_l. The agreement with experiment

(ΔC_d = 4.54%) is excellent for the corrected Mach numbers, the computed drag reduction ΔC_d being an intermediate value between 3.32% and 6.22%, which are the values corresponding to the different Mach number corrections (*Mcor*=0.839 and *Mcor*=0.841, respectively). Anyway, the computed efficiency of the bump has been improved substantially by the shift in Mach number, now determining a drag reduction instead of the initial drag increase of 3.0%.

Table 12 ADIF-3D Datum, Re=6.7x10^6, Cl=0.574

	α	C_l	C_m	C_d	C_{dp}/C_{dw}
Mcor=0.839	-0.045	0.576016	-0.129151	0.008113	0.003668(p)
Mcor=0.837	0.000	0.578538	-0.127632	0.008138	0.003656(p)
Mcor=0.835	0.000	0.575323	-0.126159	0.007298	0.003175(p)
Exp. DLR (M=0.852)	0.000	0.574	-	0.01323	-

Table 13 ADIF-3D with Bump, Re=6.7x10^6, Cl=0.574

	α	C_l	C_m	C_d	C_{dp}/C_{dw}	$\Delta Cd(\%)$
Datum(M=0.839)	-0.045	0.576016	-0.129151	0.008113	0.003668(p)	-
Exp. DLR Datum (M=0.839)	0.0	0.574	-	0.01323	-	-
Bump (Mcor=0.839)	-0.074	0.573605	-0.129734	0.007839	0.003636(p)	-3.38
Bump (Mcor=0.841)	-0.114	0.574012	-0.131601	0.007608	0.003388(p)	-6.22
Exp. DLR Bump (M=.852)	0.25	0.576	-	0.01263	-	-4.54

- Test Cases with the Intermediate Bump ($H_{bump2D} = 0.35\%c$, $H_{bump3D} = 0.3146\%c$)

The intermediate-bump-height test case was required due to the necessity of finding an intermediate solution between the advantages offered by the higher bump and those of the lower bump described above. The former works better at high angles of attack, where it alleviates buffet onset, while the latter contributes to total drag reduction at low incidences.

The test cases with the intermediate bump have been performed at three C_{l2D}-*values* (0.5, 0.6 and 0.7) and the corresponding 3D-values in order to study the efficiency of the bump dependent on lift. As shown in Figure 14 for C_l=0.5, the agreement with the 2D experimental data measured by DLR is excellent. Tables 14 to 18 illustrate the results of the 2D and 3D calculations with and without bump control.

Table 14 ADIF-2D, M=0.757, Re=8x10^6, Cl=0.7

	α	C_l	C_m	C_d	C_{dp}/C_{dw}	ΔCd (%)
w/o bump	-0.2790	0.706031	-0.196802	0.013223	0.009415(p)	-
with bump	-0.308000	0.716740	-0.201902	0.012071	0.008266(p)	-8.71
w/o bump	-0.1560	0.731050	-0.198990	0.014481	0.010683(p)	-
Exp. DLR w/o bump	0.0	0.7308	-	0.0170	-	-
Exp. DLR with bump	0.0	0.7129	-	0.0135	-	-20.6

Table 15 ADIF-2D, M=0.757, Re=8x10^6, Cl=0.6

	α	C_l	C_m	C_d	C_{dp}/C_{dw}	ΔCd (%)
w/o bump	-0.751000	0.600000	-0.188348	0.009729	0.0058795(p)	-
with bump	-0.750000	0.614767	-0.192786	0.009058	0.005208(p)	-6.87
Exp. DLR w/o bump	0.0	0.6393	-	0.0125	-	-
Exp. DLR with bump	0.0	0.6126	-	0.0103	-	-17.6

Numerical and experimental tests on ADIF-wing have shown that the bump works better at higher incidences, i.e., in the presence of strong shock waves: the higher the bump, the higher the lift coefficient at which the bump begins to be efficient in reducing total drag. For lower lift coefficients, the ADIF 2D lower bump produced an increase in lift starting at a lift coefficient of 0.4, whilst the higher bump of 0.35% chord becomes efficient for $C_l > 0.55$. An intelligent bump, which should adapt its shape and position to the external flow conditions, may represent the ideal solution. For the 0.35% c high bump, the drag reductions were about 7 to 9% in the 2D case and 2.5 to 4% in the 3D case at a lift coefficient of about 0.6 - 0.7. At lower lift coefficients, here C_l=0.5, this 2D-bump produced a drag rise of 10% - 11%.

Table 16 ADIF-2D, M=0.757, Re=8x10^6, Cl=0.5

	α	C_l	C_m	C_d	C_{dp}/C_{dw}	ΔCd (%)
w/o bump	-1.159	0.500011	-0.182206	0.008368	0.004485(p)	-
with bump	-1.147000	0.508717	-0.186120	0.009310	0.005433(p)	+11.26
Exp. DLR w/o bump	0.0	0.5259	-	0.0105	-	-
Exp. DLR with bump	0.0	0.4943	-	0.0116	-	+10.48

Table 17 ADIF-3D, M=0.842, Re=9.9x10^6, Cl=0.59

	α	C_l	C_m	C_d	C_{dp}/C_{dw}	
w/o bump	-0.073600	0.590097	-0.133750	0.007998	0.004158(p)	-
with bump	-0.106000	0.590726	-0.135586	0.007657	0.00381940(p)	-4.26

Table 18 ADIF-3D, M=0.842, Re=9.9x10^6, Cl=0.57

	α	C_l	C_m	C_d	C_{dp}/C_{dw}	$\Delta Cd(\%)$
w/o bump	-0.150000	0.572900	-0.132373	0.007674	0.003625(p)	-
with bump	-0.205000	0.564328	-0.133491	0.007488	0.003661(p)	-2.42

14.6 Conclusions

The aim of the present work, performed at CIRA within the EUROSHOCK II project, is to study active control techniques based on ventilation (suction/ blowing) and on contour deformations (bump), and to emphasise their application. The use of these techniques is mainly aimed at improving performance of aerodynamic transonic transport aircraft wings by minimising the increase in drag due to the presence of strong adverse pressure gradients, including shock waves, and to prevent buffet generated by shock-induced separation. Numerical tests have been performed for both, turbulent (RAE-5225) and laminar-type airfoils (DRA-2303), as well as for the ADIF basic airfoil and corresponding sheared wing.

The work conducted by CIRA demonstrates the capability of the numerical model to accurately simulate control effects based on ventilation. The studies performed to improve the ventilation model for the flow through the holes confirmed the equivalence of both transpiration laws tested: Poll's and the University of Karlsruhe's (non-linear) law whose coefficients have been determined by a precise calibration of the porous-plate characteristics. Other preliminary numerical tests revealed the necessity of an accurate calibration of the code by performing slight Mach number corrections or by refining the grid in the control region.

Control techniques based on ventilation used in the control of shock/boundary-layer interaction act not only on the shock wave, reducing its intensity, but also aim to improve boundary-layer characteristics [17]. Thus, the use of localised or distributed suction in shock and boundary layer control increases the resistance of the boundary layer to destabilising factors and limits the increase in the displacement and momentum thicknesses. Because of the minor friction on the porous surface, discrete suction is more efficient than distributed suction in reducing total drag. Discrete blowing generates only a minor total drag reduction compared to discrete suction, but distributed blowing permits the realisation of various control devices ("pneumatic" bumps, riblets, spoilers, etc.). In the present study, the "pneumatic" bump has been numerically tested alone or associated with discrete suction. The hybrid discrete suction/distributed blowing technique optimised with regard to intensity and location seems to be the most efficient technique for drag reduction and buffet alleviation ($\Delta Cd = -12.84\%$).

This new control device solution was inspired by porous transonic plenum walls [18]. A “multi-box” plenum underneath the airfoil surface, communicating with the external flow by means of a “multi-slot” surface, allows the separate control of the internal pressure in each box and thus the intensity of the ventilation (suction/blowing) through each slot. In fact, the device, based only on ventilation, has the advantage of easily simulating discrete (distributed) suction, blowing (“pneumatic”, flexible bumps) and their combinations at locations, intensities, and types of ventilation (constant, parabolic, etc.) optimised for each flight regime.

Another technique numerically tested in the present work is the contour bump. The bump has been investigated on various types of airfoils, laminar (DRA-2303) and turbulent ones (RAE-5225), and on the 2D and the equivalent 3D ADIF-configuration. Its effect has been studied on both, drag reduction and buffet onset. As a single technique, the contour bump seems to be the most efficient in reducing drag as well as in alleviating buffet. Thus, similar drag reductions of about 8.6% for the RAE-5225 and 9% for the DRA-2303 airfoils have been calculated. Higher values of 17% in 2D and 16% in 3D have been obtained with the highest bump for the ADIF airfoil/sheared wing ($H_{bump2D} = 0.60\%c$, $H_{bump3D} = 0.54\%c$). Numerical and experimental tests with the ADIF-wing have shown that the bump works better at higher incidences, i.e., in the presence of stronger shocks: the higher the bump, the higher the lift coefficient at which the bump begins to be efficient in reducing total drag. For lower lift coefficients (here C_l=0.5), the ADIF 2D intermediate bump produced a drag increase of about 10%. Computations and experiments have also shown that the values of the drag reduction calculated in 2D are higher than the 3D ones.

From the present analysis of both, global aerodynamic coefficients and boundary-layer characteristics, calculated for the test cases described, it is concluded that none of the active control techniques succeeds, by itself, to meet the requirements of both, reducing drag and delaying buffet onset. Only the combination of these control techniques, as, for instance, bump + slot suction, becomes here most efficient for an optimised bump geometry and intensity of suction. The corresponding drag reductions are the highest of all control techniques analysed: 9.40% for the RAE-5225 with contour bump + suction, and 12.84% for the DRA-2303 with "pneumatic bump" + suction. In fact, the hybrid control technique takes advantage of both techniques involved: the bump, in its geometric and "pneumatic" variants, represents an isentropic compression ramp if situated upstream of the shock wave, which reduces the shock intensity, while suction reduces the tendency of the flow to separate. In this way, a separate control of the shock wave and of the boundary layer is performed.

From the technological point of view, the control technique based on ventilation has the drawback of needing heavy equipment, which requires careful maintenance in spite of the fact that it has no moving parts. In exchange, the important drag reduction obtained in the case of the bump on the ADIF swept wing and the quite simple technology are suggesting the bump as a valid control device to be integrated into a transport aircraft wing.

References

[1] de Matteis P. and Dima, C., "Numerical Prediction of Shock and Boundary-Layer Control Effects on Airfoils at Transonic Speed", ECCOMAS, Athens, September, 1998.

[2] de Matteis P. and Dima C., "Numerical Investigation of the Passive Shock Control on Transonic Airfoils through an Euler/Boundary-Layer Coupling Technique", In EUROSHOCK-Drag Reduction by Passive Shock Control, Notes on Numerical Fluid Mechanics, Vol. 56, Chap. 12, pp. 151-170, Vieweg & Sohn, 1997.

[3] Dima C., de Matteis P. and de Nicola C., " Investigation of Concepts for the Control of Transonic Shock Wave/Boundary-Layer Interaction", AIDAA, Napoli, October, 1997.

[4] Dima C. and de Matteis P., "Extension of a VII Interaction Method to Airfoil Computations with Passive Control- Preliminary Test Cases", CIRA-TR-95-0002, EUROSHOCK TR AER2-92-49/2.2, 1995.

[5] Poll, D.I.A., Danks and M., Humphreys, B.E., "The Aerodynamic Performance of Laser Drilled Sheets", First European Forum on Laminar Flow Technology, Hamburg, March, 1992.

[6] Bohning, R. and Doerffer, P., "Bohning-Doerffer Law for a Transpiration Flow", U. Karlsruhe, Karlsruhe, May, 1998.

[7] Fulker J.L. and Simmons,M.J., "An investigation of active, suction, shock and boundary layer control techniques," DERA Report MSS4X/CR980817/1.0, EUROSHOCK II TR CT-95-0095/3.1, January 1999 (also see Chapter 19).

[8] Coiro, D., Amato, M. and deMatteis, P., "Numerical Prediction of Transonic Viscous Flows Around Airfoils through an Euler/Boundary-Layer Interaction Method", AIAA Paper 90-1537, June 1990.

[9] Jameson A.,"Numerical Solution of the Euler Equations for Compressible Inviscid Fluids", MAE Report No. 1643.

[10] Chima R. and Turkel E.,"Comparison of Three Explicit Multigrid Methods for the Euler and Navier-Stokes Equations", NASA T.M. 88878

[11] Cohen, C.B. and Reshotko, E., "The Compressible Laminar Boundary Layer with Heat Transfer and Arbitrary Pressure Gradient", NACA No. 1294.

[12] Olling, C.R., " Viscous-Inviscid Interaction in Transonic Separated Flow over Solid and Porous Airfoils and Cascades", Ph.D. Thesis, University of Texas (Austin), 1985.

[13] P.deMatteis and C. Dima., "EUROSHOCK II Task 2 Final Technical Report", CIRA-CR-LAS-99-198 Report, November 1999.

[14] Dima, C. and deMatteis P., "Numerical Investigation of the Boundary-Layer Control through Ventilation Slots in Transonic Airfoil Flows", IUTAM Symposium on Mechanics of Passive and Active Flow Control, Göttingen, 7 - 11 September 1998, Kluwer Academic Publishers, 1999.

[15] Arina, R., Ceresola N. and Piantà P.G., "Application of Unsteady Turbulence Model to the Numerical Prediction of Transonic Buffet", ICAS, Sorrento, September 1996.

[16] Dima, C. "Contribution to the Study of Shock/Boundary-Layer Interaction Control," Invited Lecturer at the "European Science Foundation ESF-AMIF" Symposium, Gdansk, September 1999.

[17] Dima, C. and deMatteis, P. "Effects of Shock and Boundary-Layer Control Techniques on Transonic Flows about Airfoils", AIAA Aerospace Sciences Meeting & Exhibit, Reno/NV, January 2000.

[18] Dima, C. and deMatteis, P. "A New Concept for the Adaptive Shock and Boundary-Layer Control on Transonic Flows about Airfoils ", ICAS, Harrogate, 2000.

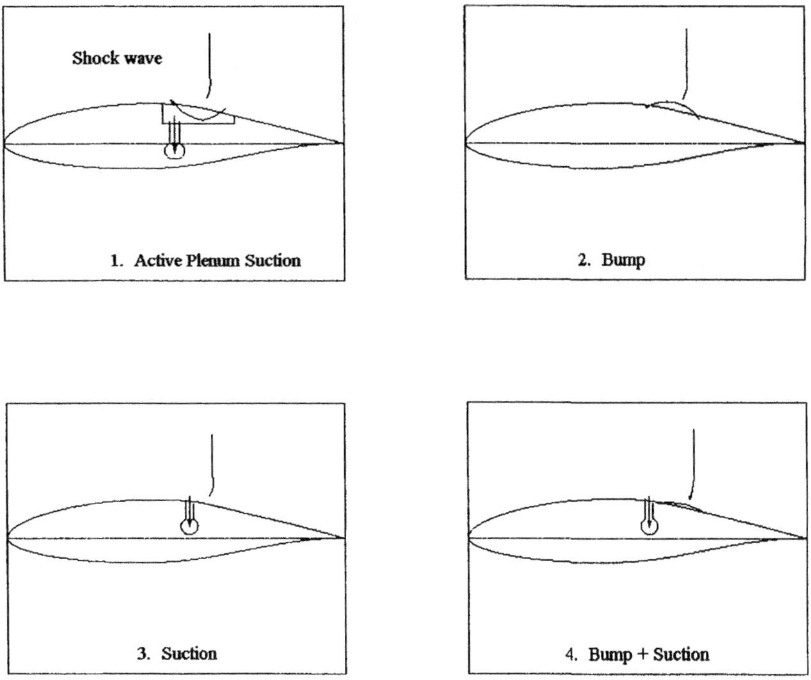

Figure 1 Control techniques for shock/boundary-layer interaction

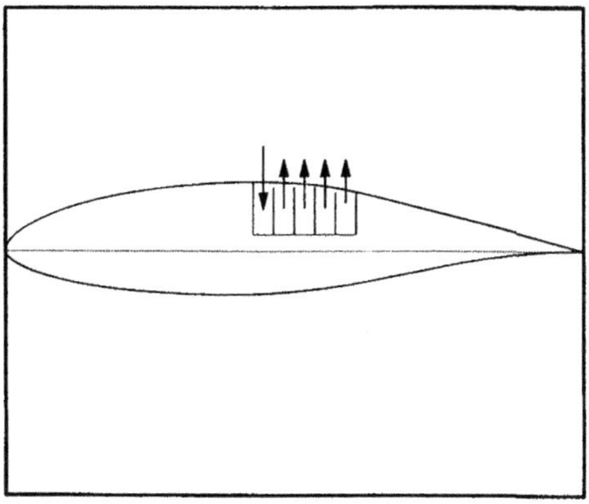

Figure 2 Sketch of the adaptive “multi-box” plenum device

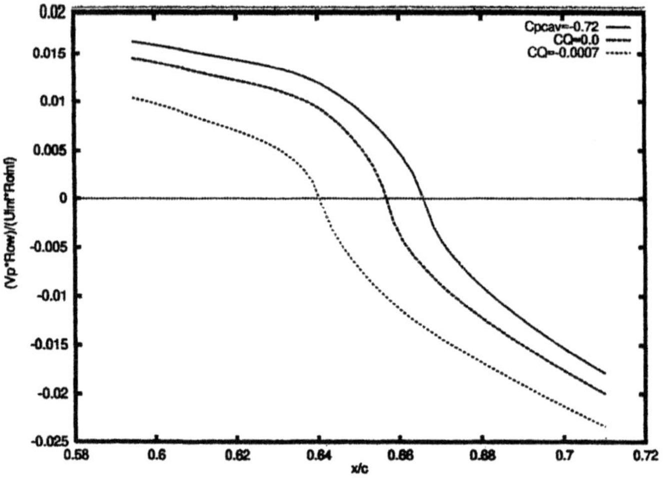

Figure 3 Non-dimensional mass flux

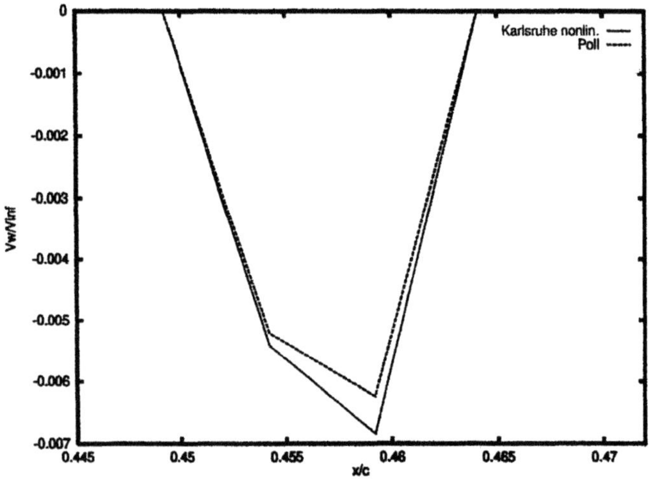

Figure 4 Control-law effect on the transpiration velocity, DRA-2303 NLF airfoil, M=0.6793, Re=19x10^6, C_N=0.60, slot-suction $C_Q = -0.00007, por = 2.48\%$

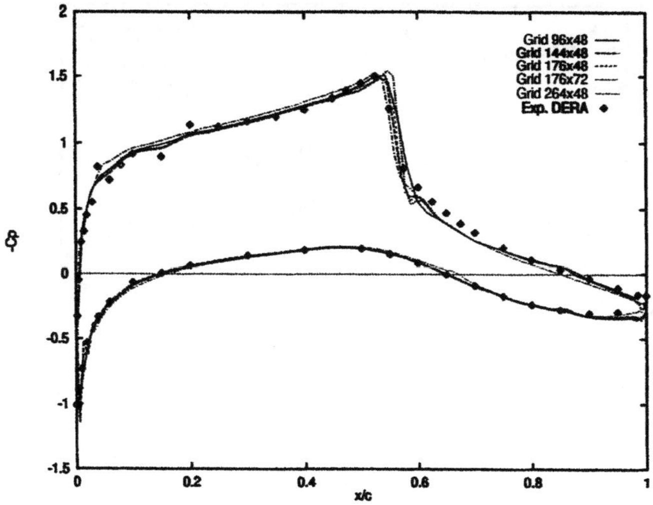

Figure 5 Grid influence on the pressure distribution on the DRA-2303 NLF airfoil

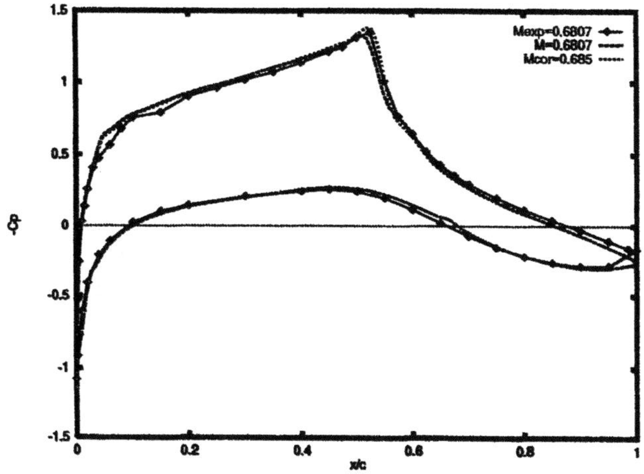

Figure 6 Mach number correction for the DRA-2303 airfoil, Re=18.89x10^6, C_N=0.6075

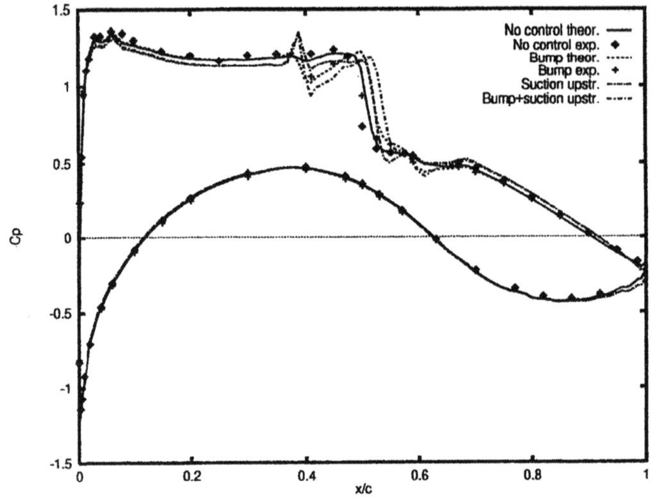

Figure 7 Active control influence for the RAE-5225 airfoil, M=0.73, Re=19x10^6, C_N=0.756

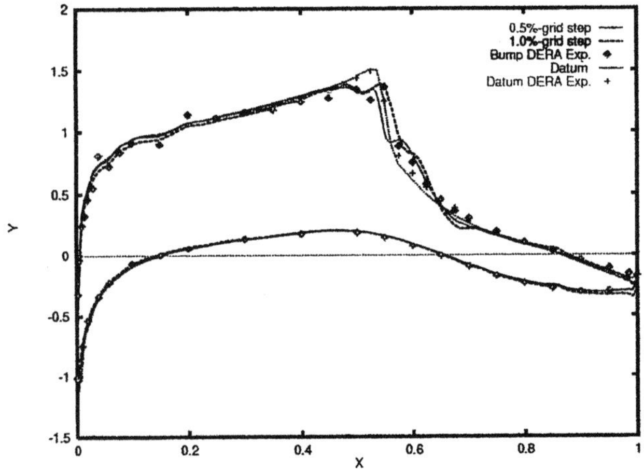

Figure 8 DRA-2303 airfoil with bump, M=0.68, Re=19x10^6, C_N=0.74

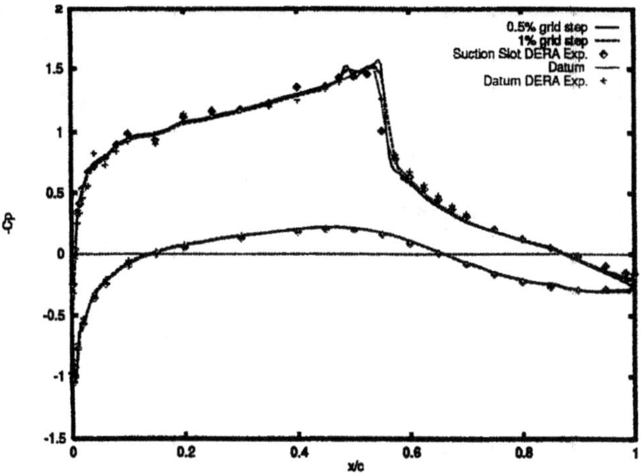

Figure 9 DRA-2303 airfoil with suction, M=0.68, Re=19x10^6, C_N=0.74

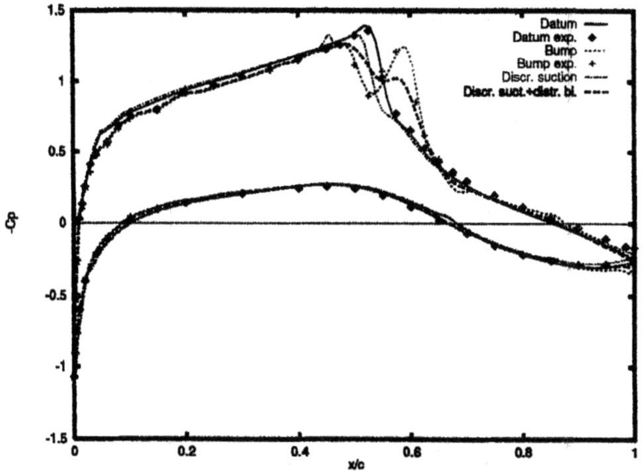

Figure 10 Influence of active control on the DRA-2303 airfoil flow
M=0.685, Re=19x10^6, C_N=0.6

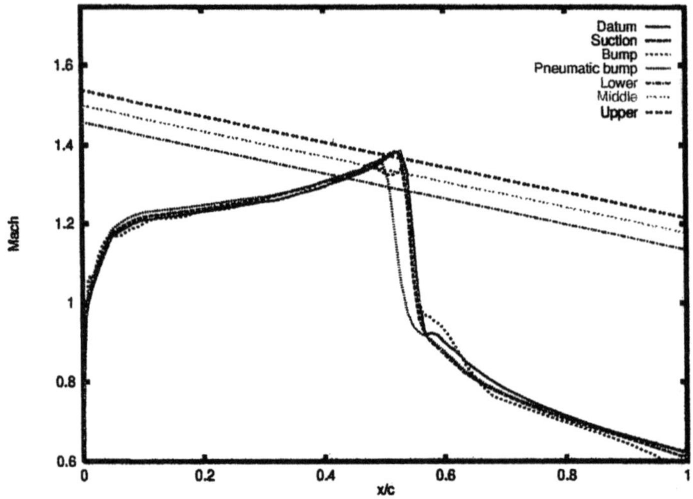

Figure 11 Buffet onset for the DRA-2303 airfoil, M=0.67, Re=19x10^6, C_N=1.05

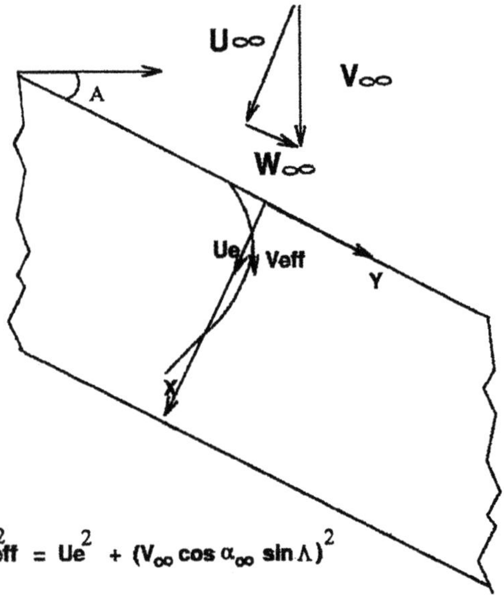

Figure 12 Quasi-3D swept wing flow

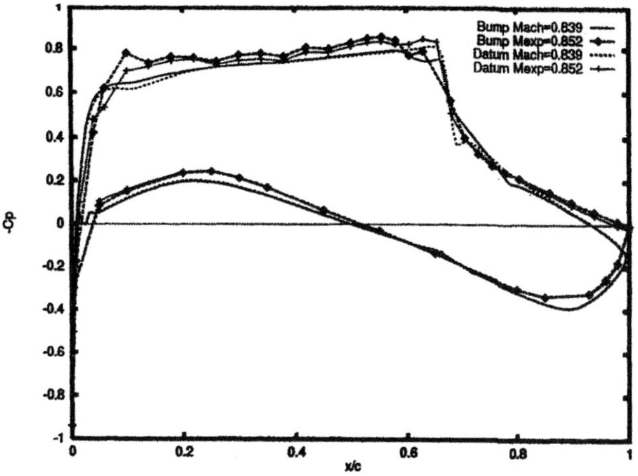

Figure 13 ADIF-3D sheared wing with and w/o bump

$$H_{bump} = 0.1573\%c, \mathrm{Re} = 6.7x10^6, C_l = 0.574$$

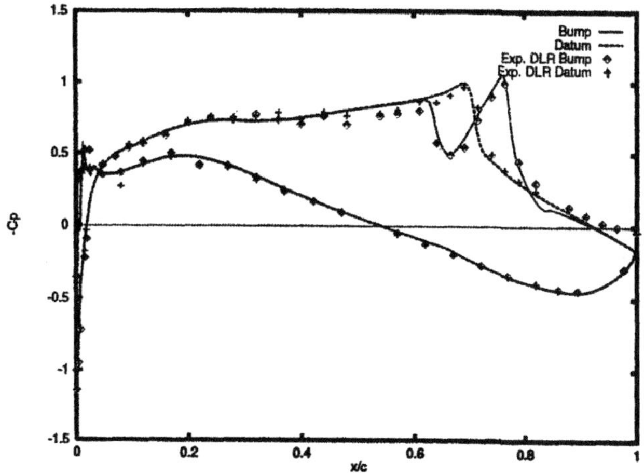

Figure 14 ADIF-2D with and w/o bump,

$$H_{bump} = 0.35\%c, M = 0.757, \mathrm{Re} = 8x10^6, C_l = 0.50$$

15 Euler/Boundary Layer Coupling to Predict Steady and Unsteady Transonic Flows Past an Airfoil with and without Shock Control

Carlo de NICOLA, Vincenzo CIRINO
Aeronautical Design Department - Piazzale Tecchio 80, 80125 Naples, Italy

Summary

A numerical method based on the Euler/boundary layer coupling has been extended to unsteady flows to investigate the control of shock wave/boundary layer interaction in transonic flow for the reduction of both, drag and buffet effects. Applications concerned the effectiveness of suction and of a bump at Mach numbers prior to and beyond buffet onset. Good agreement with experiments has been found for the pressure distributions, while the prediction of buffet onset is consistent with other methods, but quantitative differences have been found. Results clearly indicate that control devices can be helpful in reducing drag and in delaying buffet onset to higher lift coefficients or Mach numbers.

15.1 Introduction

Two main objectives have been pursued to set up a numerical tool for the calculation of transonic flows and the prediction of buffet onset in the presence of control devices for shock wave / boundary layer interaction:

- The implementation of the suction law defined by the University of Karlsruhe;
- The development of a "quasi-unsteady" coupling method starting from the steady method through the elimination of the existing Euler convergence accelerators, while the (steady) boundary layer method has not been changed.

Computations concerned all mandatory and optional test cases defined during the present research. Comparisons with Poll's law have been performed for the new transpiration law of the University of Karlsruhe.

In the next sections a short discussion will follow regarding the improvements to the numerical model, the transpiration laws that have been implemented, and the semi-empirical buffet criteria that have been used for the evaluation of buffet onset in the numerical calculations. Main results will then be presented concerning the prediction of the effect of control on drag reduction and buffet onset; an evaluation of the computing time (for a steady test case only) will be presented.

15.2 Numerical Model

The inviscid flow field is computed by solving the Euler equation on a structured grid. The algorithm is the well known and robust scheme developed by Schmidt, Jameson and Turkel [1]. The method is explicit in time, and uses a central space discretization with explicit adaptive artificial dissipation. The artificial viscosity is constructed by blending a 2nd and 4th order contribution. The 4th order background viscosity is necessary to provide stability to the scheme and avoid odd-even decoupling; a 2nd order dissipation is introduced at discontinuities by a shock sensor based on the pressure jump to locally improve the accuracy. This method is consistent in time.

Steady state calculations are performed by using different convergence accelerators: enthalpy damping, implicit smoothing, multi-grid, multilevel, residual averaging, local time stepping, local grid refinement; the use of these techniques leads to a time inconsistency of the Euler solver. The (integral) boundary layer code is based on the Cohen-Reshotko method (Stewartson transformation and Thwaites method) for the laminar part [2], and a modified Green method for the turbulent part [3].

The coupling procedure has been developed following Carter [4].

The unsteady equation of the compressible boundary layer has been solved, and the solver has been validated; when a coupling with the unsteady Euler equations was implemented, the resulting scheme showed poor quality of stability and convergence, i.e., it was unable to find an asymptotic steady state that the pseudo-unsteady scheme correctly predicts. Consequently, a quasi-unsteady flow solver has been obtained by coupling the time-consistent Euler solver to the steady boundary layer.

15.3 Transpiration Models

Two suction laws have been used for the computation of the ventilation velocity distributions.

15.3.1 Poll's law

The first control law, already implemented within the context of the EUROSHOCK (I) project, is based on Poll's formula for laser-drilled sheets [5]. This law is based on the assumptions of 1) small interference between holes and 2) laminar and incompressible pipe flow.

Poll developed a flow model, resulting in a quadratic relationship between the pressure drop across the porous plate ΔP and the mass flow rate through a hole $\dot{m} = \rho_h v_h \pi$ d^2/4: ρ is the density, v is the (mean) velocity in the hole and d is the hole diameter, while suffix "h" refers to the hole conditions. Defining

$$Y = \frac{\Delta p}{\rho}\left(\frac{d}{v}\right)^2\left(\frac{d}{t}\right)^2$$

and

$$X = \frac{\dot{m}}{\mu t}$$

the relationship that has been implemented is

$$Y = \frac{1}{K}\left[40.7X + 1.95X^2\right]$$

where v and μ are the dynamic and kinematic viscosity coefficients, t is the thickness of the porous plate, and K is a correction factor, related to the hole conicalness.

15.3.2 Bohning/Doerffer (B/D) law

The transpiration law defined at the University of Karlsruhe [6] takes into account the effect of the tangential stream along a porous plate. The B/D law has been implemented in the form

$$\frac{\Delta P}{P_w} = M_h^{\frac{1}{0.55}}\left[\left(\frac{1}{1.2}\right)^{\frac{1}{0.55}} + b\left(cM_h\right)^{\frac{1}{a}}\right]$$

with

$$c = \frac{2\tau_w}{\rho_h v_h^2};\ \text{a} = 1.52;\ \text{b} = 25$$

The suffix "w" is referred to the 'wall' conditions, and τ is the shear stress.

15.4 Buffet-onset Criteria

In order to find buffet onset three criteria have been analyzed, namely, C1 (Mach number limit), C2 (trailing edge pressure divergence) and C3 (numerical evaluation).

15.4.1 Mach-number-limit Criterion (C1)

Statistical correlations of experimental data have shown that an airfoil reaches buffet onset conditions when the Mach number distribution immediately upstream of the shock is tangent to the straight line(s) defined by

$$M_{limit} = 1.496245 - 0.322996\ x/c \pm 0.041077$$

The uncertainty in the prediction of buffet onset is contained in the third term of this equation. Buffet will not occur for conditions giving a Mach number distribution below the lower straight line, while for conditions giving a Mach number distribution above the highest one, buffet will be fully developed. The intermediate straight line

$$M_{limit} = 1.496245 - 0.322996\ x/c$$

gives an estimate of buffet onset.

Referring to the tangency of the lower, the intermediate and the higher straight lines, the respective three criteria will be termed Cl(I), Cl(II), Cl(III).

15.4.2 Trailing edge pressure divergence (C2)

This method gives an evaluation of buffet onset based on the consideration that for values of the lift coefficient, C_L, far from the buffet condition, the pressure coefficient at the trailing edge, C_p(t.e.), is nearly independent of lift, but begins to decrease close to buffet-onset conditions. The criterion assumes that buffet onset occurs when the value of C_p(t.e.) decreases with angle of attack (lift) at a rate of $\Delta Cp = -0.04$.

15.4.3 Numerical evaluation (C3)

The code, designed to calculate the boundary layer effects by steady viscous inviscid coupling, has been modified, starting from a pseudo-unsteady form, to a quasi time dependent version. In order to assure the time consistency, all convergence acceleration techniques have been eliminated, but the steady boundary layer method has not been changed.

Starting from a steady solution, very small subsequent increases of the angle of attack are usually needed in order to properly estimate buffet onset, i.e., the onset of unsteadiness.

15.5 Results

In this section, the main results of the research activity are summarized. In treating control phenomena, this section is comprised as follows:

1. Active and passive control on the VA-2 airfoil.
2. Control by slot suction on the DRA-2303 airfoil.
3. Control by a bump on the RAE-5225 and the DRA-2303 airfoils.
4. Hybrid control (bump + suction) on the RAE-5225 and DRA-2303 airfoils.
5. Buffet onset prediction and bump effects.
6. Effect of slot suction on buffet onset.
7. Code evaluation.

15.5.1 Control by active and passive cavity ventilation (VA-2 airfoil)

Two test cases (with and without active control), with a 0.5% chord step size in the cavity region, have been computed. Mach number and angle of attack corrections have been applied in order to match the shock position of the experiments. The numerical test cases have been performed using an O-type grid (320x80) coupled to the "Local Grid Refinement". To account for control, Poll's law has been used; the porosity of the plate covering the cavity was 8%, and a suction coefficient in the case of active control of C_Q = -0.00088 has been assumed.

Furthermore, a test case with passive control at the same lift coefficient, C_L, as in the experimental case, again using a porosity of 8%, has been computed.

Figure 1 shows the local suction coefficient (C_q) distributions for active and passive control; in Table 1, a summary of the results in terms of global coefficients is reported. A drag reduction of about 12.6% was determined for active as well as passive control.

Table 1 Results for the VA-2 airfoil with and w/o control

Control	M_∞	Rex10^{-6}	α	C_L	C_m	C_D	C_{Dp}
No	0.7365	2.5	2.25	0.8146	-0.1157	0.01662	0.01225
Active	0.7350	2.5	1.90	0.8341	-0.1280	0.01452	0.0098
Passive	0.7350	2.5	2.22	0.7831	-0.1091	0.01447	0.01020

15.5.2 Control by slot suction (DRA-2303 airfoil)

Besides Poll's formula, the control law defined by Bohning and Doerffer, taking into account the incoming external flow, has been implemented for further improvements of the code. For the calculations an O-type grid of 267x48 points has been used. Furthermore, the wake effect has been included up to a distance of 30% of the chord.

Two mandatory test cases have been performed for the DRA-2303 airfoil, corresponding to the data points DP 2017 (datum airfoil) and DP 2129 (control by slot suction). The test conditions are indicated in Table 2.

Table 2 Test conditions for the airfoil DRA-2303 w/o and with control by slot suction

Test	M_∞	Rex10^{-6}	C_L	Configuration
DP 2017	0.6807	18.89	0.6075	Datum airfoil
DP 2129	0.6973	18.56	0.5953	Slot suction

In the case of suction, DP 2129, the use of the nominal (experimental) porosity of 4% for the perforation covering the slot led to an inconsistency between the results obtained by imposing the cavity pressure, C_{Pcav}, or the total mass flow, C_Q; a lower porosity (2.48%) eliminated this problem and subsequent computations were performed using this porosity.

In Table 3, a comparison between experimental data and calculated results is given for the datum airfoil and the airfoil with discrete suction upstream of the shock. In the case of the latter, the Bohning – Doerffer (B/D) as well as Poll's law were employed. Both, mass flow rate, here $C_Q = 0.00007$, Case (a) in Table 3, and the cavity pressure, here $C_{Pcav} = -1.2992$, Case (b) in Table 3, respectively, were prescribed. In all instances a drag reduction is predicted due to slot suction, in tendency in agreement with experiment, although absolute levels differ. There is only a minor difference dependent on the control law employed and whether cavity pressure or suction rate is prescribed.

In Figure 2, experimental and numerical pressure distributions for the case with slot suction — and a prescribed suction coefficient in the case of the computations — are presented. The agreement between experiment and computation is quite good.

Table 3 Numerical and experimental results for the DRA-2303 airfoil w/o and with control by slot suction

	Datum DP. 2017	Suction (C_Q given) DP. 2129 (a)	Suction (C_{Pcav} given) DP. 2129 (b)
C_L (exp.)	0.6075	0.5953	0.5953
C_L (B/D)	0.6081	0.5962	0.5962
C_L (Poll)	0.6081	0.5940	0.5973
C_{Dtot} (exp.)	0.009730	0.009000	0.009000
C_{Dtot} (B/D)	0.010198	0.009719	0.009758
C_{Dtot} (Poll)	0.010198	0.009829	0.009753
C_Q (exp.)		0.00007	0.00007
C_Q (B/D)		0.00007	0.0000534
C_Q (Poll)		0.00007	0.0000414
C_{Pcav} (exp.)		-1.2992	-1.2992
C_{Pcav} (B/D)		-1.3367	-1.2992
C_{Pcav} (Poll)		-1.3837	-1.2992
C_D (B/D)		0.009522	0.009586
C_D (Poll)		0.009629	0.009596
C_{Dsuc} (B/D)		0.000094	0.000069
C_{Dsuc} (Poll)		0.000097	0.000054
C_{Desc} (B/D)		0.000103	0.000103
C_{Desc} (Poll)		0.000103	0.000103

Definitions: C_{Dsuc} = Pump drag; C_{Desc} = Excrescence drag;
$C_D = C_{Dtot} - C_{Dsuc} - C_{Desc}$

15.5.3 Control by contour bump (RAE-5225 and DRA-2303 airfoils)

Computations have been performed for the laminar-type (NLF) DRA-2303 airfoil and for the RAE-5225 turbulent airfoil, respectively, with and without bump at fixed (experimental) lift coefficients. An O-type grid with 184x32 points has been used and the wake effect has been taken into account in the calculations. The grid spacing in the bump region was 1% chord. At the conditions considered, i.e., about $M_\infty = 0.68$, $C_L = 0.74$ for the DRA-2303 airfoil, and $M_\infty = 0.7255$, $C_L = 0.756$ for the RAE-5225 airfoil, reductions in total drag due to the bump of $\Delta C_D = -18.9\%$ for the NLF airfoil and -9.8% for the RAE-5225 airfoil have been determined. Figure 3 shows, as example, the excellent agreement of the calculated pressure coefficient with the experimental data for the RAE airfoil with bump at the above condition.

In order to obtain an improved grid spacing of $\Delta(x/c) = 0.005$ in the bump region, an O-type grid of 312x80 grid points, coupled to the Local Grid Refinement, was introduced for the NLF airfoil, Figure 4. In Figure 5, a comparison between the computed pressure distributions with and without bump is depicted, while in Figure 6, a comparison between calculated and experimental pressure distributions for the datum airfoil is presented. The former clearly shows the spreading of the shock due to the bump, indicative of a wave drag reduction.

In Figure 7, the drag polars calculated with the steady code with and without bump are presented. There are two features to be recognized, one related to the grid resolution, the other to the bump characteristic: the drag reduction due to the bump at a lift coefficient of $C_L = 0.74$ is here, i.e., at a grid spacing of 0.5% chord, only about 11% — as compared to the 18% in case of the 1% chord spacing — and the bump is essentially only effective at the higher lift coefficients, increasing or hardly affecting drag at $C_L < 0.5$.

15.5.4 Hybrid control by bump and upstream suction (RAE-5225 and DRA-2303 airfoils)

Localized suction upstream of the shock in combination with a bump in the shock region has also been investigated for the two airfoils considered here at freestream conditions similar to the ones of the preceding chapter, i.e., about $M_\infty = 0.68$, $C_L = 0.74$ for the DRA-2303 airfoil, and $M_\infty = 0.7255$, $C_L = 0.756$ for the RAE-5225 airfoil. The suction coefficients were $C_Q = -0.0001$ for the DRA-2303 and $C_Q = -0.0002$ for the RAE-5225 airfoil, respectively.

A systematic reduction of the drag coefficient has been found up to an optimum rate of suction. At this condition, drag reductions of $\Delta C_D = -7.3\%$ and –12.1% for pure suction and suction plus bump, respectively, were determined for the DRA-2303 airfoil and $\Delta C_D = -5.2\%$ and -11%, respectively, for the turbulent RAE-5225 airfoil. Relevant computations have been performed by using Poll's law for the calculation of the transpiration velocity.

15.5.5 Buffet-onset prediction and bump effect

Computations have been performed for the airfoil DRA-2303 with and without bump control by using an O-type grid of 264x48 points, an adequate step size along the bump, and a constant Mach number of $M_\infty = 0.68$ at a Reynolds number of $Re = 19 \times 10^6$. The wake effect has again been included up to a distance of 30% of the chord. The three criteria C1, C2 and C3 have been used to evaluate the effect of the bump. The results are summarized in Table 4 below.

Table 4 Bump effect on the lift coefficient at buffet onset

Criterion	C_L — Bump off	C_L — Bump on
C1 (I)	0.805	0.861
C1 (II)	0.920	0.998
C1 (III)	0.996	1.070
C2	0.839	0.869
C3	0.962	0.990

Benefits related to the alleviation of buffet onset due to the presence of the bump are indicated in the table. Generally, the lift coefficients corresponding to the three criteria employed are raised due to the bump. Some further comments on the application of the three criteria related to buffet are given below.

*1. **Mach number limit:*** C1(I), C1(II) and C1(III) are related to the first, second, and third straight Mach number line as described in Chapter 15.4. Figures 8 and 9 show the Mach number distributions, with and without bump control, respectively, at angles of attack that provide tangency to these lines. The corresponding friction-coefficient distributions are plotted in Figure 10 for the case with bump. It should be noted that at the lower lift coefficient separation has not been predicted by the present code, while at the next higher lift coefficient shock-induced separation is incipient. At the highest lift coefficient, a small separation bubble is visible with trailing-edge separation being incipient. Buffet does not yet occur (due to bump control). These developments show that the present (empirical) criterion is no longer applicable in the presence of flow control.

*2. **Trailing-edge, $\Delta C_p(t.e.)$:*** It can be seen that the results obtained with the C1(I) criterion are being confirmed by using the present criterion, i.e., $\Delta C_p(\text{t.e.}) = -0.040$. In Figure 11 the lift-dependence of the trailing edge pressure coefficient, whose slope can be used as an indication of the buffet-onset conditions, is demonstrated.

*3. **Numerical evaluation:*** The unsteady version of the code has been used. Starting from a steady solution close to buffet onset at the specified Mach number ($M_\infty = 0.68$), small increases of the angle of attack are needed in order to estimate buffet onset: typically, during the present investigation, $\Delta\alpha = 0.10^{o}$. Figure 12 shows the lift coefficient C_L versus the non-dimensional time T for three values of the angle of attack: the present code clearly indicates the occurrence of buffet at $\alpha = 3.6^{o}$. At these conditions, the reduced frequency of the oscillations of the lift coefficient is k = 0.56, close to the experimental value of k = 0.49 given by DERA. In Figures 13 and 14, the pressure and skin-friction coefficient distributions, respectively, at two different values of the non-dimensional time T (corresponding to the minimum and maximum of the lift oscillation) are plotted. The oscillation of the shock and the change in the size of the separated region are indicated. The change in lift coefficient is $(\Delta C_L)_{max} = 0.015$, a rather small value, indicating that we are at the onset of buffet. It should be noted that numerical difficulties arose when the angle of attack was further increased, so the application of the method is here actually limited to the detection of buffet onset only. However, for most applications the prediction of the flow field and of the aerodynamic characteristics of an airfoil at fully developed buffet seems less important than an accurate prediction of the buffet barrier.

In concluding it should be noted that the numerical agreement of C3 with C1(1) and C2 is poor which is, of course, not surprising since the latter are empirical criteria possibly indicating conditions prior to buffet onset. Nevertheless, all criteria show a delay of buffet onset in the presence of the bump.

15.5.6 Effect of slot suction on buffet onset

Computations related to the numerical prediction of the effect of slot suction, and especially of the slot position, on buffet onset have been carried out. Computations were performed for the airfoil DRA-2303 with and without slot suction by using an O-type Grid of size 264x48, an adequate step size ($\Delta x/c$ = 0.005) over the suction slot, a constant Mach number (M_∞ = 0.68), and a constant lift coefficient (C_L = 0.80), again considering the wake effect up to 30% of the chord. Two values of the suction coefficient (C_Q =-0.00007 and -0.00014) and six positions of the slot have been analyzed by using the C1 criterion. Table 5 shows the maximum value of the upper-surface Mach number distribution (M_{max}) and the total drag coefficient, including excrescence and pump drag. The following values relate to the clean airfoil: $C_L = 0.80$, $M_{max} = 1.281$, $C_D = 0.01597$.

Buffet onset seems be independent of the slot position. Only in some cases can a small increase in M_{max} be observed. Locating the slot upstream of the shock wave reduces the drag coefficient by a few counts: for C_Q = -0.00014 and $(x/c)_{Slot}$= 0.40 and for C_Q = -0.00007 and $(x/c)_{Slot}$ = 0.50, this reduction is equal to 4 and 3.4 drag counts (ΔC_D = 0.0004 and 0.00034), respectively, i.e., about 2.4%.

Table 5 Effect of slot suction on buffet onset and drag, DRA-2303 airfoil, $M_\infty = 0.68$, $C_L = 0.80$

Slot Position	C_Q = -0.00007		C_Q = -0.00014	
x/c	M_{max}	C_D	M_{max}	C_D
No slot (Datum)	1.281	0.01597	1.281	0.01597
0.40	1.283	0.01594	1.291	0.01558
0.45	1.280	0.01588	1.284	0.01564
0.50	1.282	0.01563	1.293	0.01573
0.60	1.281	0.01617	1.291	0.01651
0.70	1.282	0.01636	1.282	0.01625
0.80	1.283	0.01715	1.283	0.01704

15.5.7 Code evaluation

For a code evaluation, five test cases have been evaluated, all related to the DRA-2303 airfoil, covering the datum airfoil and the control concepts investigated. Details of the specified conditions and the results are summarized below and in Table 6.

- Computer: PENTIUM PRO
- Clock rate: 200 MHz
- All computations made with single precision
- O-type grid of size 264x48 with the number of points on the body being N_B = 264 and the number of points in the normal direction being N_N = 48.
- Convergence eriterion: $\Delta C_L < 10^{-4}$

Table 6 Computer-code evaluation in terms of CPU time per number of points

Test Cases DRA-2303 with	Time steps	CPU [sec]	CPU/NB	CPU/(N_BxN_N) x 100
1 – No control	630	432	1.64	3.42
2 – Bump	612	420	1.59	3.31
3 - Slot suction	570	405	1.53	3.19
4 – Active cavity	492	336	1.27	2.65
5 – Passive cavity	789	532	2.01	4.19

References

[l] Jameson, A., Schmidt, W. and Turkel, E. "Numerical Solution of the Euler Equation by Finite Volume Methods Using Runge-Kutta Time-Stepping Schemes", AIAA Paper 81-1259, 1981.

[2] Stewartson, K. "Correlated Incompressible and Compressible Boundary Layers", Proc. Roy. Soc., London, Ser. A, Vol. 200, pp. 213-228, 1949.

[3] Green, J. E., Weeks, P. J. and Brooman, J. W. F. "Prediction of Turbulent Boundary Layers and Wakes in Compressible Flow by a Lag Entrainment Method", A.R.C. R&M 3791, 1977.

[4] Carter, J. E. "A New Boundary Layer Inviscid Interaction Technique for Separated Flows", AIAA Paper 79-1450, 1979.

[5] Poll, D., Danks, M. and Humphreys, B. "The Aerodynamic Performance of Laser Drilled Sheets", First European Forum on Laminar Flow Technology, Hamburg, Germany, March 1992, Paper 92-02-02.

[6] Bohning, R. and Doerffer, P. "Effect of Tangential Flow Along the Porous Plate on the Transpiration Flow", in preparation (Also see Chapter 12).

[7] Bidello, M. "Optimization of Methods for Viscous Flows for the Development of a New Airfoil Technology", Master's Thesis, University of Naples, 1993

[8] Lazzaro, M. "On the Control of the Shock Wave - Boundary Layer Interaction", Master's Thesis, University of Naples, 1992.

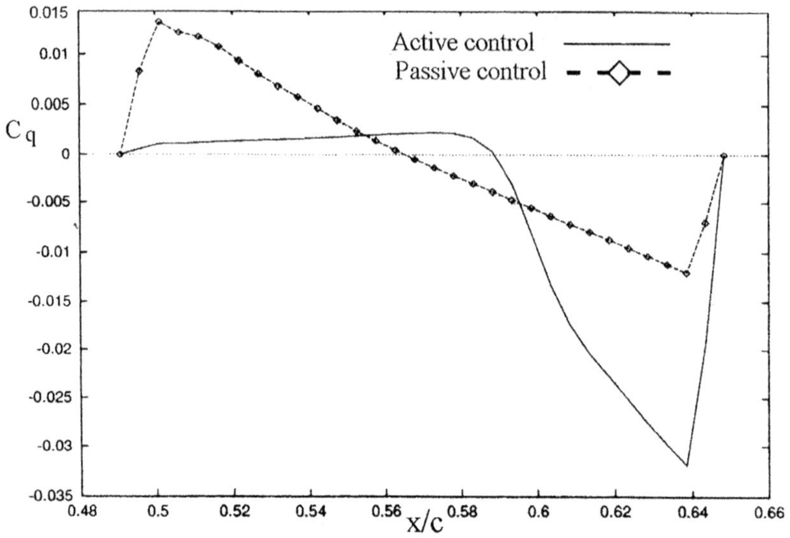

Figure 1 Local cavity suction distributions for active (Cq = -0.00088) and passive control; Airfoil VA-2, $M_\infty = 0.735$, Re = 2.50×10^6, $\alpha = 1.90^\circ$ (active) and 2.22° (passive)

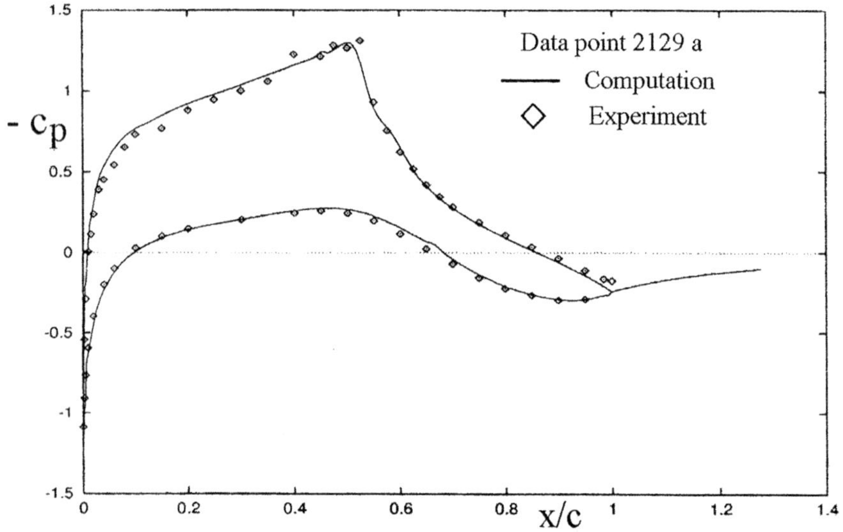

Figure 2 Numerical and experimental pressure distributions; Airfoil DRA-2303, $M_\infty = 0.6973$, Re = 18.56×10^6, $C_L = 0.596$, $C_Q = 0.00007$ (DP 2129a)

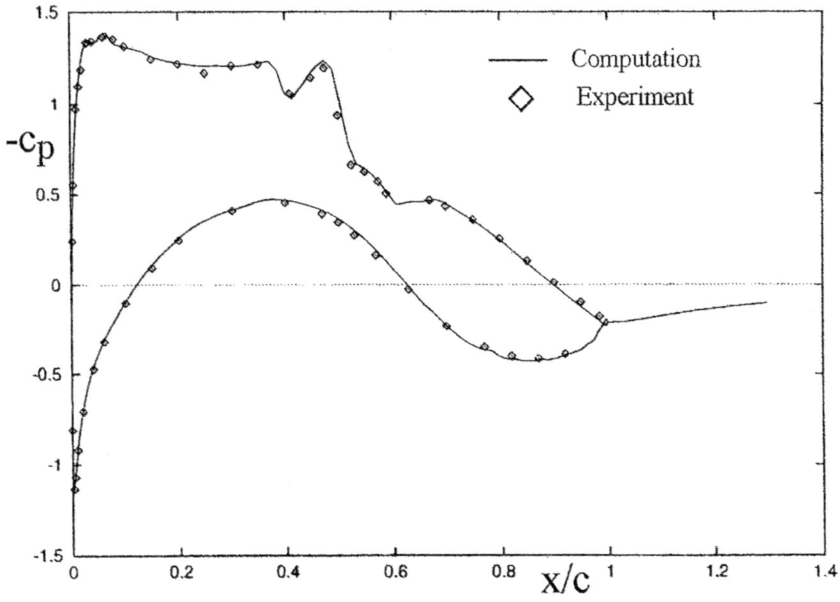

Figure 3 Airfoil RAE-5225 with bump: comparison between experimental and calculated pressure distributions, $M_\infty = 0.7255$, $Re = 18.8x10^6$, $C_L = 0.756$

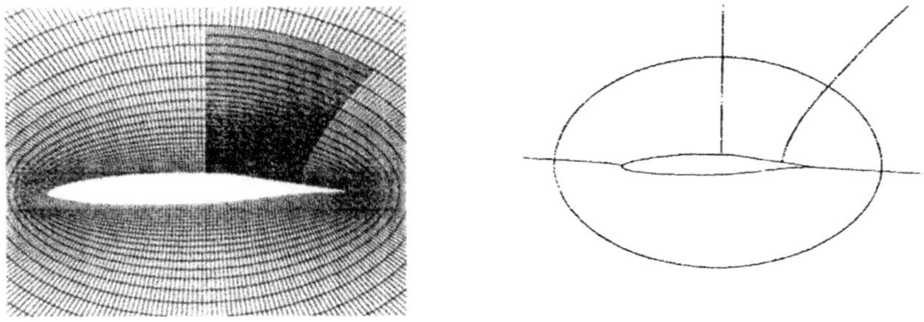

Figure 4 DRA-2303 airfoil: 0-type grid of size 312x80 and the topology related to the "Local Grid Refinement"

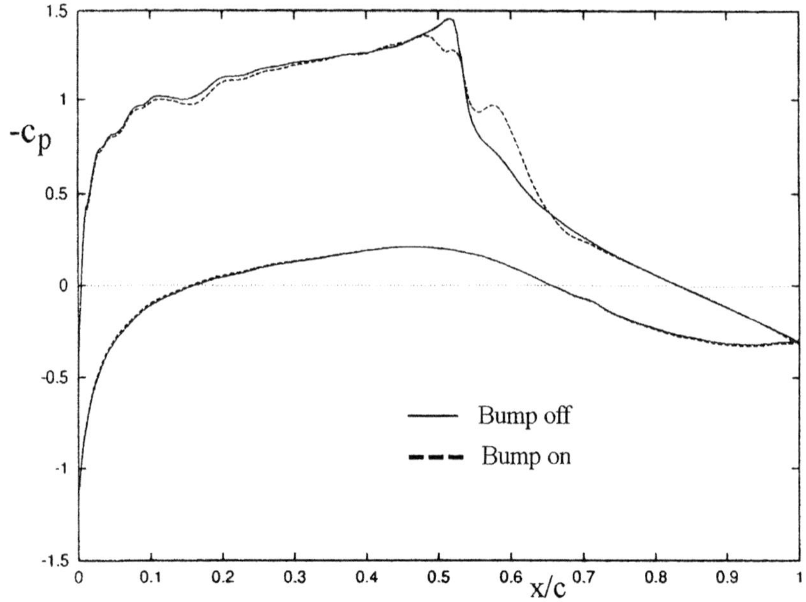

Figure 5 DRA-2303 airfoil: computed pressure distributions with and without bump, $M_\infty = 0.68$, $Re = 18.9 \times 10^6$, $C_L = 0.74$

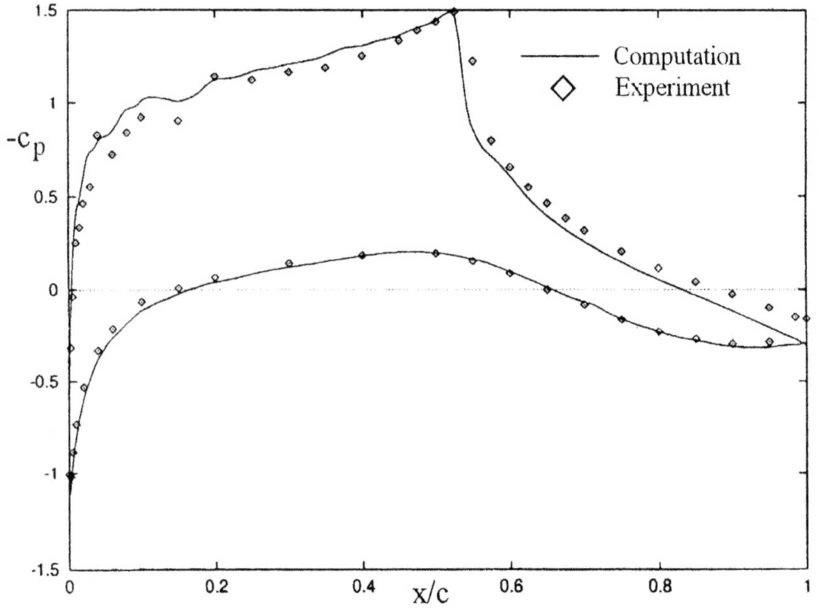

Figure 6 DRA-2303 airfoil: experimental and computed pressure distributions for the datum airfoil, $M_\infty = 0.681$, $Re = 18.9 \times 10^6$, $C_L = 0.7467$

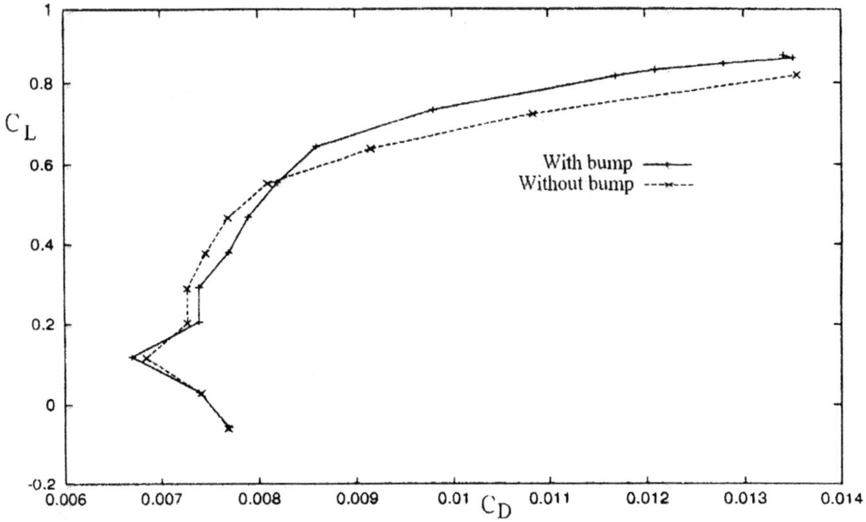

Figure 7 DRA-2303 airfoil: drag polars with and without bump, $M_\infty = 0.68$, $Re = 18.9 \times 10^6$

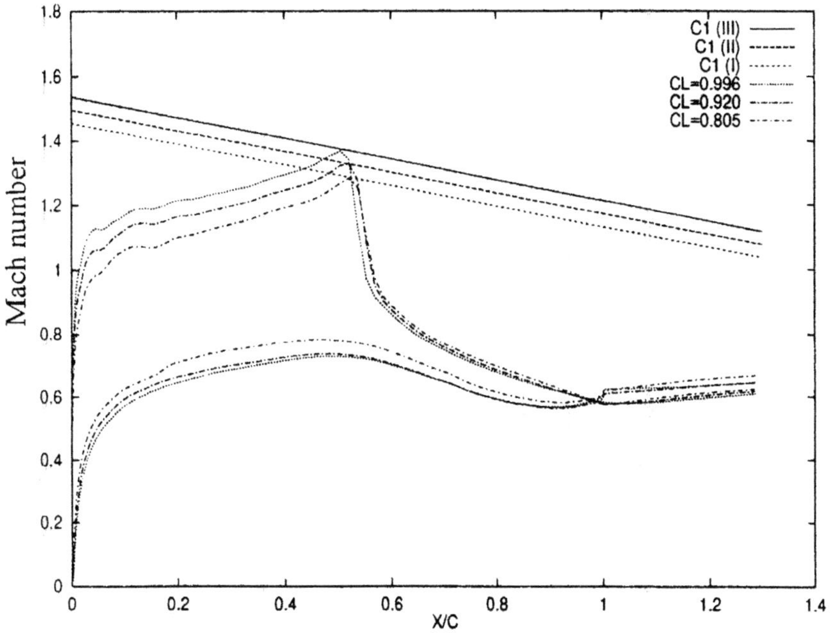

Figure 8 DRA-2303 airfoil without bump: local Mach number vs. x/c with application of the C1 criterion, $M_\infty = 0.68$, $Re = 18.9 \times 10^6$

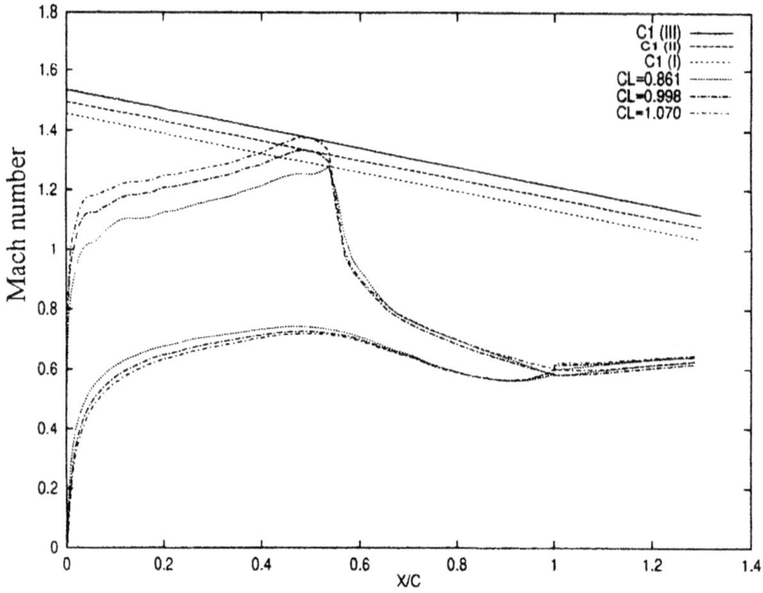

Figure 9 DRA-2303 airfoil with bump: local Mach number vs. x/c with application of the C1 criterion, $M_\infty = 0.68$, Re = 18.9×10^6

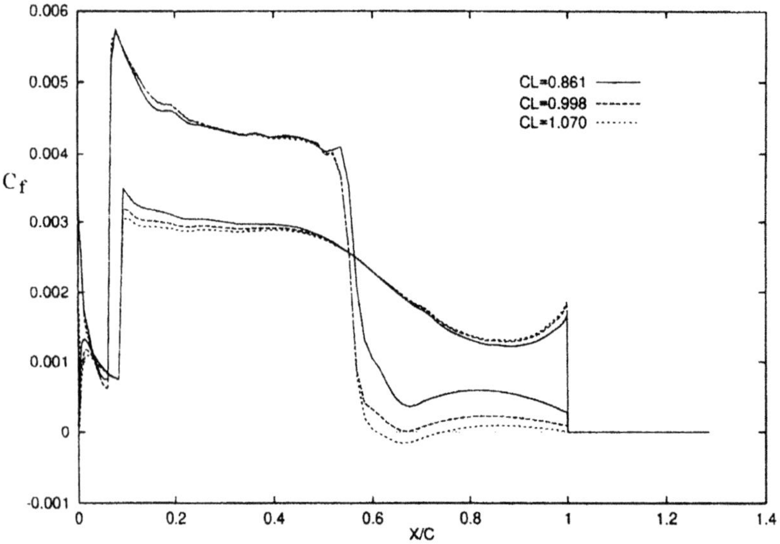

Figure 10 DRA-2303 airfoil with bump: local skin friction C_f vs. x/c corresponding to Figure 9, $M_\infty = 0.68$, Re = 18.9×10^6

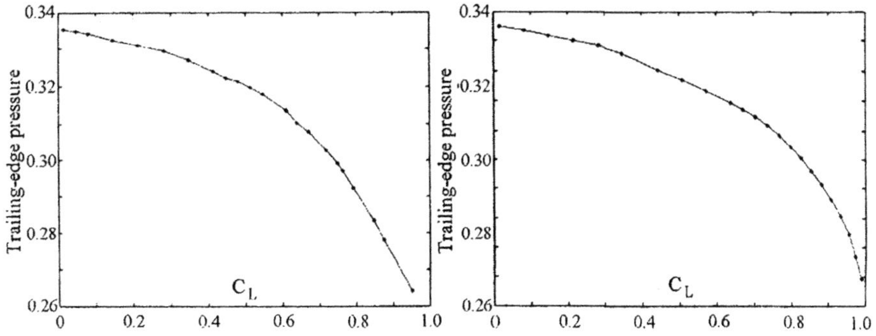

Figure 11 DRA-2303 airfoil without (left) and with (right) bump: trailing edge pressure vs. C_L and application of the C2 criterion, $M_\infty = 0.68$, Re = 18.9×10^6

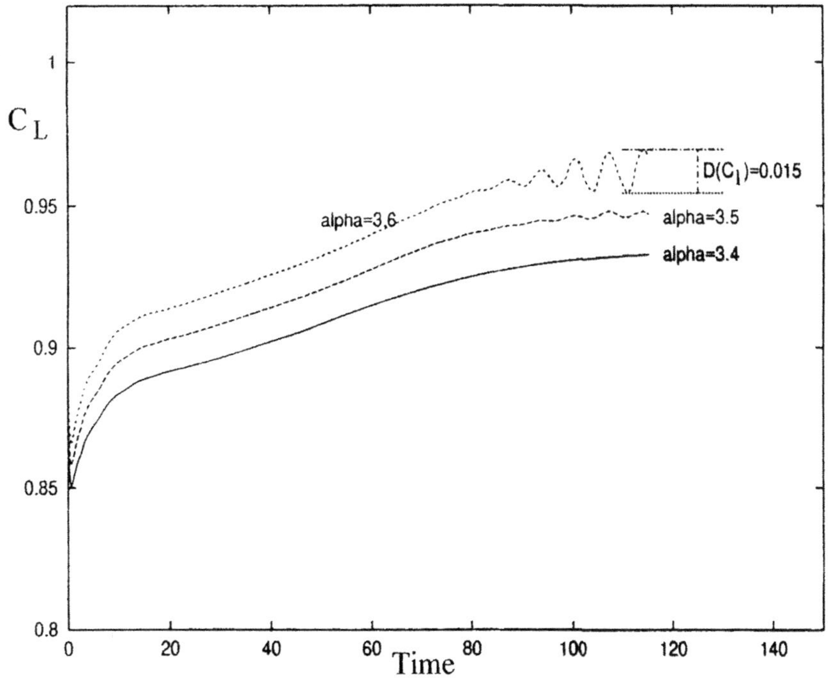

Figure 12 DRA-2303 airfoil without bump: C_L vs. time, application of the C3 criterion (unsteady calculations), $M_\infty = 0.68$, Re = 18.9×10^6

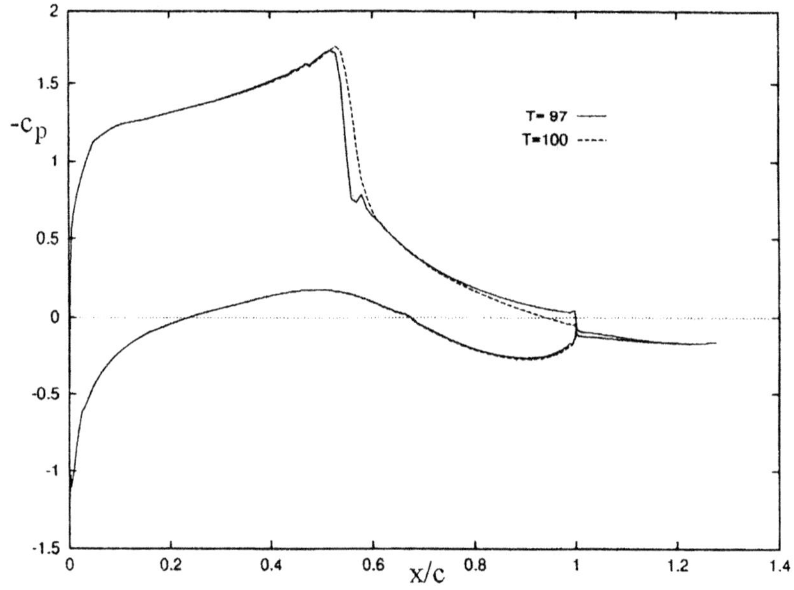

Figure 13 DRA-2303 airfoil without bump: unsteady calculations; numerical Cp distributions at maximum and minimum C_L, $M_\infty = 0.68$, Re = $18.9x10^6$

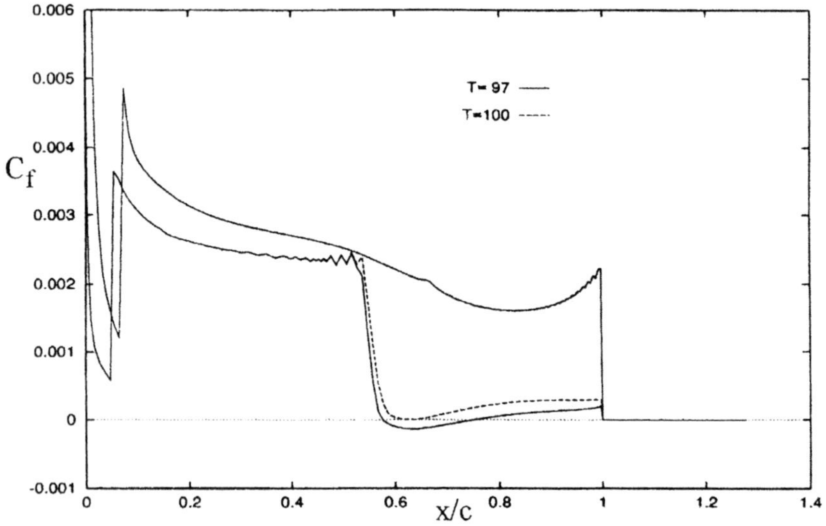

Figure 14 DRA-2303 airfoil without bump: unsteady calculations; numerical C_f distributions at maximum and minimum C_L, $M_\infty = 0.68$, Re = $18.9x10^6$

16 2D NUMERICAL INVESTIGATIONS OF SHOCK AND BOUNDARY LAYER CONTROL TECHNIQUES

N. Caballero
INTA, Fluid Dynamics Department
Ctra. Torrejón Ajalvir, Km 4.5, 28850 Madrid, SPAIN

Summary

The work developed at INTA in the frame of the EUROSHOCK-II programme as contribution to TASK-2 has focused on the numerical simulations of flows around airfoils with different devices used to control the shock boundary layer interaction phenomenon.

The numerical investigations have been carried out in two phases. In a first step, steady calculations were performed in order to understand how the control devices affect the main flow and to determine their effect in terms of drag reduction. In a second phase, unsteady calculation have been performed in order to investigate the influence of control on buffet onset.

Two airfoils have been considered for the computations: the laminar DRA-2303 and the turbulent RAE-5225. Both airfoils were tested at DERA and hence experimental results were available allowing to make comparisons between the predicted numerical results and those obtained experimentally.

16.1 Introduction

Different types of shock wave boundary layer interaction control techniques (*SBLIC*) can be considered. When no additional power is added to control the phenomenon it is said to have passive control; on the other hand control devices that require some additional energy are termed as active. A combination of both is called hybrid control.

The work programme of TASK-2 was concerned with the numerical modeling and simulation of the effect of these different types of control.

The idea behind the use of control mechanisms is to modify the conditions of the flow in the vicinity of the shock wave where the strong adverse pressure gradient causes a thickening of the boundary layer and consequently a total drag increase or it may even trigger unsteady flow behavior if the boundary layer does not sustain the pressure jump and separates. Thus, the aim of using control devices is to avoid the undesired effects mentioned that usually occur when flying in the transonic regime at off-design conditions.

When thinking of controlling shock wave boundary layer interaction in a passive way the first idea, already discussed in the EUROSHOCK-I programme, is to try to take advantage of the pressure jump across the shock. This is the basic idea behind the use of a cavity placed underneath the foot of the shock separated from the outer flow by a perforated plate. The cavity allows the fluid in the high pressure zone to recirculate towards the low pressure region. As a result the fluid is sucked from the region where the boundary layer is weak and tends to separate, and blown out in front of the shock inducing compression waves in the incoming flow and therefore causing a splitting and weakening of the shock wave.

Results from previous investigations have shown a beneficial effect of this passive control system in terms of a wave drag reduction, but also an unacceptable increase of total drag. The latter is essentially due to the extra viscous drag induced by the thickening of the boundary layer as a consequence of the blowing.

Then, if the origin of the higher total drag is the high level of blowing at the foot of the shock, the next step should concern a technique to reduce that level. The cavity becomes an active control device by sucking part of the fluid from the outer flow.

At this point it seems clear that the positive effect of blowing in front of the shock is to split of the shock. Another idea to get a similar shock splitting effect has been investigated. Basically it consists of collocating a bump at the foot of the shock. The bump forces the incoming flow to compress as it was done by the upstream blowing in the cavity case, but now the boundary layer is no more highly perturbated and hence the viscous drag does not increase. Bumps can be thought of as passive control mechanisms although the idea of actively inflating or deflating such surfaces can be considered as well.

The devices described control the shock wave boundary layer interaction region, but the control of the boundary layer far from that zone could also reduce drag and/or postpone buffet onset. In general, boundary layer control may be realized using small slots. Fluid is sucked or blown through the surface that separates the slot from the outer flow. This surface can be, as in the cavity case, a perforated plate, but also the slot can be open meaning that no physical separation surface exists between the hole and the outer flow. The location of the slots can be in front or behind the shock. When the slot is upstream of the shock and suction takes place, it is expected that a stronger boundary layer arrives at the shock and better resist the pressure jump. When blowing through this slot, the same effect of splitting the shock that has been already mentioned for the cavity case could be obtained. The suction slot behind the shock tends to reduce the thickness of the boundary layer thickened by the shock.

Apart from investigating individually all the devices mentioned, also the possibility of using combinations of devices has been considered. The main motivation is an attempt to alleviate the secondary effect induced by a single device through the simultaneous usage of suitable combinations. These are the hybrid control devices that also have been considered in the present work.

16.2 Numerical Tool

The numerical simulations have been carried out using a 2D Navier-Stokes Reynolds averaged solver. It is a time accurate solver that uses the implicit scheme of Beam & Warming of second order in time [1]. The spatial derivatives are discretized with central second order finite differences formulas.

The turbulence model is the Baldwin-Lomax eddy viscosity model. Some modifications were needed to take into account the transpiration velocity through perforated surfaces.

16.2.1 Turbulence Model

Modifications in the basic turbulence model are needed since in the interaction region the flow is fully turbulent and one has to take into account the surface mass transfer as well as the longitudinal pressure gradient that cannot be neglected in this region.

The Baldwin-Lomax [2] turbulence model originally implemented in the code has been changed to take into account the new physical features by means of a generalization of *Van Driest's* [3] wall damping function:

$$D = 1 - \exp\left(-\frac{y^+}{A^+}\right) \tag{1}$$

with the original values of

$$A^+ = 26 \qquad y^+ = y\frac{u_\tau}{\nu}$$

u_τ being the friction velocity

$$u_\tau = \sqrt{\frac{\tau_w}{\rho}} \; .$$

The value of A^+ has been modified to include the two effects mentioned. According to the work of *Bushnell et al.* [4], A^+ can be written as:

$$A^+ = \frac{26}{(\tau^+)^n} \tag{2}$$

The value of τ^+ is found, as proposed by *R. Bur* [5], by solving the y-wise integral of the simplified x-wise momentum equation

$$\frac{\partial \tau}{\partial y} - \frac{\rho \bar{v}_p \tau}{\mu} = \frac{d\bar{p}}{dx} \tag{3}$$

with v_p being the transpiration velocity.

The location where τ^+ is evaluated is the edge of the viscous sublayer defined empirically as $y_s^+ = 11.8$. So, finally, the value of A^+ (assuming $n = \frac{1}{2}$) is:

$$A^+ = 26[exp(11.8v_p^+) - \frac{p^+}{v_p^+}\{exp(11.8v_p^+) - 1\}]^{-\frac{1}{2}} \tag{4}$$

with

$$v_p^+ = \frac{v_p}{\bar{u}_\tau} \qquad p^+ = -\frac{d\bar{p}}{dx}\frac{\nu}{\bar{\rho}\bar{u}_\tau^3} \quad .$$

The calculation of v_p^+ and p^+ may cause problems in the region behind the shock where probably separation occurs ($\tau_w \leq 0$). To avoid this situation a different velocity scale has been considered in this zone: the friction velocity is taken at the position where the absolute value of the shear stress reaches its maximum value.

16.2.2 Transpiration Velocity

The transpiration velocity through the perforated surface has been modeled by two different laws. Both consider the flow through a single hole as if it were alone so no interference by the presence of the other holes is considered.

The first law assumes under a certain hypothesis a developed laminar pipe flow through the hole. Then the velocity through the hole is a function of the pressure drop across the perforated plate. The second law is based on dimensional analysis and establishes a dependency of the Mach number at the hole on the pressure drop and the shear stress at the wall.

Poll model

The first model used to predict the transpiration velocity along the perforated wall is based on the work of *Poll et al.* [6].

The basic assumption consists in considering the flow through an individual hole as a fully developed laminar pipe flow. Also the effects of compressibility are considered to be small enough to be neglected.

Then, the velocity profile at the exit plane is assumed to be parabolic and the pressure drop through the pipe can be written as:

$$\frac{\Delta P}{0.5\rho v^2} = 2.24 + \frac{64}{R_d}\frac{e}{d} \tag{5}$$

with v being the velocity at the exit plane , e/d the ratio of plate thickness to hole diameter, and R_d the Reynolds number based on hole diameter.

The latter relation can be worked out to give:

$$Y = [40.7 * X + 1.95 * X^2] \tag{6}$$

with X proportional to the mass flux, $\dot{m}$, and Y proportional to the pressure difference:

$$X = \frac{\dot{m}}{\mu e} \qquad \dot{m} = \rho\pi\frac{d^2}{4}v \tag{7}$$

$$Y = \frac{\Delta P d^2}{\rho\nu^2}(\frac{d}{e}^2) \tag{8}$$

with ρ being the density at the exit, μ the molecular viscosity and ν the kinematic viscosity.

Laser drilling does not produce perfect holes. The spacing can be uniform but the holes are not strictly circular. As a direct result of the burning procedure, the holes are bigger on the side closest to the laser and the diameter varies in an irregular way with depth. This means that it is not possible to specify the diameter of an individual hole accurately, but one must rather define a nominal diameter which can only be determined by a statistical survey of a large number of holes.

Consequently, it is necessary to specify an equivalent diameter in equation (6) introducing a factor K defined as:

$$K = \frac{effective\ \ diameter}{"measured"\ \ diameter} \tag{9}$$

Then equation (6) becomes:

$$Y = \frac{1}{K}[40.7 * X + 1.95 * X^2] \tag{10}$$

The characteristics of the perforated plates and flow conditions of the airfoils investigated in EUROSHOCK II give values of the transpiration velocity through the holes of the order of $30m/s$ and a Re_d of around 100. Those values support the fully developed laminar "pipe" flow assumption and expression (10) can be used for the computations.

Bohning&Doerffer model

In TASK 1 Bohning and Doerffer from the University of Karlsruhe have developed a new law for the transpiration velocity through a perforated plate.

The main points of this new model are the introduction of the effect of an outer tangential flow in the process of sucking fluid into a cavity (tangential flow does not influence blowing) and the particular treatment of the plate porosity.

The proposed law for the transpiration velocity through a perforated surface is:

$$M_h = 1.2 \left(\frac{dP}{P_0} \right)^{0.555} \tag{11}$$

where M_h stands for the Mach number in the hole.

When tangential flow is present and suction takes place the proposed model is modified with the addition of a term that takes into account the presence of shear stress at the wall. In this case the transpiration velocity through a hole can be obtained from the expression:

$$\frac{dP}{P_0} = (M_h)^{\frac{1}{0.55}} \left[\left(\frac{1}{1.2} \right)^{\frac{1}{0.55}} + 25 \left(\frac{\tau_0}{0.5 \rho_h u_h^2} M_h \right)^{\frac{1}{1.52}} \right] \tag{12}$$

As far as the porosity is concerned, Bohning&Doerffer have suggested an expression for determining the so called effective porosity:

$$\frac{dP}{P_0} = 0.063 \frac{\dot{m}/A}{Por} \tag{13}$$

where $\dot{m}$ is the mass flow rate in $[kg/sec]$, A the corresponding flow area in $[m^2]$, dP the pressure drop across the surface and Por the effective porosity in %.

The effective porosity is usually lower than the prescribed value given by the manufacturer of the plate and takes into account the possible irregularities that can exist. Once the effective porosity is determined, its value has to be introduced into the computations that make use of the formula proposed by Bohning&Doerffer (Eq. 12) for the transpiration velocity.

16.3 Use of Bumps as Control Devices

In this section some aspects of the flow around airfoils with bumps as control devices are discussed. How these mechanisms act on the steady flow around the wing section and what the results are in terms of drag reduction are the main issues considered.

16.3.1 DRA-2303 Airfoil

The original airfoil has been modified with an asymmetric bump of .0025c height located on the upper surface between $x/c = 0.5$ and 0.68. In Fig. 1 (a) it is possible to notice the presence of the bump comparing the original section with the new one.

For the computations of this airfoil a C-type grid with 300x70 points has been used and the outer boundary was located at ten chords distance from the airfoil. In Fig. 1 (b) a selected region of the grid is shown.

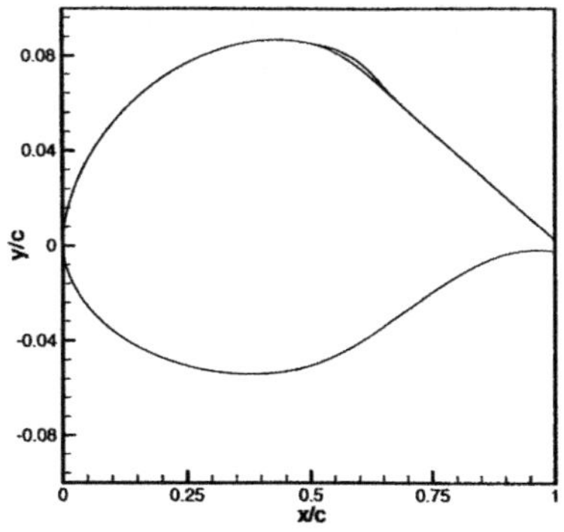

(a) Airfoil geometry.

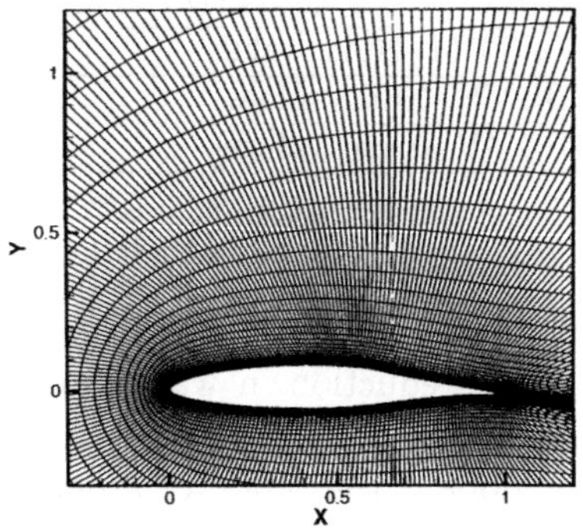

(b) 300x70 C-grid.

Figure 1: DRA-2303 airfoil.

The effect of the bump on the main flow has been investigated for the following flow conditions: $M = 0.68$, $Re = 18.8E6$ and transition fixed at 5% of the chord on both, the upper and the lower surface.

In Fig. 2 the predicted pressure coefficient distribution is compared with the experimental data for several lift coefficients. The figures on the left correspond to the datum airfoil while on the right the bump section data are plotted. Each couple of figures corresponds to simulations performed for the same angle of attack. Figures 2 (a) and 2 (b) show the results of computing the flow around the airfoil at an incidence of $\alpha = 1.0$. figures 2 (c) and 2 (d) to $\alpha = 1.75$, and figures 2 (e) and 2 (f) to $\alpha = 2.25$. Remarkable is the good agreement between experimental and computed data.

The main effect induced by the presence of the bump is the split of the shock, preferably into an isentropic compression followed by a weak shock. At low incidences the flow undergoes first a compression when it reaches the bump and then expands up to the main shock, Figs. 1 (a) and (b).

The pressure jump across this shock is probably similar compared to the one produced without bump and hence the wave drag is not noticeably.

When increasing the angle of attack the two compressions become closer until they coalesce resulting initially in considerable drag reductions as also indicated in Table1.

In Fig. 3 the iso-mach contours show clearly the effect of the bump on the flow field as previously described. Again on the left the datum airfoil results

Table 1: DRA-2303: ORIGINAL vs BUMP section $C_l \approx 0.7$.

DRA-2303	α	Cn	Cm	Cd	ΔCd
DATUM -Exp. values (D.P.275)-	1.890	0.7131	-0.1003	0.0125	
BUMP -Exp. values (D.P.1638)-	1.890	0.7027	-0.0983	0.0110	-12.0%
DATUM -Num. values-	1.750	0.7255	-0.1047	0.0117	
BUMP -Num. values-	1.750	0.6793	-0.0936	0.0096	-17.9%
BUMP -Num. values-	2.000	0.7251	-0.0945	0.0100	-14.5%

are given and on the right the corresponding bump-section flow data.

The reduction in wave drag together with the unchanged viscous drag produces a global reduction in drag. Table1 shows the effect of the bump on the global coefficients for a particular point of the polar. These data correspond to the pressure coefficients and iso-mach contours plotted in Figs. 2 (c)-(d) and 3 (c)-(d), respectively. For the same incidence the lift decreases slightly while the total drag is drastically reduced. The computed drag coefficients show a reduction of the order of 15% when comparing cases with the same lift coefficient $Cl = 0.725$.

16.3.2 RAE-5225 Airfoil

The original airfoil has been modified introducing a circular arc bump of $.00175c$ height located between the coordinates $x/c = 0.4$ and 0.6. Again, in the left picture of Fig. 4 the geometry of the airfoil with bump is shown.

As in the laminar airfoil case, the computations for the RAE-5225 airfoil have been performed using a C-type grid with 300x70 points and the outer boundary located at ten chords from the airfoil. In the right picture of Fig. 4 a region of this grid is shown.

The results computed corresponding to the flow conditions $M = 0.725$, $Re = 18.7E6$ and transition fixed at 5% of the chord will be discussed.

Fig. 5 shows the iso-mach lines for a lift coefficient of 0.7 for both, the datum airfoil and the airfoil with bump. The original shock wave is splitt into two shocks and the total drag increases with respect to the no bump case.

In this modified airfoil and for the mentioned lift coefficient of 0.7 the compression of the flow at the beginning of the bump is realized through a first shock wave. After this compression the flow is accelerated becoming again supersonic and a second shock wave occurs down of the position of the shock on the original airfoil.

The Mach number in front of each of the shocks described is higher than

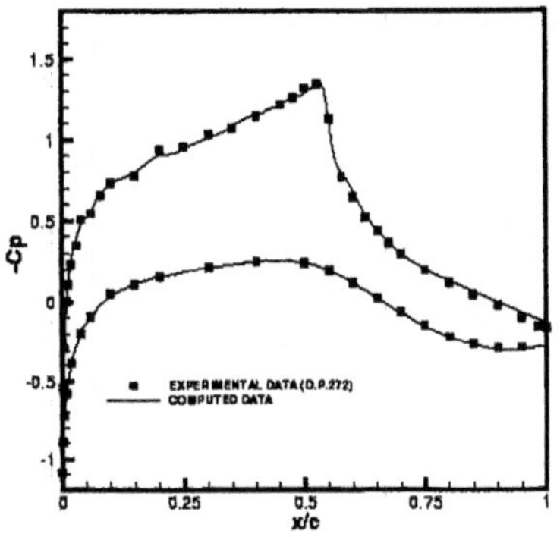

(a) DATUM $C_l = 0.61$.

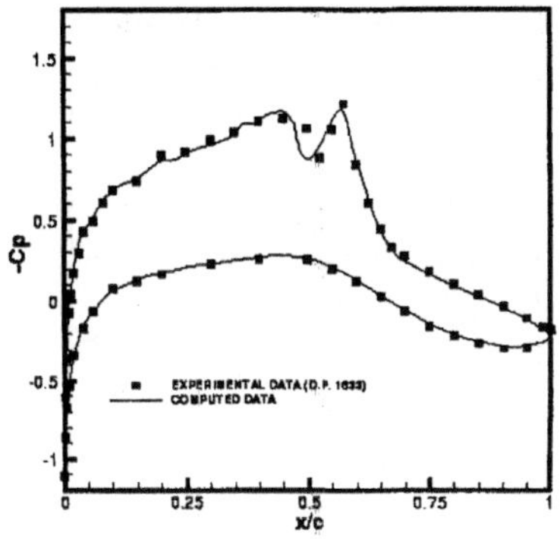

(b) BUMP $C_l = 0.55$.

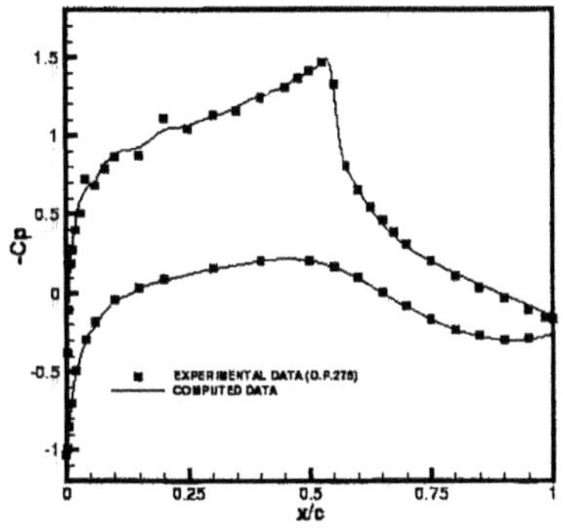

(c) DATUM $C_l = 0.72$.

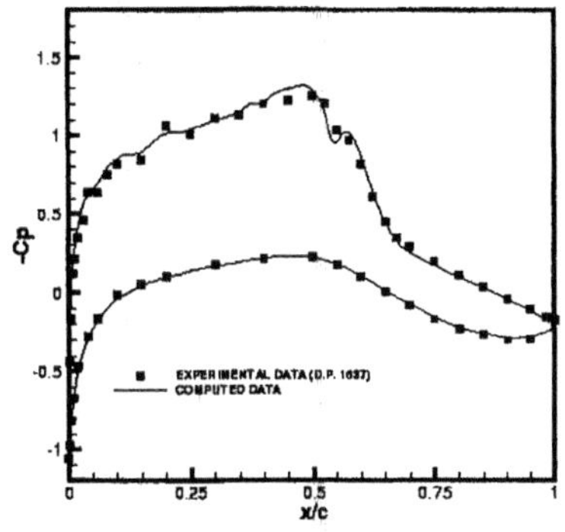

(d) BUMP $C_l = 0.68$.

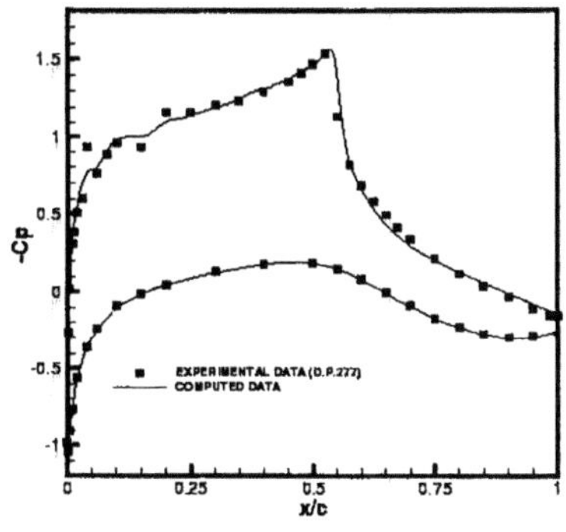

(e) DATUM $C_l = 0.79$.

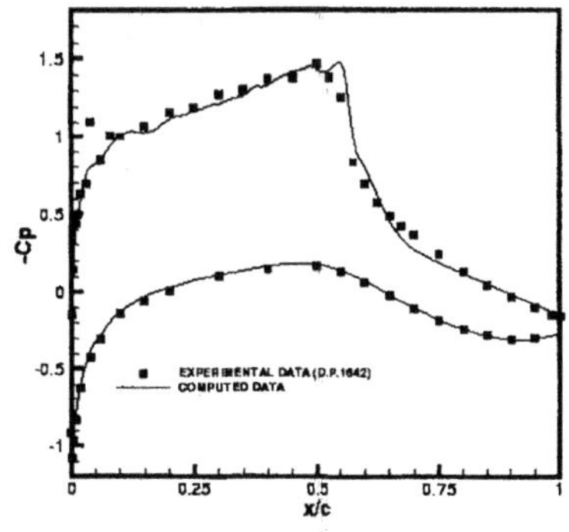

(f) BUMP $C_l = 0.77$.

Figure 2 DRA-2303: Pressure coefficients, M = 0.68.

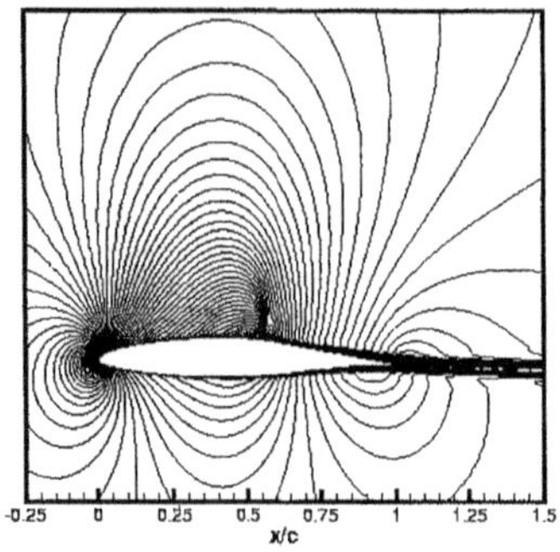

(a) DATUM $C_l = 0.61$.

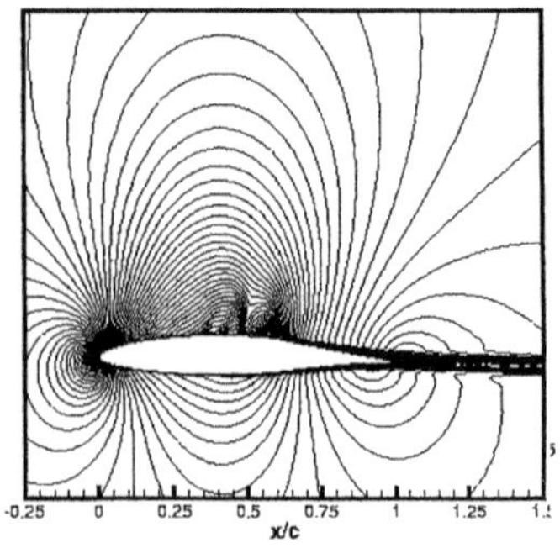

(b) BUMP $C_l = 0.55$.

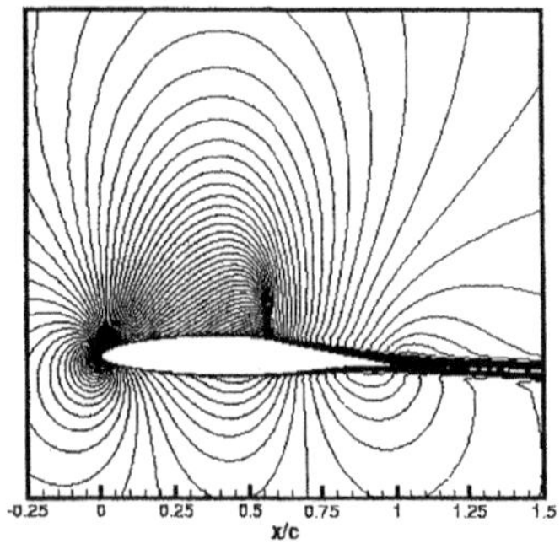

(c) DATUM $C_l = 0.72$.

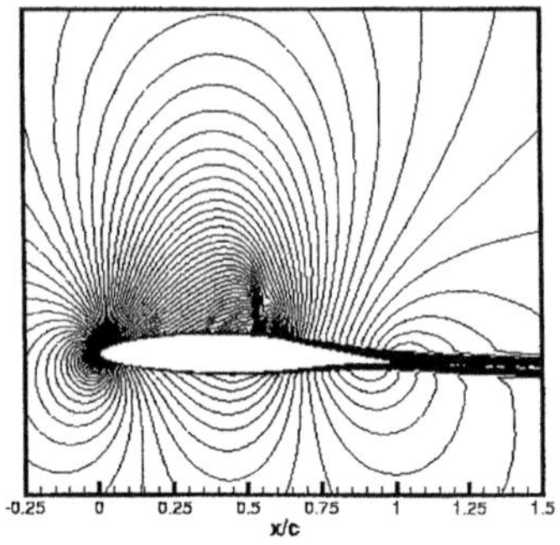

(d) BUMP $C_l = 0.68$.

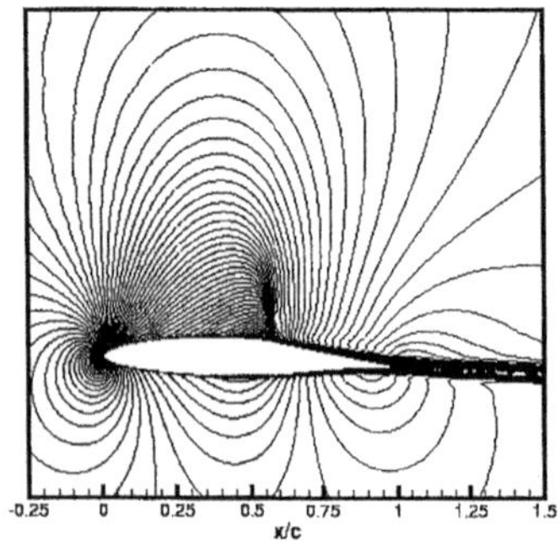

(e) DATUM $C_l = 0.79$.

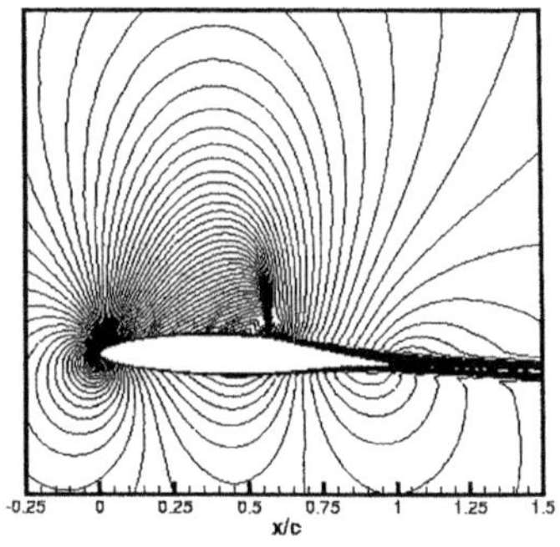

(f) BUMP $C_l = 0.77$.

Figure 3 DRA-2303: Iso-Mach contours, M = 0.68.

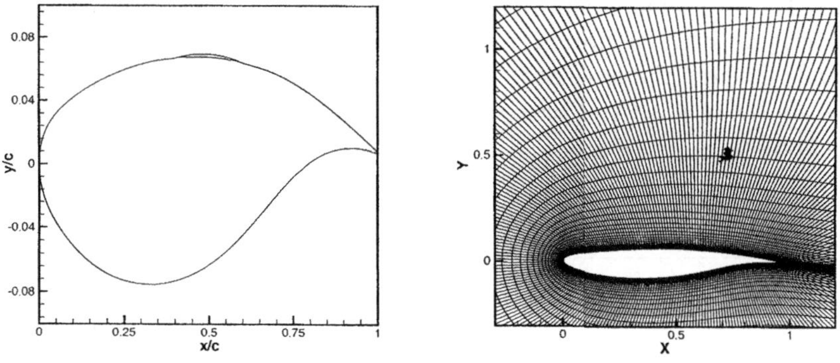

(a) Airfoil geometry. (b) 300x70 C-grid.

Figure 4: RAE-5225 airfoil.

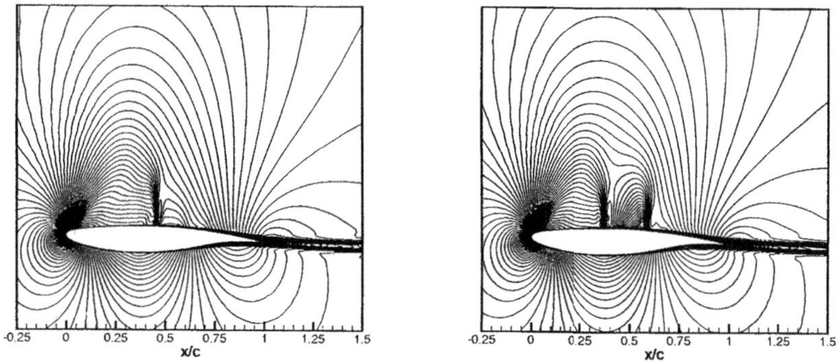

(a) DATUM section. (b) BUMP section.

Figure 5: RAE-5225: Iso-Mach contours, $C_l \approx 0.7$, $M = 0.725$.

the one in the no-bump case, hence the shock strength is considerably increased with respect to that of the original section. As a secondary effect, the presence of the bump causes here a thicker boundary layer, wich is basically due to the very high pressure jump through the second shock. Therefore, the viscous drag is also higher than for the original airfoil.

Only when increasing incidence will the effect of the bump become positive. The first compression is clearly reduced and a lambda structure develops associated with an important reduction in wave drag. This can be seen in Fig. 6 and 7 where the iso-mach contours and the pressure coefficient distributions of the original and the section with bump, corresponding to $C_l = 0.75$, are plotted.

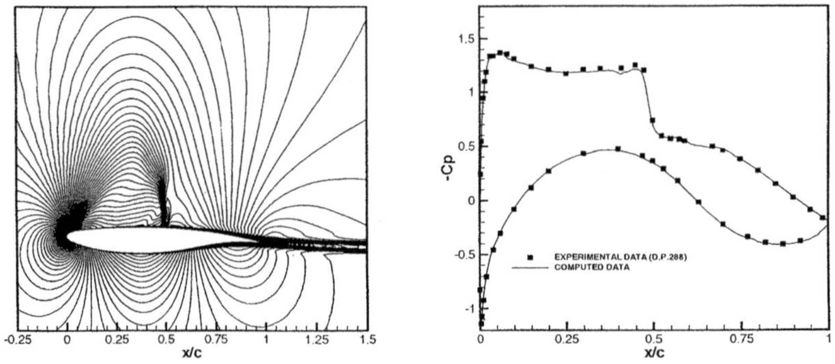

(a) Iso-Mach contours. (b) Pressure distribution.

Figure 6: RAE-5225: DATUM SECTION, $C_l = 0.75$, $M = 0.725$.

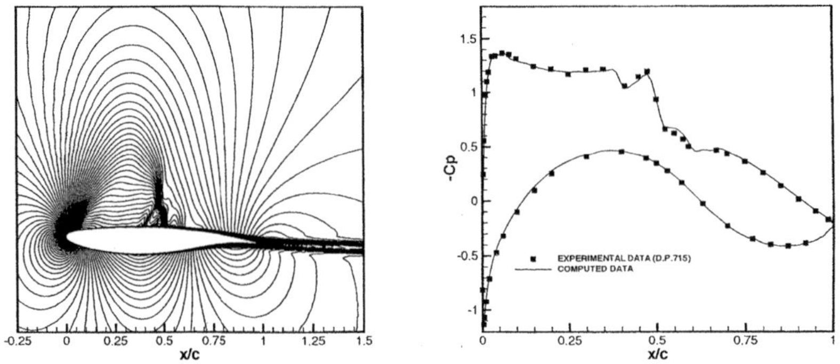

(a) Iso-Mach contours. (b) Pressure distribution.

Figure 7: RAE-5225: BUMP SECTION, $C_l = 0.75$, $M = 0.725$.

In Table2, the computed global coefficients for this last case show a reduction in total drag of about 9% while the experimental data give even higher reductions.

Table 2: RAE-5225: ORIGINAL versus BUMP section $C_l \approx 0.75$.

RAE-5225	α	Cn	Cm	Cd	ΔCd
DATUM -Exp. values (D.P.288)-	2.959	0.7550	-0.0918	0.0119	
BUMP -Exp. values (D.P.714)-	2.958	0.7433	-0.0898	0.0098	-17.6%
DATUM -Num. values-	2.750	0.7488	-0.0984	0.0118	
BUMP -Num. values-	2.750	0.7507	-0.0988	0.0108	-8.5%

The reason for this lift-dependent behavior is found in the features of the supercritical airfoils. The transonic regime for this type of airfoils is characterized by a shock wave that moves downstream and increases its intensity when increasing the angle of attack. Therefore, the relative position of the shock wave on the original section with respect to the bump location varies with lift.

In the first case, $C_l \approx 0.7$, the shock on the original airfoil is located at the beginning of the bump, while in the case of $C_l \approx 0.75$ the shock has moved downstream towards the center of the bump. This difference causes the different behavior. It is important to note that, in the first case, the shock locations on the airfoil bump are not the same as on the original section, while at the higher lift the shock varies slightly in position with respect to the original shock.

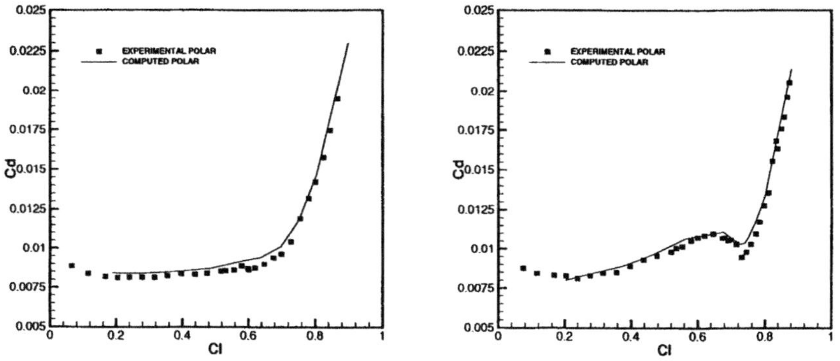

(a) Datum section polar. (b) Bump section polar.

Figure 8: RAE-5225: Experimental and computed polars, $M = 0.725$.

The comparison between experimental and computed results is also depicted in Fig. 8 where the polars for the datum and the airfoil with bump are plotted.

Fig. 9 shows the computed polars for both airfoils. Clearly seen is the range of lift coefficients where the bump section has a negative effect with respect to the original airfoil and also the range where the bump effect becomes positive.

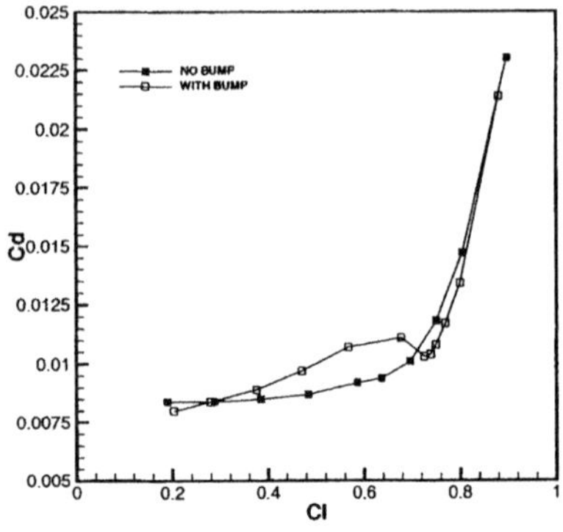

Figure 9: RAE-5225: Computed polar, $M = 0.725$.

16.4 Use of Cavities as Control Devices

The use of cavities separated from the outer flow by a porous wall as control devices has been investigated for the laminar DRA-2303 airfoil. The cavity, according to the data delivered by DERA, has been located in the upper surface from $x/c = 0.475$ to $x/c = 0.625$. The plate characteristics have been fixed to: thickness $1.0mm$, porosity 4%, hole diameter $0.076mm$; the airfoil chord c was $635mm$.

In the following discussion the parameter C_q denotes the non dimensional mass flow rate through the porous plate:

$$C_q = \frac{q}{\rho_\infty U_\infty c} \tag{14}$$

with q being the net mass flow through the perforated plate and c the airfoil chord.

With the previously described cavity the case of passive as well as active control has been considered. When comparing the computed results with the experimental data for any of these controls, some discrepancies have been observed as far as mass flow rate and cavity pressure are concerned.

This is illustrated in Fig. 10 (a) where the pressure distribution for the passive case corresponding to $C_l = 0.6$ is compared to the experimental data corresponding to the point 2199. The two computed curves, obtained using Poll's law for the estimation of the transpiration velocity, correspond to a zero mass flow rate through the perforated plate and the experimental pressure coefficient in the cavity, respectively, as input. According to these numerical results no passive control takes place when the pressure in the cavity is $(Cp)_{cav} = -0.946$ since some blowing occurs ($C_q = 0.0002$). In order to achieve passive control, the pressure in the cavity should be $(Cp)_{cav} = -1.025$, i.e., much lower than the experimental value. The same results are obtained when using the model of Bohning&Doerffer with very similar values for C_q and $(Cp)_{cav}$. Figure 10 (b) depicts the computed transpiration velocity non-dimensionalized by the value of the sound speed at infinity. non-dimensional variable).

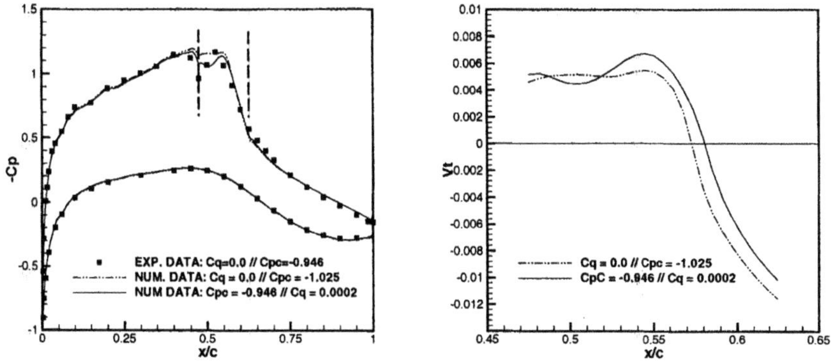

(a) Pressure distribution (b) Velocity transpiration

Figure 10: DRA-2303 WITH CAVITY: $C_q = 0.0$, (Poll's law), $M = 0.68$ $Cl = 0.6$.

Figure 11 presents the results corresponding to active control (experimental $C_q = -.00009$). Similar comments as those made for passive control also applied here. Again, the experimental pressure coefficient C_{pc} gives a net blowing and a slightly higher compression in front of the shock.

With respect to the global coefficients it can be concluded that the use of cavities does not improve the performances of the airfoil. Experimentally and numerically a decrease in lift and a considerable increase in total drag have been determined, although wave drag is reduced, when the cavity is activated either as passive or active control.

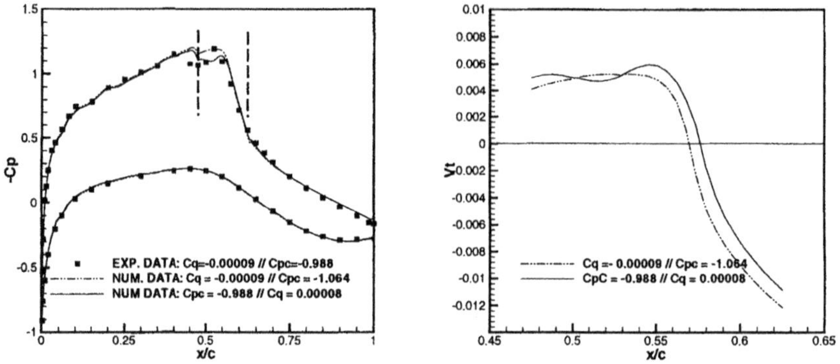

(a) Pressure (b) Velocity transpiration

Figure 11: DRA-2303 WITH CAVITY: HIGH SUCTION (Poll's law), $M = 0.68$, $Cl = 0.6$.

16.5 Slot upstream of the Shock

16.5.1 Open Slot

One of the techniques used to actively control the boundary layer consists of sucking through an open slot located upstream of the shock. This means to modify the boundary layer that will undergo the high pressure jump across the shock wave. The aim of such control is to reduce viscous drag and possibly avoid shock induced separation if it occurred in the no-control case.

Some numerical investigations have been performed for the two airfoils mentioned, i.e., the DRA-2303 and the RAE-5225, with and without bumps, at two levels of suction. For the cases with bump plus suction there were no experimental data available.

16.5.2 DRA-2303

For the laminar airfoil a 1%c wide slot has been located at $x/c = 0.475$ to 0.485 on the original airfoil as well as on the bump configuration. The flow conditions are the same as in section 16.3.1, i.e., $M = 0.60$, $Re = 18.8E6$.. Two different suction levels have been investigated, a high suction coefficient of $C_q = -0.0003$ and a lower one of $C_q = -0.00003$.

The computed pressure distributions for two different lift coefficients can be seen in Fig. 12, where the effect of sucking through the slot is compared with the corresponding no-control case.

In the presence of suction the flow expands when it reaches the slot and rapidly compresses at the end of the slot. The shock wave is not very much altered. Only when the suction coefficient is high, the shock intensity becomes slightly higher than it was for the datum airfoil. Therefore the effect of sucking upstream of the shock can even be negative in terms of a wave drag reduction.

The iso-mach lines in Fig. 13 the iso-mach lines show the flow field around the airfoil at the same lift coefficients as in Fig. 12 for the low suction coefficient. The flow pattern show the expansion already described followed by a lambda shock structure.

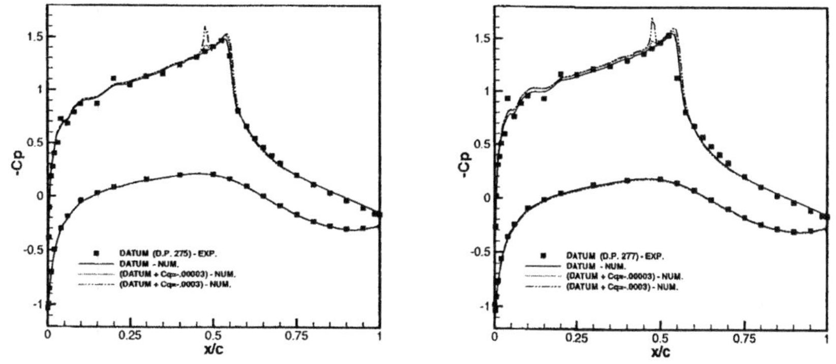

(a) $C_l \approx 0.7$. (b) $C_l \approx 0.8$.

Figure 12: DRA-2303 DATUM + OPEN SLOT: Pressure coefficients, $M = 0.68$.

Even if the wave drag increased, a total drag reduction is achieved due to the decrease of the viscous drag contribution.

The effect of increasing the suction intensity seems to be negative, at least when some level is reached. This is due, as already mentioned, to the negative effect of increasing the shock strength.

When slot suction is used together with the bump, the flow structure does not change very much with respect to the previously described suction on the original section. The slot is located slightly ahead of the bump so the flow is again accelerated upstream of the slot and then compressed at the rear part of the slot.

The low level of suction, $C_q = -0.00003$, does not change the flow field very much as can be seen in Fig. 14 where the pressure distribution for lift

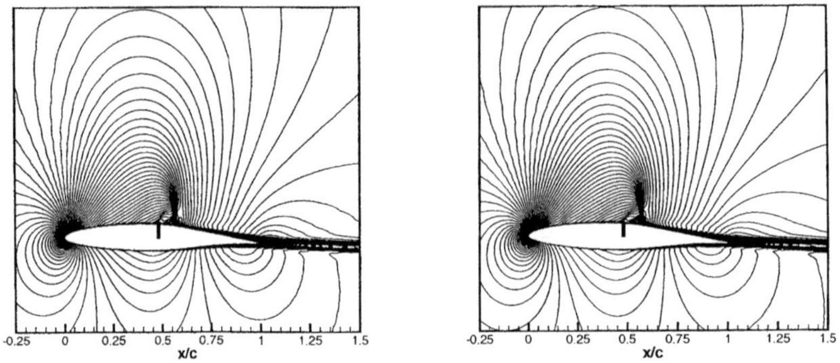

(a) $C_l = 0.73, C_q = -0.00003.$ (b) $C_l = 0.81, C_q = -0.00003.$

Figure 13: DRA-2303 DATUM + OPEN SLOT: Iso-Mach contours, $M = 0.68$.

coefficients of $C_l \approx 0.7$ and $C_l \approx 0.8$ are plotted and compared with the respective pressure distribution, for the airfoil with bump only.

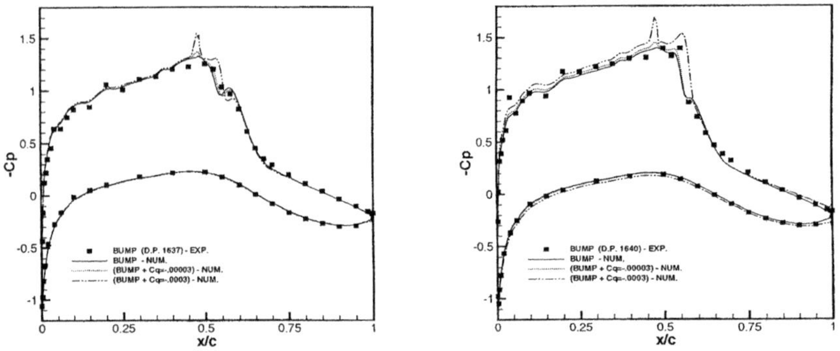

(a) $C_l \approx 0.7.$ (b) $C_l \approx 0.8.$

Figure 14: DRA-2303 BUMP + OPEN SLOT: Pressure coefficients, $M = 0.725$.

On the other hand, the higher suction rate causes a very different behavior depending on the incidence (lift): when low angles of attack are considered, sucking with $C_q = -0.0003$ in front of the shock results in a total drag reduction of the order of 5%. The high level of suction causes a negative effect, i.e., an increase in drag when applied at high incidences. This increment in total drag can be explained by the increase in wave drag due to the presence of a stronger shock compared to the one on the airfoil with bump only. This is shown in Fig. 14 (b), where the dot-dashed line represents the pressure distribution around the airfoil with bump and high suction.

16.5.3 RAE-5225

For the turbulent airfoil RAE-5225 a slot with a width of 1% of the chord has been located at $x/c = 0.385$ to 0.395. As in section 16.3.2 the flow conditions were $M = .725$ and $Re = 18.7E + 6$.

Datum and bump section have been tested with two different suction coefficients, $C_q = -0.00006$ and a higher value $C_q = -0.0006$.

For the airfoil it has been observed that the use of slot suction can have negative consequences in that an increase of total drag is obtained depending on the lift coefficient. The influence of the suction intensity on the flow depends also on the outer inviscid flow. If the shock intensity is not very high the total drag reduction increases with suction rate , as can be seen in Table3. In Fig. 15 (a) the pressure distributions corresponding to the airfoil without and with suction at a lift coefficient of the order of $C_l \approx 0.6$ are plotted.

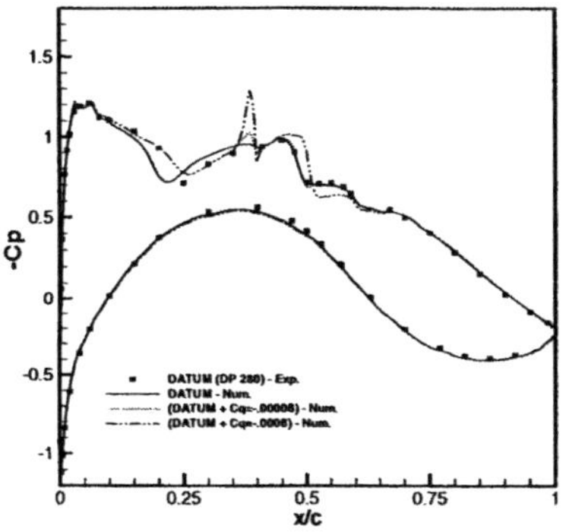

(a) $C_l \approx 0.6$.

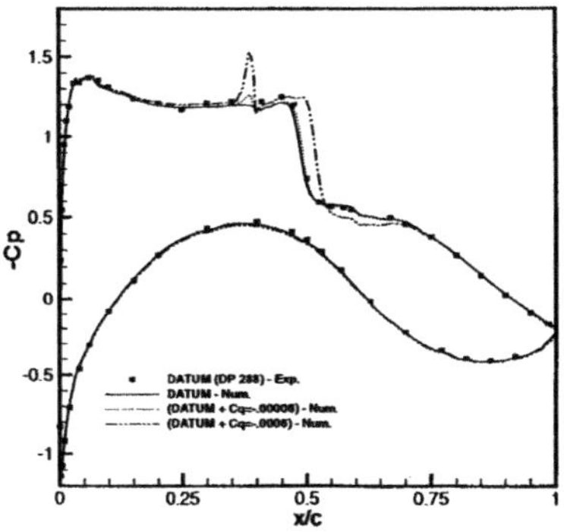

(b) $C_l \approx 0.75$.

Figure 15: RAE-5225 DATUM SECTION + OPEN SLOT: Pressure coefficients, $M = 0.725$.

The shock position does not change when sucking with the lower intensity but when suction is increased, the shock moves downstream. Hence, the lower C_q does not influence very much either the local variables or the global coefficients but the higher C_q increases lift and reduces total drag.

Table 3: RAE-5225: ORIGINAL SECTION vs OPENED SLOT $C_l \approx 0.6$.

RAE-5225	α	Cn	Cm	Cd	ΔCd
DATUM -Exp. values (D.P.280)-	2.169	0.5795	-0.0930	0.0086	
DATUM -Num. values-	2.000	0.5828	-0.1019	0.0092	
DATUM + ($C_q = -.00006$)	2.000	0.5871	-0.1023	0.0091	-1.1%
DATUM + ($C_q = -.0006$)	2.000	0.6109	-0.1030	0.0086	-6.5%

When increasing the angle of attack the shock becomes stronger and the effect of suction is negative. Again, the lower C_q does not affect the flow noticeably but the higher C_q shifts the shock, as for the lower C_l value, and increases its intensity as can be seen in Fig. 15 (b).

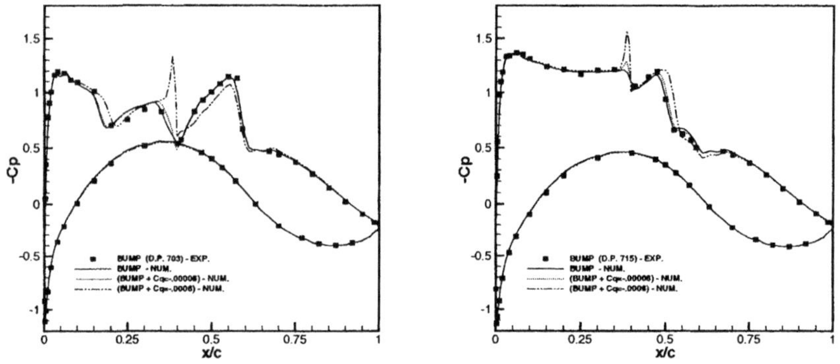

(a) $C_l \approx 0.6$.   (b) $C_l \approx 0.75$.

Figure 16: RAE-5225 BUMP + OPENED SLOT: C_p, $M = 0.725$

The flow around the airfoil with bump suffers a first compression in the vicinity of the bump, then expands and again compresses as already seen in Section 16.3.2.

The effect of sucking fluid through the slot, located at the beginning of the bump, depends on the flow field. If there are two compression shocks and suction takes place at the foot of the first shock, a low C_q does not affect very much the outer flow as can be seen in the pressure distribution for

$C_l \approx 0.6$ Fig. 16 (a), and in the corresponding flow field pattern (iso-mach lines), Fig. 17 (a). In this same case, the high suction coefficient causes a quite different behavior. The two shocks become a lambda shock and the total drag is drastically reduced. In the Table4, global coefficients show a maximum drag reduction of around 10%.

Table 4: RAE-5225: BUMP SECTION versus OPENED SLOT $C_l \approx 0.6$.

RAE-5225	α	Cn	Cm	Cd	ΔCd
BUMP -Exp. values (D.P.703)-	2.153	0.5800	-0.0977	0.0105	
BUMP -Numerical values-	2.000	0.5679	-0.1031	0.0107	
BUMP+($C_q = -.00006$)	2.000	0.5708	-0.1026	0.0106	-0.9%
BUMP+($C_q = -.0006$)	2.000	0.5880	-0.1022	0.0096	-10.2%

When the incidence increases and the flow around the airfoil without suction shows a lambda structure, see Fig. 7 (a) in Section 16.3.2, the effect of sucking is no longer positive. In Fig. 16 (b) the pressure distribution for $C_l \approx 0.75$ is plotted, the low suction coefficient does not affect very much the local pressure coefficient but sucking with a higher intensity causes a stronger shock and a net increment in total drag.

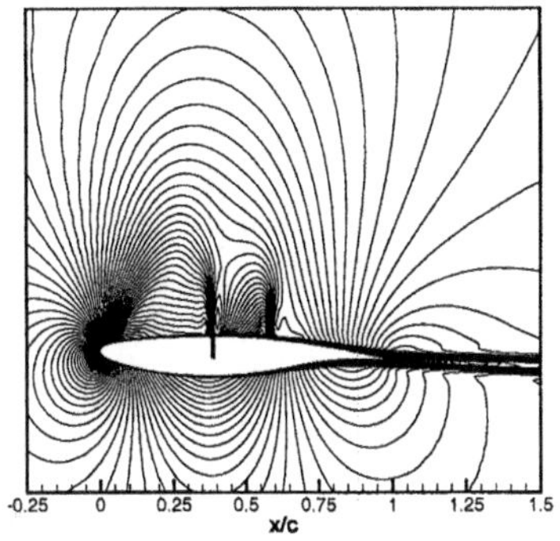

(a) $C_q = -0.00006$

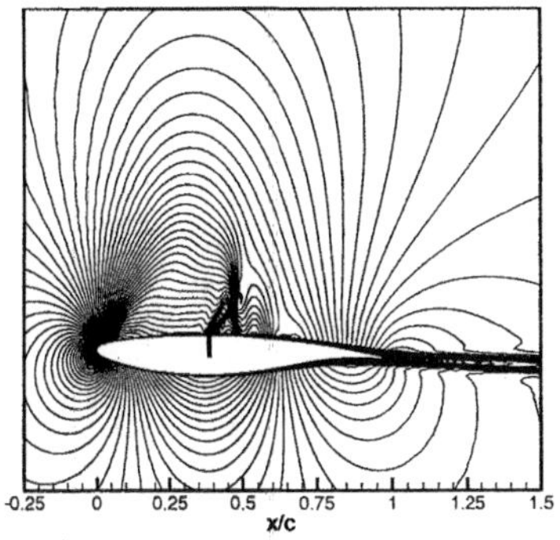

(b) $C_q = -0.0006$

Figure 17: RAE-5225 with BUMP + SUCTION: $C_l \approx 0.6$, $M = 0.725$.

16.5.4 Slot covered by a Perforated Plate

In this section the case of using a suction slot separated from the outer flow by a porous plate, positioned as in the previous section in front of the shock, will be discussed.

The case that has been investigated is the DRA-2303 airfoil with a slot of width $0.01c$ located on the upper surface at $x/c = .45$ to $.46$. The flow conditions are the same as in the previous NLF cases, i.e., $M = 0,68$, $Re = 18.8E6$ (see Section 16.3.1).

As in the case of sucking through the cavity, some discrepancies between the experimental data and the numerical results have been found. Basically, the difference lays in the mass flow rate through the slot and the corresponding pressure inside the cavity. The computed pressure is always higher than the experimental one, either using Poll's law or the Bohning&Doerffer model.

In order to try to clarify that fact, it is useful to observe the experimental data corresponding to a lift coefficient around $C_l = 0.6$. In Fig. 18 these data are plotted for various suction rates. The horizontal lines correspond to the associated experimental pressure coefficients in the cavity.

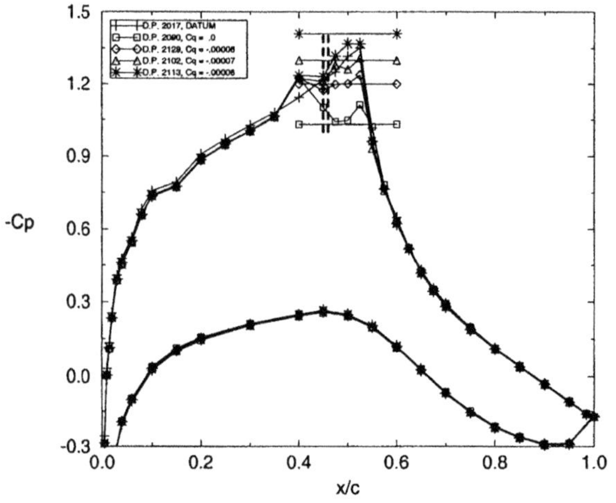

Figure 18: DRA-2303 + DISCRETE SUCTION: Exp. data $\alpha = 1.285$ $M = 0.68$.

The most significant case is the set of values corresponding to passive control, $C_q = 0$, data point 2090. The figure shows a level of pressure inside the cavity higher than that for the datum airfoil at the slot position. That would mean that blowing took place which is also indicated by the pressure coefficient on the airfoil in the vicinity of the slot in the case of D.P. 2090.

To understand the origin of the discrepancies a set of calculations has been performed corresponding to the experimental data points shown in fig. 18: 2090 ($Cq = 0.0$ and $(Cp)_{cav} = -1.03$), 2129 ($Cq = -0.00006$ and $(Cp)_{cav} = -1.30$), 2102 ($Cq = -0.00007$ and $(Cp)_{cav} = -1.41$) and 2113 ($Cq = -0.00008$ and $(Cp)_{cav} = -1.20$).

The computations have been carried out also with the aim of comparing the Poll and Bohning-Doerffer transpiration velocity models. According to the different assumptions for each model, the cases should be of relevance to show the differences between the two laws since suction takes place in the presence of an outer supersonic tangential flow.

In Fig. 19 the net mass flow rate is plotted versus the pressure coefficient in the cavity. The experimental points are joint by vertical dashed lines to the predicted values when fixing the pressure in the cavity equal to the experimental ones, and by horizontal lines to the predicted values when the mass flow rate is fixed. The Bohning & Doerffer law has been used with the given porosity factor of 4% and also with the effective porosity of 2.48% recommended by the authors.

Starting with the passive control case, D.P.2090, Fig. 19 shows a numerically predicted blowing when the pressure in the cavity is set equal to the experimental value. That behavior is predicted for the two transpiration velocity models considered. It is noticeable that the value of net mass flow computed with the B/D law and an effective porosity of 2.48% matches with the corresponding Poll's law and the porosity factor of 4%, while the B/D law for a porosity of 4% predicts a higher level of blowing.

The predicted cavity pressure is always lower than the corresponding experimental value when the value of the net mass flow rate of zero is prescribed. In this case the three computed points are coincident as they should.

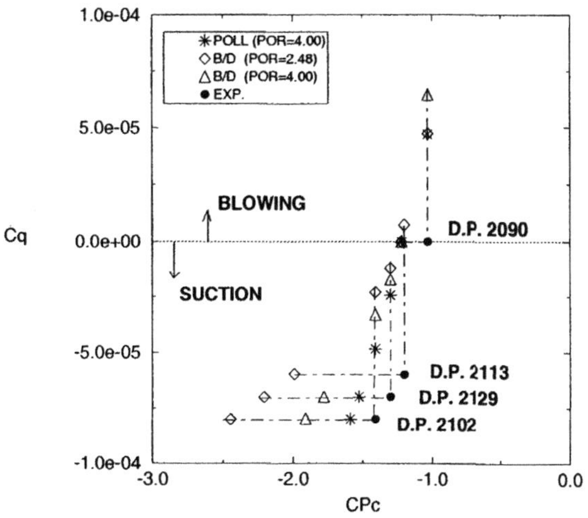

Figure 19: DRA-2303 + DISCRETE SUCTION: Mass flow coefficient.

When considering active control the situation is quite similar. The experimental cavity pressure gives always a computed suction coefficient lower than the corresponding experimental value, and if the suction level is imposed the computed cavity pressure is lower. In these cases there are some differences between the predicted values employing either Poll's law and the B/D law. The computed data show a higher suction coefficient when using Poll's model. This agrees with the assumptions made by Bohning and Doerffer, namely that the tangential flow should hinder suction as the computations show.

So from these cases it can be judged that the Poll and B/D models, respectively, predict similar values when there is no suction and the porosity factor is used properly. If suction takes place in the presence of a tangential flow, both models give different results since only one of them take this effect into account.

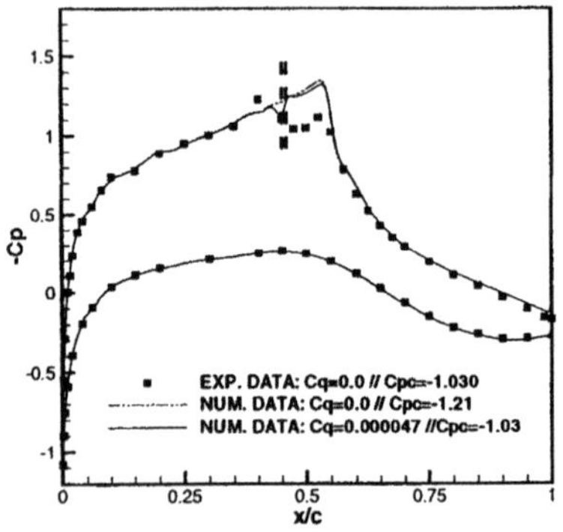

(a) $C_q = 0.0$, Poll law.

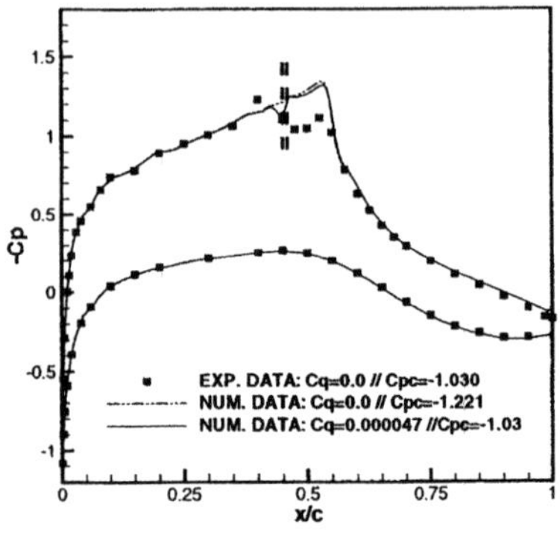

(b) $C_q = 0.0$, B&D law.

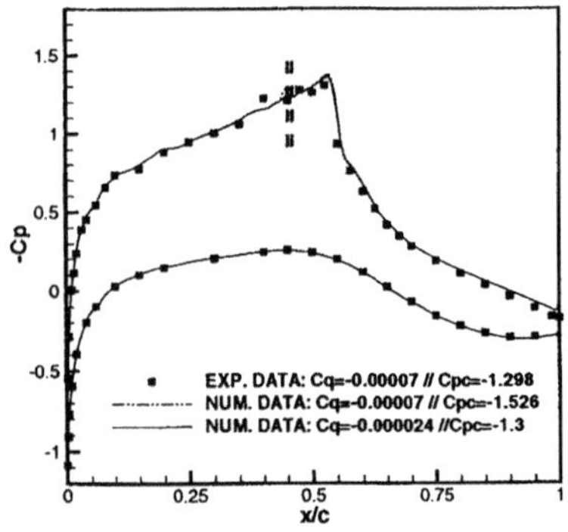

(c) $C_q = -.00007$, Poll law.

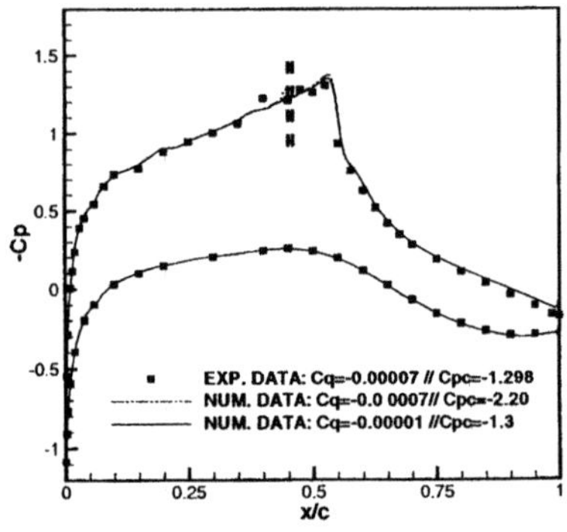

(d) $C_q = -.00007$, B&D law.

Figure 20: DRA-2303 + DISCRETE SUCTION: $C_l \approx 0.6$.

Fig. 20 shows the pressure distributions around the airfoil for the cases described above at a lift coefficient of the order of 0.6. The pictures on the left correspond to the simulations performed with Poll's model and those on the right correspond to the ones employing the Bohning-Doerffer law.

In Fig. 20 (a) and (b) the computed pressures corresponding to passive control are plotted and compared with the experimental data. When imposing zero net mass flow (dashed line), the pressure distribution is basically the same as for the closed-surface model (see Fig. 2 in Section 16.3.1). When blowing is predicted, the compression at the slot has the same level as the experimental one, but inmediately behind the slot the flow expands again while the experimental data showinitially a further increase in pressure.

In Figs. 20 (c) and (d) the predicted pressure distributions corresponding to a medium suction level show good agreement with the experimental data. It can be observed that the slot suction does not much affect the shock wave. So any change in total drag is due to the reduction of viscous drag. Similar results are found when sucking with higher intensity.

16.6 Hybrid Control

The investigations of the effect of hybrid control have been performed adding slot suction to the configuration already discussed in Section 16.4, i.e., a cavity located at the foot of the shock for the laminar airfoil case. The slot of width 1%c has been located behind the shock $x/c = 0.65$ to 0.66. The effect of the slot is mainly to remove part of the perturbed boundary layer, reducing its thickness and thus reducing viscous drag.

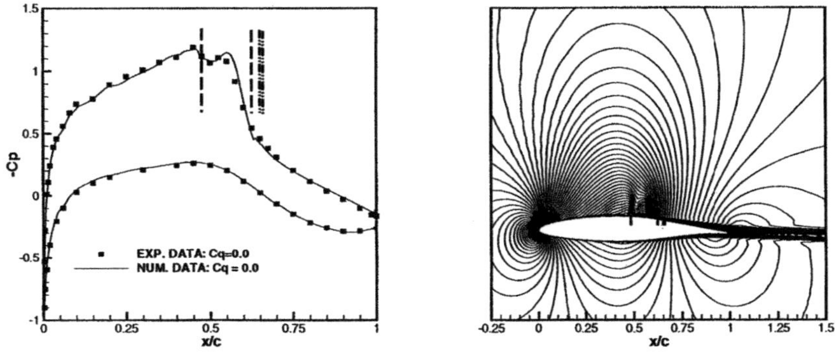

(a) Pressure distribution. (b) Iso-Mach contours.

Figure 21: DRA-2303 + HYBRID CONTROL: $C_q = 0.0, C_l = 0.6$.

The simulations have been performed using two levels of suction through the slot and also with zero net mass flow. The latter can be considered as passive control.

Some benefits have been observed compared to the case of single cavity control but still, the net total drag is higher than for the original airfoil.

Fig. 21 (a) shows the pressure distribution for the no-suction case that corresponds to the experimental data point 2291. The numerical results fit very well with the experiments. In Fig. 21 (b) the corresponding iso-mach contours are plotted. The flow structure does not differ much from that of the single cavity.

16.7 Unsteady Behavior

One of the important issues of the use of control devices is the influence that such elements can have on buffet onset.

Unsteady calculations have shown that the use of bumps can reduce or even eliminate buffet conditions. But it is not easy to determine buffet onset and the effect of control on buffet onset.

Different criteria have been used traditionally in the industry to estimate buffet onset. Here, we will predict buffet onset based on one of the criteria given by ALENIA. This criterion relates the Mach number in front of the shock to a linear function of the Mach number dependence on the distance from the leading edge. The function is

$$M_{limit} = 1.496245 - 0.322996x \pm 0.041077 \tag{15}$$

and the criterion establishes that buffet onset is reached when the Mach number just upstream of the shock is tangent to that function.

Results for the DRA-2303 airfoil are shown in the Fig. 22 for the datum airfoil and the airfoil with the bump.

It is clearly seen how according to this criterion the use of a bump postpones the appearance of buffet conditions from a lift coefficient of 0.86 to a cofficient of 0.95.

Some numerical investigations have also been performed to study the effect of slot suction on buffet onset.

As mentioned in Section 16.5.1, sucking in front of the shock can alleviate shock induced separation. Therefore it was expected that the use of slots as control devices could delay the appearance of buffet. To study the influence of these control devices on buffet, we have followed the strategy of applying suction through a 1%c wide open slot to a completely developed unsteady flow on the laminar DRA-2303 airfoil at an incidence of $\alpha = 4.8$. Two main

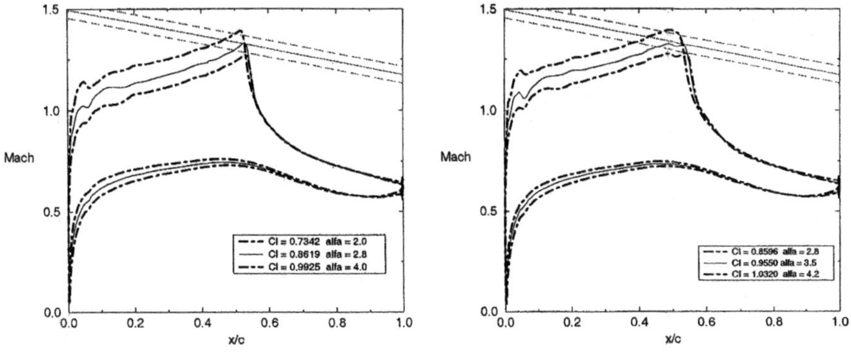

(a) ORIGINAL SECTION. (b) BUMP SECTION.

Figure 22: DRA-2303: BUFFET ONSET

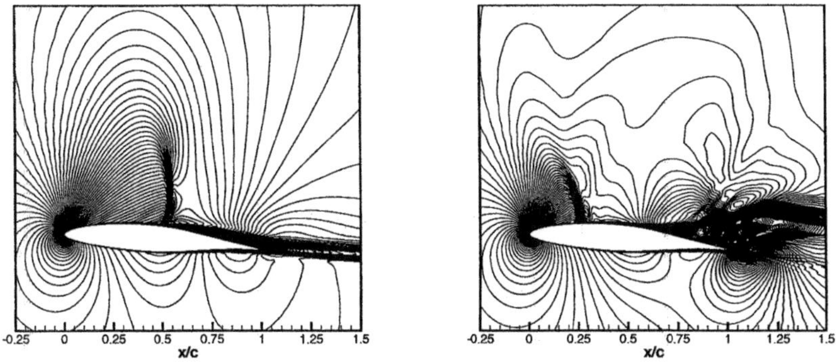

Figure 23: DRA-2303: Unsteady state: no control, $\alpha = 4.8$.

parameters can be freely chosen in order to obtain the desired effects: the position of the slot and the suction intensity.

In the developed unsteady flow, Fig. 23, the shock wave travels over the upper surface from an upstream position of minimum lift to a maximum lift position downstream. The slot locations investigated are between these extreme shock positions, namely, $x/c = 0.45 - 0.46$, $x/c = 0.49 - 0.50$ and $x/c = 0.55 - 0.56$. Two suction coefficients have been investigated, a low value $C_q = -0.00007$ and a higher one $C_q = -0.0007$. For the low level of suction no relevant effect has been found with respect to buffet onset. The unsteady flow does not suffer any change and buffet onset remains at the same limits as if without control.

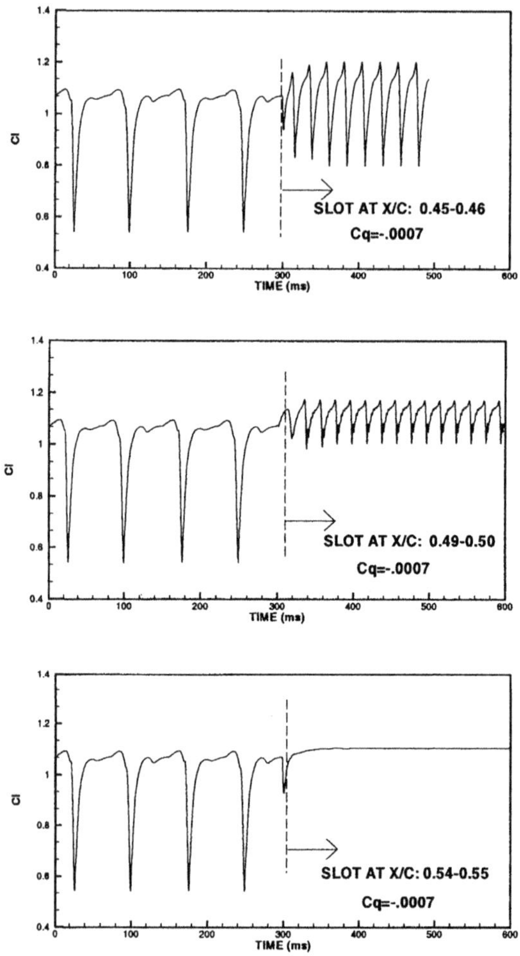

Figure 24: DRA-2303: Histograms w/o control $\alpha = 4.8$.

When increasing suction the situation changes drastically. In Fig. 24 the lift coefficient history is plotted for the different slot locations. The first part of the three histograms corresponds to the no-control case and it is followed by the computed control case. The results are quite relevant since a completely absence of unsteadiness is predicted when the slot is located at $x/c = 0.54$ to 0.55.

In Fig. 25 the iso-mach lines correspond to the maximum and minimum lift coefficient of the flow field around the airfoil when the suction slot is located at $x/c = 0.45$ to 0.5. Even if the histogram (see Fig. 24) shows an oscillatory behavior of lift, the shock position does not change.

When moving the slot towards what seems to be the fixed shock position, the unsteadiness vanished completely and the steady flow field pattern is that shown in Fig. 26.

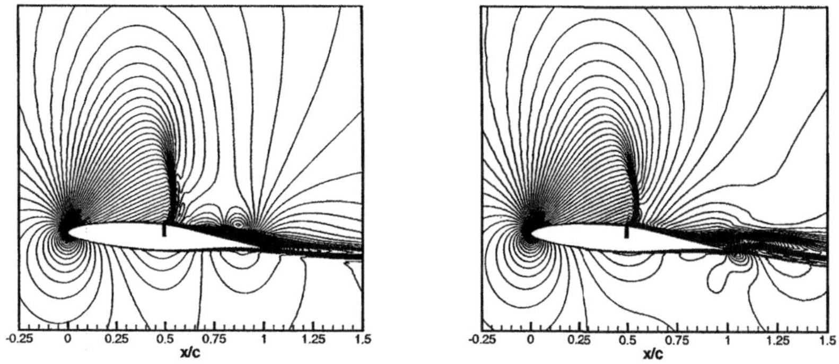

Figure 25: DRA-2303: Unsteady state, slot at x/c: 0.49-0.50, $\alpha = 4.8$.

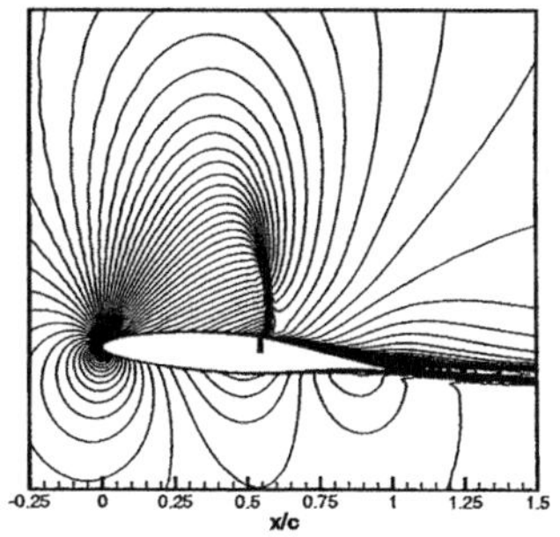

Figure 26: DRA-2303: Steady state, slot at x/c: 0.54-0.55,$\alpha = 4.8$.

16.8 Conclusions

Numerical investigations performed for two different airfoils, laminar and turbulent sections, have shown the possibility of controlling the shock wave boundary layer interaction phenomenon that takes place in the transonic regim when flying at design and off-design conditions.

From the various control devices investigated, it can be concluded that the use of properly designed bumps may give excellent results in terms of total drag reduction. On the other hand, the use of cavities, in a passive fashion, gives very good results in terms of wave drag reduction but it is accompanied by an important increase in viscous drag. The different means of reducing the induced viscous drag, such as sucking fluid from the cavity or including suction slots, seem to be efficient only when very high suction intensities are applied and, therefore, they do not appear to be useful in wing applications.

With respect to the second goal of the work, the computations show a delay in buffet onset when control mechanisms are used. The most outstanding result shows how well developed buffet on an airfoil can even disappear when using an appropriate suction slot location.

Finally it is important to note the good agreement found between numerical and experimental results. That shows, once more, the advantage of using numerical tools in order to predict the flow behavior and, as a consequence, be able to reduce the number of expensive wind tunnel tests.

References

[1] BEAM, R.M. and WARMING, R.F. "An Implicit Finite-Difference Algorithm for Hyperbolic Systems in Conservation-Law Form". *Journal of Computational Physics, Vol 22*, (1976).

[2] BALDWIN, R.S. and LOMAX, H. "Thin Layer Approximation and Algebraic Model for Separated Turbulent Flows". *AIAA Journal, Vol. 99*, (1978).

[3] VAN DRIEST, E. R. "On Turbulent Flow near a Wall". *Journal of the Aeronautical Sciences, Vol. 23*, (1956).

[4] BUSHNELL, D.M., CARY, A.M.and HARRIS, J.E. "Calculation Methods for Compressible Turbulent Boundary Layers". *NASA, SP-422*, (1977).

[5] BUR, R. "Passive control of a shock wave/turbulent boundary layer interaction in a transonic flow". *Rech. Aerosp.-n. 1992-6.*

[6] POLL, D. I. A. , DANKS, M. and HUMPHREYS, B.E. "The Aerodynamic Performance of Lasser Drilled Sheets". In *First European Forum on Laminar Flow Technology,Hamburg*, March 1992, Paper 92-02-028.

17 Prediction of Transonic Airfoil/Wing Flow and Buffet with Control Using a Time-accurate Viscous-Inviscid Interaction Approach

J.C. Le Balleur, P. Girodroux-Lavigne, H. Gassot

ONERA
BP.72, 92322 Châtillon, France

Summary

Different concepts of shock wave and/or boundary layer control have been numerically investigated with the 2D and 3D time-accurate strong viscous-inviscid interaction codes VIS15 and VIS25 within Task 2 of EUROSHOCK II. The 2D code VIS15 has been used to study the influence of bumps, discrete suction, hybrid control, and passive and active control through perforated walls in the case of different laminar and turbulent airfoils. The time-accurate 2D investigation has also been performed with the unsteady VIS15 code for studying the effect of several of these control devices on the suppression or damping of buffet. The 3D-unsteady code VIS25, developed at ONERA as an extension of the 2D VIS15 code, has finally been used to study sweep effects on the flow with and without bump in the case of the ADIF sheared wing.

17.1 Introduction

Passive control of shock/boundary-layer interaction and buffet perforated surfaces was previously numerically investigated within the EUROSHOCK (I) project [1], using the strong viscous-inviscid interaction "VIS" (*V*iscous-*I*nviscid *S*olvers) approach developed at ONERA [2], [3], [4], [5], [6], [7], [8], [9], [10], [11], [12] in both, steady and time-accurate forms. It was found, in agreement with the experimental results, that passive control leads, in general, to a decrease of wave drag but an increase of total drag. It was also shown that the main interest of passive control appears at off-design conditions by shifting buffet onset to higher Mach numbers and lift coefficients.

In order to overcome the penalties encountered with passive control at design conditions, different other concepts of shock and/or boundary layer control have been studied within the EUROSHOCK II project [13], [14], [15]. The 2D time-consistent strong coupling VIS15 code [8], [9] has first been used to investigate the effects of bumps, discrete control through a small suction slot located upstream of the shock, and hybrid control (bump and suction slot) in the case of the turbulent RAE-5225 airfoil and the laminar DRA-2303 airfoil. New computations of passive and active control by ventilation via a porous plate/cavity arrangement located at the wall below the shock have also been carried out in the case of the turbulent VA-2 airfoil.

The capability of the time-accurate VIS15 code to discriminate without any change between steady solutions and unsteady solutions with shock-induced separation has been used to investigate the effect of the bump and of discrete suction on buffet onset. The occurrence of buffet, directly predicted by the VIS15 unsteady computations, has been compared with two empirical steady criteria provided by Alenia. VIS15 computations in the buffet regime, providing a direct full description of the unsteady pressure signals, have also been performed and compared to the experimental results of DLR in the case of a laminar airfoil with and without passive control by ventilation through a perforated wall.

During this work, the new control law proposed in Task 1 by the University of Karlsruhe, which takes into account the incoming tangential flow, has been introduced into the VIS15 code as an alternative to Poll's law implemented during EUROSHOCK (I).

In order to study the effect of a bump in three dimensions, both, 2D and fully-3D "VIS" computations over an infinite swept wing have been perormed in the case of the ADIF sheared wing. The sweep effect has been investigated by comparing the 3D-VIS computations for the sheared wing geometry with 2D-VIS calculations for the corresponding basic ADIF airfoil. The 3D calculations have been achieved using the fully three-dimensional unsteady code VIS25 [16], which is a direct extension of the 2D VIS15 code, and which has been adapted for infinite swept wing conditions.

In all 2D and 3D VIS computations, appropriate fine grids, as required, have been used with streamwise grid-clusterings, both, in the control zone and in the shock region, in order to minimize the numerical dissipation and to effectively capture numerically a description of the physics of the streamwise small length scales that are induced by shock boundary layer interaction.

17.2 Numerical Methods

17.2.1 Time-consistent Viscous-Inviscid Interaction Approach in 2D and 3D

The 2D-unsteady code VIS15 and 3D-unsteady code VIS25, developed, respectively, for transonic airfoils or wings, have the common methodology and the similar coding of the Viscous-Inviscid-Solvers "VIS", initiated around 1979 in the VIS05 code [2], [4]. The advantage of the VIS methodology, which splits the numerical scheme into two viscous and inviscid coupled schemes, is to have a much lower numerical viscosity for a given grid than a single-scheme Navier-Stokes approach, and to give access to the finer grids necessary to resolve the physics of Shock Boundary Layer Interaction at high Reynolds numbers, thus minimizing the cost and the uncontrolled numerical truncation of the equations.

The viscous-inviscid splitting is based on Le Balleur's "Defect-Formulation Theory" for the full Navier-Stokes (NS) equations [2], [3], [5], [4], that replaces the single-field "NS" by a double "VII" field with full overlay ("Defect-NS" plus interacting "Pseudo-inviscid"). At present, only a thin-layer approximation of this theory is solved by the codes VIS15 and VIS25. Nevertheless, the thin-layer approximations used are neither wall-prescribed as in boundary layer theory, nor simply grid-prescribed as they are in the "Thin-layer Navier-Stokes" equations. With the Defect Formulation theory they are better governed by the aerodynamic field itself, the thin-layer truncations being performed in the "Displacement Reference Frame" [2], [3], [6], thus extending the validity of the thin-layer equations to massively separated flows.

The viscous part is a "Hybrid Field/Integral" method (FD/IBL), including the full unsteady viscous terms. It is solved by a marching thin-layer numerical technique with non-linearly implicit schemes, Le Balleur [2], [3], [4], in direct/inverse modes. At each viscous station, the method discretizes parametric turbulent velocity profiles, designed for attached or deeply separated flow [4], [5], [3], [2], in the direction normal to the local interacting inviscid streamlines. The turbulence is computed either with an algebraic model (mixing length plus velocity profiles [3], [2], [5]) or with an original 2-equation model "k - u'v', forced by the velocity profiles" [4], [5], [3], [2], for the Reynolds stresses and the entrainment.

The "Semi-implicit" time-consistent coupling algorithm first defined in 2D, Le Balleur, Girodroux [8], then extended to 3D [16], is used in the VIS15 and VIS25 codes as a substitute for the "Semi-inverse" algorithm [7], [2], [3] used in VIS05 with full potential. This algorithm is well adapted to the alternate direction implicit ADI-scheme of the inviscid TSP part of VIS15 and VIS25. The full "time-consistent coupling" is obtained by discretizing and converging the VII-coupling at each new time step, iterating the viscous and inviscid parts. This VII convergence within the time step is necessary to take into account the full viscous upstream influence at a given time-step (fully time-parabolic system, equivalent to a NS system). This permits to compute the strong viscous interaction phenomena, such as shock-boundary layer interaction and separation, in unsteady flows with the same properties as a full Navier-Stokes solver.

The time-consistent code VIS15 can be used to compute buffet type flows over fixed airfoils, where the unsteady flow is self-induced by separation, as well as unsteady shock-induced separated flows over airfoils undergoing forced oscillations (pitch or flap oscillations). An important point is that the fully time-consistent code VIS15 is used exactly in the same way independent of the fact whether the flow is steady or unsteady. This makes it possible for the code to discriminate between a steady or an unsteady solution and so to compute the buffet boundary. The 3D code VIS25 has been developed for the computation of buffet over wings, and some preliminary interesting results have already been obtained.

17.2.2 Self-adaptive Grids and Mesh Refinements

The technique of streamwise self-adaptive VII grids, first initiated in the 2D-steady code VIS05 [4], [2] and then implemented in the unsteady code VIS15, has been used for all the steady state calculations. Even if it is now possible in the code VIS15 to self-adapt the streamwise mesh at each time-step for unsteady computations, this possibility has not been adopted for the buffet flow configurations, which were all computed on a frozen mesh. The grid nodes distribution of the viscous part of the VII computation is also self-adaptive in the normal direction, between the wall and the edge of the boundary layer, both, to the local thickness and to the maximum gradient of the velocity profiles. In the tangential direction, for which the viscous and inviscid grids have coincident nodes at the wall and wake-center line, streamwise mesh clusterings are introduced not only in the shock region but also in the region of control, when this control zone is different from the shock region (for example in the case of a suction slot upstream of the shock).

At the shock, the refinement is generally such that $\Delta x \sim 0.3\delta$ to $\Delta x \sim 0.5\delta$, where δ is the local incoming boundary layer thickness. Such grids at the shock are not devoted to improve the inviscid shock treatment, but are just reaching the minimal Δx step size required for actually discretizing at the scale of the "free-interaction viscous process" [17] which governs the physics and the longitudinal extent of the shock boundary layer interaction.

This minimal grid refinement is believed to be necessary, in any method, for making calculations without a dominant effect of artificial viscosity on shock boundary layer interaction, especially in view of closely studying the effects of control in the neighborhood of the shock boundary layer interaction.

17.2.3 Implementation of the Control Law of Karlsruhe University

The new control law defined by Drs. Bohning and Doerffer (also see Chapter 12) has been implemented in the 2D code VIS15 as an alternative to Poll's law already included into the code during Euroshock (I).

A first linear version of this control law has been proposed, giving a relationship between the pressure drop across the porous plate, Δp, and the Mach number in the holes, M_h:

$$\frac{\Delta p}{p_o} = M_h^{\frac{1}{0.55}} \left(\frac{1}{1.2} \right)^{\frac{1}{0.55}} \qquad (1)$$

In this formula, p_o is the stagnation pressure on the inflow side of the flow. For suction, p_o is equal to the static pressure at the wall p_w, whereas for blowing, because the flow velocity in the cavity is very small, one may assume that the stagnation pressure is equal to the cavity pressure p_c.

It has been shown that a good correlation with the experimental data is provided by this linear law for blowing. However, for suction, a better formula is provided by the following non-linear law which takes into account the incoming outer tangential flow, i.e.,

$$\frac{\Delta p}{p_o} = M_h^{\frac{1}{0.55}} \left[\left(\frac{1}{1.2} \right)^{\frac{1}{0.55}} + b \left(\frac{2\,\tau_o}{\rho_h q_h^2} M_h \right)^{\frac{1}{a}} \right] \qquad a = 1.52, \; b = 25 \tag{2}$$

where τ_o is the shear stress of the incoming flow and ρ_h, q_h are, respectively, the density and the velocity in the holes. Since we have

$$\rho_h q_h^2 = \frac{\gamma M_h^2 p_{oh}}{\left(1 + \frac{\gamma - 1}{2} M_h^2\right)^{\frac{\gamma}{\gamma - 1}}}$$

and assuming, moreover, that the stagnation pressure in the hole p_{oh} is equal to the stagnation pressure p_o on the inflow side of the flow, the non-linear law may be written as:

$$\frac{\Delta p}{p_o} = M_h^{\frac{1}{0.55}} \left[\left(\frac{1}{1.2} \right)^{\frac{1}{0.55}} + b \left(\frac{2\,\tau_o \left(1 + \frac{\gamma - 1}{2} M_h^2\right)^{\frac{\gamma}{\gamma - 1}}}{\gamma M_h p_o} \right)^{\frac{1}{a}} \right]. \tag{3}$$

For given values of the pressure drop, stagnation pressure and shear stress, Formula (1) or (3), allows to compute the Mach number in the holes, M_h. The mean value of the ventilation velocity, q_w, is then deduced from the following relation for the mass flow rate

$$\dot{m} = A\;\rho_w q_w = A_h\;\rho_h q_h \qquad \eta_P = \frac{A_h}{A}, \tag{4}$$

where A_h and A are, respectively, the surface of the holes and the total (effective) surface of the porous region, and η_P is the porosity factor:

$$q_w = \eta_P \frac{\rho_h}{\rho_w} q_h = \eta_P \frac{1}{\rho_w} \frac{M_h p_o \sqrt{\frac{\gamma}{RT_o}}}{\left(1 + \frac{\gamma - 1}{2} M_h^2\right)^{\frac{\gamma + 1}{2(\gamma - 1)}}}, \tag{5}$$

again assuming that the stagnation pressure and temperature in the holes are equal to their values p_o and T_o on the inflow side of the stream.

Both, the linear and non-linear versions of the control law have been implemented in the 2D-unsteady code VIS15 code.

17.3 Steady Computations in 2D Flow

17.3.1 Investigation of Different Control Concepts

17.3.1.1 RAE-5225 and DRA-2303 with/without bump

The possibility to obtain drag reductions using a bump has first been investigated with the code VIS15 in the case of the laminar DRA-2303 airfoil and the turbulent RAE-5225 airfoil. An asymmetric bump located at x/c = [0.485:0.685], with a 0.25% chord relative thickness, has been used on the laminar airfoil. The symmetric bump of the turbulent airfoil, which has a relative thickness of 0.175% chord, is of circular-arc type and located at x/c = [0.4:0.6]. The 2-equation Le Balleur *k-u'v'* turbulence model has been used for all computations. Fine grids have been adopted for the calculations: the H-type grid used for the inviscid part of the computation is made of 300x100 points (with 187 streamwise nodes on each side of the airfoil, 50 nodes normal to the upper and lower surfaces) extending from -10 to 10 chords horizontally and vertically. The C-type grid used for the viscous part of the computation contains 490x49 nodes around the airfoil and its wake and is self-adaptive to the viscous layer thickness and to the local velocity gradient. Grid clusterings in chordwise direction are used in the regions of strong viscous interactions and in the control zones. This allows a chordwise concentration of 40 points in the bump region, with a minimal mesh size of 0.35% chord, comparable to the local boundary layer thickness of about 0.8% chord.

Figure 1 displays the Mach number field computed in the case of the turbulent RAE-5225 airfoil with bump. The pressure distributions with and without bump are shown, for the same airfoil, in Figure 2. An overall good agreement is found with the DERA experimental results (also see Chapter 19). The pressure levels are, however, slightly overestimated upstream of the shock on the upper surface of the airfoil, probably due to the TSP inviscid approximation of the VIS15 solver. The fine grid adopted in the bump region allows a good prediction of the relative pressure changes with and without bump.

Table 1, which summarizes the global aerodynamic coefficients, shows that a reduction of the total drag is predicted due to the bump for both airfoils, as in the experiments. This reduction of drag, which seems mainly be due to a decrease of the wave drag, is, however, found to be smaller in the computations.

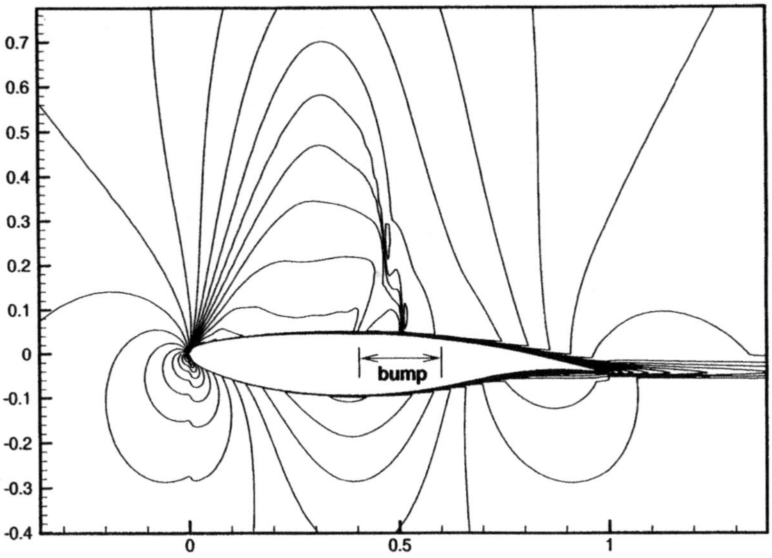

Figure 1 RAE-5225 airfoil. Mach number field with bump (VIS15 computation). M=0.73, C_L=0.756, Re=19x10^6

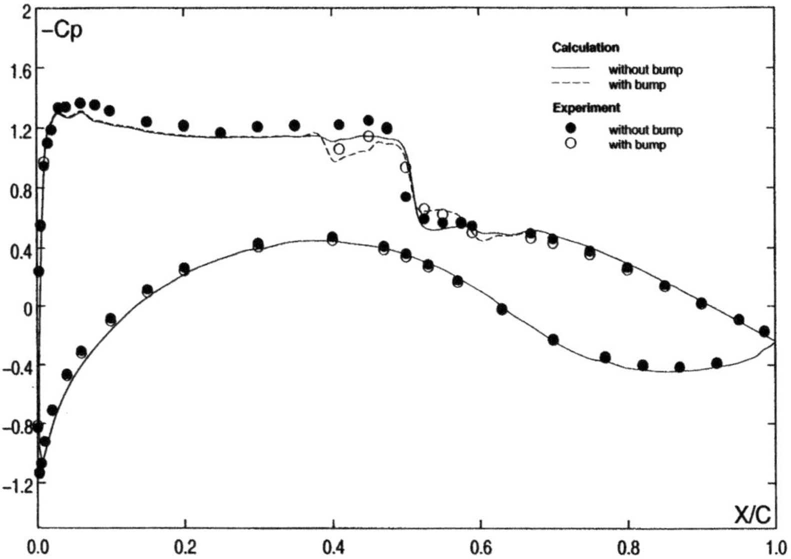

Figure 2 RAE-5225 airfoil. Pressure distributions with and without bump (VIS15 computation). M=0.73, C_L=0.756, Re=19 10^6.

Table 1 DRA-2303 and RAE-5225 airfoils with and without bump.

	Mach	Re	alpha	CL	CM	CD_{tot}	CD_{wave}
DRA-2303 w/o bump	0.6873	18981944.	1.9987	0.74677	-0.1010	0.01173	0.00244
DERA Experiment	0.6829	18982500.	2.0970	0.74721	-0.1006	0.01346	
DRA-2303 with bump	0.6845	18964728.	2.0015	0.73939	-0.1005	0.01069	0.00141
DERA Experiment	0.6801	18965200.	2.0980	0.73988	-0.0989	0.01138	
RAE-5225 w/o bump	0.7318	18812002.	2.8952	0.75783	-0.1002	0.01346	0.00408
DERA Experiment	0.7271	18811500.	2.9590	0.75539	-0.0918	0.01192	
RAE-5225 with bump	0.7302	18742444.	2.8994	0.75644	-0.0996	0.01321	0.00383
DERA Experiment	0.7255	18742300.	3.0160	0.75721	-0.0903	0.01029	

17.3.1.2 Influence of bump, discrete suction, and hybrid control

Different concepts of shock and boundary layer control have then been studied for the two airfoils: a bump, discrete control using suction through a small slot located upstream of the shock, and hybrid control. For each airfoil, four computations have been performed (baseline case without control, bump alone, suction alone, hybrid control by bump plus suction) for the same Mach number and lift coefficient. The total mass flow in the case of slot suction is $C_Q = -0.0001$ for the laminar airfoil and $C_Q = -0.0002$ for the turbulent one.

A grid sensitivity study has first been performed in the case of hybrid control which combines the bump and suction. Two different grids have been considered: a coarse grid with 150x100 nodes for the inviscid part (Δx=0.50% in the suction slot and Δx_{max}=1% in the bump region) and a fine grid with 300x100 nodes for the inviscid part (Δx=0.125% in the suction slot and Δx_{max}=0.5% in the bump region). It has been shown that the computations with the coarse grid underestimate the decrease of the displacement thickness downstream of the slot and so overestimate the viscous drag. A larger spreading of the shock is also predicted by the coarse-mesh calculation with, as a consequence, a lower estimation of wave drag. This demonstrates the necessity to use small streamwise space steps both, in the suction slot and in the shock region.

The fine grid (Figure 3), which includes 9 streamwise nodes in the small suction slot region whose width is only 1% of the airfoil chord, has been used for the final computations.

Figure 4 displays the Mach number fields predicted by the VIS15 code in the case of the RAE-5225 airfoil without control, with bump alone, discrete suction control alone, and hybrid control. The effect of suction on the pressure distributions is mainly to move the shock slightly downstream, both, with and without bump, Figure 5. The other effect of suction, also shown in Figure 5, is to increase the level of skin-friction and to reduce the displacement thickness in the suction zone and downstream up to the trailing-edge.

Tables 2 and 3 show the global coefficients obtained for both airfoils, an estimate of the pump drag in the case of suction being included for the laminar airfoil.

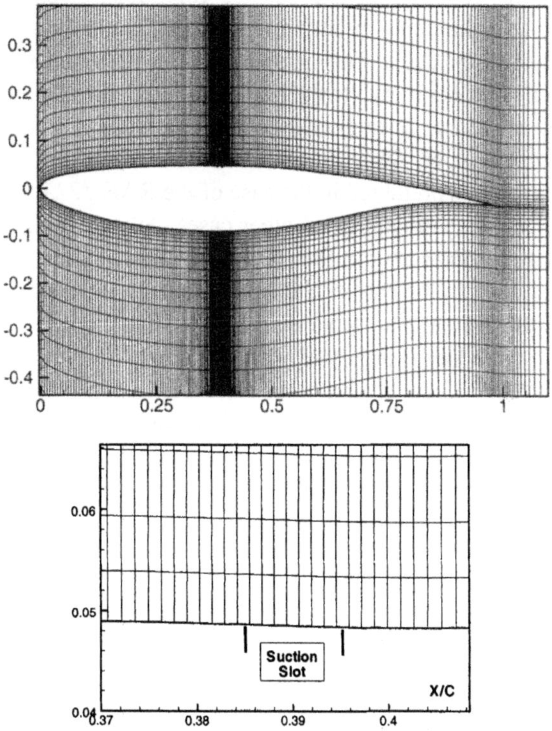

Figure 3 View of the inviscid part of the grid and details in the suction slot region. RAE-5225 airfoil, M=0.7318, Re=18.81x10^6, C_L=0.756.

Table 2 RAE-5225 airfoil. Influence of bump and local suction (VIS15 calculation). M=0.7318, Re=18.81x10^6, C_L=0.756, C_Q=-0.0002.

	alpha	CL	CM	CD_{tot}	CD_{wave}	CD_{visc}
Datum	2.9100	0.75606	-0.1020	0.01337	0.00401	0.00935
Bump	2.8990	0.75596	-0.1029	0.01291	0.00360	0.00931
Suction	2.7350	0.75692	-0.1068	0.01269	0.00474	0.00795
Hybrid	2.7170	0.75612	-0.1081	0.01073	0.00304	0.00769

Table 3 DRA-2303 airfoil. Influence of bump and local suction (VIS15 calculation). M=0.6873, Re=18.98x10^6, C_L=0.74, C_Q=-0.0001.

	alpha	CL	CM	CD_{tot}	CD_{wave}	Cd_{visc}	CD_{pump}
Datum	2.0000	0.74050	-0.1022	0.01127	0.00227	0.00901	0.00000
Bump	2.0000	0.74022	-0.1027	0.01086	0.00172	0.00915	0.00000
Suction	1.9300	0.74126	-0.1040	0.01135	0.00291	0.00833	0.00011
Hybrid	1.9300	0.74022	-0.1043	0.01103	0.00244	0.00848	0.00011

The bump leads, for both airfoils and the conditions given, to a decrease in total drag due to its dominating effect on wave drag. Suction leads in all cases to a reduction of wake drag, C_{dvisc}. The larger reduction observed in the case of the RAE-5225 airfoil is probably due to the higher value of the suction coefficient C_Q.

The level of wave drag strongly depends on the value of the total mass flow removed through the slot. With the adopted values for the parameter C_Q, a decrease of wave drag is predicted in the case of the RAE-5225 airfoil with hybrid control, but an increase is found in all other cases (ruling out the bump). The net effect is a decrease of total drag in the case of the turbulent airfoil, hybrid control giving the best results. In the case of the laminar airfoil, the decrease of wake drag is cancelled by the increase of wave drag and, adding, moreover, pump drag, a small increase of total drag is predicted with suction.

The influence of the level of suction has also been studied. Using higher values of total mass flow (C_Q=-0.0003 for the laminar airfoil and C_Q=-0.0006 for the turbulent airfoil), the increase of wave drag becomes much larger, leading then, with or without bump, to an increase of total drag for both airfoils.

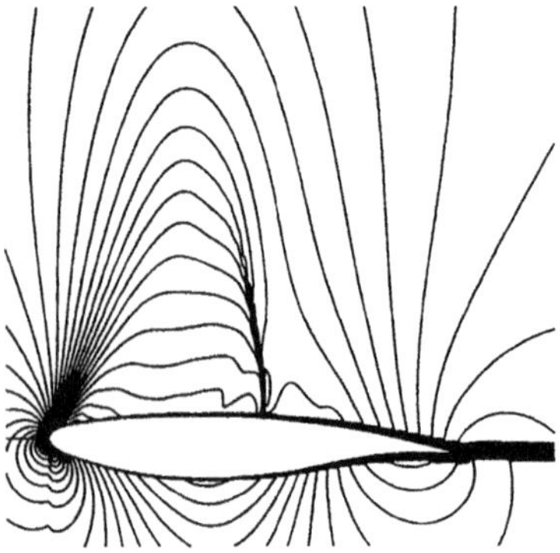

a) No Control

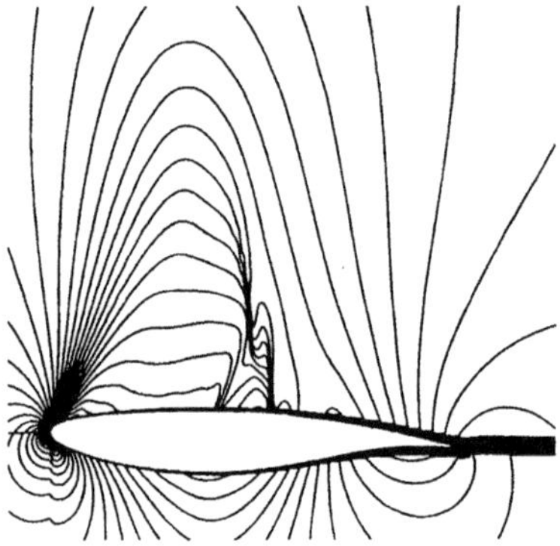

b) Bump

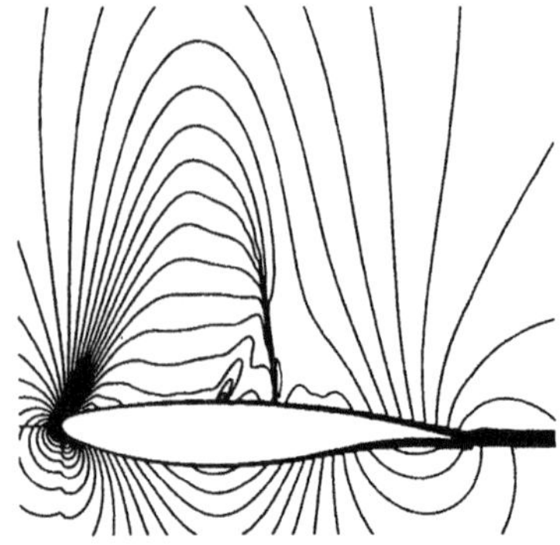

c) Discrete Suction

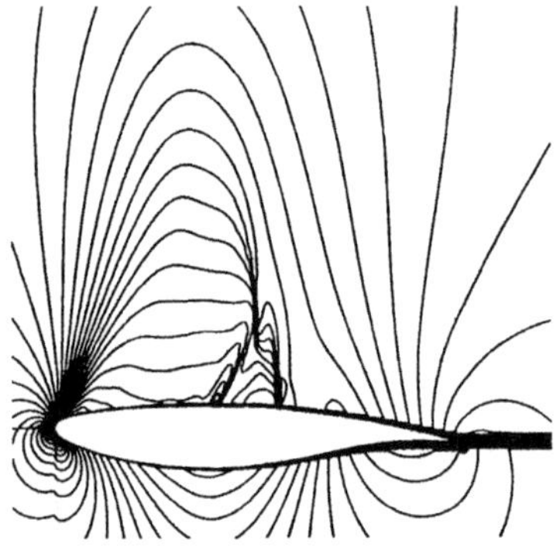

d) Hybrid control (Bump and Suction)

Figure 4 Mach number fields computed for different concepts of control (VIS15 code). RAE-5225 airfoil, M=0.7318, Re=18.81×10^6, C_L=0.756.

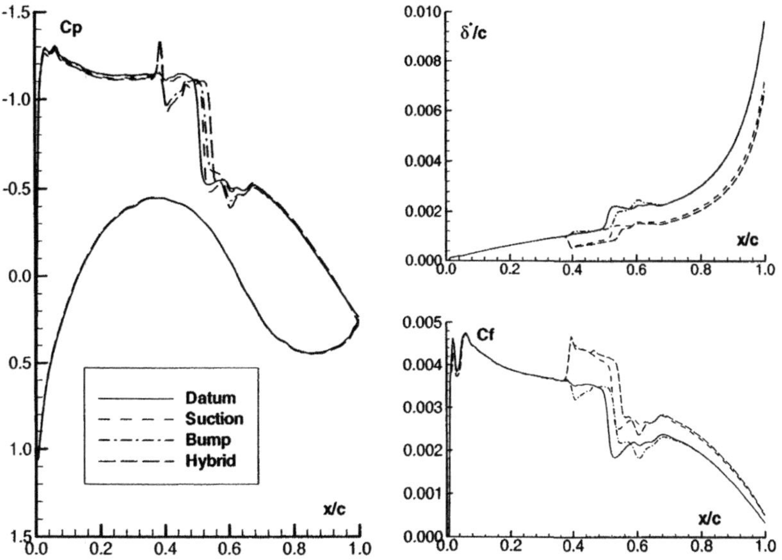

Figure 5 Influence of bump, discrete-suction, and hybrid control (VIS15 computation). Pressure, upper-surface displacement thickness, and skin-friction distribution. RAE-5225 airfoil, M=0.7318, Re=18.81×10^6, C_L=0.756.

17.3.1.3 VA-2 airfoil with active and passive control (perforated wall area)

Computations for the VA-2 airfoil have been performed with active and passive control. The Mach number and the incidence have been corrected for the baseline case (without control) in order to match the experimental lift and shock position. These values of Mach number and angle of attack have then been used for the computations with passive or active control. About 40 streamwise nodes have been placed in the porous region, as shown by the ventilation velocity distribution in Figure 6 (lower right-hand plot). This corresponds to a space-step Δx between 0.3% and 0.4% of the chord, smaller than the local boundary layer thickness in this region ($\delta/c \approx 0.7\%$). For the active control case, computed prescribing the total mass flow C_Q, a reduction of the experimental total mass flow has been applied in order to avoid convergence problems (C_Q=-0.00033 instead of C_Q=-0.00088). A rather good agreement is found between the calculated and experimental pressure distributions, Figure 6.

Table 4 shows that, for a prescribed incidence, the computed lift and moment coefficients are increased by active control and decreased by passive control, as in the experiments. The predicted drag coefficients are, however, smaller than in the experiment. The computed total drag coefficient is only slightly decreased by passive control and increased in the case of active control. This increase may perhaps be due to the value of the net mass flow being different in the computation and in the experiment. Also pump drag is not included in the experimental drag with active control.

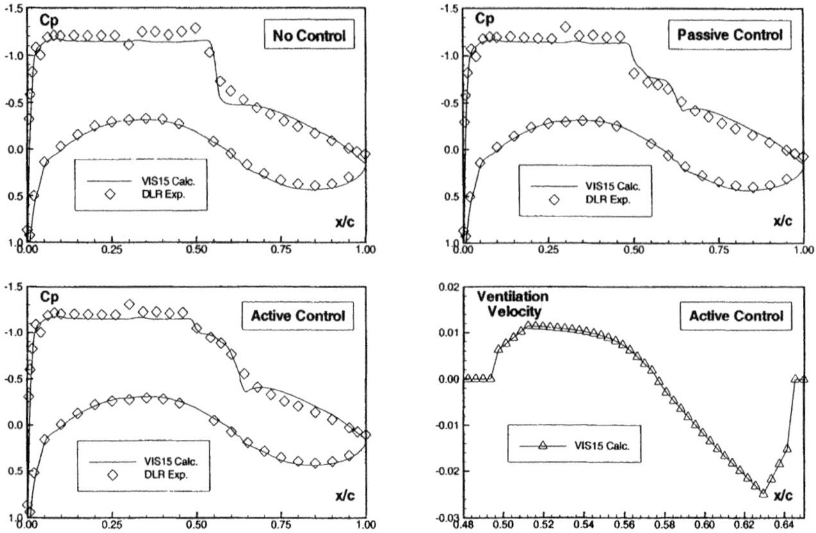

Figure 6 VA-2 airfoil. Pressure distributions w/o control, with passive and active control (0.49<x/c<0.65). Ventilation velocity with active control. M=0.734, $\alpha = 1.665^{o}$, Re=2.5×10^{6}.

Table 4 VA-2 airfoil without control and with active and passive control.

	α	CL	CM	CD_{tot}	CD_{wave}	CD_{visc}	CD_{pump}
Calc. No control	1.665	0.8136	-0.1388	0.01505	0.00570	0.00935	
Exp. No control	*2.000*	*0.8136*	*-0.1269*	*0.02740*			
Calc. Passive cont.	1.665	0.8010	-0.1371	0.01501	0.00498	0.01004	0.00000
Exp. Passive cont.	*2.000*	*0.7837*	*-0.1217*	*0.02320*			
Calc. C_Q=-0.00033	1.665	0.8190	-0.1391	0.01599	0.00568	0.00966	0.00065
Exp. C_Q*=-0.00088*	*2.000*	*0.8296*	*-0.1299*	*0.02000*			

17.3.2 DRA-2303 Airfoil with Discrete Suction Upstream of the Shock

It was decided, after this first phase devoted to the investigation of different control concepts, to focus on discrete suction control in the case of the laminar DRA-2303 airfoil. Polar computations have been performed with the unsteady code VIS15 in order to study the influence of discrete suction for a Mach number of M_{∞}=0.69 and a Reynolds number of Re = 19×10^{6}. The angle of attack has been progressively increased by steps, starting at $\alpha = -1.8^{o}$, to about $\alpha = 4^{o}$, covering the range of incidences investigated in the DERA experiments (also see Chapter 19). A fine grid with strong streamwise mesh refinements both, in the shock region and over the suction slot, located between x/c=0.45 and x/c=0.46 (Δx/c=0.001), has been used for these polar computations. Poll's law, with a prescribed total mass flow C_Q=-0.00007, has been employed for the calculation of control by discrete suction.

The lift evolution versus angle of attack, Figure 7, is rather well predicted by the code VIS15. The computations with and without control overestimate, however, slightly the lift for the higher angles of attack. Unsteady buffet solutions are also predicted by the computations, the onset of buffet corresponding to α=3.7° both, with and without discrete suction. Computed viscous drag, wave drag and total drag are also shown in Figure 7. With control, the calculations predict a decrease of total drag for lift coefficients lower than C_L=0.7. This drag reduction is mainly due to a decrease of the viscous drag. A very small increase of wave drag is generally also observed due to suction.

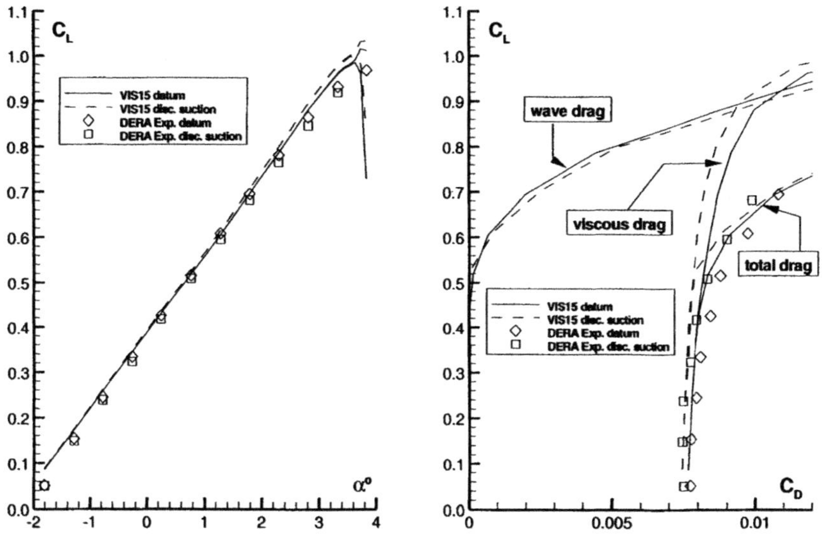

Figure 7 Influence of discrete suction (Poll's law with prescribed total mass flow rate). Lift versus incidence and lift versus drag. DRA-2303 airfoil, M=0.69, Re=19x10^6.

The mandatory test case with discrete suction, corresponding to a lift coefficient of C_L=0.6075 (experimental data point 2129), has been computed, prescribing either the cavity pressure or the total mass flow rate, respectively, in the slot, and using the different control laws implemented in the VIS15 code. An example of the pressure distribution and the suction velocity in the slot is shown in Figure 8, using the linear law of Karlsruhe.

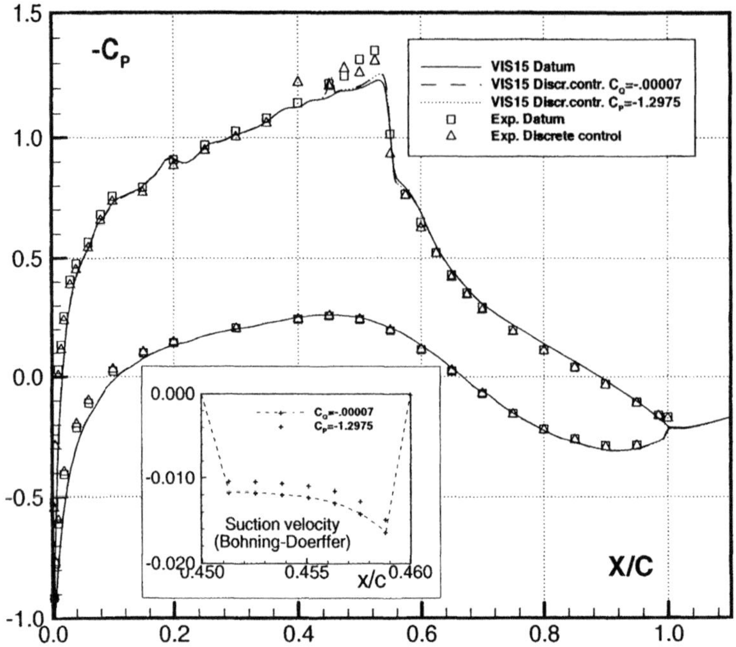

Figure 8 Influence of discrete suction control. Wall pressure and suction velocity. DRA-2303 airfoil, M=0.69, Re=19x10^6, C_L=0.6075.

Table 5 compares, in the case of a prescribed pressure, C_p=-1.2975, in the slot, the drag reductions predicted by the VIS15 code, using Poll's law and the two versions (linear and non-linear) of the law of Karlsruhe University. For the latter, both, the nominal porosity of 4% and an effective porosity of 2.48%, determined by Bohning and Doerffer, have been used. The best fit with Poll's law is obtained using the non-linear version of the law of Karlsruhe University with the porosity of 4%. The contribution of the non-linear term, which takes into account the tangential flow, is roughly as important as the linear term, as expected for transonic flows. The drag reduction with respect to the no-control case remains lower than in the experiments, i.e., 7.5%, ranging from 2.5% to 5%.

Table 5 Influence of the control law (VIS15 computation). DRA-2303 airfoil with discrete suction control, M=0.69, Re =19x10^6.

Law	Porosity	CQ	Drag reduction
Poll	4.00%	-0.000035	3.5%
Karlsruhe linear	4.00%	-0.000066	4.9%
Karlsruhe non-linear	4.00%	-0.000035	3.5%
Karlsruhe linear	2.48%	-0.000047	4.0%
Karlsruhe non-linear	2.48%	-0.000024	2.5%

The influence of the slot location on drag reduction and on buffet onset has also been studied with the code VIS15. Six slot positions have been investigated. Figure 9a shows that, for C_L=0.6075, the drag reduction becomes very small when the slot is close to the shock, and that the highest drag reductions are obtained when the slot is far upstream of the shock. Concerning buffet onset, the angle of attack has, for each slot position, been increased up to unsteady conditions. A small decrease of the angle of attack corresponding to buffet onset is predicted by the unsteady computations when the slot is moved from its nominal position to the shock location. Then buffet onset is shifted to higher angles of attack when the slot is located downstream of the shock, Figure 9b.

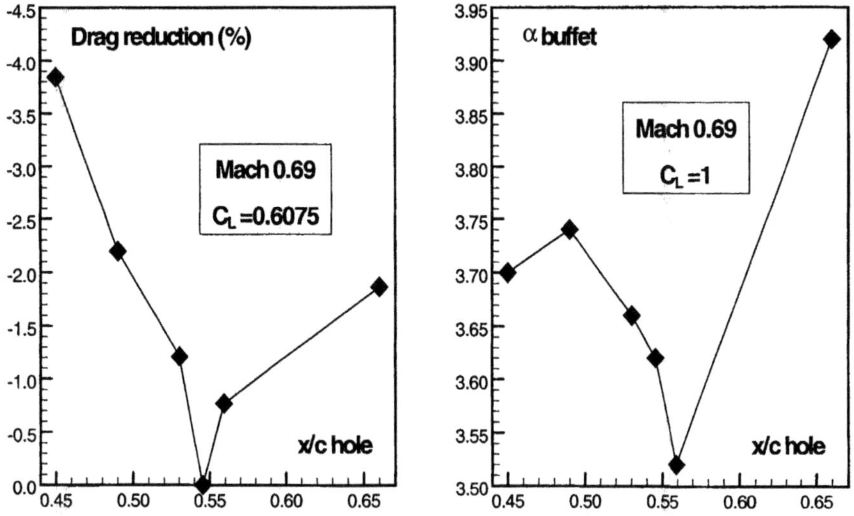

a) Drag reduction at C_L=0.6075 b) Onset of buffet

Figure 9 Effect of slot location on drag reduction and buffet onset (VIS15 computations). DRA-2303 airfoil, M=0.69, Re=19x10^6, (Poll's law, C_Q=-0.00007).

17.4 Unsteady Computations in 2D Flow

17.4.1 Influence of a Bump on Buffet Onset

The unsteady code VIS15 has been used to study time-accurately the onset of buffet in the case of the laminar DRA-2303 airfoil with and without bump. The angle of attack has been progressively increased by steps up to buffet conditions. These α-sweeps have been started at an angle of attack of α=2.5° with and without bump. Figure 10 displays the envelope of the lift coefficient versus angle of attack corresponding to the minimum and maximum (unsteady) lift coefficient predicted for each incidence; these two values are identical if the flow is steady and differ when the flow is unsteady. The gap between the two curves gives an idea of the unsteadiness of the buffet phenomenon.

The main effect of the bump is to postpone buffet onset to higher lift values and slightly higher angles of attack. For $\alpha > 3.6^o$ without bump and $\alpha > 3.8^o$ with bump, a fast growing of the flow unsteadiness is observed. The amplitude of the lift variation is, however, smaller, at a given angle of attack, in the case of the airfoil with bump. It can also be seen in Figure 10 that the computed lift coefficients corresponding to the onset of buffet are not far from the experimental maximum lift values.

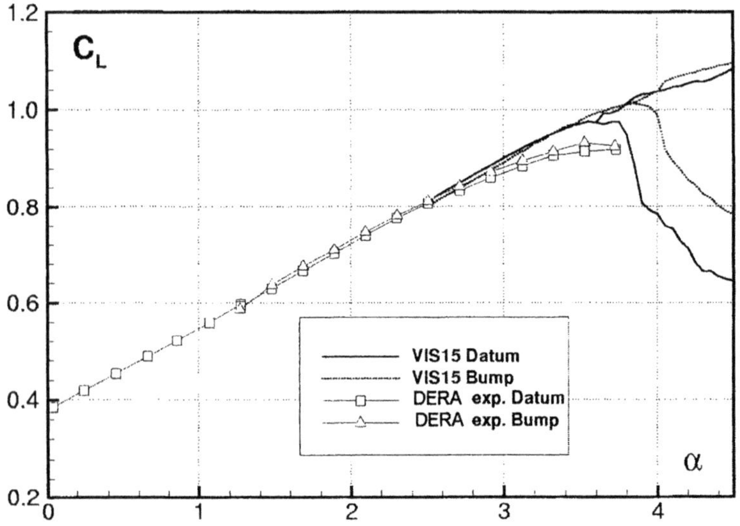

Figure 10 Unsteady prediction of buffet onset using α -sweeps (lift envelopes, VIS15 computation). DRA-2303 airfoil without control and with bump, M=0.69, Re =19x10^6.

Systematic comparisons have been made between the empirical steady criteria C1 and C2, provided by ALENIA, and the actual occurrence of unsteady solutions in the viscous computation, (criterion C3), Figure 11. Criterion C1 is based on the Mach number just upstream of the shock, and criterion C2 is based on the pressure divergence at the trailing-edge of the airfoil.

The angles of attack and lift coefficients, for which the VIS15 computations satisfy these two empirical criteria, are compared in Table 6 to the real onset of buffet in the unsteady calculations (criterion C3) for the datum airfoil and for the airfoil with bump, respectively. The three criteria C1 (top), C2 and C3 predict the onset of buffet for very similar aerodynamic conditions α and C_L in the case of the airfoil with bump. For the datum airfoil, the computations satisfy criteria C2 and C3, but the onset of buffet is predicted too early, i.e., for lower values of α and C_L, using criterion C1.

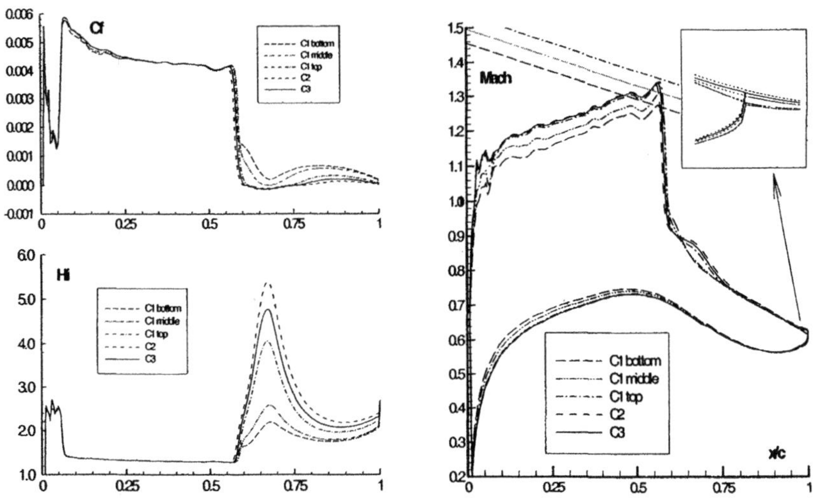

Figure 11 Buffet onset prediction. Criteria C1 (M_{shock}),C2 (ΔCp_{TE}), C3(unsteady VIS15 solution)
DRA-2303 airfoil with bump, M=0.687, Re=19x10^6.

Table 6 Evaluation of different industrial criteria for buffet onset (VIS15 calculations). DRA-2303 airfoil with/without bump, M=0.69, Re =19x10^6.

	Without Bump			With Bump		
Criterion	Mach	α	C_L	Mach	α	C_L
C1 bottom	0.690	2.25	0.7643	0.687	2.95	0.8826
C1 middle	0.690	2.55	0.8197	0.687	3.25	0.9368
C1 top	0.690	2.95	0.8922	0.687	3.65	0.9965
C2	0.690	3.55	0.9763	0.687	3.80	1.0111
C3	0.690	3.55	0.9763	0.687	3.75	1.0069

17.4.2 Cancellation of Buffet Using Discrete Suction Control

The possibility to alleviate buffet oscillations using discrete suction control has also been studied with VIS15. A computation without suction has first been performed for α=3.8°, an angle of attack which is well into the buffet domain, as shown in Figure 12 by the large lift variations versus time and in Figure 13 by the instantaneous velocity fields and pressure distributions at 4 time steps within a buffet period. Then strong suction has been applied, using the same values as INTA (see Chapter 16) for the total mass flow, C_Q=-0.0007, and for the slot location, x/c=[0.54:0.55]. The buffet oscillations are completely damped, Figure

12, with the other effect of this strong suction being an increase of lift due to a more downstream shock location compared to the no-control case.

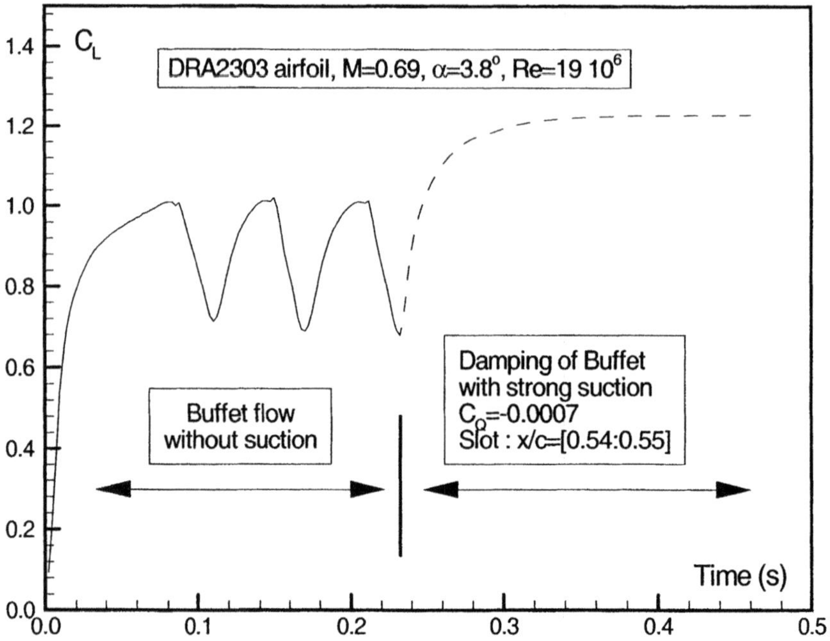

Figure 12 Cancellation of buffet using discrete-suction control with strong suction (VIS15 computation). M=0.69, α =3.8°, Re=19x10^6.

17.4.3 Damping of Buffet by Passive Control through a Perforated Wall

The possibility to dampen buffet oscillations by passive control through a perforated surface in the shock region has been experimentally tested at DERA in the case of the DRA-2303 airfoil (data points "747" without control and "1303" with passive control; see References [1.8] and [1.10] of Part A, Chapter 1). The capability of the unsteady code VIS15 to describe the unsteady pressure signals in the buffet regime has been used to confirm these experimental results. A computation without control has first been performed at flow conditions close to the nominal experimental conditions, for which strong buffet is predicted. Some examples of the time evolution of the pressure at different locations on the upper surface of the airfoil are shown in Figure 14.

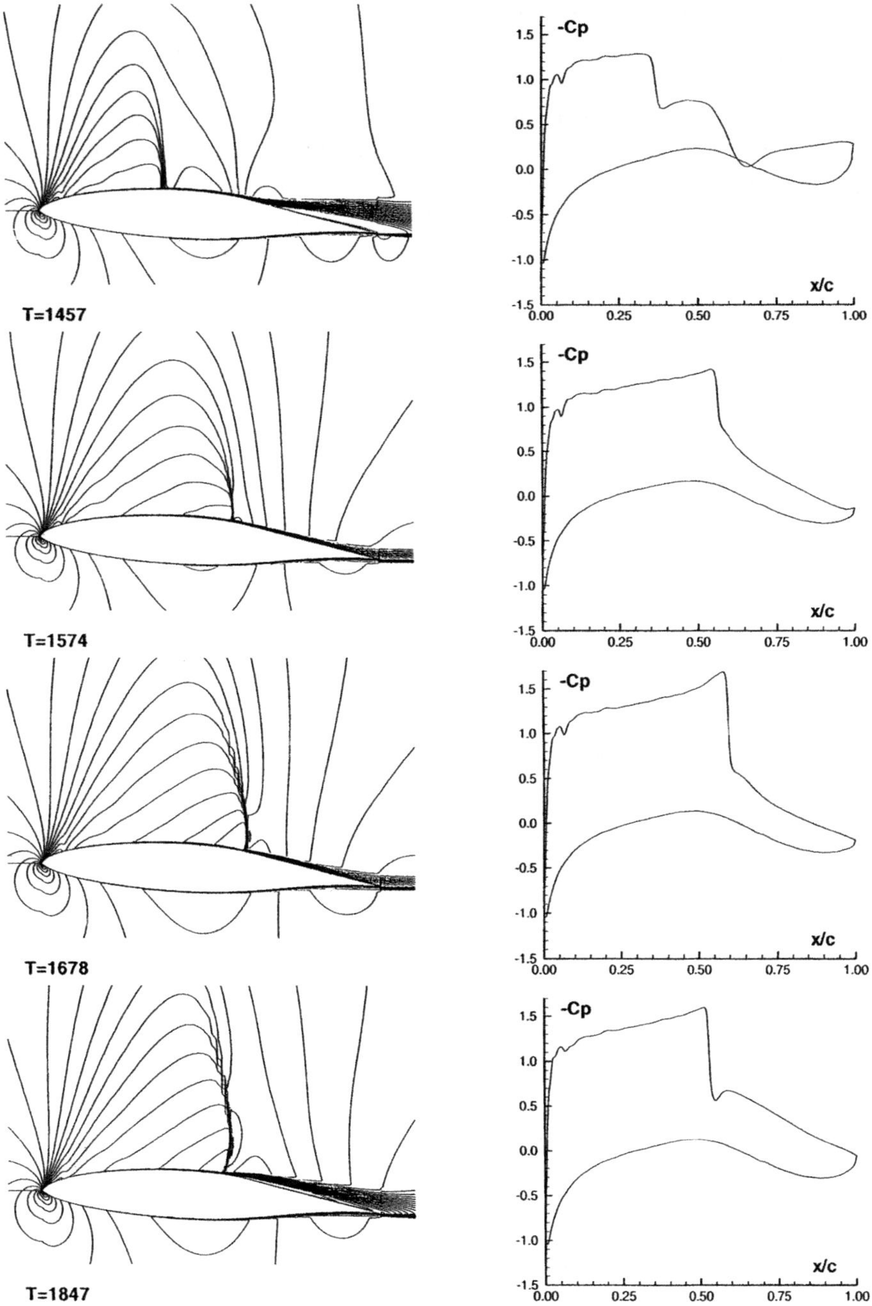

Figure 13 VIS15 buffet calculation. DRA-2303 airfoil, M=0.69, α=3.8°, Re=19x10^6
Instantaneous velocity fields and pressure distributions at the wall.

Even if the computation somewhat overpredicts the amplitude of the shock motion, the computed mean value of the pressure over a buffet cycle and the root mean square value (RMS) of the pressure fluctuations reproduce very well the experimental behavior. A rather complete damping of the buffet oscillations is found in the numerical simulation with passive control at exactly the same aerodynamic conditions, Figure 15. In the experimental results, this damping is only partially observed, as shown by the RMS levels in the shock region, but the angle of attack in the experiment is not identical and higher with control than without control.

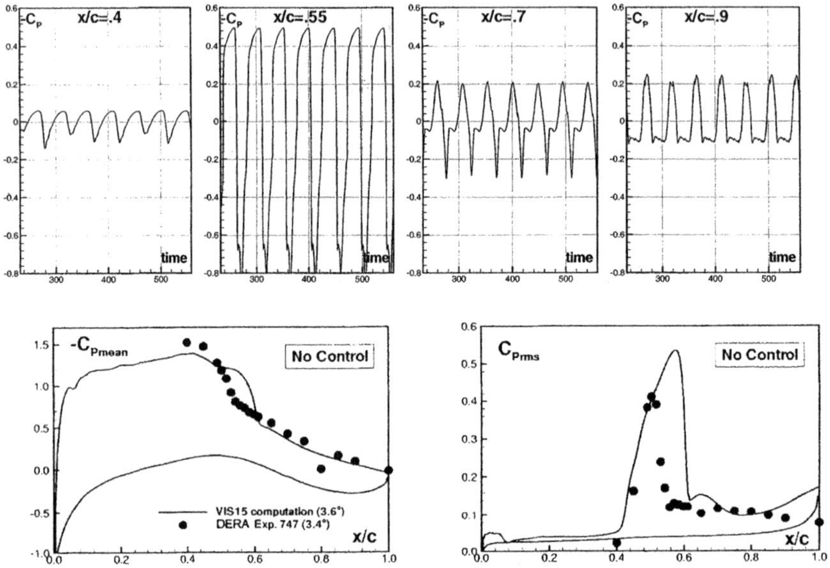

Figure 14 Pressure signals at different locations on the upper surface (VIS15 code). Mean value of the pressure distributions over one buffet cycle and RMS pressure fluctuations. DRA-2303 airfoil without control, M=0.69, α=3.6°, Re=19x10^6.

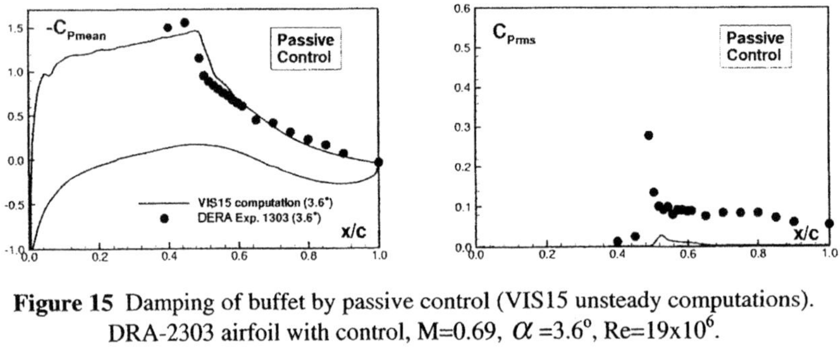

Figure 15 Damping of buffet by passive control (VIS15 unsteady computations). DRA-2303 airfoil with control, M=0.69, α=3.6°, Re=19x10^6.

17.5 ADIF Swept Wing 2D and 3D Calculations with and without Bump

The effect of sweep on the efficiency of control has been experimentally studied at DLR within Task 3 in the case of the turbulent ADIF sheared wing. Contoured end plates have been designed to realize infinite swept-wing conditions. 2D experiments have first been performed with the basic airfoil, with and without control, as reference for the final sheared-wing 3D experiments.

A number of 2D and 3D computational test cases have been defined for different bump heights and lift coefficients, Table 7. The VIS15 code has been used for the 2D calculations, and the fully three-dimensional unsteady VIS25 wing code (direct extension in 3D of the unsteady VIS15 code) has been adapted for the computation of the infinite swept-wing configurations. The Mach numbers in Table 7 are corrected values provided by DLR for the calculations, the experimental Mach number being slightly higher.

Table 7 2D and 3D ADIF sheared-wing test cases (defined in Task 2).

	2D conditions				3D conditions					
	Mach	Re	C_L	h_{bump}	Mach	Re	C_L	h_{bump}	X_{crest}	
case 1	0.755	$8.0\ 10^6$	0.755	0.600%	0.840	$9.9\ 10^6$	0.61	0.54%	74.5%	mandatory
case 2a	0.757	$8.0\ 10^6$	0.5	0.175%	0.842	$9.9\ 10^6$	0.404	0.1573%	76%	mandatory
case 2b	0.757	$8.0\ 10^6$	0.5	0.350%	0.842	$9.9\ 10^6$	0.404	0.3146%	76%	mandatory
case 3a	0.757	$8.0\ 10^6$	0.6	0.175%	0.842	$9.9\ 10^6$	0.485	0.1573%	76%	optional
case 3b	0.757	$8.0\ 10^6$	0.6	0.350%	0.842	$9.9\ 10^6$	0.485	0.3146%	76%	optional
case 4a	0.757	$8.0\ 10^6$	0.7	0.175%	0.842	$9.9\ 10^6$	0.565	0.1573%	76%	mandatory
case 4b	0.757	$8.0\ 10^6$	0.7	0.350%	0.842	$9.9\ 10^6$	0.565	0.3146%	76%	optional

17.5.1 Numerical Results for 2D and Corresponding 3D Conditions

All the mandatory test cases have first been computed using the corrected Mach number and adapting the angle of attack in order to match the experimental lift. The 2D and 3D computed global coefficients are summarized in Table 8.

With the bump, a drag reduction of about 10% has been determined for test case 1. Figure 16 compares the 2D and 3D pressure distributions also showing the skin-friction lines predicted by the code VIS25 on the upper surface of the wing. The 3D computation provides both, the streamwise (cf^1, δ^1) and cross-flow (cf^2, δ^2) components of the viscous fields, Figure 17. The large height of the present bump leads to a double shock system with a strong expansion in the bump region.

Test cases 2a and 4a with the small bump, and 2b with the medium-sized bump, have also been computed both, in 3D (code VIS25) and 2D (code VIS15). The 2D pressure distributions are in good agreement with the experimental results for the lower lift cases 2a and 2b, Figure 18. For the higher lift, case 4a, the shock is located slightly more downstream than in the experiment.

Table 8 ADIF sheared wing. Predicted global coefficients in 2D (VIS15) and 3D (VIS25) computation.

		Mach	α	CL	CD_{wave}	CD_{visc}	CD_{tot}	(CD) %
Case 1	2D No Bump	0.755	-0.1260	0.75542	0.00533	0.01016	0.01549	
Case 1	2D Bump 0.60%	0.755	-0.1260	0.76708	0.00355	0.01043	0.01398	-9.7%
Case 1	3D No Bump	0.840	0.1800	0.61018	0.00590	0.01056	0.01646	
Case 1	3D Bump 0.54%	0.840	0.0714	0.61088	0.00491	0.01006	0.01497	-9.0%
Case 2	2D No Bump	0.757	-1.2799	0.5259	0.00082	0.00795	0.00877	
Case 2	*Exp. No Bump*	*0.765*	*-1.0000*	*0.5259*			*0.01050*	
Case 2a	2D Bump 0.175%	0.757	-1.3980	0.5016	0.00050	0.00797	0.00847	- 3.4%
Case 2a	*Exp. Bump 0.175%*	*0.765*	*-1.0000*	*0.5017*			*0.00930*	*- 11.4%*
Case 2b	2D Bump 0.350%	0.757	-1.5246	0.4943	0.00314	0.00775	0.01089	+24.2%
Case 2b	*Exp. Bump 0.350%*	*0.765*	*-1.0000*	*0.4943*			*0.01160*	*+10.5%*
Case 2	3D No Bump	0.842	-0.8700	0.4247	0.00072	0.00902	0.00974	
Case 2a	3D Bump 0.1573%	0.842	-0.9850	0.4054	0.00036	0.00903	0.00939	- 3.6%
Case 2b	3D Bump 0.3146%	0.842	-1.0600	0.3999	0.00100	0.00892	0.00992	+1.8%
Case 4	2D No Bump	0.757	-0.4363	0.7307	0.00492	0.00838	0.01330	
Case 4	*Exp. No Bump*	*0.765*	*0.0900*	*0.7308*			*0.01700*	
Case 4a	2D Bump 0.175%	0.757	-0.5090	0.7091	0.00290	0.00831	0.01121	-15.7%
Case 4a	*Exp. Bump 0.175%*	*0.765*	*0.0000*	*0.7090*			*0.01340*	*-21.2%*
Case 4	3D No Bump	0.842	0.0700	0.5909	0.00543	0.01047	0.01589	
Case 4a	3D Bump 0.1573%	0.842	0.0000	0.5799	0.00402	0.01029	0.01431	-9.9%

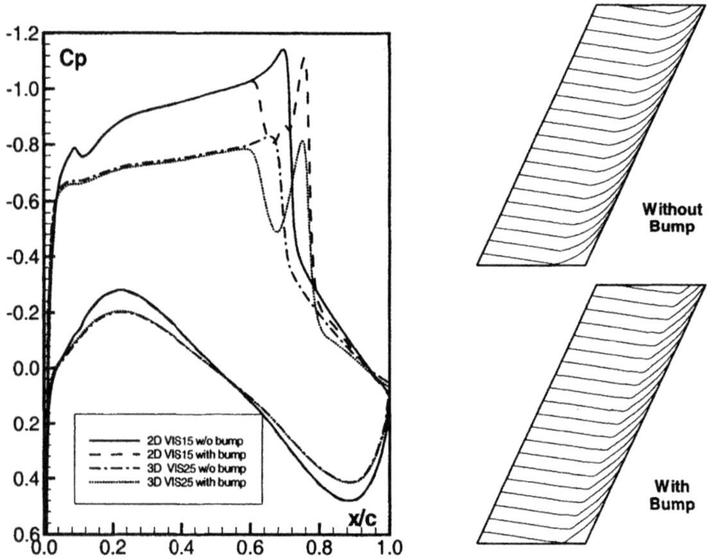

Figure 16 ADIF airfoil/ sheared wing with/without bump. Test case 1. Pressure distributions predicted by 2D-VIS15 and 3D-VIS25 computations, and 3D skin-friction lines on the upper surface.

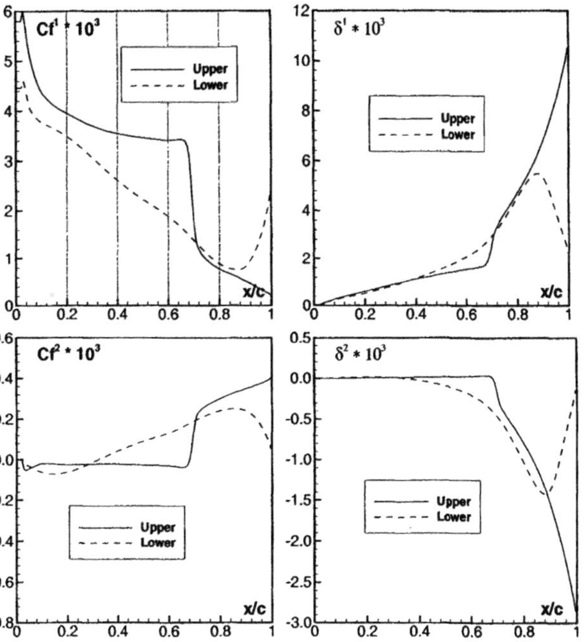

Figure 17 Streamwise and cross-flow components of skin-friction and displacement thickness. ADIF wing, test case 1 (VIS25 3D computation).

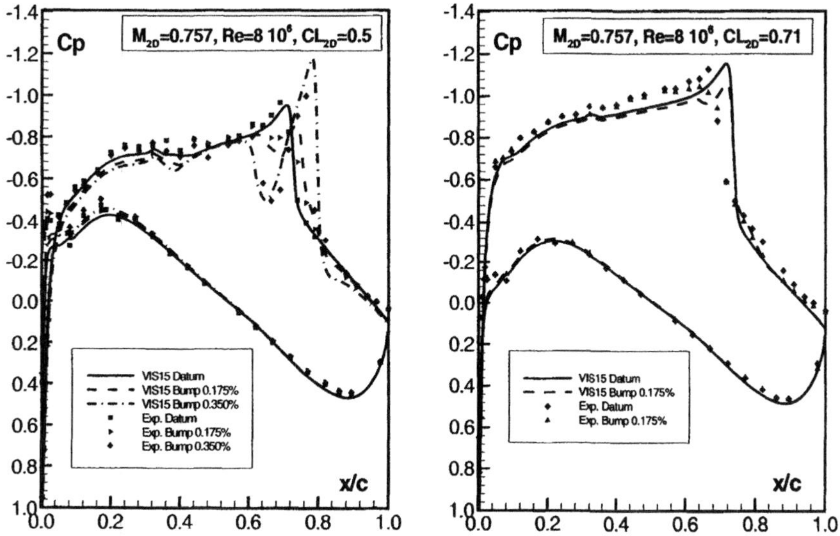

Figure 18 ADIF airfoil. 2D computations with / without bump. Test cases 2a, 2b and 4a.

The computations predict, as the experimental results, at the present lift coefficient ($C_L \approx 0.50$), a decrease of total drag with the small bump and an increase of total drag with the medium-sized bump. The 2D and 3D computations underpredict, however, the drag reduction with respect to the experiments for test case 2a (-3.4% in the 2D computation and -3.6% in 3D versus -11.4% in the experiments). For test case 2b, a larger increase of drag (24.1%) than in the experiments (10.5%) has been found in 2D due to the strong expansion over the medium-sized bump, also present in the experimental results. In 3D, this increase of total drag is only 1.8%. For test case 4a, a decrease in drag is obtained both, in the computations (-15.7% in 2D and -9.9 % in 3D) and in the 2D experiments (-21.2%).

However, these results have been found to be very sensitive to small variations of the Mach number, as shown by the drag polars of Figure 19 computed with the 2D code VIS15 for M=0.757 and for the slightly higher Mach number of M=0.762, the latter being closer to the experimental nominal value of M=0.765.

In both cases, the computations, as the experiments, predict a better performance with the small bump at the lower lift coefficients while the large bump is more effective at higher lift. In the case of M=0.757, the computed limits of effectiveness of the bumps are $C_L \approx 0.5$ (small bump) and $C_L \approx 0.65$ (large bump), these values being higher than in the DLR experiments ($C_L \approx 0.4$ for the small bump, $C_L \approx 0.55$ for the large bump). This explains the very small decrease of total drag observed for test case 2a at C_L=0.5. In the case of M=0.762, the computed limits of efficiency are shifted to lower lift values, rather very close to the experimental results, the relative drag variations for C_L=0.5 being now -8% with the small bump and +16% with the large bump.

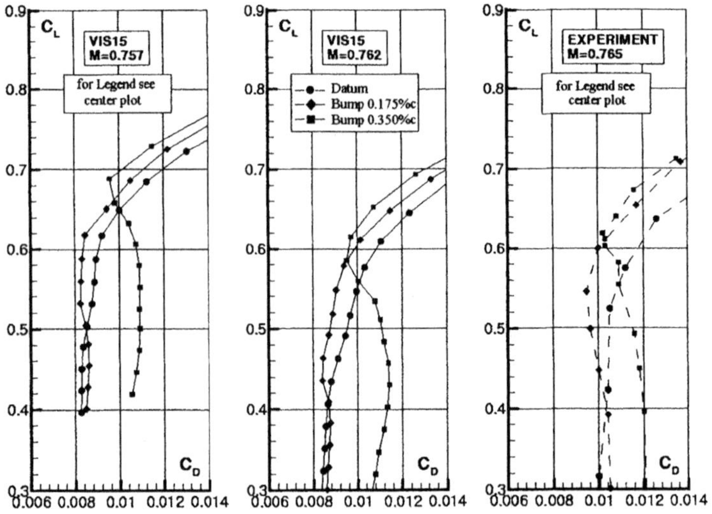

Figure 19 ADIF airfoil. Computed and experimental drag polars. Influence of the Mach number. VIS15 2D computations.

17.5.2 Final 3D Experimental Cases and Corresponding Numerical Results

The final 3D experiments were only performed towards the end of the Euroshock II project. There were some changes (bump location, Reynolds number, transition location, suggested corrected Mach number lower than 0.842) with respect to the originally scheduled experiments. As a consequence, the previous 3D computations, performed in parallel with the 2D calculations, were carried out for conditions different from the experimental ones. The final 3D experimental conditions for the new final 3D computations are shown in Table 9 below.

Table 9 Final ADIF wing 3D experiments.

	Mach	Re	C_L	h_{bump}	X_{crest}
case 2b	0.837	$6.7x\ 10^6$	0.405	0.3146%	71.5%
case 3a	0.837	$6.7x\ 10^6$	0.475	0.1573%	71.5%
case 4a	0.837	$6.7x\ 10^6$	0.587	0.1573%	71.5%
case 4b	0.837	$6.7x\ 10^6$	0.576	0.3146%	71.5%

Since the most important point is to have the "correct" location of the shock with respect to the bump, it has been decided, in these final 3D computations with the unsteady code VIS25, to adapt the angle of attack in order to match the shock location, instead of the lift coefficient, for the datum test cases without bump. Then, the test cases with bump have been run at exactly the same conditions. The corrected Mach number used in all the computations is M=0.837.

Without bump, the experimental shock location is reached in the computations using a slightly higher angle of attack than in the experiments (α=-0.25° instead of α=-0.50° for test case 3, α=0.25° instead of α=0.00° for test case 4). The effect of both bumps considered is then fairly well predicted by the VIS25 computations, as shown by the pressure distributions in Figures 20 and 21. In the case of the lower lift configuration (3a) with the small bump, it can be seen that the shock is located more upstream than without bump, but it is followed by a strong expansion both, in the computational and in the experimental results. A rather different behavior is observed both, with the small bump (test case 4a) and the medium-sized bump (test case 4b) at the higher lift coefficient, with a weakening of the strong shock predicted without bump.

As in 2D, polar computations have been performed in 3D in order to investigate the limits of efficiency of the two bumps. Very similar trends are found by the computations and the experimental results, as shown by the drag polars in Figure 22. With the small bump, an increase in drag of about 10 counts has been found for lift coefficients lower than 0.52, whereas a decrease of drag, higher in the computations than in the experiments, is predicted above this lift value. As in the experiments, the limit of efficiency is also shifted to higher lift coefficients with the medium-sized bump, but this limit is overestimated as is the drag increase or decrease with respect to the experimental results.

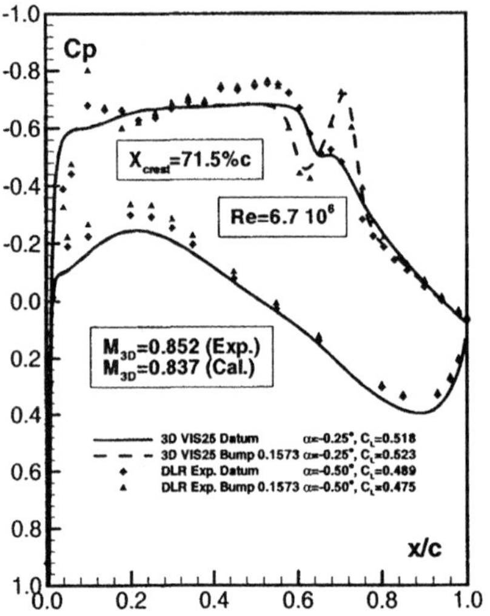

Figure 20 ADIF sheared wing. VIS25 3D-computations. Test case 3a.

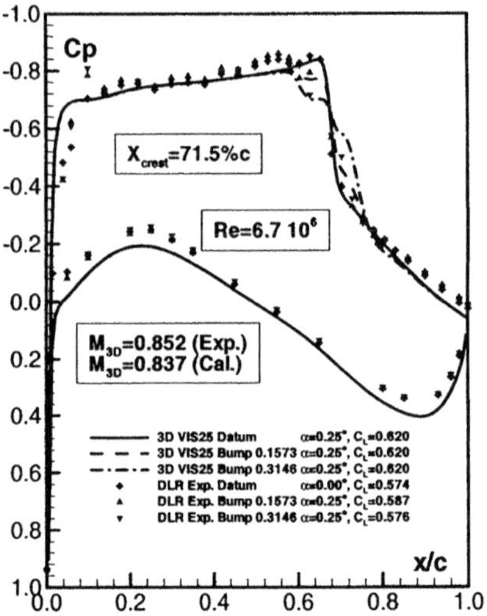

Figure 21 ADIF sheared wing. VIS25 3D-computations. Test cases 4a and 4b.

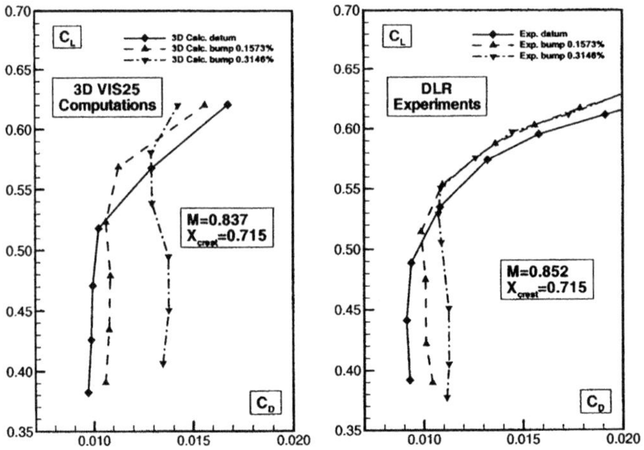

Figure 22 ADIF airfoil. Computed and experimental drag polars. VIS25 3D computations.

17.6 Conclusion

As an extension of the numerical work performed during EUROSHOCK (I) both, in viscous steady flow (VIS05 code) and viscous unsteady flow (VIS15 code) to study passive flow control over transonic airfoils, the 2D and fully-3D time-consistent Viscous-Inviscid Interaction (VII) codes VIS15 and VIS25 have been used in EUROSHOCK II for further numerical investigations of transonic flow control.

The 2D unsteady code VIS15 has been utilized in the investigation of several different concepts of shock and/or boundary layer control (bump, discrete slot suction, hybrid control, active and passive control by ventilation). The capability of the time-accurate VIS15 code to discriminate without any change in the method between a steady and an unsteady (buffet) solution has been used to study not only the effect of the control devices on drag variation at steady near-design conditions, but also to study the effect of several of these control devices on buffet onset and buffet damping at off-design conditions.

The fully-3D unsteady code VIS25, which has been developed as extension of the 2D VIS15 code, has been adapted for infinite swept wing calculations and used to compute the effect of sweep in the case of the ADIF sheared wing with and without bump.

The necessary level of very fine streamwise grids has been reached in these VII calculations with adaptive streamwise strong clustering of the mesh in the shock area and with a prescribed clustering in the slot region in the case of discrete suction control. Such adaptive streamwise grid clustering, conjugated with the normal grid adaptation of the VII approach, is necessary to actually discretize at the longitudinal scale of the "free interaction viscous process" [17] which governs the physics and the extent of shock boundary layer interactions.

As far as 2D steady conditions are concerned, it has been found that a decrease of wave and total drag is always obtained with the bump at design conditions, both, in the case of the turbulent RAE-5225 and the laminar DRA-2303 airfoil. Using discrete control by suction through a small slot located upstream of the shock, a reduction of the viscous drag together with an increase of the wave drag is predicted for both airfoils, the net effect being, however, a decrease of total drag for optimal levels of suction. The highest drag reduction has been obtained by hybrid control with both bump and discrete suction, at least for the turbulent airfoil.

Polar computations up to buffet conditions in the case of the laminar DRA-2303 airfoil with control by discrete suction have shown that a drag reduction is obtained for lift coefficients lower than about 0.7. This decrease of drag is, however, slightly smaller than determined in the DERA experiments, and ranges from 2.5% to 5% depending on the control law used for the computation of the suction velocity.

As far as the 2D unsteady computations and buffet control are concerned, time-accurate buffet control configurations have been computed with success by the 2D unsteady VIS15 code, increasing progressively the angle of attack. The main effect of bump control and of discrete-suction control, at moderate levels of suction, is to shift the onset of buffet to slightly higher lift coefficients, whereas a complete damping of buffet has been predicted, either using passive control with porous walls, or using discrete-suction control at the foot of the shock at high levels of suction.

Finally, as far as the 3D investigation of drag reduction and flow control is concerned, 2D and 3D polar computations have been successfully performed by the 2D and fully-3D unsteady codes VIS15 and VIS25 in the case of the ADIF airfoil and the ADIF swept wing with and without bump. Numerical 2D and 3D results have shown, in agreement with the experimental results, that a drag reduction is obtained with a bump for lift coefficients higher than the one marking the "limit of efficiency" of the bump. This limit of efficiency is shifted to higher lift coefficients when the height of the bump is increased. Numerical results also indicated this limit to be rather sensitive to the freestream Mach number.

References

[1] Le Balleur, J. C., Girodroux-Lavigne, P. and Gassot, H., "Development of Viscous-Inviscid Interaction Codes for Prediction of Shock Boundary Layer Interaction Control (SBLIC) and Buffet over Airfoils". In: EUROSHOCK – Drag Reduction by Passive Shock Control. Notes on Numerical Fluid Mechanics, Vol. 56, Chap. 15, pp. 221-244, Vieweg & Sohn, 1997.

[2] Le Balleur, J. C., "Viscous-inviscid calculation of high-lift separated compressible flows over airfoils and wings". AGARD-CP-515, Paper 26, Symposium AGARD /FDP on High-lift Aerodynamics, Banff (Canada), 5-8 October 1992, (or ONERA TP 1992-184).

[3] Le Balleur, J. C., "New possibilities of viscous-inviscid numerical techniques for solving viscous flow equations with massive separation". Proceed. Fourth Symp. Numerical and Physical Aspects of Aerodynamic Flows, Long-Beach, USA, Jan. 16-19, 1989, Selected papers, Chap. 4, p. 71-96, Cebeci (Ed.), Springer-Verlag, 1990. (or ONERA TP 1989-24).

[4] Le Balleur, J. C., "Strong matching method for computing transonic viscous flows including wakes and separations. Lifting airfoils." La Recherche Aérospatiale 1981-3, p.21-45, English and French editions, March 1981.

[5] Le Balleur, J. C., "Numerical viscous-inviscid interaction in steady and unsteady flows." Proceed. 2nd Symp. "Numerical and Physical Aspects of Aerodynamic Flows", Long-Beach, 1983, Chap. 13, p. 259-284, T. Cebeci (Ed.), Springer-Verlag, (1984), or ONERA TP 1983-8.

[6] Le Balleur, J. C., "Viscous-inviscid interaction solvers and computation of highly separated flows." Symposium ICASE, NASA Langley, USA, (July 9-10 1985), Proceedings Studies of Vortex Dominated Flows, Chap. 3, p. 159-192, M. Hussaini and M. Salas, Springer-Verlag 1987, or ONERA TP 1986-4.

[7] Le Balleur, J. C., "Viscous-inviscid flow matching: Numerical method and applications to two-dimensional transonic and supersonic flows." La Recherche Aérospatiale 1978-2, p. 67-76, French or English transl. ESA-TT-496, March 1978.

[8] Le Balleur, J. C. and Girodroux-Lavigne, P., "A semi-implicit and unsteady numerical method of viscous-inviscid interaction for transonic separated flows." La Recherche Aérospatiale 1984-1, p.15-37, English and French editions, Jan. 1984.

[9] Le Balleur, J. C. and Girodroux-Lavigne, P., "A viscous-inviscid interaction method for computing unsteady transonic separation." Proceedings of 3rd Symp. Numerical and Physical Aspects of Aerodynamic Flows, Long-Beach, USA, January 21-24, 1985, Cebeci (Ed.), Springer-Verlag 1987, or ONERA TP 1985-5.

[10] Le Balleur, J. C. and Girodroux-Lavigne, P., "Prediction of buffeting and calculation of unsteady boundary layer separation over airfoils." Proceed. IUTAM Symp. Boundary Layer Separation", London, August 26-28, 1986, p. 19-35, Editor F.T. Smith, S.N. Brown, Springer-Verlag 1987, or ONERA TP 1986-95.

[11] Le Balleur, J. C. and Girodroux-Lavigne, P., "Viscous-Inviscid Strategy and Computation of Transonic Buffet. Proceed." Symp. IUTAM Transsonicum III, Göttingen, May 24-27, 1988, Springer-Verlag 1988, or ONERA TP 1988-111.

[12] Girodroux-Lavigne, P. and Le Balleur, J. C., "Time-consistent computation of transonic buffet over airfoils." Proceed. of the 16th ICAS Congress, Jerusalem, Israel, August 28-September 2, 1988, or ONERA TP 1988-37

[13] Le Balleur, J. C., Girodroux-Lavigne, P. and Gassot, H., "Prediction of Transonic Airfoil/Wing Flow with Control using a Time-Accurate Viscous-Inviscid Interaction approach." EUROSHOCK-II Contract BRPR-CT95-0076, ONERA TASK 2, First Annual Report, May 1997.

[14] Le Balleur, J. C., Girodroux-Lavigne, P. and Gassot, H., "Prediction of Transonic Airfoil/Wing Flow with Control using a Time-Accurate Viscous-Inviscid Interaction approach. Brite-Euram EUROSHOCK-II Contract BRPR-CT95-0076, ONERA TASK 2, Second Annual Report, April 1998.

[15] Le Balleur, J. C., Girodroux-Lavigne, P. and Gassot, H., "Prediction of Transonic Airfoil/Wing Flow with Control using a Time-Accurate Viscous-Inviscid Interaction approach." Brite-Euram EUROSHOCK-II Contract BRPR-CT95-0076, ONERA TASK 2, Third Annual Report, April 1999.

[16] Le Balleur, J. C. and Girodroux-Lavigne, P." Calculation of fully three-dimensional separated flows with an unsteady viscous-inviscid interaction method." Proceed. Fifth Symposium on Numerical and Physical Aspects of Aerodynamic Flows, California State University, Long-Beach, USA, January 13-15, 1992, and ONERA TP 1992-01.

[17] Chapman, D.R., Kuehn, D.M. and Larson, H.K., "Investigation of separated flows in supersonic and subsonic streams with emphasis on the effect of transition." NACA Report 1356 (1958).

18 Steady and Unsteady Numerical NS-Calculations for the DRA 2303 Airfoil with and without Bump

W. Geissler

German Aerospace Centre, DLR
Institute of Aerodynamics and Flow Technology
Bunsenstr.10, 37073 Göttingen, Germany

Summary

Numerical calculations on the basis of the 2D time-accurate Navier-Stokes equations have been carried out for the DRA-2303 airfoil to investigate the steady as well as the unsteady transonic flow about this airfoil with and without a bump. The following work has been carried out and analysed in detail: ① calculations of the steady lift-, drag- and pitching moment distributions for the datum airfoil and the airfoil with bump, ② implementation of new turbulence models and the investigation of their effect on buffet prediction, ③ determination of buffet onset by several onset criteria, and ④ calculation of well-established buffet oscillations and comparison with experimental data. As far as experimental data were available, comparisons with the calculations have shown that agreement was achieved in most cases. As far as steady drag calculations are concerned, at least the right tendencies have been predicted, although the absolute values differed from the experimental data.

18.1 Introduction

The present numerical code, based on the time-accurate 2D Navier-Stokes equations, has repeatedly been applied for steady as well as unsteady calculations of the shock boundary layer interaction problem during the EUROSHOCK (I) project [1][2]. This project concentrated on passive shock control utilising a cavity underneath the shock wave. The cavity was covered by a perforated surface allowing the flow to circulate through the holes of the perforation. The numerical code has been modified to take into account the perforated airfoil surface and the flow through the cavity by introducing different control laws which have been developed within other tasks of the EUROSHOCK (I) project.

One of the main outputs of these investigations was that an increase in total (steady) drag for the laminar airfoil sections investigated, using passive control, was predicted in agreement with the experimental findings: the reduction in wave drag due to passive control was overcompensated by a corresponding increase in viscous drag. However, improvements were determined in first studies with respect to the buffet boundary of the airfoils by using active control, i.e., the net inflow into the cavity was increased to simulate increased suction through the perforated surface. The result was a remarkable shift of the buffet boundary to higher incidences.

It has been shown in earlier investigations [3], that improvements can be achieved by local airfoil-shape modifications with so-called bumps. If a bump is placed close to the foot of the shock, the shock strength is reduced. The effect on drag is that not only wave drag but, in addition, viscous drag may be reduced. The question arises whether a bump does also favourably affect buffet. To answer this question, a numerical code is necessary which allows time-accurate calculations utilising, for instance, the complete the Navier-Stokes (NS) equations [4].

It has already been found in [2] that turbulence modelling plays an important role with respect to the prediction of steady as well as unsteady shock induced separation. With the simple algebraic Baldwin-Lomax turbulence model, used in the previous studies, the separation point location was predicted too far downstream compared to experimental data. Therefore, new, more sophisticated models have been attached to the code [5] and partly been investigated in the present study.

18.2 Numerical Code

The computer code applied in the present study is an Approximate Factorisation Implicit solution procedure of the 2D time-accurate Navier Stokes equations as originally proposed by Beam and Warming of NASA. This code discretises the complete equations and is, therefore, not restricted to thin layer approximations. The grid topology used for the calculations is the C-grid topology with grid lines attached to the airfoil as well as to the outer boundary allowing the airfoil to move or deform with corresponding deformations of the grid. This feature has not been applied in the present study but has been used in other, helicopter related, unsteady flow investigations [6]. In future studies dynamic bumps will be investigated and then the option of a deforming airfoil surface becomes important.

Special emphasis has been placed on the development and implementation of new turbulence and transition models into the code. In addition to the standard algebraic Baldwin-Lomax turbulence model, the Johnson-King model [7] as well as the Spalart-Allmaras model [8] have been implemented during this study. In addition, k-ω models in different versions are available [9].

In addition to turbulence modelling, transition modelling has been improved: fixed transition as well as the Chen-Tyson transition model with different transition onset criteria are available in the code [10]. The different turbulence models and transition options can, of course, also be combined.

18.3 Steady Calculations

Mandatory flow cases have been defined in Task 2 of EUROSHOCK II to calculate steady pressure distributions for the DRA-2303 datum airfoil as well as for the airfoil with bump. Corresponding experimental data were made available by DERA.

Figure 1 shows as a typical example of both, pressure and skin friction distributions for the DRA-2303 airfoil without bump (upper figures) and with bump (lower figures). Experimental results were only available for the pressure distributions and have been included in Figure 1. The tables show, in addition, the force and moment coefficients obtained in both, calculation and experiment.

To eliminate possible effects of wind tunnel walls, the calculated lift coefficient has been matched to the experimental value. This was achieved by a simple linear interpolation procedure between two calculated incidences.

Figure 1 shows a very good agreement between the calculated and measured pressure distributions for both, datum and bump test cases with the details in the bump region also matching quite well. The skin friction distributions (only calculated) do not show separation on the airfoil surfaces. As far as the force coefficients are concerned, a considerable drag reduction has been achieved in the case with bump: the experimental reduction was as high as 15.45% compared to 9.22% in the calculations. The trends are clearly matching although some differences occur in the level of drag with the higher C_D-values being obtained in the experiments. The reason for these discrepancies still remains to be investigated.

The second part of the numerical studies with the present Navier-Stokes code, i.e., the investigation of active and passive control by a cavity/perforated plate arrangement and hybrid control, combining passive flow control by a cavity with slot suction downstream, has been carried out by INTA — who have obtained a version of the present Navier-Stokes code having been introduced to the code during an EU-Trainee program. INTA has extended the code to take into account the additional control devices. The corresponding results are discussed in the INTA contribution, Chapter 16.

18.4 Turbulence Modelling

Of major concern in numerical studies including separation in its various forms is the modelling of turbulence. The use of turbulence models is unavoidable for practical applications. A large number of models is available today which can be categorized into algebraic models, e.g., the Baldwin-Lomax model, in 1/2-equation models with the solution of an *Ordinary Differential Equation* for the unknown eddy viscosity, e.g., the Johnson-King model, in one-equation models, e.g., the Spalart-Allmaras model and various 2-equation models, i.e. k-ω and k-ε models in their different forms. The present study still uses the Baldwin-Lomax (BA) model in both, steady as well as unsteady applications but, in addition, some results will also be shown for more sophisticated models, i.e., the Johnson-King (JK) and Spalart-Allmaras (SA) models. In particular the SA-model has shown considerable improvements compared to the simpler models without a major increase in computing time. A result obtained with the SA-model will be discussed in Chapter 18.6.

In Figure 2 unsteady results are shown for both, the BL- and the JK-turbulence models. In both cases the incidence of the DRA-2303 airfoil has been successively increased until unsteady responses of the flow solution signal the occurrence of buffet. The upper left graphs in Figure 2 show C_L versus time above the calculated buffet boundary with the lift showing a well established buffet oscillation. In the case with bump (right graph), the lift remains completely steady. The incidence for both calculations is $\alpha=5.5^{\circ}$.

Using the Johnson-King model for the same flow case and applying the same numerical procedure, the buffet oscillations already occur at $\alpha=5.0^{\circ}$ in the datum airfoil case. The amplitude of the buffet oscillation is reduced compared to the corresponding case with the BL-model. Calculations with bump (lower right graph) employing the JK-model still show some oscillations which, however, clearly dampen to a steady value.

From these results it can be concluded that a shift of the buffet boundary due to the bump to higher incidences is predicted by both turbulence models; however, the incidence differs such that the J-K model reacts more sensitive with respect to buffet onset. No experimental data were available for these flow cases to compare with.

18.5 Buffet Boundary Investigations

Of major concern during the present study was the investigation of buffet and buffet onset. Several buffet-onset criteria have been defined and were applied during the present study, viz.,

- C1: Tangent line to the upper surface Mach number distribution (ALENIA, Figure 3)
- C2: Increase of the pressure divergence at the airfoil trailing edge with lift or Mach number (ALENIA)
- C3: Appearance of flow unsteadiness during numerical calculations (Figure 4)

The C1-criterion has been defined by ALENIA and specifies a line in the Mach number - x/c- plot which is tangent to the Mach number distribution upstream of the shock (see Fig.3). C2 defines buffet onset by an increase of the trailing edge pressure by a specified amount, Δc_p=0.04. This criterion has also been defined by ALENIA. C3 finally defines buffet onset, as has already been pointed out in Chapter 18.4, by successively increasing the incidence until unsteady buffet oscillations occur.

Table 1 includes lift coefficients applying the three criteria for the datum DRA-2303 airfoil. Table 2 includes the corresponding results for the airfoil with bump. Figure 3, upper plots (datum airfoil) and lower plots (airfoil with bump) shows pressure, local Mach number and skin friction distributions for the airfoil. The upper surface Mach number curves are tangent to the line defined by ALENIA characterizing the C1 buffet onset criterion. The corresponding C_L values are included in Tables 1 and 2, respectively, as C1 (middle). It can be seen in Figures 3 that the airfoil with bump has buffet onset at a slightly higher incidence (lift), i.e., the bump has a stabilizing effect on the flow.

In both, datum and bump cases the corresponding skin friction coefficients show a shock-induced separation area of limited downstream extent. It was found that with increasing incidence the bubble extends beyond the airfoil trailing edge. If this occurs, well defined buffet oscillations are commencing.

Table 1 Buffet onset on the DRA-2303 airfoil w/o bump, M=0.68, Re=$19x10^6$

	C1(middle)	C2	C3
DLR	0.939	1.1079	1.130*)

*) Baldwin-Lomax turbulence model, $\alpha = 5.0^o$

Table 2 Buffet onset on the DRA-2303 airfoil with bump, M=0.68, Re=$19x10^6$

	C1(middle)	C2	C3
DLR	0.985	1.156	1.194*)

*) Baldwin-Lomax turbulence model, $\alpha = 5.5^o$

The trailing edge criterion (C2) assumes buffet onset if the trailing edge pressure is reduced by Δc_p=0.04, here when increasing the angle of attack. The results applying this criterion for both, datum and bump cases are included in Tables 1 and 2 under C2. This criterion also shows a shift of buffet onset to higher lift. The tendency of the stabilizing effect of the bump can also be observed in this case.

Time-accurate numerical calculations are necessary to investigate the third criterion (C3). The incidence of the airfoil is successively increased until the flow is no longer steady but shows non-periodic unsteady behaviour. Figures 4 shows lift, drag and pitching moment distributions versus time (Nr. of Time Steps). The dotted curves in this figure show steady data close to the start of periodic buffet oscillations, the latter indicated by the solid curves in Figure 4. The trends which have already been found with the two empirical buffet onset criteria (C1,C2), can also be observed for the unsteady criterion: for the datum airfoil buffet oscillations start at about $\alpha=5^o$. The airfoil with bump shows buffet oscillations to start at a slightly higher incidence of α=5.5°. The corresponding lift coefficients for the C3 criterion are also included in Tables 1 and 2. In this case, the C_L values are the ones shortly before periodic oscillations occur.

It has been pointed out in Chapter 18.4 that turbulence modelling plays an important role in the numerical calculation: qualitatively similar results may still differ quantitatively if the calculations are carried out with different turbulence models. Note that Tables 1 and 2 include the results for the numerical calculations obtained with the BL turbulence model.

Figure 5 shows lift, drag and pitching moment coefficients versus incidence obtained from a pitch-up motion with a very slow pitching rate (A=0.005 deg/s). The pitch-up motion is a ramp-type motion, i.e., the time dependency of the incidence is linear, here with a very small slope to avoid unsteady effects. The advantage of this procedure is that one needs to carry out only one unsteady sweep to get first information about the start of unsteadiness and therefore about the buffet boundary. Figure 5 (left graphs) shows results for the datum airfoil. The unsteadiness starts at about α=5.8°. In the case with bump (Figure 5 right graphs) buffet onset is shifted to the higher incidence of α=6.1°. This result coincides with the findings obtained with the different buffet-onset criteria discussed above.

18.6 Buffet on an 18%-thick Circular Arc Airfoil

Figure 6 shows results of both, buffet onset and buffet decay by successively increasing (decreasing) the Mach number in steps of ΔM=0.01. From the Mach number of well established buffet oscillations, the Mach number is reduced stepwise until buffet decay occurs. In agreement with experimental data [11], a hysteresis effect is observed: buffet decay occurs at a considerably lower Mach number than buffet onset. Also the characteristic data, like oscillation amplitude and frequency, are in very close agreement with the measured data of McDevitt (see table in Fig.6).

The table included in Figure 6 indicates comparisons between calculation and experiment for the Mach number at buffet onset and buffet decay, respectively, as well as for the buffet frequency, taking the wind tunnel model chord into account [11]. These data show the hysteresis effect of buffet onset and decay due to the stepwise Mach number increase and reduction. The agreement between calculated and measured buffet frequencies is very remarkable.

For the calculations, a 353x65 C-grid has been used; the Spalart-Allmaras turbulence model has been applied in the calculations just discussed. Additional calculations, using the same grid but the Baldwin-Lomax turbulence model, show qualitatively very similar results. However, details of the shock oscillation as well as amplitude and frequency are slightly different.

It is important to note that prior to buffet onset (see Fig.6) a small-amplitude high-frequency oscillation occurs. This phenomenon has been observed to be caused by alternating separation at the airfoil trailing edge. These oscillations do not affect the shock wave before buffet onset. In the presence of well-established buffet oscillations, the high-frequency oscillations disappear but are re-established after buffet decay. The experimental data also show the high-frequency oscillations.

This result is very encouraging and seems to put the SA-model into a favourable position. However, limitations have been found very recently for a case where the SA-model has been applied to the investigation of buffet onset on a supercritical airfoil (OAT15A [12]). In this case, the procedure of successively increasing the incidence until the buffet boundary is reached failed: the calculation did not show any buffet up to high incidences. The corresponding calculations with the BL-model, however, clearly indicated well established buffet. The reason for this discrepancy is not yet known.

18.7 Conclusions and Future Activities

Numerical calculations have been carried out, employing a time-accurate 2D Navier-Stokes code, to determine steady as well as unsteady flow behaviour for the DRA-2303 airfoil with and without a bump. In steady flow, the calculations were able to predict a drag reduction due to the bump in agreement with experimental data, although absolute drag values differed by some amount. The reason for this discrepancy is not yet known and has to be studied further.

Concerning unsteady flow behaviour, an increase in lift at buffet onset due to the bump was predicted by the computations for the DRA-2303 airfoil using three criteria, viz., C1, based on an empirical value for the shock-upstream Mach number, C2, based on trailing-edge pressure divergence, and C3, based on the onset of unsteadiness in the computations. Buffet onset and buffet decay were numerically investigated for an 18%-thick circular-arc airfoil. It was found, in good agreement with experiment, that a hysteresis effect exists which places buffet onset at a noticeably higher Mach number than buffet cessation.

In addition to the standard Baldwin-Lomax turbulence model, further turbulence models, i.e., the Johnson- King and Spalart-Allmaras model as well as some 2-equation models, have been implemented into the code. The results show that the former three all predict buffet onset well (only compared for criterion 3), however, difference occur in details of the shock oscillation i.e., in the form of the time dependence, as well as in amplitude and frequency.

The present Navier-Stokes code has already been prepared to take into account dynamic airfoil deformations. In the future, the code will be used to investigate buffet control by means of dynamic bumps. It is envisaged to operate with a double-bump system, i.e., one bump located in the shock region and a second bump located close to the leading edge in the region of the sonic line. This combination of two bumps and their dynamic deformations, taking into account phase shifts between the bump motions, is assumed to be a powerful tool for controlling buffet. The aim of this investigation is buffet control for fixed wing applications but is also of interest for helicopter rotor blades. In case of the latter, dynamic bumps may influence the motion and strength of the shock wave with a potential to reduce high-speed impulse noise on the advancing rotor blade.

It is judged to be very important to combine numerical calculations with comprehensive and reliable wind tunnel experiments. Similar to the numerical procedures are experimental investigations in the unsteady separated flow regime very sensitive to any disturbance. Turbulence and transition modelling as well as time-accuracy are still candidates for improvements of the numerical codes; similarly are wind tunnel wall effects on both, steady and unsteady flows, and effects of turbulence in the incoming flow to be investigated experimentally.

In future projects these complicated problems need to be solved in order to get further improvements of steady as well as unsteady flow control devices.

References

[1] Stanewsky, E., Délery, J., Fulker, J. and Geissler, W. (Eds.), "EUROSHOCK-Drag Reduction by Passive Shock Control", Notes in Numerical Fluid Mechanics, Volume 56, Verlag Vieweg & Sohn, 1997.

[2] Geissler, W., "Modelling of Shock/Boundary Layer Interaction Phenomena", ECCOMAS '96, Paris, France, Sept. 1996.

[3] Ashill, P.R. and Fulker, J., "A Novel Technique for Controlling Shock Strength of Laminar-Flow Airfoil Sections", First European Forum on Laminar Flow Technology, Hamburg, Germany, March 1992.

[4] Geissler, W., "Instationäres Navier-Stokes Verfahren für beschleunigt bewegte Profile mit Ablösung (Unsteady Navier-Stokes Code for Accelerated Moving Airfoils with Separation)", DLR Research Report DLR-FB 92-03, 1992 (in German).

[5] Geissler, W. and Ruiz-Calavera, L.P, "Transition and Turbulence Modelling for Dynamic Stall and Buffet", 4th International Symposium on Engineering Turbulence Modelling and Measurements, Frantour & CCAS, Porticcio-Ajaccio, Corsica, France, May 24-26, 1999.

[6] Geissler, W. and Sobieczky, H., "Unsteady Flow Control on Rotor Airfoils", Proceedings of the AIAA 13th Applied Aerodynamics Conference, 19 - 22 June, 1995, San Diego, Ca, USA, pp. 1-9.

[7] Müller, A., "Numerische Simulation stationär angestellter und harmonisch schwingender Profile mit Hilfe eines erweiterten Halbgleichungs - Turbulenzmodells", Dissertation at the Institut für Aerodynamik und Gasdynamik der Universität Stuttgart, 1999.

[8] Spalart, P.R. and Allmaras, S.R., "A One-Equation Turbulence Model for Aerodynamic Flows", AIAA Paper, AIAA-92-0439, 1992.

[9] Wilcox, D.C., "Reassessment of the Scale Determining Equation for Advanced Turbulence Models", AIAA Journal, Vol.26, No. 11, pp. 1299-1310, Nov. 1988.

[10] Geissler, W., Chandrasekhara, M.S., Platzer, M. and Carr, L.W., "The Effect of Transition Modelling on the Prediction of Compressible Deep Dynamic Stall", 7th Asian Congress of Fluid Mechanics, Dec. 8-12, 1997, Chennai (Madras), India

[11] McDevitt, J.B., "Supercritical Flow About a Thick Circular Arc Airfoil", NASA TM-78549, 1979.

[12] Sobieczky, H., Geissler, W. and Hannemann, M., "Expansion Shoulder Bump for Wing Section Viscous/Wave Drag Control" in Proceedings of the IUTAM Symposium on Mechanics of Passive and Active Flow Control (Eds. G.E.A. Meier and P.R. Viswanath), pp. 29-34, Kluwer Academic Publishers, 1999.

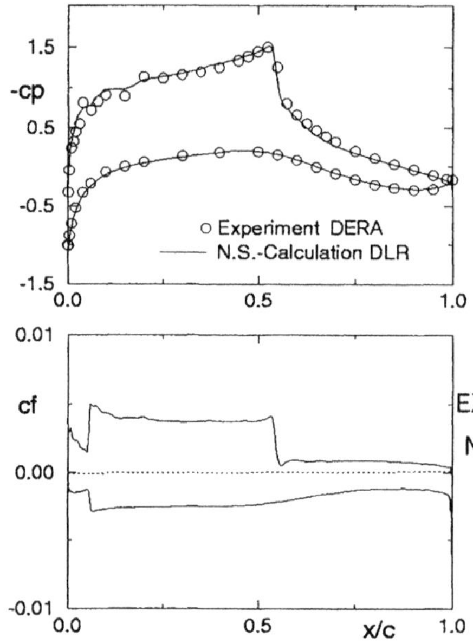

DRA-2303 NLF Datum

M=0.68
Re=19x10^6

EUROSHOCK II

	α	CL	CM	CD
EXP	2.09700	0.747208	-0.1006	0.01346
NS	2.250	0.7513	-0.09195	0.01226

Navier-Stokes Calculations with:
353x65 C-Grid

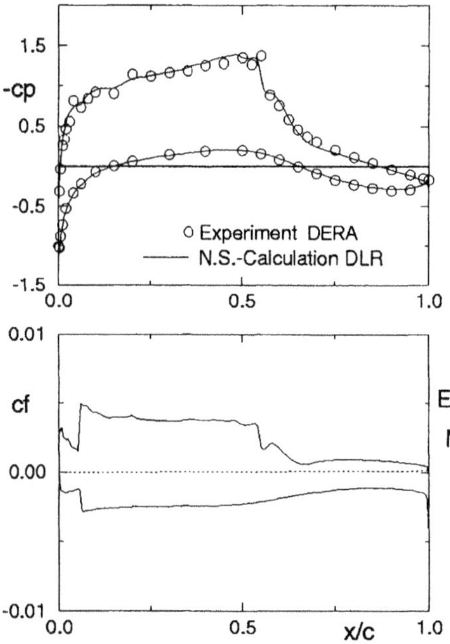

DRA-2303 NLF Datum

M=0.68
Re=19x10^6

Drag Reduction: 15.45% Exp. 9.22% N.S.

	α	CL	CM	CD
EXP	2.098	0.73988	-0.0989	0.01138
NS	2.150	0.7454	-0.0937	0.011134

Navier-Stokes Calculations with:
353x65 C-Grid

Figure 1 Steady flow about the DRA-2303 airfoil with and without bump

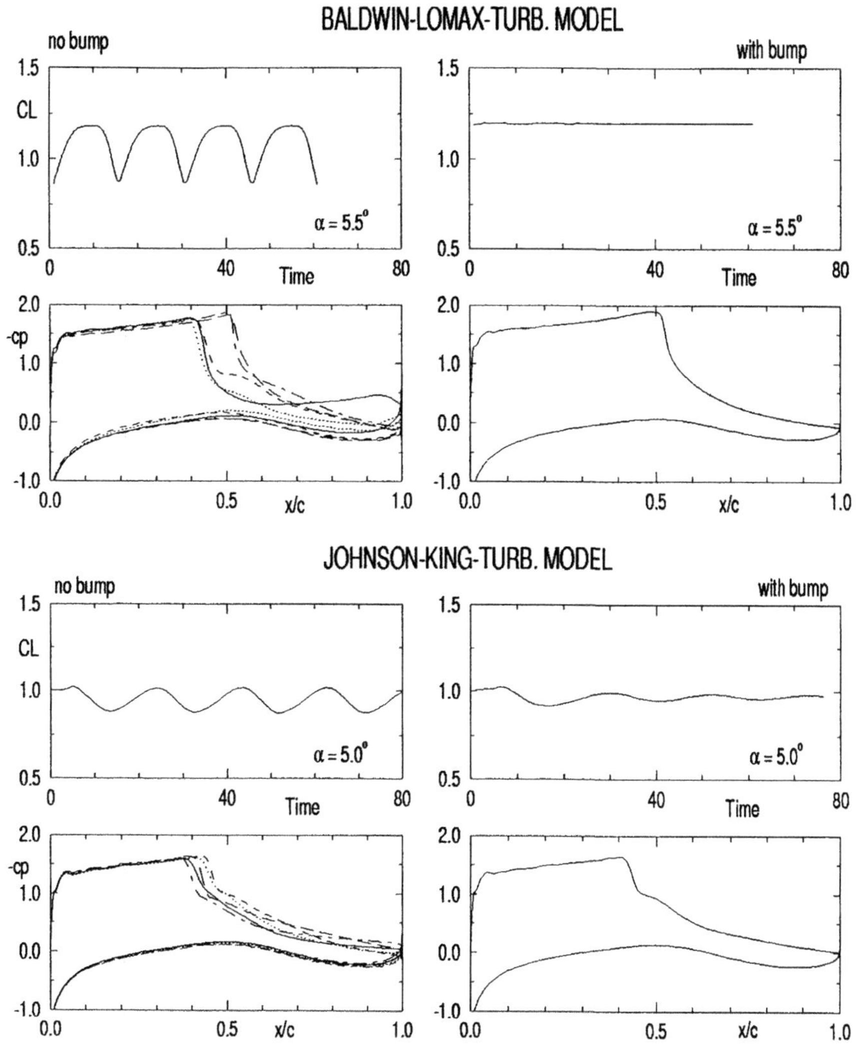

Figure 2 Buffet boundary without and with bump, M=0.68, Re=19x10[6]
Baldwin-Lomax versus Johnson-King turbulence model

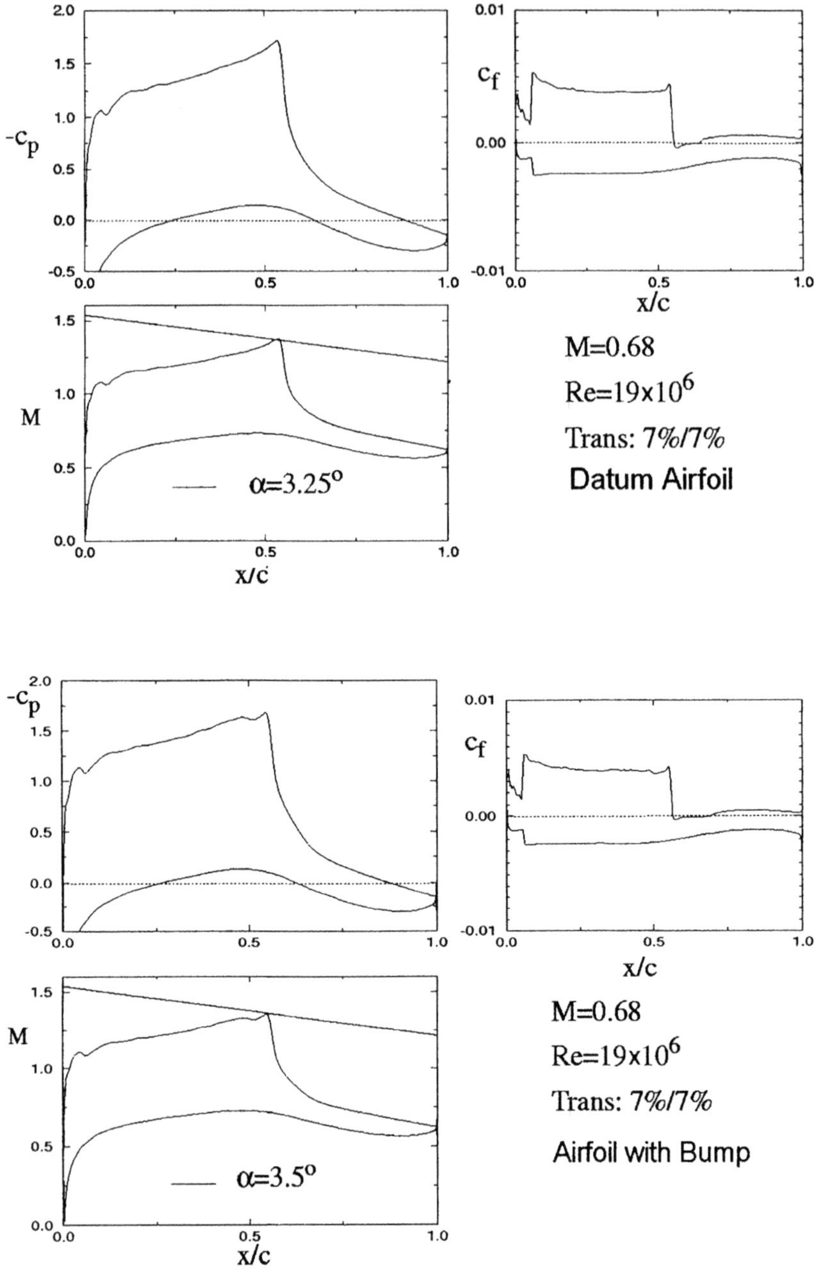

Figure 3 Application of the C1 Buffet-onset Criterion — Tangent to the upper surface Mach number distribution

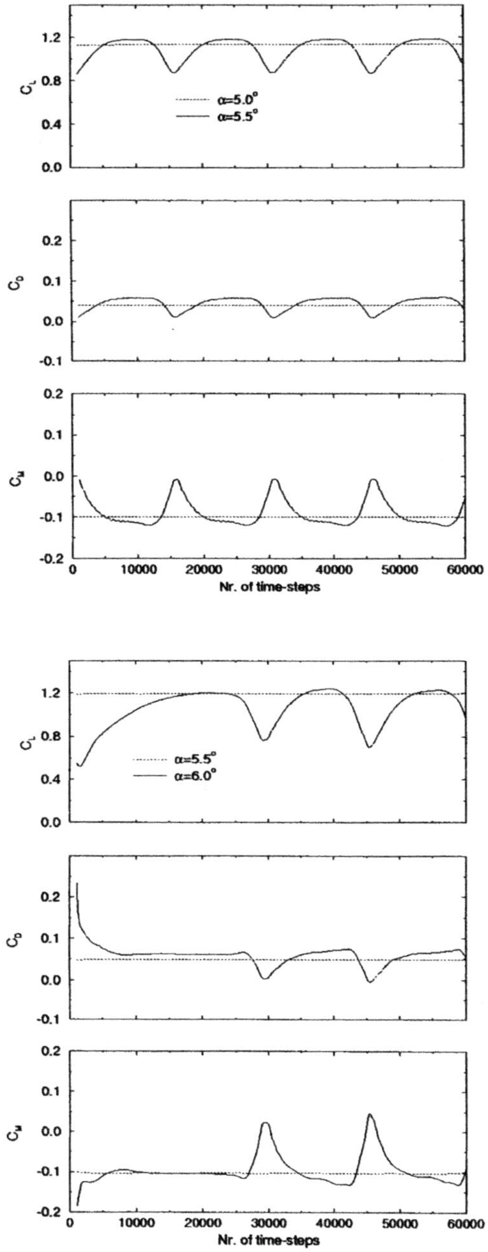

Upper Figures:
DRA-2303 Airfoil (Datum)

Lower Figures:
Airfoil with Bump

Solid Curves in Figures Indicate Force and Moment Coefficients just Above Buffet Onset:
Established Periodic Time Histories

Dashed Curves in Figures Indicate Force and Moment Coefficients just Below Buffet Onset
Time Independent (Steady) Results

M=0.68
Re=19x10^6
Transition: 7%/7%

Determination of Buffet Onset by Successive Increase of Incidence in NS-Code.
(Numerical Determination of Buffet Onset)

Figure 4 Application of the C3 Buffet-onset Criterion — Onset of unsteadiness

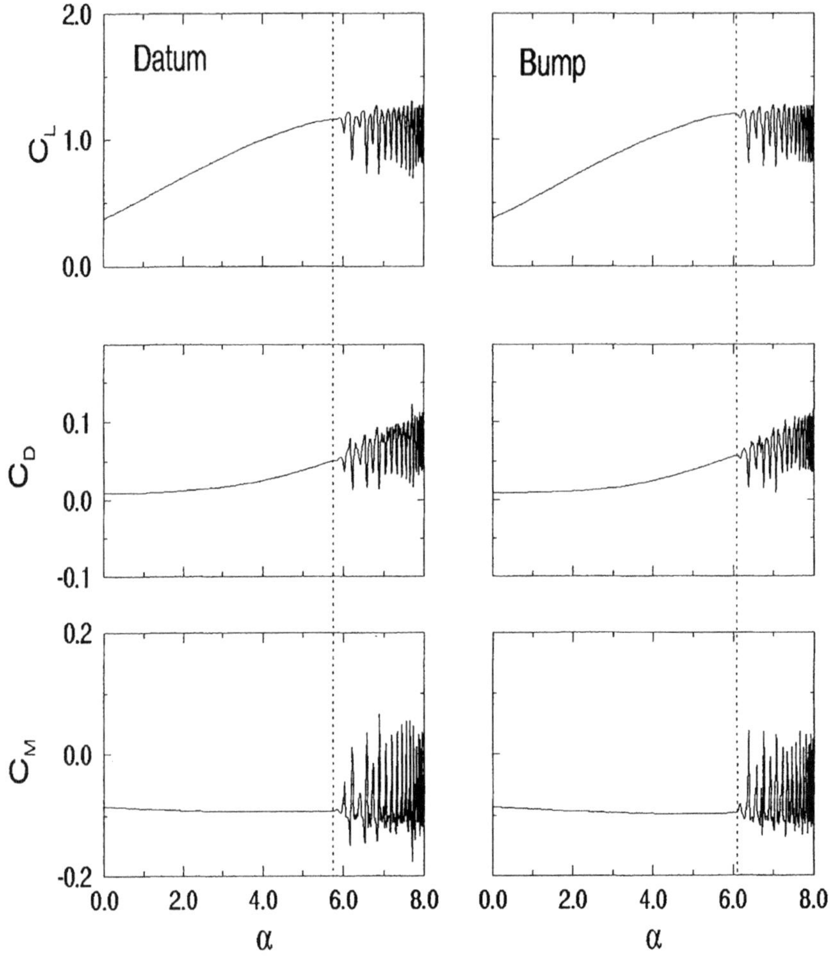

M=0.68, Re=19x10^6, Transition: 7%/7%

Figure 5 Buffet investigations with the DRA-2303 airfoil using ramp motion at a pitch-rate of A=0.005 (Baldwin-Lomax turbulence model)

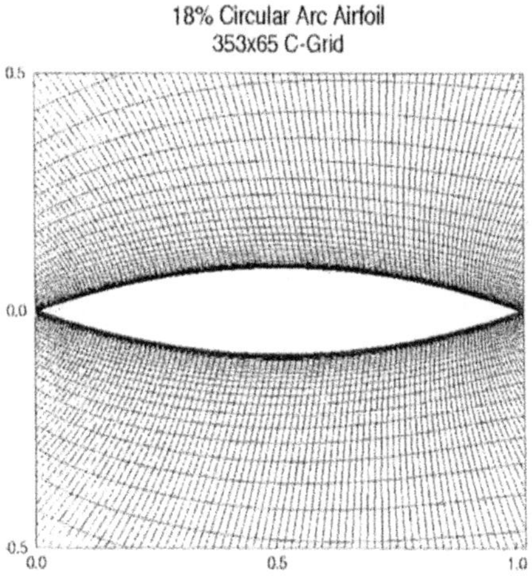

High density grid (left)

Buffet Onset (lower left)

Buffet Decay (lower right)

Mach number has been changed successively in steps of $\Delta M=0.01$.

High Frequency Oscillations of Lift Curves due to Periodic Trailing Edge Separation.

	Calc.	Exper.
Buffet Onset	M=0.75	M=0.76
Buffet Decay	M=0.72	M=0.73
Buffet Frequ.	183Hz	187Hz

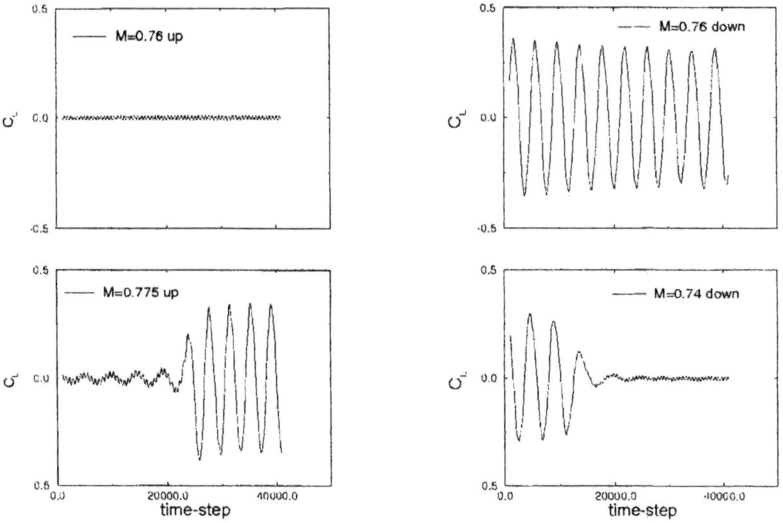

Figure 6 Buffet calculations for an 18%-thick circular-arc airfoil (Spalart - Allmaras turbulence model)

19 An Investigation of Active, Suction, Shock and Boundary Layer Control Techniques

J. L. Fulker and M. J. Simmons
Technology Dept., Aerodynamics and Hydrodynamics Centre.
DERA Bedford, Clapham, Beds., MK41 6AE, UK.

Summary

This report describes experimental and theoretical studies at high subsonic speeds on a two-dimensional airfoil of novel drag reduction systems. The control systems are all 'active' in the sense that they utilise suction to remove some of the mass flow within the boundary layer. The systems considered were:

1) Active Suction: 'Passive' control by means of a porous region on the surface in the region of the shock wave is very effective at reducing wave drag by controlling the development of the shock wave, but this is at the expense of an increase in viscous drag mainly due to the flow through the surface. In the 'active' variant of this system a small amount of air is removed from the plenum with the aim of reducing the streamwise growth of the boundary layer thickness in the control region, thus arriving at a nett drag reduction.

2) Discrete Suction: The effect of applying suction locally, for example by a narrow porous strip or slot, will be to reduce the boundary-layer momentum thickness across the strip. If the strip is placed on the airfoil or wing upper surface, where the local velocity is higher than that of the free-stream, the effect will be 'magnified', so that the drag reduction will be greater than that of the equivalent flat plate at the same free-stream speed.

3) Hybrid Suction: This system combines the benefits of 'Passive' control and the potential of Discrete Suction, by reducing wave drag in the passive control region and eliminating the viscous drag increase by a narrow suction strip downstream.

In assessing the potential benefits of these 'active' suction systems it is necessary to allow for the internal or 'pump' drag to arrive at a nett drag for the system.

The theoretical calculations and experiments show that Discrete Suction is very effective at reducing drag. The current configuration of Active Suction does not reduce drag, but theoretical parametric studies have revealed a potential control system, which could be examined by a simple modification to the existing system. Hybrid Suction is shown to be in-effective at reducing drag.

19.1 Introduction

The onset of shock waves on the wings of an air vehicle marks the start in the deterioration of the performance of the aircraft, as, for example, typified by the range and sustained turn radius. This is due to the onset of a rapid rise in drag associated with the appearance of shock waves, on the wings of aircraft, at high subsonic speeds. The rate at which the drag rises with either Mach number or angle of incidence depends on the type of wing section. For example, the drag rise can be severe for modern designs with relatively far-aft shock waves, the latter in order to ensure either increased lift or laminar flow over large parts of the wing surface. Thus, for these designs, the limiting effect of shock waves on aircraft performance can be serious. Similarly for a given lift, increasing rear camber has the beneficial effect of reducing wave drag for airfoils and wings with strong shock waves. Unfortunately, this benefit is negated by an increase in viscous drag associated with the rapid growth of the upper surface boundary layer in the strong adverse pressure gradients. Methods for controlling shock strength and the interaction between the shock wave and the boundary layer need to be established and if successful, developed.

Various techniques have been proposed or can be considered for controlling shock strength as follows:

i) 'Passive' control: In this method[1,2,3,4], the wing surface is ventilated in the region of the shock wave, the porous surface having a common plenum beneath it. A natural recirculation occurs between the region downstream and upstream of the shock wave. The resulting outward displacement of the streamlines upstream of the shock wave causes compression waves, similar to those found on bumps or ramps, which weaken the main shock wave and reduce wave drag. However, this reduction in wave drag is accompanied by an increase in viscous drag[1].

ii) 'Active' control: Two families of control methods can be considered here: a) Distortion or deflection of the wing surface in the region of the shock wave by 'bumps' or 'ramps'. The upstream face of the ramp induces compression waves that weaken the shock wave in the flow-field, thereby reducing wave drag[1,5,6]. b) In the case of passive control it was inferred that the increase in viscous drag is caused by the thickening of the boundary layer in the region of the porous surface. This suggests the possibility of an active form of control where the passive recirculation effect is retained but a certain amount of air is removed from the plenum in an attempt to reduce the boundary-layer thickness in the control region.

In cases where airfoils or wings have been designed for low wave drag the effect of increases in rear camber is disadvantageous; for such wings/airfoils alternative approaches need to be considered. The most obvious alternative is laminar flow control. The benefits of this approach are well understood, however, there are many problems to be overcome before it can be realised in practice. It will be some time before these problems are solved so it is worth considering if boundary-layer suction can be used to reduce the drag of classical turbulent airfoils and wings. Limited studies have been made with turbulent airfoils equipped with suction slots, but the emphasis of this work has been on control of shock/boundary-layer interactions rather than on drag reduction[7].

The effect of applying suction locally, for example by a porous strip or slot, will be to reduce the boundary-layer momentum thickness across the strip. If the strip is placed on the airfoil or wing upper surface, where the local velocity is higher than that of the free-stream, the effect will be 'magnified', so that the drag reduction will be greater than that of the equivalent flat plate at the same free-stream speed. Thus for modern airfoil sections with relatively high local velocities on the upper surface at the design condition, the drag reduction could be significant.

This investigation was performed to study the effectiveness of 3 different systems for drag reduction as follows:

a) Active suction (Fig. 1a): the removal of some of the air from the plenum chamber beneath a porous surface placed in the region of the shock wave.

b) Discrete suction (Fig. 1b): local suction at a narrow, porous strip.

c) Hybrid suction (Fig. 1c): a combination of 'passive' control to reduce wave drag and discrete suction, downstream of the control region, to reduce viscous drag.

In assessing the potential benefits of 'active' suction systems it is, of course, necessary to allow for the internal or 'pump' drag.

This Report describes the results of an initial experimental investigation, supplemented by theoretical investigations, of the systems, described above, on an airfoil model in the 8ft x 8ft Subsonic-Supersonic Wind Tunnel at DERA Bedford. The aim of the work is to investigate the potential for drag reduction and buffet improvements with such systems at Reynolds numbers approaching full scale.

After the experimental techniques are described in Section 2, details of the theoretical methods are given in Section 3 and data from the experiments are presented in Section 4. Salient features of the effect of the control systems on the upper surface flow are presented in Section 5, together with comparisons with theoretical results. Concluding remarks of the initial study are given in Section 6 and recommendations for further work in Section 7.

19.2 Model Details and Measurements

19.2.1 The Model

The layout of the model in the Wind Tunnel is illustrated in Figure 2 and a photograph is shown in Figure 3.

The model was mounted firmly to a rotating mechanism on the starboard side of the working section and within a free rotating bearing on the port side.

The model consists of a main spar with detachable leading edge (0 - 0.17c) and trailing edge (0.7 - 1.0c) sections, manufactured from high-tensile steel. The datum airfoil section was designed to be representative of a natural laminar-flow section with long runs of favourable pressure gradient on both the upper and lower surfaces extending to 0.5c close to the design conditions (a freestream Mach number, M, of 0.68 and a lift coefficient of 0.5). On the upper surface of the main spar there is a removable panel between 0.39c and 0.69c allowing control systems to be inserted. An insert was also manufactured to form the original profile in order to obtain a datum configuration.

The control inserts had perforated surfaces between 0.475c and 0.625c (Active Suction), 0.45c and 0.46c (Discrete Suction) and 0.475c and 0.625c followed by 0.65c and 0.66c (Hybrid Suction), with an open area ratio of 4% based on the local area. The perforations were formed by laser drilling with a nominal diameter of 0.076mm (0.003ins). A calibration of the control surface by the method suggested by Poll et al[11] allowed a value of the so called K modifier to be derived. In this case the value was found to be equal to 1.0 for flow both into and out of the plenum, based on an extensive survey of the surface for various mass flows. Thus, although it is not possible to specify the diameter of an individual hole accurately, there is a nominal hole diameter which can be determined for a large number of holes, which in this case is identical to that specified.

Suction could be applied to the plenum chamber beneath the perforated surfaces by evacuating a secondary plenum in the form of pipes laid into the bottom of the cavity.

Ordinates of the datum airfoil are given in Reference 1. All of these ordinates were obtained from an inspection in March 1993 and are accurate to within ±0.02mm (0.00003c).

The airfoil and inserts are fitted with surface static pressure holes at three spanwise stations (0.25, 0.5 and 0.75 span) each having 35 stations on the upper surface and 22 on the lower surface. All the static holes in the wing are of 0.5mm diameter and are drilled normal to the surface. Provision is also made for measuring static pressure in the control cavity (plenum) when the control inserts are in place.

19.2.2 Measurements

The static pressures at the model surface, wake rake pressures and static pressures on the working section walls were measured using ten electronic pressure scanning modules having a working range of ±1.02 bar for measuring static pressures and ±0.34 bar for total pressures. Total and static pressures were measured on a wake rake of pitot and static tubes 1.75 chord lengths downstream of the model trailing-edge, Figures 2 and 3.

The overall accuracy of the pressure coefficients, allowing for uncertainties in transducer calibration and dynamic and static pressures, is estimated to be ±0.002 at the test conditions.

Normal force and pitching-moment coefficients were determined by appropriate integration of the surface static pressures around the airfoil.

Geometric angles of incidence were measured by a digital encoder attached to the half-model balance housing, supporting one end of the model, the setting for zero angle of incidence having previously been determined using an electro level meter. The estimated accuracy of the setting is ±0.005°. The model was loaded before the tests to a maximum of 740Nm about its flexural axis. No detectable movement could be recorded; hence it has been assumed that aeroelastic distortion of the model is negligible.

No corrections have been applied to the wake pitot readings for displacement effect. Sectional drag was inferred from the pitot and static pressure measurements in the wake, using the method described in Ref. 8.

19.2.3 Boundary-layer Transition Trips

Narrow bands of sparsely distributed ballotini, cemented to the model by epoxy resin, fixed boundary layer transition. Transition was fixed at 5% chord on both surfaces of the model, the band width in each case being 2.5mm and the diameter of ballotini used being 0.1mm - 0.13mm.

19.2.4 Test Conditions

Tests with transition fixed on both surfaces were performed at a freestream Mach number of 0.68 at a nominal Reynolds number of $R_C = 19 \times 10^6$ for each configuration.

The 8ft x 8ft Wind Tunnel has very accurate control of total pressure and total temperature during a run. Hence variations in these parameters during the acquisition of data are expected to be small and to exert negligible effects on the measurements taken.

A correction to the Mach number for the blockage effect of the model has been applied. A complete description of the method is given in Ref 9 but, briefly, it relies on linear theory to calculate:

a) the effect of the model and its images beyond the tunnel walls on velocity increments at two points on the roof and corresponding points on the floor, in both cases on the tunnel centre line and close to the model centre of volume, and

b) the arithmetic mean value of the blockage increment. The ratio of this increment to the arithmetic mean of the increments in a) is combined with the mean of the measured changes in static pressure at the four points relative to those in the empty tunnel to infer blockage at the model.

The correction to angle of incidence due to tunnel-wall interference is determined using linear theory[10], together with the lift coefficient inferred from the measured model pressures, to determine the strength of the vortex simulating airfoil circulation. The vortex is placed at the centre of pressure, which is also inferred from measured pressures.

Unless the 'nozzle' Mach number is adjusted, corrected Mach number increases with angle of incidence owing to changes in the blockage effect of the model, were kept to within ±0.002 of the required value during testing.

19.3 Calculation Methods

A version of the transonic airfoil code BVGK, BVGK6E, has been used to calculate the flow characteristics for the Discrete Suction cases. Based on the viscous-inviscid interaction procedure, this method combines a numerical solution of the exact potential flow equation for the equivalent inviscid flow with an integral method for solving the shear flow equations. The turbulent shear layers are calculated using an inverse procedure suitable for flows with separation. 'Higher-order' terms are included in BVGK[9] to allow for effects that are known to be important for flows close to or with separation.

The suction strip is situated in the turbulent boundary layer on the airfoil upper surface and its effect is modelled by a decrement in boundary-layer momentum thickness at the position of the leading edge of the strip, Δθ, where the symbol Δ refers to a decrement due to suction. This decrement is related to the mass flow per unit span into the strip, m, (from the momentum equation) by:

$$\frac{\Delta\theta}{c} = \frac{m}{(\rho_s U_s)c},$$

where ρ and U are density and stream speed, c is airfoil chord and suffix s refers to conditions at the edge of the boundary layer at the position of the strip. Thus suction coefficient is given by:

$$C_Q = \frac{m}{\rho_\infty U_\infty c} = \frac{\Delta\theta}{c}\frac{\rho_s U_s}{\rho_\infty U_\infty}, \qquad (3\text{-}1)$$

where suffix ∞ refers to conditions far upstream of the airfoil. In the code, the shape parameter, $H = \delta^*/\theta$, is maintained constant across the strip and this method has been shown to give good predictions of drag magnification factors for airfoils with excrescencies[9].

In this report the notation follows that used in Ref. 9, which should be read in conjunction with this report.

Reducing the boundary-layer momentum thickness by suction will result in an increase in skin friction coefficient over the suction strip. If the suction strip is modelled with infinitesimal width, the drag penalty due to the increase in skin friction will be negligible. However, in reality the suction will be applied over a small distance and therefore will introduce a drag penalty which needs to be taken into account in the calculation method.

The effect of the increase in local skin friction is to reduce the effective suction coefficient as derived in Appendix A such that:

$$C_{Q_{eff}} = C_Q \frac{2\overline{H}}{(\overline{H} + H_1)} \tag{3-2}$$

or

$$C_{Q_{eff}} = -C_Q \left\{ \left(1 - \frac{2\overline{H}}{(\overline{H} + H_1)} \right) \left(0.16 \log_{10} R_\theta - 0.07 \right) - 1 \right\} \tag{3-3}$$

A typical value of $\overline{H}=1.3$ gives C_{Qeff} equal to 28% of the nominal C_Q, from Equation 3-1, using Model A[12] for skin friction in boundary layers with suction, whereas using Model B, Equation 3-3 gives C_{Qeff} as 61% of the nominal C_Q. It is thus possible to allow for the effect of skin friction by simply reducing the C_Q and hence the decrement in momentum thickness.

A second method has been used to calculate the Active and Discrete Suction cases, again based on the transonic airfoil code BVGK, BVGK(P), but in this case the porous surface is modelled directly by allowing for surface transpiration[12, 13]. The use of this method allows the effect of variations in the width of the porous region to be modelled explicitly.

In both cases and for the experimental results an allowance has been made for internal pump drag, where appropriate, and the method used is described in Appendix B.

19.4 Experimental Data

The airfoil model was tested over a range of angles of incidence from a value closely corresponding to that for zero lift to an angle above which trailing-edge pressure decreases rapidly with angle of incidence (see later), for all control configurations. The data for M=0.68 is representative of data obtained in the Mach number range M=0.65 to 0.75. An allowance for pump drag is only applied to cases where a nett drag reduction is obtained in order to illustrate the potential of the system, in all other cases no correction is applied.

19.4.1 Discrete Suction

The variation of normal-force coefficient with angle of incidence is shown in Figure 4. It can be seen that the effect of suction is small. For the no-suction case, $C_Q=0$, the effect is to reduce normal force by approximately $C_N=0.04$, at a constant angle of incidence, at the higher values of normal-force coefficient. As the suction coefficient is increased the decrement in normal-force coefficient is restored; at the maximum suction coefficient (C_Q=0.0009) the normal force curve is almost indistinguishable from the datum (solid) airfoil case. This is in contrast to previous theoretical investigations[14], where the effect of discrete suction was to increase lift for a fixed angle of incidence. The latter was due to the direct effect of increasing rear camber due to a reduced boundary-layer thickness and rearward movement of the shock wave on a turbulent airfoil/wing. However, for the present type of airfoil (laminar design) this is not the case and there is little evidence of camber increase or shock movement; this is probably a feature of this type of airfoil design where the favourable pressure gradient is ensuring that the boundary layer is kept thin. The benefits of discrete suction are therefore likely to be greater for turbulent designs than for laminar ones.

Figure 5 shows the variation of drag coefficient with normal-force coefficient. It can be seen that for the no-suction case drag is increased, however, as suction coefficient is increased, drag is reduced. The reduction in drag can be seen to be slightly larger for higher values of normal-force coefficient.

The variation in trailing-edge pressure with normal-force coefficient is shown in Figure 6. The effect of increasing levels of suction is to increase the normal-force coefficient for trailing-edge pressure divergence (the point at which trailing-edge pressure decreases rapidly as flow separation moves forward from the trailing edge). The effect of having no suction is to decrease the normal-force coefficient for trailing-edge pressure divergence.

Typical pressure distributions are shown in Figure 7. One observes that the pressure distributions are identical everywhere, except in the region just ahead of the upper surface shock wave. The effect of switching the suction system off is to introduce a mild compression ahead of the shock wave. However, there is a suggestion of a rapid re-expansion of the flow immediately ahead of the shock potentially giving rise to a shock at least as strong as that for the datum case. This is conjecture which can only be resolved by increasing the density of pressure tappings in this region, however, in view of the higher drag of this configuration (Figure 5) compared to the datum, it would appear that this can only be associated with an increase in wave drag. For the cases with suction applied it can be seen that for low values of suction the effect is a small reduction in shock strength; as the suction rate increases this trend is reversed until at the maximum suction rate, C_Q=0.00009, there is a small increase in shock strength.

19.4.2 Active Suction

The variation of normal-force coefficient with angle of incidence is shown in Figure 8 for the Active Suction configuration (also see Figure 1a). It can be seen that for high values of normal-force coefficient (C_N>0.7) there is a considerable loss of normal-force for this configuration, although it is to be noted that the trend with increasing levels of suction is to move towards datum levels.

Figure 9 shows the variation of drag coefficient with normal-force coefficient. For the no-suction case drag is increased significantly, which is to be expected based on earlier studies of Passive Control[1]. The use of suction indicates that cutting the passive recirculation effect can reduce the drag penalty associated with Passive Control.

The variation in trailing-edge pressure with normal-force coefficient is shown in Figure 10. The effect of introducing this control system is to reduce the normal-force coefficient for trailing-edge pressure divergence for the no-suction case and, as suction is applied, the indications are that some of the reduction is restored.

Typical pressure distributions are shown in Figure 11. Again the pressure distributions are identical except in the vicinity of the upper surface shock wave. In the case of no suction the pressure distribution is typical of that found in these (cavity-control) cases (this case being essentially Passive Control[1]). There is some recompression of the flow ahead of the shock wave thus weakening it but this is accompanied by an increase in boundary layer thickness behind the shock as indicated by the lower pressures in this region. The effect of suction is to maintain the reduction in shock strength but there is a suggestion of an increase in the pressures downstream of the shock wave as suction is increased.

19.4.3 Hybrid Suction

The variation of normal-force coefficient with angle of incidence is shown in Figure 12 for the Hybrid Suction configuration (also see Figure 1c). It can be seen that for high values of normal-force coefficient (C_N>0.7) there is a considerable loss of normal-force for this configuration, although losses in this case are not as great as for the Active Suction configuration (Figure 8). The effect of varying suction from C_Q=0 to 0.0001 is negligible in this case.

Figure 13 shows the variation of drag coefficient with normal-force coefficient. For the no-suction case drag is again increased significantly, this being essentially a Passive Control case. The use of suction indicates that the drag penalty associated with Passive Control can be reduced at low values of normal-force coefficient (C_N<0.3), but for higher values the effect of suction on reducing the excess drag is small.

The variation in trailing-edge pressure with normal-force coefficient is shown in Figure 14. The effect of introducing this control system is to reduce the normal-force coefficient for trailing-edge pressure divergence for the no-suction case and, as suction is applied, the indications are that some of the reduction is restored.

Typical pressure distributions are shown in Figure 15. Again the pressure distributions are identical except in the vicinity of the upper surface shock wave. In all cases there is some recompression of the flow ahead of the shock wave thus weakening it but this is accompanied by an increase in boundary layer thickness behind the shock as indicated by the lower pressures in this region. The effect of suction is to maintain the reduction in shock strength but there is a suggestion of an increase in the pressures downstream of the shock wave as suction is increased; the length of this region is reduced compared to the Active Suction case.

19.5 Discussion

It is worth studying in detail the experimental data described in Section 19.4 with the aim of identifying the important features of the flows associated with the various control systems, also where possible comparing the results of the measurements with theoretical calculations.

19.5.1 Discrete Suction

Figure 5 shows that for all values of normal-force coefficient the effect of applying suction is to reduce drag, whereas the no-suction case, C_Q=0, increases drag significantly, probably due to the aerodynamic roughness of the porous surface. The reduction in drag with suction coefficient appears to be monotonic, suggesting that for higher values of suction, larger drag reductions should be possible. The highest value of suction coefficient, C_Q=0.00009, is currently limited by the experimental set up, where, although high mass flows are potentially available, the mass flow removed from the boundary layer is restricted by the available volume of the ducting arrangement in and adjacent to the model. The use of a slot in place of the porous strip may also allow an increase in suction rate by removing the possibility of choking the flow in the holes.

Considering the pressure distributions, Figure 7, the effect of low values of suction is to weaken the shock wave compared to the datum case and for its position to remain unchanged. These are indirect effects caused by the tendency for suction to make the boundary layer downstream of the suction-strip thinner and consequently, to increase the effective rear camber and rear loading. A suction peak is introduced in the upper surface pressure distribution upstream of the suction-strip position, while a trough is induced downstream of the strip. In these cases, the flow is locally close to sonic conditions where the suction is applied. At the highest suction, C_Q=0.00009, the shock strength is slightly increased relative to the datum. This suggests a small increase in wave drag, which is similar to the effect that is found when rear camber is increased, indicating that the effect of suction is to increase effective rear camber, as noted before. However, this small increase in wave drag is more than off-set by the reduction in viscous drag, but will need to be borne in mind if higher values of suction are to be considered.

Clearly the effect of suction is to cause changes in the boundary-layer development downstream of the suction strip. The skin-friction coefficient will be increased aft of the strip by suction and, consequently, it will be larger than that for the datum case at the trailing-edge. This implies that the effect of suction is to reduce the tendency for trailing-edge separation, an observation that is confirmed by the fact that suction improves the normal-force coefficient for trailing-edge pressure divergence, Figure 6. It is worth noting that this system is primarily a boundary-layer control device and therefore independent of flight condition.

As noted earlier the true effectiveness of suction systems can only be assessed once allowance is made for internal or 'pump' drag. The variation of drag coefficient with normal-force coefficient is shown in Figure 16 for the maximum suction rate, with and without allowance for pump drag, calculated using the method described in Appendix B. The value of pump drag calculated by the method is approximately half of the drag reduction.

Figure 17 shows a comparison between experiment and theory of the variation of drag coefficient with normal-force coefficient for various models of the control system using BVGK6E. Considering first the datum case, it can be seen that the theory underestimates drag by between $0.0002C_D$ and $0.0004C_D$; this is consistent with the increment in drag due to the roughness effect of the transition trips[12]. For the cases with suction, the effect of the various models of skin friction is clear. If the effect of suction on skin friction is ignored, C_{qact}, it can be seen that the reduction in drag is overestimated, indicating that for finite width suction strips the effect on skin friction has to be considered. If Model A[12] is used and C_{qeff} is derived from Equation 3-2, the curve labelled C_{qeff1} is derived. This model is clearly pessimistic and included here for the sake of completeness. It does however have the advantage that it is simple to apply, since it only requires a knowledge of $\overline{H}$ at the position of the suction strip from the datum case. If Model B[12], Equation 3-3, is used much better agreement with experiment is achieved (labelled C_{qeff2}). Again the difference between experiment and theory is consistent with the roughness effect of the transition trips. Although slightly more difficult to implement than Model A, Model B has the advantage that it is the same model as used in BVGK(P), therefore adding to consistency in comparisons between the methods, and is consequently used for all further comparisons.

The effect of varying the suction rate is shown in Figure 18. It can be seen that the incremental reduction in drag is well predicted (the drag values shown here are absolute values, i.e., no allowance has been made for pump drag for clarity). The trend with increasing suction suggests that it would be worth attempting to obtain higher rates, if possible, to see if there is any physical limit. One limit is clearly the state when all of the boundary layer is removed, however, to do this would involve an unacceptable pump drag penalty. Another option that needs considering is to replace the porous strip with a narrow slot, which would remove the throttling effect of the holes.

Figure 19 shows a comparison of drag between BVGK6E and BVGK(P) for the maximum suction case compared with the datum. The differences between the two methods are small. This comparison has been included because BVGK(P) is more rigorous in its treatment of porous surfaces, confirming the validity of the modelling of the suction system used in BVGK6E. However, it has the disadvantage that it is not as robust as BVGK6E. Nevertheless, it is the only method available once the chordwise extent of porosity is increased and the secondary flow to and from the plenum develops.

19.5.2 Active Suction

Figure 8 shows that for high values of normal-force coefficient, (C_N>0.7) there is a significant loss of normal force associated with the application of suction. Although there is a suggestion that as the rate of suction is increased the loss in normal-force is restored, it is clear that very high rates of suction will be necessary to restore normal-force to the datum levels.

Figure 9 shows that for all values of normal-force coefficient and suction rate, drag is always greater than for the datum case. Compared to the no-suction case, C_q=0, only modest reductions in drag are achieved as suction is increased, again suggesting that high rates of suction will be necessary to reduce the drag to even the datum level. However, it has been shown that whilst the effect of suction is to reduce viscous drag, this is at the expense of an increase in wave drag[12]. Therefore it is not clear that even if higher rates of suction could be achieved, a nett drag reduction (after allowance for pump drag) can be obtained with this control arrangement.

The reduction in normal-force coefficient for trailing-edge pressure divergence is shown in Figure10; it is consistent with the loss of normal-force shown in Figure 8. This is evidence that the thickening of the boundary layer in the control region is still excessive, and that the 'passive' re-circulation effect needs to be reduced to a minimum.

Considering the pressure distributions, Figure 11, the effect of the introduction of control is to generate a recompression into the upper surface flow starting from a point some way ahead of the control region. This is followed by a re-expansion to the final shock wave. It would appear that the shock strength is reduced compared to that of the datum case, although the effect of increasing suction suggests an increase in shock strength compared to the no-suction case. Clearly the pressures downstream of the shock wave are reduced, suggesting a considerable thickening of the boundary layer in this region, and this extends right to the trailing edge, where even here are indications of reduced pressures. However, the effect of increasing suction does appear to increase pressures in this region indicating a reduction in boundary-layer thickness compared to the no suction case. The pressure distributions, therefore, provide indirect evidence that the effect of this form of control is to potentially increase wave drag and reduce viscous drag What is not clear is whether there is an optimum suction rate which will lead to a nett drag reduction.

Figure 20 shows a comparison of the variation in drag coefficient with normal-force coefficient between experiment and theory (BVGK(P)) for the datum and maximum suction case. For the datum case theory again underestimates drag as discussed previously. However, for the suction case theory underestimates the drag penalty, even if a similar allowance is made for the drag due to trip roughness. Although BVGK(P) does not predict the drag penalty exactly it can still be considered a useful tool for carrying out a parametric study of control systems in an attempt to identify an optimum.

Using BVGK(P), the effect of varying the length of the control region in different ways is shown in Figures 21 to 23, with all other parameters fixed. The effect on drag of reducing the length of the control region whilst keeping the start position fixed is shown in Figure 21. It can be seen that the drag of the section is always greater than that of the datum for control cases where the end of the control region is greater than x/c = 0.58. The optimum is obtained when the suction strip is as narrow as possible, in other words when the system reduces to the discrete suction case. Figure 22 shows a similar trend. In this case where the end of the control region is fixed and the start varied, the optimum is discrete suction. In Figure 23 the length of the control region is varied by changing the start and end position; again the optimum is discrete suction, in this case with the control region centred on the shock position. It is clear from Figures 21 to 23 that, provided the suction strip is reduced to a minimum width, the actual position is unimportant, at least for positions between x/c = 0.47 and 0.63. In these cases an absolute drag reduction of the order of 10% is obtained, 7% if pump drag is accounted for. In view of these results it worth considering a modification to the Active Suction insert to represent this optimum arrangement.

19.5.3 Hybrid Suction

Figure 12 shows that for high values of normal-force coefficient (C_N>0.7) there is, again, a significant loss of normal force associated with this control system, although it is to be noted that the loss is not as great as that for the Active Suction system (Figure 8). There appears to be no discernible effect of suction, in that the suction on and off cases are essentially indistinguishable.

The variation in drag coefficient with normal-force coefficient is shown in Figure 13 for varying suction rates. In this case the secondary suction control region behind the 'passive' control region is attempting to reduce the additional viscous drag generated by passive re-circulation. It can be seen that the effect of increasing suction is to reduce the drag relative to the suction-off case. However, it is clear that in order to obtain significant drag reductions it would be necessary to resort to very high suction rates which in turn results in a larger pump drag penalty. It is therefore debatable as to whether a nett drag reduction can be achieved with this arrangement, another factor being that at high suction rates the flow will choke in the holes of the porous surface, rendering any further increases in suction futile. This would suggest the possibility of either increasing the porosity or replacing the porous surface with a slot, but at the high suction rates needed it would appear unlikely that such a system would be viable.

The effect of suction on trailing-edge pressure divergence, Figure 14, would appear to be marginal for this configuration; again it is clear that high levels of suction would be necessary to restore trailing-edge pressure to the level of the datum section.

Typical pressure distributions, Figure 15, show that the flow in the 'passive' control region is identical, as would be expected. The classical effect of 'passive' control is seen whereby the flow is recompressed from a point ahead of the start of the control region and the supercritical flow is terminated by a shock wave which is much weaker than that of the datum section. By inference the wave drag in the control case is therefore significantly reduced. Downstream of the shock wave the pressures are lower than the datum case, as was noted for Active control. In the present case, however, the pressures downstream of the second, active, control region are increased and are similar to those of the datum case. This suggests that the effect of the active control region is to reduce the boundary-layer thickness in this region albeit not to or below the thickness of the undisturbed, datum section.

No theoretical calculations have been performed for this control configuration due to the lack of a suitable method, although it is possible to envisage how this could be done. A hybrid method based on features of the BVGK6E and BVGK(P) methods could be developed, utilising the ability of BVGK(P) to model 'passive' control allied to the BVGK6E modelling of discrete suction.

19.6 Conclusions

This report has described an initial experimental and theoretical assessment of novel techniques for reducing the drag of airfoils. The main conclusions arising from the study are as follows:

a. The use of Discrete Suction leads to a nett reduction in drag, after allowance is made for pump drag, for all values of suction coefficient. Examination of the results suggests that larger reductions are possible if the suction rate could be increased. The effectiveness of the system is also independent of Mach number and lift coefficient, therefore making it useful at all flight conditions.

b. Examination of the comparisons of predicted and measured drag polars show that satisfactory predictions are made using both variants of the BVGK method. In the case of the simplified model utilised by BVGK6E it is essential that the increase in skin friction due to suction is correctly accounted for.

c. Active Suction is ineffective at reducing drag for the configuration tested, the suppression of the passive re-circulation effect being insufficient. However, the data obtained has allowed a validation of the BVGK(P) method to be made. This has allowed a simple parametric study to be made which has identified a control arrangement that should allow significant nett reductions in drag.

d. Predictions of the increase in drag due to Active control by BVGK(P) are in reasonable agreement with measurement for the conditions shown. This suggests that the method may be used with some confidence to calculate the effects of active suction at Mach numbers close to the design value and at high Reynolds numbers typical of flight conditions.

e. The Hybrid Suction system is unable to overcome the development of the thickening boundary layer associated with the blowing portion of the 'passive' control region. It is probable that in order to reduce the boundary layer thickness sufficiently to yield a reduction in drag the rate of suction required would be such that the benefits in terms of wave drag reduction would be nullified by the pump drag.

References

[1] Fulker, J. L. and Simmons, M. J., "An experimental investigation of passive shock/boundary-layer control on an airfoil," in: Notes on Numerical Fluid Mechanics, Vol. 56, EUROSHOCK-Drag Reduction by Passive Control, pp. 379-400, 1997.

[2] Krogmann, P. and Thiede, P, "Transonic shock-boundary layer control", ICAS Paper 84-2.3.2, September 1984.

[3] Thiede, P and Krogmann, P, "Improvement of transonic airfoil performance through passive shock/boundary-layer interaction control", IUTAM Symposium, Palaiseau, France, 1985, Ed. J. Délery, Springer, Berlin, 1986.

[4] Thibert, J. J., Reneaux, J. and Schmitt, V, "ONERA activities on drag reduction", Paper ICAS 90-3.6.1, September 1990.

[5] Ashill, P. R., Fulker, J. L. and Shires, A., "A novel technique for controlling shock strength of laminar-flow airfoil sections", Paper presented to the First European Forum on Laminar Flow Technology, Hamburg, Germany, March 1992.

[6] Fulker, J. L., Ashill, P. R. and Simmons, M. J., "Study of simulated active control of shock waves on airfoil sections", DRA Unpublished, 1993.

[7] Krogmann, P. and Stanewsky, E., "Effects of local boundary layer suction on shock-boundary layer interaction and shock-induced separation", Paper AIAA-84-0098, 1984.

[8] Lock, C. N. H., Hilton, W. F. and Goldstein, S., "Determination of profile drag at high speeds by pitot traverse method", ARC R&M 1970, 1940.

[9] Ashill, P. R., Wood, R. F. and Weeks, D. J., "An improved semi-inverse version of the viscous Garabedian and Korn method", RAE Technical Report TR 87002, 1987.

[10] Ashill, P. R., "Calculation and measurements of transonic flows over airfoils with novel rear sections", Proceedings of the 16th Congress of ICAS, Jerusalem, Israel, Paper No. 88-3.10.2. September 1988.

[11] Poll, D. I. A., Danks, M. M. and Humphreys, B. E., "The aerodynamic performance of laser drilled sheets", Paper presented to the First European Forum on Laminar Flow Technology, Hamburg, Germany, March 1992.

[12] Ashill, P. R. and Gaudet, I. M., "Assessment of a CFD method for transonic airfoils with shock control", DRA Unpublished, 1996.

[13] Gaudet, I. M., "A user's guide to the DRA program BVGK(P)", DRA Unpublished, 1996.

[14] Hackett, K. C., "A theoretical investigation of the effect of boundary-layer suction on the aerodynamic performance of a turbulent wing", DRA Unpublished, 1995

[15] Pankhurst, R. C. and Gregory, N., "Power requirements for distributed suction for increasing maximum lift", ARC CP82, 1952.

List of Symbols

c	airfoil chord length, distance between airfoil leading and trailing edges along airfoil reference axis
C_D	drag coefficient
C_f	local skin-friction coefficient, $2\tau_w/\rho_{iw}U^2_{iw}$
C_N	normal-force coefficient
C_p	static pressure coefficient
C_{pte}	trailing-edge static pressure coefficient
C_Q	mass-flow coefficient, $= m/\rho_\infty U_\infty c$, taken positive when air is removed from the plenum (CQ in Figures)
C_{qeff}	effective mass-flow coefficient
EIF	Equivalent Inviscid Flow
H	boundary-layer shape parameter, $=\delta^*/\theta$
$\overline{H}$	transformed shape parameter
H_1	mass-flow shape parameter, $= (\overline{\delta} - \delta^*)/\theta$
K	hole calibration constant
M	Mach number
m	mass flow withdrawn from plenum chamber per unit span
p	pressure
R_c	Reynolds number based on free-stream density and speed and airfoil chord
R_θ	Reynolds number based on boundary-layer momentum thickness and local EIF quantities at the wall
RVF	Real Viscous Flow
s	distance around airfoil surface in streamwise direction
U,W	flow velocities in the s and z directions, respectively
x	distance along airfoil chord downstream of leading edge
z	distance normal to and from airfoil surface
α	angle of incidence (also referred to as ALPHA in Figures)
δ	boundary layer thickness
$\overline{\delta}$	equivalent boundary layer thickness
θ	boundary-layer momentum thickness
Δ	decrement in
ρ	density
σ	$\rho_w W_w/\rho_{iw}U_{iw}$

Suffixes

c	beneath the suction surface position in the plenum chamber
i	Equivalent Inviscid Flow
s	solid surface
w	wall or airfoil surface
*	sonic conditions when applied to C_p
∞	freestream conditions

Appendix A

Derivation of effective mass flow coefficient for discrete suction cases

The streamwise momentum equation in zero pressure gradient can be written as:

$$\frac{d\theta}{ds} = \frac{C_f}{2} + \sigma ,$$

where

$$\sigma = \frac{\rho_w W_w}{\rho_{iw} U_{iw}} ,$$

and the suffix iw refers to wall or airfoil surface quantities in the Equivalent Inviscid Flow, EIF, and w refers to wall quantities in the Real Viscous Flow, RVF. Therefore,

$$\frac{d\theta}{ds} = \frac{C_f}{2} + C_Q \frac{c}{l} \frac{\rho_\infty U_\infty}{\rho_{iw} U_{iw}} , \qquad \text{(A-1)}$$

since

$$C_Q = \frac{\rho_w U_w l}{\rho_\infty U_\infty c} ,$$

where l is the width of the porous region and c is the airfoil chord.

The skin friction increases in the region of porosity and so Equation (A-1) can be written as:

$$\frac{d\theta}{ds} = \frac{C_f}{2} + \frac{\Delta C_f}{2} + C_Q \frac{c}{l} \frac{\rho_\infty U_\infty}{\rho_{iw} U_{iw}} ,$$

Integrating this expression with respect to s and considering the increment in momentum thickness we have

$$\Delta\theta = \frac{\Delta C_f}{2} l + C_Q c \frac{\rho_\infty U_\infty}{\rho_{iw} U_{iw}} ,$$

but taking "Model A" of skin friction from Ref 12, i.e,

$$\frac{\Delta C_f}{2} = -\sigma\left(1 - \frac{2\overline{H}}{\overline{H}+H_1}\right),$$

we obtain:

$$\Delta\theta = -C_Q\left(\frac{\rho_\infty U_\infty}{\rho_{iw} U_{iw}}\right)\left(1-\frac{2\overline{H}}{\overline{H}+H_1}\right)+C_Q\frac{\rho_\infty U_\infty}{\rho_{iw} U_{iw}},$$

hence

$$\frac{\Delta\theta}{c} = C_Q\left(\frac{\rho_\infty U_\infty}{\rho_{iw} U_{iw}}\right)\frac{2\overline{H}}{\overline{H}+H_1},$$

and

$$C_{Q_{eff}} = C_Q\frac{2\overline{H}}{(\overline{H}+H_1)}.$$

If Model B from Ref. 12 is used, we obtain:

$$C_{Q_{eff}} = -C_Q\left\{\left(1-\frac{2\overline{H}}{(\overline{H}+H_1)}\right)\left(0.16\log_{10} R_\theta - 0.07\right)-1\right\}$$

Appendix B

Allowance for internal pump drag in total drag coefficient

According to Pankhurst and Gregory[15], if

a. the efficiency of the suction pump is taken to be equal to the efficiency of the main propulsion system of the aircraft,
b. the duct losses are negligible, and
c. there is a single internal chamber or plenum into which the boundary-layer air flows through the suction strip,

the internal pump drag coefficient, C_{DP}, may be expressed as:

$$C_{DP} = C_{PC}C_Q\,. \qquad \text{(B-1)}$$

with

$$C_{PC} = \frac{(P_\infty - p_c)}{\frac{1}{2}\rho_\infty U_\infty^2},$$

where P is the total pressure and suffix ∞ refers to conditions in the external flow far upstream. Suffix c refers to conditions in the plenum chamber beneath the suction strip.

If, in addition, the pressure losses through the strip into the plenum are ignored, then for area suction through the strip it is reasonable to assume that $p_c = p_s$ and

$$C_{PC} = \frac{(P_\infty - p_s)}{\frac{1}{2}\rho_\infty U_\infty^2} = \frac{(P_\infty - p_\infty)}{\frac{1}{2}\rho_\infty U_\infty^2} - C_{PS};$$

here suffix s refers to conditions in the external flow just above the strip.

According to Equation 3-1, Equation B-1 may be written as

$$C_{DP} = C_{PC}\frac{\Delta\theta}{c}\frac{\rho_s U_s}{\rho_\infty U_\infty}.$$

Thus the total drag coefficient, including allowance for internal pump drag, may be expressed as:
where ΔC_D is the decrement in drag coefficient allowing for internal pump power,

$$C_D = C_{D0} - \Delta C_D = C_{D0} + \Delta C_{DE} - C_{PC}\frac{\Delta\theta}{c}\frac{\rho_s U_s}{\rho_\infty U_\infty}, \quad \text{(B-2)}$$

ΔC_{DE} is the decrement in drag coefficient due to the effect of suction on the external flow and C_{D0} is the drag coefficient of the airfoil without suction. Noting that

$$\Delta C_{DE} = -2m_D\frac{\Delta\theta}{c},$$

it is possible to rewrite Equation B-2 as

$$C_D = C_{D0} - \frac{2\Delta\theta}{c}\left(m_D - \frac{C_{PC}}{2}\frac{\rho_s U_s}{\rho_\infty U_\infty}\right).$$

It follows that, within the framework of the assumptions described, suction leads to a reduction in drag provided that

$$m_D > \frac{C_{PC}}{2}\frac{\rho_s U_s}{\rho_\infty U_\infty}.$$

For the flows considered in this report, this is found to be the case.

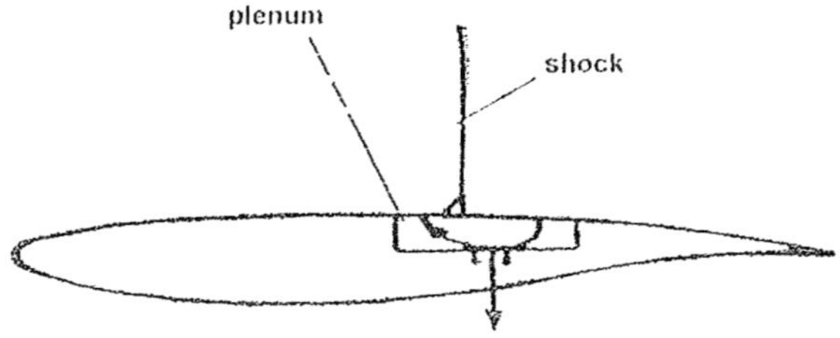

Fig 1a Aerofoil section with Active Suction

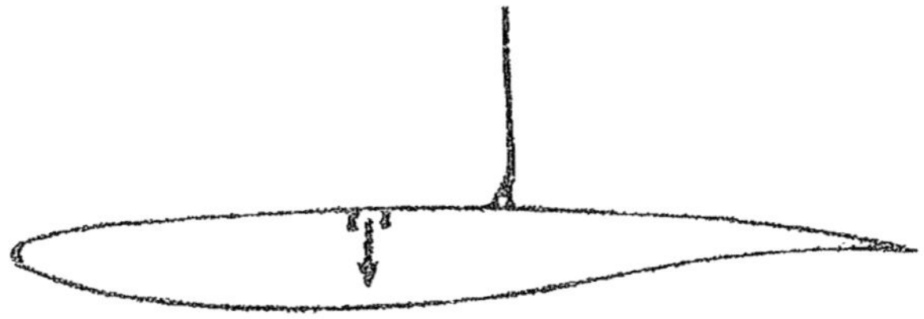

Fig 1b Aerofoil section with Discrete Suction

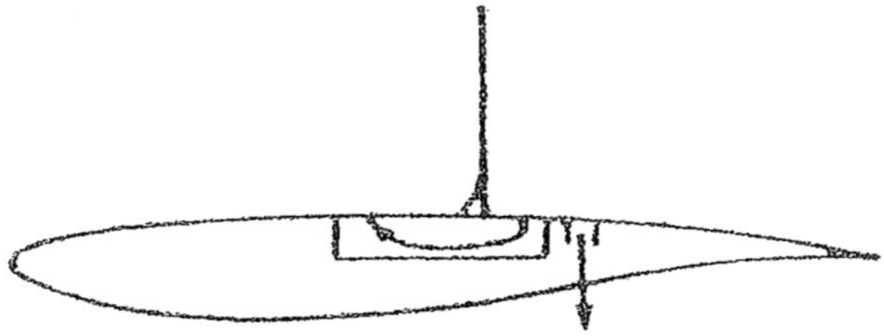

Fig 1c Aerofoil section with Hybrid suction

Figure 1 Schematics of control systems

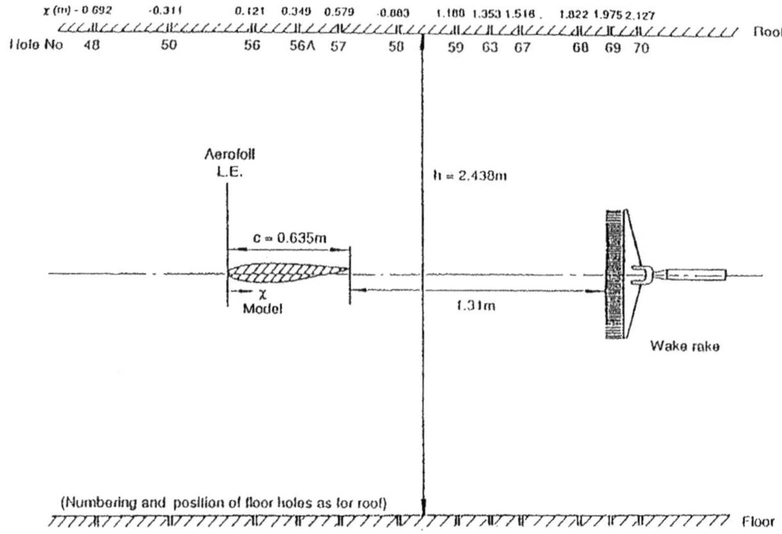

Figure 2 Layout of model and wake rake in the working section

Figure 3 Photograph of the model in the working section

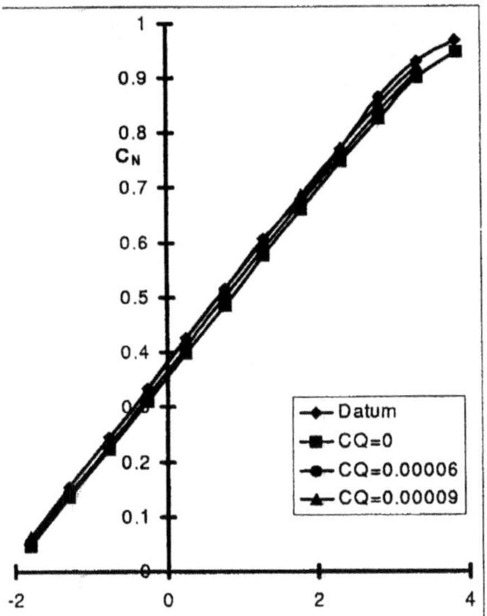

Figure 4 Variation of normal-force coefficient with angle of incidence for Datum and Discrete-suction configurations, $R_c = 19 \times 10^6$, M = 0.68

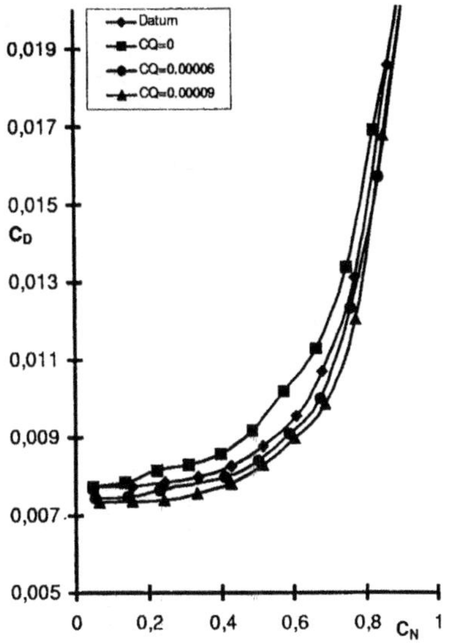

Figure 5 Variation of drag coefficient with normal force coefficient for Datum and Discrete-suction configurations, $R_c = 19 \times 10^6$, M = 0.68

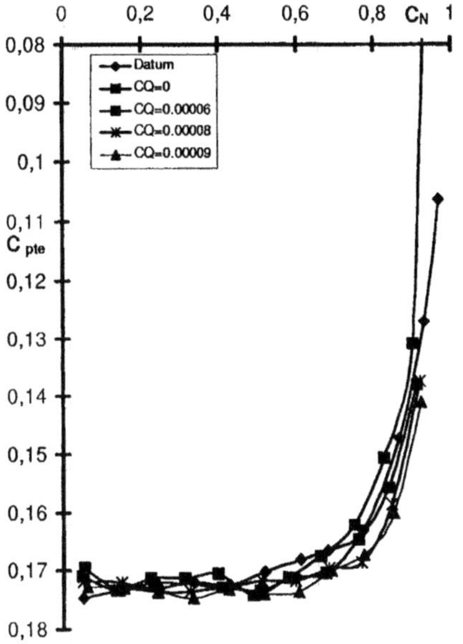

Figure 6 Variation of trailing-edge pressure with normal force coefficient for Datum and Discrete-suction configurations, $R_c = 19 \times 10^6$, M = 0.68

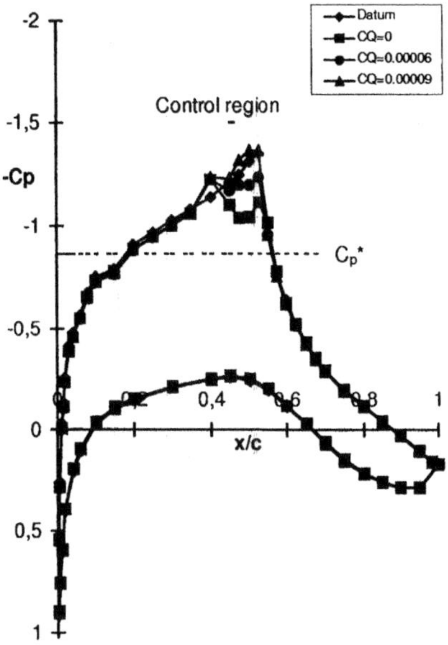

Figure 7 Comparison of pressure distributions for Datum and Discrete-suction configurations, $R_c = 19 \times 10^6$, M = 0.68, C_N=0.6

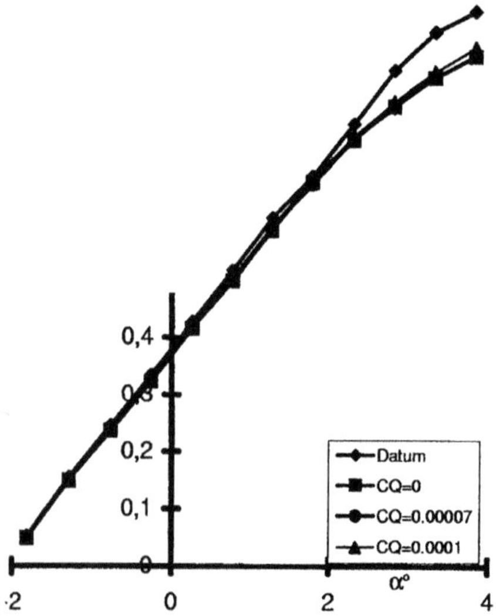

Figure 8 Variation of normal-force coefficient with angle of incidence, Datum and Active configurations, $M = 0.68$, $R_c = 19 \times 10^6$

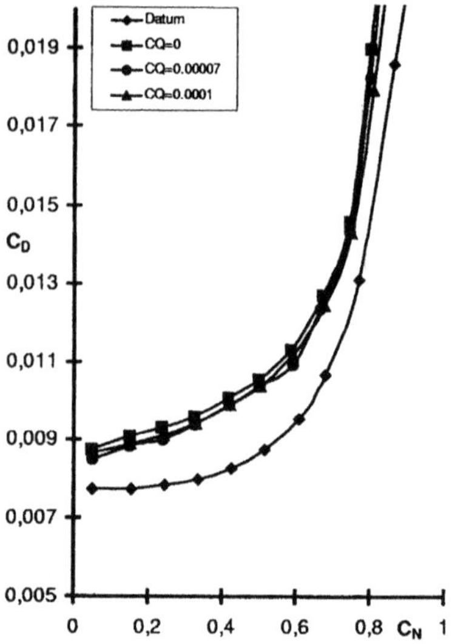

Figure 9 Variation of drag coefficient with normal-force coefficient, Datum and Active configurations, $M = 0.68$, $R_c = 19 \times 10^6$

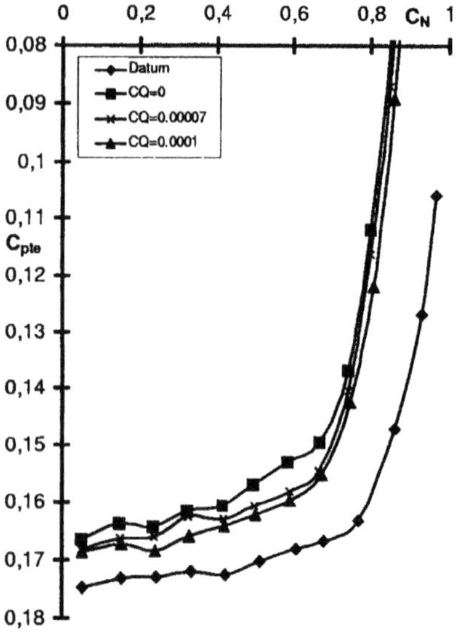

Figure 10 Variation of trailing-edge pressure with normal-force coefficient, Datum and Active configurations, $M = 0.68$, $R_c = 19 \times 10^6$

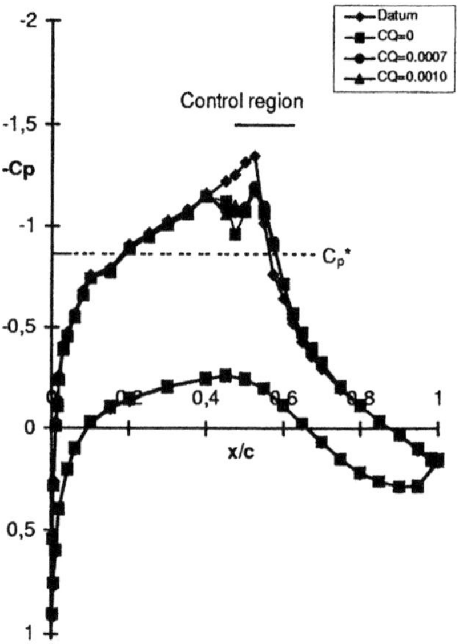

Figure 11 Comparison of pressure distributions, Datum and Active configurations, $M = 0.68$, $R_c = 19 \times 10^6$, $C_N = 0.6$

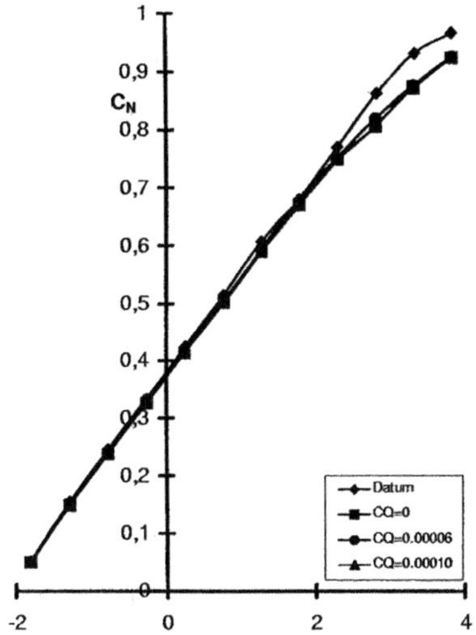

Figure 12 Variation of normal-force coefficient with angle of incidence, Datum and Hybrid configurations, $M = 0.68$, $R_c = 19 \times 10^6$

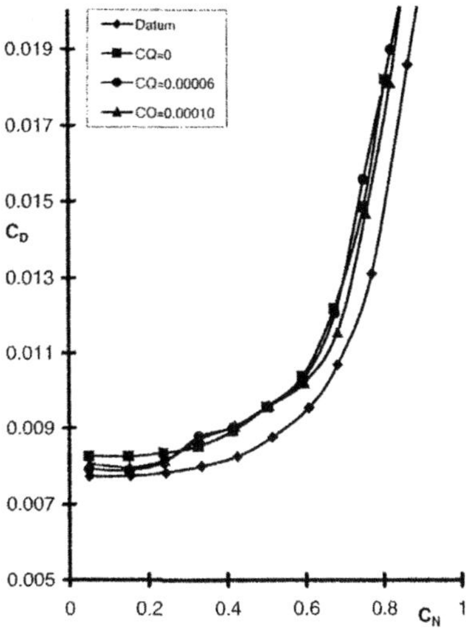

Figure 13 Variation of drag coefficient with normal-force coefficient, Datum and Hybrid configurations, $M = 0.68$, $R_c = 19 \times 10^6$

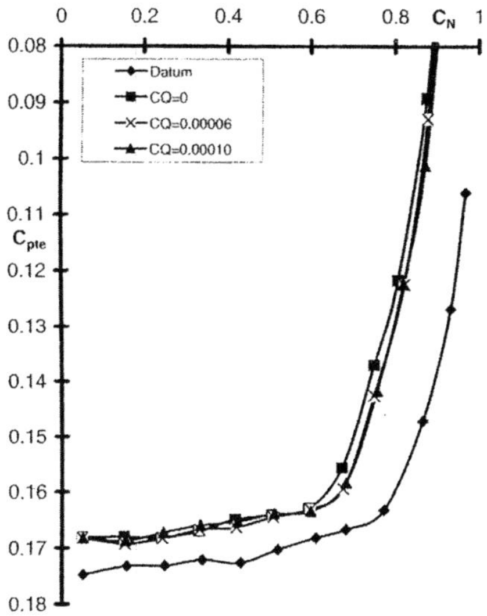

Figure 14 Variation of trailing-edge pressure with normal-force coefficient, Datum and Hybrid configurations, $M = 0.68$, $R_c = 19 \times 10^6$

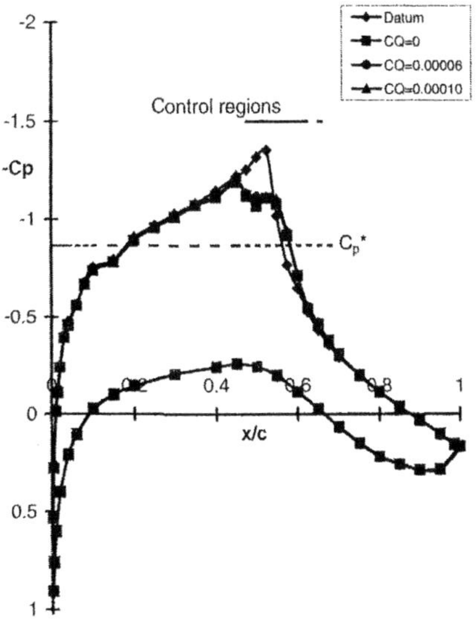

Figure 15 Comparison of pressure distributions, Datum and Hybrid configurations, $M = 0.68$, $R_c = 19 \times 10^6$, $C_N = 0.6$

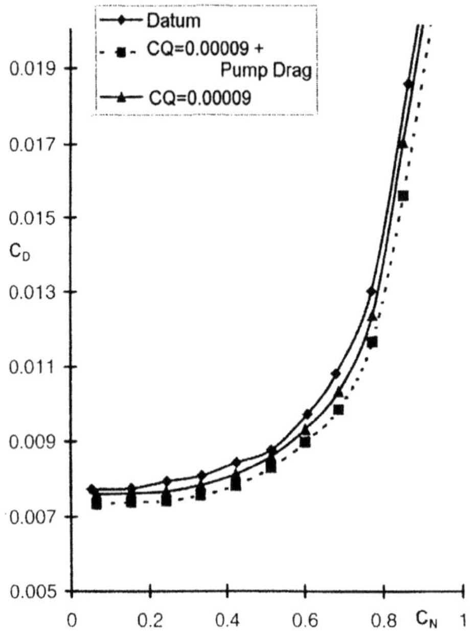

Figure 16 Variation of drag coefficient with normal-force coefficient, Datum and Discrete-suction configurations, M = 0.68, $R_c = 19 \times 10^6$, $C_Q = 0.00009$ + correction for pump drag

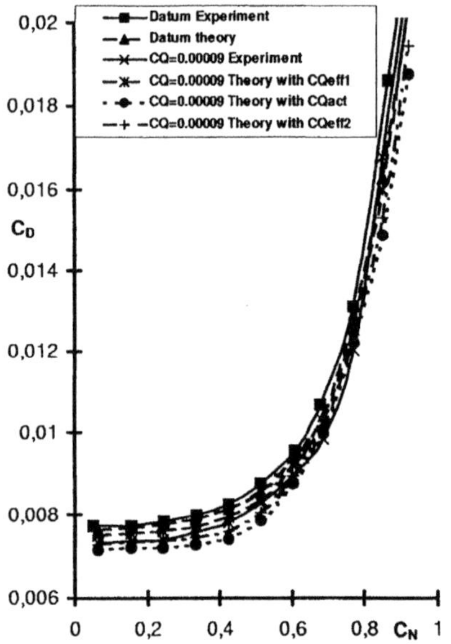

Figure 17 Variation of drag coefficient with normal-force coefficient, Datum and Discrete-suction configurations, M = 0.68, $R_c = 19 \times 10^6$, $C_Q = 0.00009$, comparison between experiment and theory

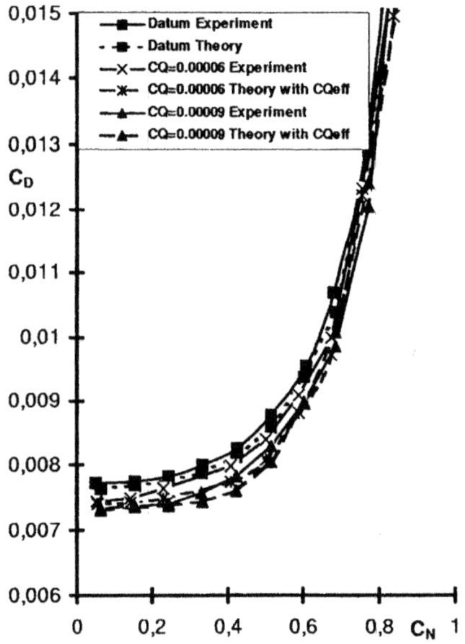

Figure 18 Variation of drag coefficient with normal-force coefficient, Datum and Discrete-suction configurations, M = 0.68, $R_c = 19 \times 10^6$, varying C_Q, comparison between experiment and theory

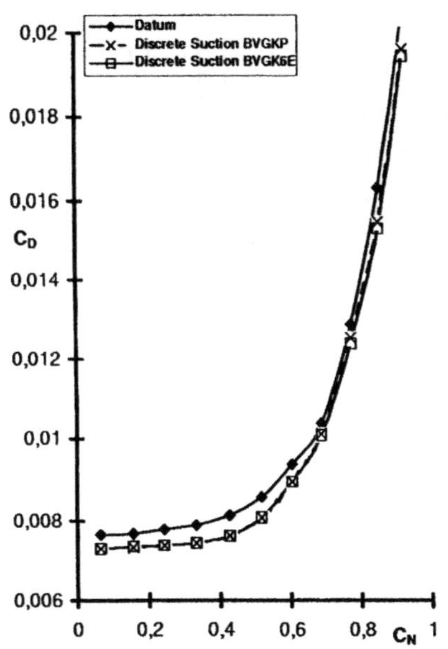

Figure 19 Variation of drag coefficient with normal-force coefficient, Datum and Discrete-suction configurations, M = 0.68, $R_c = 19 \times 10^6$, $C_Q = 0.00009$, comparison between BVGK6E and BVGK(P)

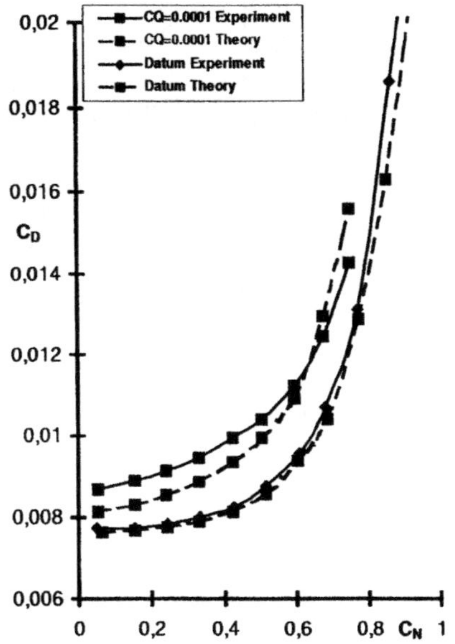

Figure 20 Variation of drag coefficient with normal-force coefficient, Datum and Discrete-suction configurations, M = 0.68, $R_c = 19 \times 10^6$, $C_Q = 0.0001$, comparison between experiment and theory

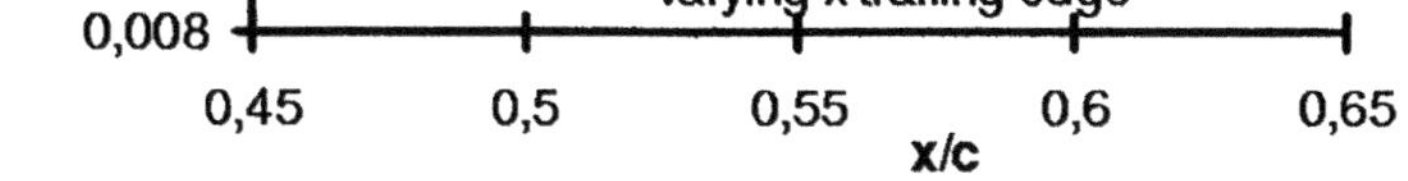

Figure 21 Variation of drag coefficient with control surface length, M = 0.68, $R_c = 19 \times 10^6$, $C_Q = 0.00014$, varying trailing-edge position.

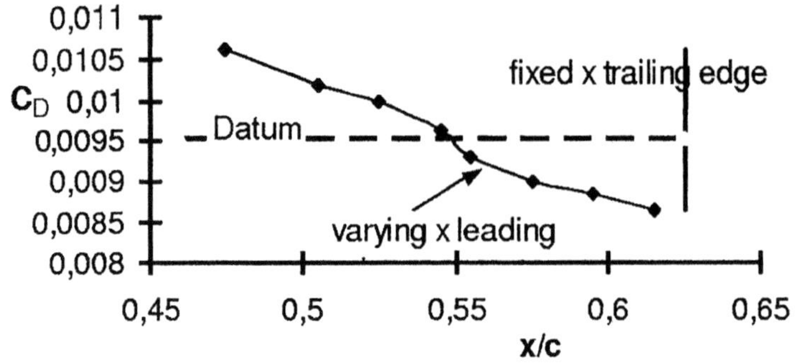

Figure 22 Variation of drag coefficient with control surface length, M = 0.68, $R_c = 19 \times 10^6$, $C_Q = 0.00014$, varying leading-edge position.

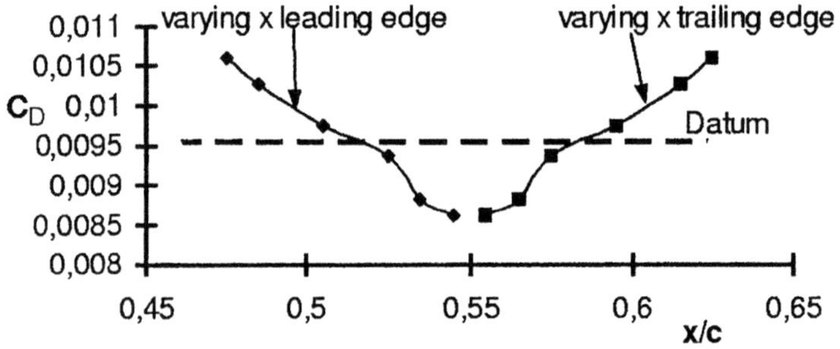

Figure 23 Variation of drag coefficient with control surface length, M = 0.68, $R_c = 19 \times 10^6$, $C_Q = 0.00014$, varying leading- and trailing-edge position.

20 Experimental Investigation of the Transonic Airfoils ADIF and DA LVA-1A and the ADIF Sheared Wing with and without Control

H. Rosemann and J. Birkemeyer

DLR Institute of Aerodynamics and Flow Technology
Bunsenstrasse 10, D-37073 Göttingen

Summary

Employing shock and boundary layer control can be assumed to have a large potential for improving flight performance of transonic transport aircraft in terms of cruise drag, hence speed and/or fuel consumption, and with respect to the drag-rise and buffet boundaries. An extensive wind tunnel program on airfoils and an infinitely-swept sheared wing was, therefore, carried out investigating the effect of shock and boundary layer control on the improvement of airfoil and wing performance. The airfoils considered were the laminar-type design DA LVA-1A and the A340-type airfoil ADIF; the sheared wing studied was based on the ADIF airfoil. Control schemes investigated included passive control via a perforated plate/cavity arrangement, discrete slot suction, control by contour bumps, and hybrid control, viz., a passive cavity and a bump, respectively, combined with slot suction. All control schemes were investigated with the sheared wing, while for the airfoils only bump control was applied. It was found that shock and boundary layer control by a bump and by discrete slot suction can greatly reduce drag, also when applied in combination, while control involving a passive cavity always increased drag due to its dominating negative effect on viscous drag. When passive cavity control was combined with discrete slot suction, a minor drag reduction, especially at low lift coefficients, could be achieved. All devices considered have the potential of increasing the buffet boundary.

20.1 Introduction

Mostly experimental investigations concerned with the control of shock waves for the purpose of drag reduction and the enhancement of the drag-rise and buffet boundaries of transonic transport aircraft wings have been performed at DLR since 1979. The investigations commenced with single-slot suction at the foot of the shock to avoid shock-induced separation at off-design conditions [1], followed by the application of double-slots and perforated surfaces, both with underlying cavities and suction from these cavities, mainly for drag reduction and the avoidance of separation [2, 3]. These early studies were carried out with the turbulent airfoil VA-2 designed by EADS-Airbus D.

Concerning single-slot suction, it was found that this type of control was very effective in increasing maximum lift — hence positively affecting the buffet boundary — and reducing drag at high lift conditions. Control by double slots and perforated surfaces with underlying cavities in the shock region shifted the buffet boundary to higher Mach numbers and/or lift coefficients, but also drag was reduced over a wider range of freestream conditions. The effect was, qualitatively, even achieved without external suction, i.e., by passive cavity control. This is demonstrated in Figure 1 by the range parameter L/D (lift to drag ratio). Clearly, an increase in L/D over a considerable range of flow conditions is indicated, which is mainly caused by a reduction in total drag due to control.

Inspired by the success of passive shock control for the turbulent airfoil VA-2 and the renewed interest in laminar-wing transport aircraft, where relatively strong shock waves may already be present on the wing upper surface at cruise conditions, passive control by a perforated plate/cavity arrangement was investigated by DLR for the laminar-type airfoil DA LVA-1A, also designed by EADS Airbus D [4], within the EUROSHOCK (I) project [5, 6]. As is well documented in [6], and in agreement with the findings of other investigations within this project, e.g., results obtained by DERA for the laminar-type airfoil DRA-2303 [7], total drag was always increased by the application of passive cavity control in spite of a strong reduction in wave drag. The increase in total drag was traced to the dominating increase in viscous drag caused by the re-circulating flow in the cavity region and the subsequent amplification of the local increase of the momentum and displacement thicknesses by the strong rear adverse pressure gradients prevailing on the upper surface of the airfoil.

Due to the negative results related to drag, measurements with passive cavity control on the turbulent VA-2 airfoil were repeated with the results, however, confirming the earlier measurements, although the drag reductions observed were less pronounced. Analyzing the data for the turbulent airfoil VA-2 and the laminar airfoil DA LVA-1A lead to the conclusion that the thickness of the boundary layer and the magnitude of the skin friction coefficient upstream of the control region may have an important influence on the efficiency of control and the increase in viscous drag. Both parameters may differ greatly between a laminar and a turbulent airfoil.

The results of the EUROSHOCK (I) investigation lead to the consideration of active control methods designed to keep the boundary layer growth over and downstream of the interaction/control region low, possibly completely avoiding an increase in viscous drag due to control. Control candidates considered for investigation by the EUROSHOCK (I)-Consortium (see [5]) were active cavity control with part external suction, discrete single-slot suction, contour modifications in the shock region (bumps) and combinations of control schemes (hybrid control) such as a passive cavity in the shock region in conjunction with active control by discrete slot suction downstream of the passive cavity, or a bump in combination with slot suction upstream.

The airfoil/wing configurations investigated by DLR in conjunction with various control schemes were the laminar-type airfoil DA LVA-1A, the A340-type airfoil ADIF, and an infinitely-swept (sheared) wing, based on the ADIF airfoil, the latter to study the effect of sweep on control effectiveness. The investigations were performed within the present (EUROSHOCK II) project. In the following chapters we will briefly describe the airfoil/wing/control configurations studied and the experimental facilities utilized, and then discuss essential results of the present investigation.

20.2 Experimental Setup and Wind Tunnel Models

Experiments at various Mach numbers and associated angle of attack sweeps were carried out with the A340-type airfoil ADIF and the laminar-type airfoil DA LVA-1A, respectively, in the no-control datum configuration and with control by contour bumps of different heights. The measurements were performed in the Cryogenic Ludwieg-tube wind tunnel of DLR (KRG) [8]. Tests with the 26-degree infinitely-swept wing (sheared wing) were performed in the 1x1 Meter Transonic Wind Tunnel Göttingen (TWG) [9]. Control mechanisms considered here included contour bumps, similar to the 2D configuration, a bump in conjunction with discrete suction upstream, pure discrete slot suction, and passive ventilation via a perforated plate/cavity arrangement, also in conjunction with discrete slot suction.

20.2.1 Wind Tunnel Characteristics

The Cryogenic Ludwieg-tube of DLR Göttingen (KRG) is a short-duration facility with a test time of up to one second consisting of a 130-meter long tube, a contraction section with a contraction ratio of 3.6, an adaptive-wall test section, a combination of second throat for Mach number control and quick-opening valve, and a dump tank, Figure 2. The test section has a cross section of $0.40 \times 0.35 m^2$ and a length of 2m which allows to obtain interference-free results for the model chords investigated here, i.e., $c \approx 0.18m$. The performance characteristics of the tunnel, which uses gaseous nitrogen as test gas, are summarized in Figure 2.

The 1 x 1 Meter Transonic Wind Tunnel Göttingen (TWG) is a closed-circuit continuous tunnel with a cross-section of $1 \times 1 m^2$. There are three independent test sections available: a perforated test section, 6% open with 60° slanted holes, for the Mach number range 0.40 to 1.3, an adaptive-wall transonic test section for the Mach number range 0.3 to 0.95, and a supersonic test section for Mach numbers of 1.4 to 2.2. The total pressure can be adjusted between 0.6 and 1.6 bar allowing for some Reynolds number variation. The present investigation was performed in the adaptive-wall test section to reduce wall-interference effects; the side walls of the test section were, for the present sheared-wing tests, contoured, as outlined below, to generate infinite swept-wing flow.

20.2.2 Airfoil/Wing Characteristics and Wind Tunnel Models

20.2.2.1 ADIF and DA LVA-1A airfoils

The **ADIF airfoil** is an A340-type modern airfoil with a thickness of about 12% chord. It has a moderately accelerating flow on the upper surface upstream of the shock at near-cruise conditions. The shock position varies slightly with changing freestream conditions, which is typical of airfoils with only moderate pressure gradients on the upper surface following the initial strong acceleration around the leading edge (see, e.g., Figure 6). The cryogenic wind tunnel model of the airfoil is depicted in Figure 3 [10]: the model has a span of 400 mm and a chord of 175 mm and is inserted between the wind tunnel side walls; the model instrumentation, mainly consisting of static and dynamic pressure orifices, is similar to the one for the sheared wing model, Figure 4. In all tests, a wake rake was used for drag determination. Various bumps, optimized in a parametric study using a 2D coupled Euler/boundary layer method [10, 11], and a discrete suction device upstream of the bump, which can be employed independently of and in conjunction with the bump, are available. The shape of the bumps tested is asymmetric, in that the initial ramp is 12% chord in length while the closure is 8% chord. Two bumps of 0.175%c and 0.35%c height, respectively, were investigated (see lower two bumps in the corresponding insert to Figure 3). Note that here only bump control will be discussed.

The **DA LVA - 1A** airfoil is a transonic laminar-type airfoil with a thickness of 12% chord. It was designed by EADS-Airbus D to have natural laminar flow on the upper and lower surfaces up to 50% of the chord at a Mach number of M_{∞} = 0.73, a lift coefficient of C_L= 0.4 and a Reynolds number of $Re_c = 20x10^6$ [4]. The pressure distribution is characterized by a moderate-strength shock wave occurring already at design conditions with strongly accelerating flow on the upper surface up to the shock to sustain laminar flow. An essential feature of this airfoil, especially important for shock control, is that with increasing lift coefficient, as the shock grows stronger, the position of the shock remains rather unchanged. The model design, similar to the ADIF models, provides an exchangeable insert to accommodate various optimized bumps (see Chapter 21). Here, two different bump contours with heights of h/c = 0.2% and h/c = 0.4%, respectively, located between x/c = 0.59 and x/c = 0.79 were investigated. An asymmetric shape was used, similar to the bump contour of the ADIF airfoil. The model was equipped with two chordwise pressure plotting sections with a total of 64 pressure orifices and two dynamic pressure transducers positioned downstream of the shock control region to determine pressure fluctuations, hence buffet behavior. Thermocouples are located on the inside of the model wall to determine transition location.

20.2.2.2 Sheared wing

The **Sheared-wing** model, Figure 4, has been designed to study sweep effects on shock control [12]. The wing has a chord length of 400mm and a sweep angle of 26°. On the upper surface of the wing there is a removable insert between 0.575c and 0.84c, similar to the ADIF airfoil model. It allows the contour of the datum wing to be exchanged with a bump contour or a perforated surface with a cavity underneath. Three different bump contours with the same asymmetric shape and the same length of 20% chord have been investigated with the crest (point of maximum thickness) locations and the height of the bumps being as follows:

1) bump Ia: $x_{crest}/c = 0.715$, $h_{bump}/c = 0.1573\%$
2) bump Ib: $x_{crest}/c = 0.715$, $h_{bump}/c = 0.3146\%$
3) bump II: $x_{crest}/c = 0.760$, $h_{bump}/c = 0.1573\%$

The perforated plate / cavity arrangement was placed between $x/c = 0.60$ and $x/c = 0.75$; the porosity of the plate was 8% and the diameter of the holes 0.3mm. A suction slot was located at $x/c = 0.55$ and, in case of the ventilation insert, an additional perforated strip of 1% chord length was placed at $x/c = 0.80$.

The model was equipped with conventional pressure orifices and with dynamic pressure transducers for buffet detection, Figure 4. As was mentioned before, the effective wind tunnel side walls are contoured according to stream surfaces to achieve infinitely swept-wing flow conditions.

20.3 Results of the Airfoil Experiments

20.3.1 ADIF Sheared-wing Reference Airfoil

Experiments with the ADIF airfoil were carried out in the DLR Göttingen Cryogenic Ludwieg Tube (KRG) with and without contour bumps [10]. The results serve, among others, as a reference for the investigation of sweep effects and to verify the design of the surface 'bumps', which were optimized in a parametric study using, as mentioned above, a two-dimensional coupled Euler/boundary layer method (Drela code [11]). Based on the numerical study and the wind tunnel experiments, a number of bump contours were chosen for the sheared wing and further optimized using a three-dimensional Navier-Stokes method [12].

In the present experiments, angle of attack sweeps were performed at Mach numbers of $M_\infty = 0.735, 0.755, 0.765$ and 0.775 at a Reynolds number of $Re_c = 8 \times 10^6$ with transition fixed at 30% chord on the upper surface and 7% chord on the lower surface — generally denoted 30/7 in the figures — to simulate flight Reynolds number conditions. Tests were also performed with free transition and the correspondingly thinner boundary layers to investigate, in addition, the influence of the boundary layer displacement thickness on the effectiveness of the contour bumps.

Typical results for the bumps, here at $M_\infty = 0.765$, show drag reductions of up to 20 drag counts ($\Delta C_D = 0.0020$) at a lift coefficient of approximately $C_L \leq 0.66$ and up to 35 counts at higher lift coefficients, Figure 5. This amounts to drag reductions of 17% and 23%, respectively. Also noticeable in Figure 5 is that the lower bump shows better performance at the lower lift coefficients while the higher bump is more effective at higher lift, i.e., in the presence of stronger shock waves, but even causes an increase in drag at low lift coefficients due to a mismatch between bump location/height and shock strength/location.

The spreading of the shock due to the contour change, causing the reduction in drag, is demonstrated by the pressure distributions shown in Figure 6 which also indicates — by the wake total pressure distributions — that not only wave drag but also viscous drag appears to be reduced by the presence of the bumps. The data also indicate that an adaptive bump seems to be required to fully exploit the bump potential. It should be noted that the behavior of the flow due to the bumps was well predicted by the Drela code [10, 11].

20.3.2 DA LVA-1A Airfoil

In the test series with the DA LVA-1A airfoil, employing shock control by contour bumps, angle of attack sweeps have been performed at three Mach numbers, viz., $M_\infty = 0.76$, 0.77, and 0.78, at a Reynolds number of $Re_c = 6 \times 10^6$ and transition fixed at 48% chord on both airfoil surfaces to simulate high Reynolds number conditions.

Results for the datum airfoil and the lower bump in the form of drag polars are presented in Figure 7. The polars show, as for the ADIF airfoil, significant drag reductions in the lift range of increasing wave drag, commencing here, in comparison to the ADIF airfoil, already at low lift coefficients ($C_L \approx 0.1 - 0.2$). The results are similar for all Mach numbers investigated; furthermore, there is no pronounced minimum in drag in the case of bump control as can be observed for the ADIF polars in Figure 5 between $0.5 \leq C_L \leq 0.6$. Only small gains in drag reduction, compared to the lower bump, are achieved with the higher bump, which are, in addition, restricted to the lift coefficients $0.3 < C_L < 0.45$; at higher lift coefficients the higher bump is less effective while producing additional drag at low lift coefficients of $C_L < 0.1$.

Pressure distributions corresponding to a freestream Mach number of $M_\infty = 0.77$ and a lift coefficient of $C_L \approx 0.315$ are compared in Figure 8 for the datum airfoil and the airfoil with the small bump. The spreading of the shock is clearly indicated. The wake profile shows that wave drag as well as viscous drag are reduced by the bump resulting here in a drag reduction of 12 drag counts ($\Delta C_D = 0.0012$) or roughly 12%.

Figure 9 shows the Mach number dependence of lift and drag at a constant angle of attack of $\alpha = 1.0°$. One observes that drag is reduced and lift increased due to the bump almost over the entire range of Mach numbers investigated. The beneficial effect of the bump in increasing the drag-rise boundary is also clearly indicated.

20.4 Results of the Sheared-wing Experiments

Tests were mainly performed at a Mach number of $M_\infty = 0.852$, the design Mach number for the contoured side walls, and a Reynolds number of $Re_c = 6.7 \times 10^6$. Due to a contamination of the leading edge, transition had to be fixed at 10% chord on the suction side. On the pressure surface transition was fixed at 15% of the model chord.

Since a swept adaptation of the roof and floor is not practicable in a 2D adaptive-wall test section, the best approach for adaptation has been investigated prior to the tests employing 3D Navier-Stokes simulations of the wind tunnel flow [12]. The best agreement with the pressure distribution for the swept wing in free air was achieved with wall contours corresponding to the shape of the center streamline in an unbounded flow around the model. Therefore, in the experiments, the pressure distributions at the centerline of the upper and lower wind tunnel walls were used for the adaptation process. In the following chapters, we will first consider the effectiveness of the side walls at their design condition, then the effect of passive cavity ventilation with and without additional single-slot suction applied, and finally control by a contour bump, also with and without additional slot suction.

20.4.1 Effectiveness of the Side-wall Contour

The purpose of the contoured liners is to produce a quasi two-dimensional flow field. As the liner contour depends on the wing geometry and the flow conditions, the contour is correct only for the design point, $M_\infty = 0.852$, $\alpha = 0^\circ$. The pressure distributions at the three spanwise stations S1, S2 and S3 (also see Figure 4) indicate a rather small spanwise gradient, Figure 10. At off-design conditions, the wind tunnel flow differs slightly from the infinite-span swept-wing flow: mainly the sweep angle of the shock differs from the geometrical sweep angle of the wing, and the trailing edge pressure is not constant in spanwise direction. The influence of the different means of shock control on drag reduction is, therefore, mainly judged at the design point of the sidewalls.

20.4.2 Passive Ventilation without and with Discrete Slot Suction

The effect of passive cavity ventilation has been investigated in detail within the EUROSHOCK (I) project [5]. It was found that wave drag is reduced substantially via the displacement effect due to the ventilation (see Figure 1 for a sketch of the passive cavity). However, for the laminar-type airfoils investigated, a large increase in viscous drag has been found to overcompensate the wave drag reduction, thus leading to an increase in total drag. Since passive ventilation essentially caused an increase in the boundary layer thickness parameters over and downstream of the interaction/control region leading to the dominating increase in viscous drag, a combination of passive ventilation with discrete boundary layer suction downstream or upstream of the passive control region has been considered in the present investigations. This type of hybrid control is expected to reduce the increase in viscous drag and hence reduce total drag.

Figure 11 confirms that the application of passive ventilation, here applied to the sheared-wing flow, leads to a strong increase of total drag, e.g., at the design point of the sidewall contour this increase amounts to ΔC_D = 11%. If the ventilation is coupled with boundary layer suction upstream of the interaction region, viscous drag is reduced — and wave drag possibly increased — resulting in a situation where any further decrease in viscous drag due to increased suction rates is exactly compensated by the wave drag increase and a further total drag reduction cannot be achieved. A drag reduction is, however, accomplished by suction at the lower lift coefficients where shock waves, hence wave drag, are of minor influence.

Figure 12 compares pressure distributions for the datum airfoil and the airfoil with passive control and with additional upstream slot suction at various suction rates at an angle of attack of $\alpha = 0^{\circ}$. Firstly, it should be noted that at the constant angle of attack lift is decreased due to passive control, however, increases slightly as suction is applied. The wake profiles in Figure 12 show that due to the higher lift in case of the datum airfoil drag is also higher. In the case of control, drag is highest for passive control but is reduced when suction is applied. Clearly indicated in the surface pressure distributions is the spreading of the shock due to control, hence the reduction in wave drag which can also be deduced by considering the wake profiles. Also discernable in the wake profiles is the increase in wave drag when activating slot suction. The reduction in viscous drag due to suction is indicated by the improved pressure recovery over the rear of the airfoil upper surface and the narrowing of the wake.

If boundary layer suction is applied downstream of the passive ventilation, a small gain in drag reduction is attained, Figure 13, increasing as the lift coefficient is reduced. No clear tendencies concerning the influence of the suction rate on drag reduction can be found at the higher lift coefficients, while at lower lift drag decreases with the suction rate. As is shown in the surface and wake pressure distributions, Figure 14, the drag reduction below the "passive-cavity-drag" is mainly due to a reduction in viscous drag indicated by the improved upper-surface pressure recovery and the reduced width of the wake. The higher drag of the datum airfoil at the constant angle of attack ($\alpha = 0^{\circ}$) is again due to the higher lift coefficient at these conditions.

20.4.3 Control by Contour Bumps with and without Slot Suction

20.4.3.1 Control by contour bumps

The control of shock boundary layer interaction by contour bumps was already discussed extensively in conjunction with the ADIF and DA LVA-1A airfoils, Chapter 20.3. For the present sheared wing, the influence of the bump height on drag reduction is demonstrated in Figure 15, where the drag polars of the datum wing and the wing with the bumps Ia (height 0.1573% chord) and Ib (height 0.3146% chord), respectively, are shown. Both bumps result at the design lift coefficient of the side walls, $C_L = 0.574$ ($\alpha = 0^o$), in a drag reduction of $\Delta C_D = 8\%$. It should be noted that this drag reduction is considerably less than the drag reduction of the ADIF airfoil at corresponding freestream conditions; the reason for the apparent reduced effectiveness of the bump will be discussed below. While at lift coefficients of $C_L > 0.55$ the sensitivity of drag reduction to the bump height seems to be negligible, leads the higher bump to a higher increase in drag at low lift coefficients, as was also observed for the basic ADIF airfoil.

At the design point of the side walls both bumps convert the strong shock of the datum airfoil into a nearly isentropic recompression, Figure 16, indicating the reduction in wave drag; however, the improved upper-surface pressure recovery shows that also viscous drag seems to be reduced, a trend also confirmed by the wake profiles.

20.4.3.2 Control by bump plus discrete upstream slot suction

The effect of a contour bump, here the lower bump with $h_{bump}/c = 0.1573\%$, in combination with upstream slot suction is demonstrated in Figure 17 by the drag polars. As already seen in the previous chapter, the bump alone results, at the design point of the contoured sidewalls, in a drag reduction of 8%. Adding upstream suction increases the drag reduction to 12% and, raising the suction coefficient to $C_Q = -0.00015$, to 22%. The corresponding surface and wake pressure distributions in Figure 18 indicate that the bump reduces wave drag due to the spreading of the shock as well as viscous drag. Adding suction mainly further reduces viscous drag which is again indicated by the better upper-surface pressure recovery over the rear of the wing but also by the narrowing of the wake. Suction does not seem to increase wave drag, as it would in the absence of the bump, which is due to the higher effectiveness of the bump in the presence of the thinner boundary layer now approaching the interaction region [12].

Regarding the comparison of the bump effectiveness in case of the airfoil and the sheared-wing addressed above, the following must be considered: in the case of the ADIF swept-wing experiments, transition had, as mentioned above, to be fixed on the upper surface at 10% chord; this lead to a larger boundary layer thickness upstream of the interaction region than in the case of the ADIF airfoil, where high Reynolds number flight conditions (thin boundary layer) were simulated by duplicating the displacement thickness upstream of the shock by fixing transition at 30% chord. If one now regards the boundary layer suction upstream of the control region in case of the sheared wing as a means to simulate higher Reynolds numbers, one may determine the effect of sweep on drag reduction due to the bump by a comparison of the two configurations at similar upstream boundary layer conditions, Figure 19: it follows that, at the design point of the sidewalls ($C_L \approx 0.57$) — only here is a comparison valid — the drag reduction in the case of the swept wing is slightly lower than the one for the corresponding airfoil, viz., 13% as compared to 15%, which is essentially due to the reduced contribution of wave drag to total drag in the case of the sheared wing.

20.4.4 Effect of Bump and Cavity-Ventilation Control on Buffet

The smaller bump (Ia), just considered, which has been designed to reduce wave drag, is not able to postpone the buffet boundary to higher lift coefficients. Bump II with $x_{crest}/c = 0.76$, $h_{bump}/c = 0.157$, i.e., the small bump placed just downstream of the shock, does not reduce the shock strength at all, but it is very effective in shifting the buffet boundary to a higher lift coefficient by its positive effect on separation. This is indicated in Figure 20 where the dependence on lift of the root-mean-square (rms) value of the upper-surface pressure at 90% chord is depicted. Also noticeable is the (slight) shift in buffet onset by passive ventilation, similar to the corresponding two-dimensional results. The buffet boundary is here defined according to Criterion C3, onset of (major) unsteadiness [12].

20.5 Conclusions and Future Work

The experimental results for the transonic airfoils and wings considered here show that the application of shock control in the form of bumps significantly reduces total drag. Analysis of the experimental data shows that this is a result of a large reduction in wave drag due to the bump while having only a minor, but frequently positive effect, on viscous drag. This behavior is especially beneficial for laminar-type airfoils (or wings) since strong shock waves tend to form at a nearly fixed chordwise position following the long favorable pressure gradient necessary to maintain natural laminar flow. However, the results have also shown that an adjustment of the bump in height is needed for optimum performance gains over a given range of lift coefficients (or Mach numbers for that fact).

For airfoils and wings of turbulent design, the experiments suggest that similar benefits can be obtained, however, a complication for this class of wings is the potential, large chordwise movement of the shock wave. In this case an adaptive bump, not only in height, but also in chordwise direction is needed for optimum gains.

The ADIF sheared-wing experiments have shown that sweep effects are actually minor with the drag reductions observed only slightly less than in the case of the airfoil. Furthermore, combining discrete slot suction upstream of the interaction region with bump control has an additive positive effect on drag reduction since here not only wave drag but also viscous drag is being considerably reduced. A bump placed downstream of the shock, i.e., downstream of the position optimized for drag reduction, was found to be very effective in increasing the buffet boundary, another reason for an adaptive bump.

The use of control by cavity ventilation does not appear to be a suitable solution to the problem of drag reduction in the presence of strong shock waves, even if passive cavity ventilation is combined with discrete slot suction either upstream or downstream of the passive cavity. It was found or confirmed that pure passive cavity control always increased drag: applying discrete slot suction upstream of the passive cavity just nullified the increase in drag caused by passive cavity ventilation bringing drag back to the no-control level; downstream suction resulted in a small drag reduction of 3%, an amount that would most certainly be nullified by pump drag, as would any increase in the suction rate meant to achieve higher drag reductions.

Future work concerning the application of a bump should essentially consider the structural integration of adaptive bumps into existing and/or new wing designs, turbulent or laminar. Especially for the former, emphasis should be placed on how the incorporation of a bump may lead to the design of thicker wings with the benefit of reduced structural weight, hence reduced costs and time to market. Furthermore, other flow control mechanisms, aiming at drag reduction and the avoidance of separation, such as, for instance, sub-boundary layer vortex generators, mass-less air jets, and possibly trailing-edge mini flaps, should be considered in combination with bumps. Concerning control involving cavity ventilation, emphasis should be placed on applications where drag reduction is not the main driver, such as, for instance, in the case of supersonic air intakes where the avoidance of separation and shock oscillations as well as the establishment of the best possible flow uniformity are of prime interest.

References

[1] Krogmann, P. "Pressure Distribution and Boundary Layer Measurements on a Transonic Airfoil with Single-slot Suction", DFVLR (DLR) Technical Report IB 222 81 A 06, 1981 (in German)

[2] Stanewsky, E. and Krogmann, P. "Transonic Drag Rise and Drag Reduction by Active/Passive Boundary Layer Control", AGARD Report No. 723, Lecture Series Aircraft Drag Prediction and Reduction, July 1985

[3] Thiede, P., Krogmann, P. and Stanewsky, E., "Active and passive shock/boundary layer interaction control on supercritical airfoils", AGARD-CP-365, Brussels, 1984

[4] Dargel, G. and Thiede, P., "Design and Manufacture of the Cryogenic Laminar-type Airfoil Model DA LVA-1A with Passive Shock Control", in EUROSHOCK – Drag Reduction by Passive Shock Control, Notes on Numerical Fluid Mechanics, Volume 56, Friedr. Vieweg & Sohn Verlagsgesellschaft mbH, Braunschweig/ Wiesbaden, 1997, pp. 319 - 329

[5] Stanewsky, E., Délery, J., Fulker, J. and Geissler, W. (Ed.), "EUROSHOCK – Drag Reduction by Passive Shock Control", Notes on Numerical Fluid Mechanics, Volume 56, Friedr. Vieweg & Sohn Verlagsgesellschaft mbH, Braunschweig/ Wiesbaden, 1997

[6] Rosemann, H., Knauer, A. and Stanewsky, E., "Experimental Investigation of the Transonic Airfoils DA LVA-1A and VA-2 with Shock Control", in EUROSHOCK – Drag Reduction by Passive Shock Control, Notes on Numerical Fluid Mechanics, Volume 56, Friedr. Vieweg & Sohn Verlagsgesellschaft mbH, Braunschweig/ Wiesbaden, 1997, pp. 355 -378

[7] Fulker, J. L. and Simmons, M. J., "An experimental investigation of passive shock/boundary-layer control on an aerofoil", in EUROSHOCK – Drag Reduction by Passive Shock Control, Notes on Numerical Fluid Mechanics, Volume 56, Friedr. Vieweg & Sohn Verlagsgesellschaft mbH, Braunschweig/Wiesbaden, 1997, pp. 379 - 400

[8] Rosemann, H., "The Cryogenic Ludwieg-tube Tunnel at Göttingen", in AGARD-R-812, Special Course on Advances in Cryogenic Wind Tunnel Technology, 1997

[9] Binder, B., Riethmüller, L., Tusche, S. and Wulf, R., "Upgrading the Transonic Wind Tunnel Göttingen", in Yearbook of the DGLR, Volume 1, 1992

[10] Knauer, A., "Performance Improvement of Transonic Airfoils Through Contour Modifications in the Shock Region", DLR Research Report 98-03, January 1998, and Ph.D. Thesis University Hannover

[11] Drela, M., "A User's Guide to MSES 2.95", MIT Computational Aerospace Sciences Laboratory, 1996

[12] Birkemeyer, J., "Drag Minimization on a Transonic Wing by Ventilation and Adaptive Contour Bump", DLR Research Report 1999-28, July 1999, and Ph.D. Thesis University Hannover

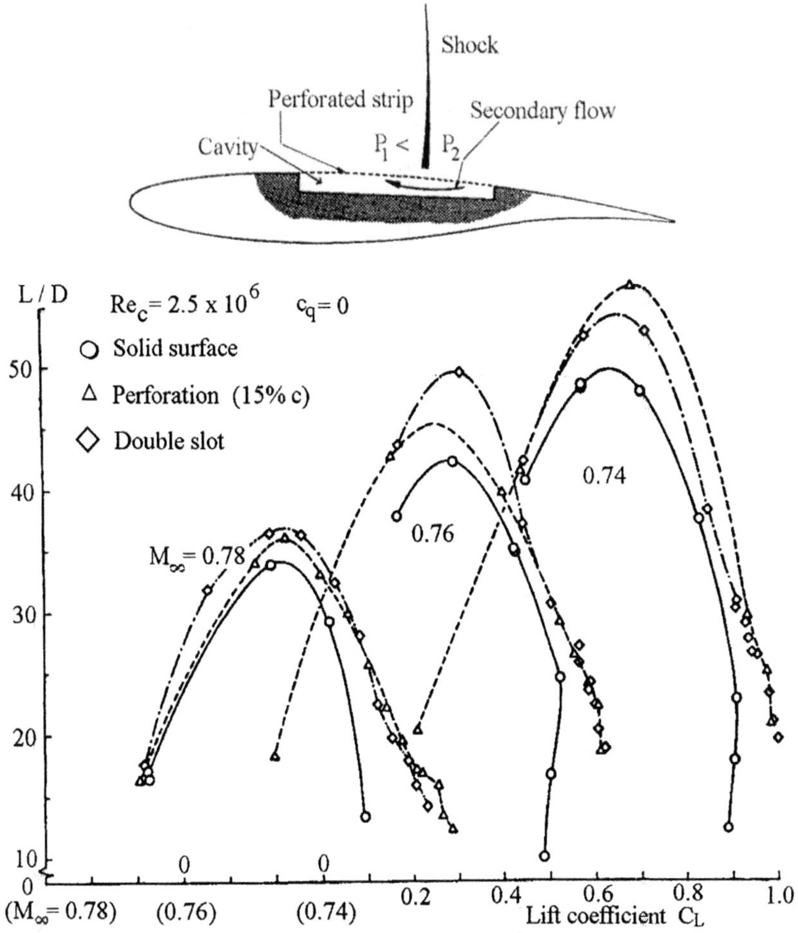

Figure 1 Effect of passive shock control on the range parameter L/D
Airfoil VA-2 [2]

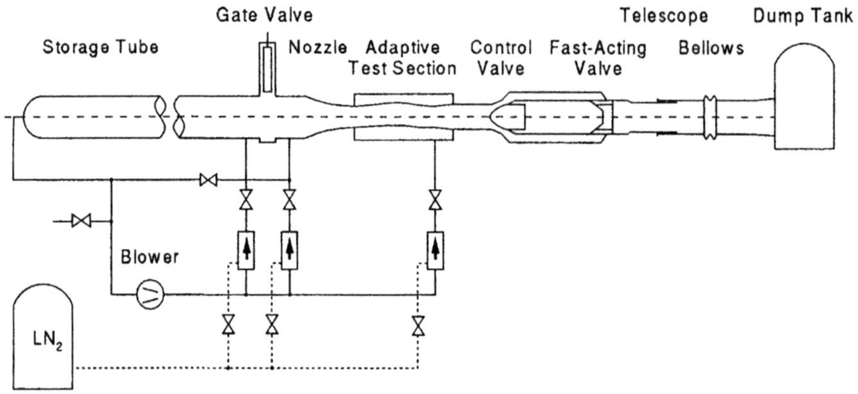

Tube	Diameter	0.8 m	Max. total pressure	10 bar
	Length	130 m	Temp. range	100 - 300 K
	Charge pressure	12.5 bar	Mach number	0.25 - 0.95
Test section	Cross section	0.40x0.35 m^2	Max. Reynolds No.	70x10^6
	Length	2.0 m	Run time	0.6 to 1.0 sec
	Model chord length	0.18 m		

Figure 2 Schematic and characteristics of the Cryogenic Ludwieg-tube [8]

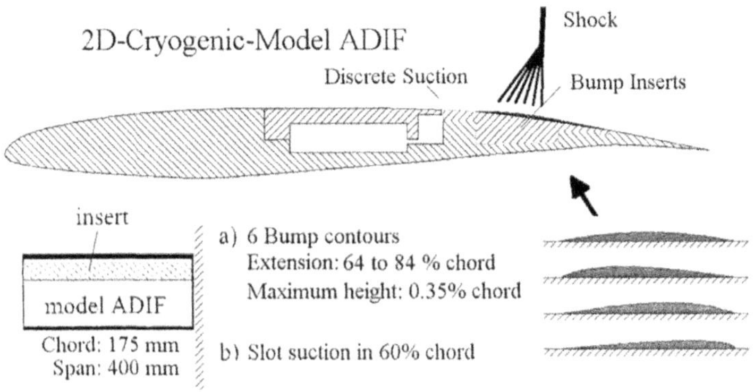

Figure 3 Wind tunnel model of the ADIF airfoil with bump inserts and suction slot [10]

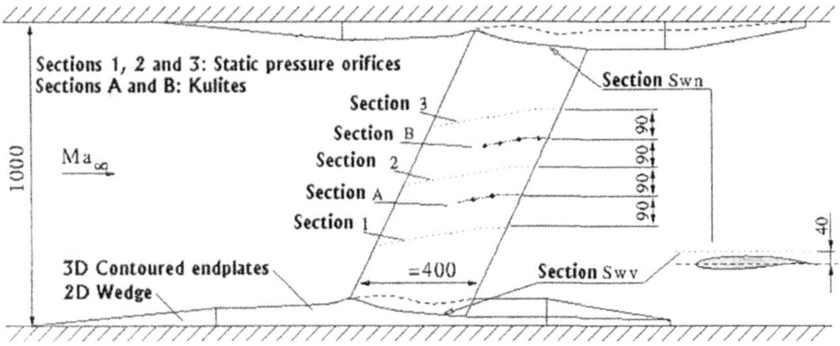

Figure 4 Test setup and instrumentation of the ADIF swept-wing configuration [12]

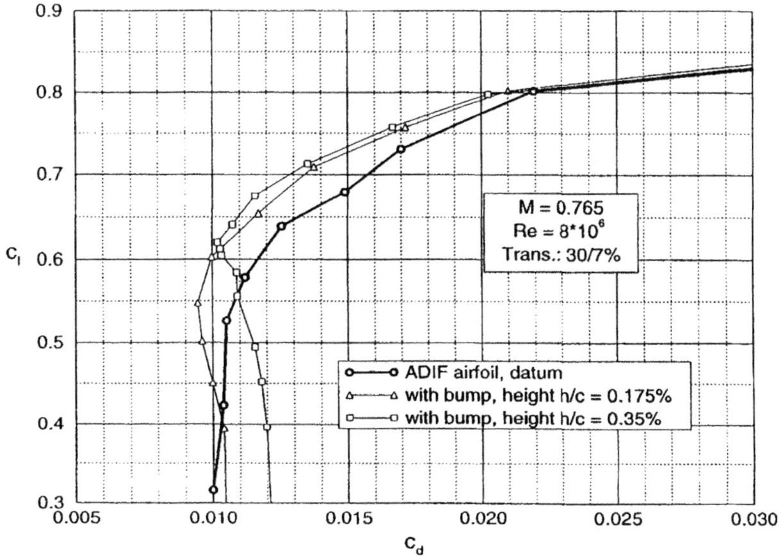

Figure 5 Drag polars for the ADIF airfoil without and with bump control
$M\infty = 0.765$, $Re_c = 8x10^6$ [10]

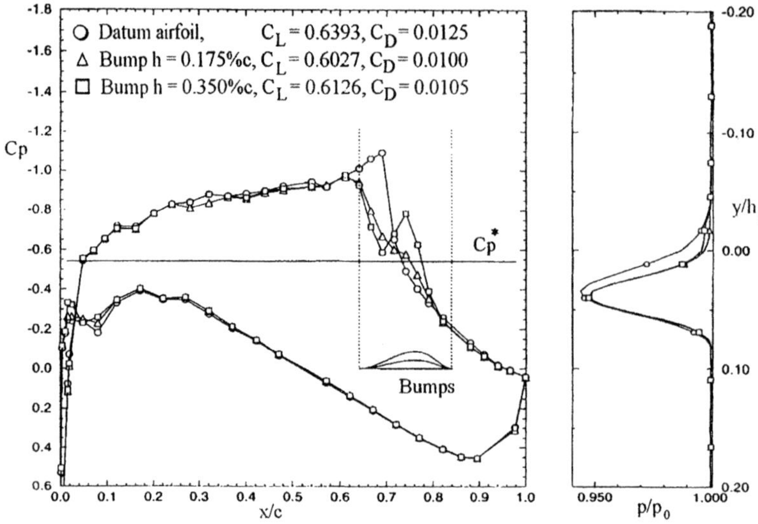

Figure 6 Pressure distributions and wake profiles for the ADIF airfoil without and with bump control, $M\infty = 0.765$, $Re_c = 8x10^6$, $x_{tr}/c = 30/7\%$, $C_L \approx 0.610$ [10]

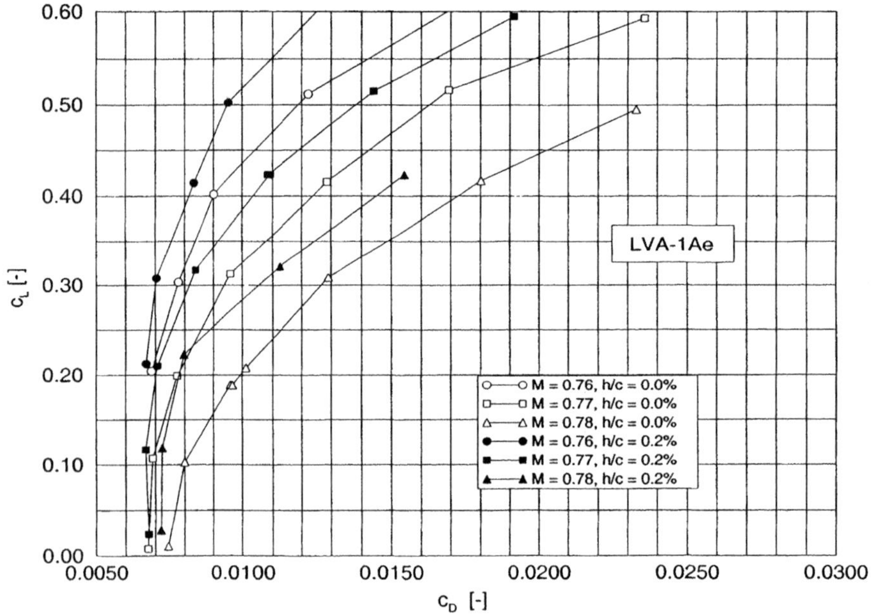

Figure 7 Drag polars for the LVA-1Ae airfoil with and without shock control $M_\infty = 0.76$, 0.77 and 0.78, $Re_c = 6 \times 10^6$, $x_{tr}/c = 48/48\%$

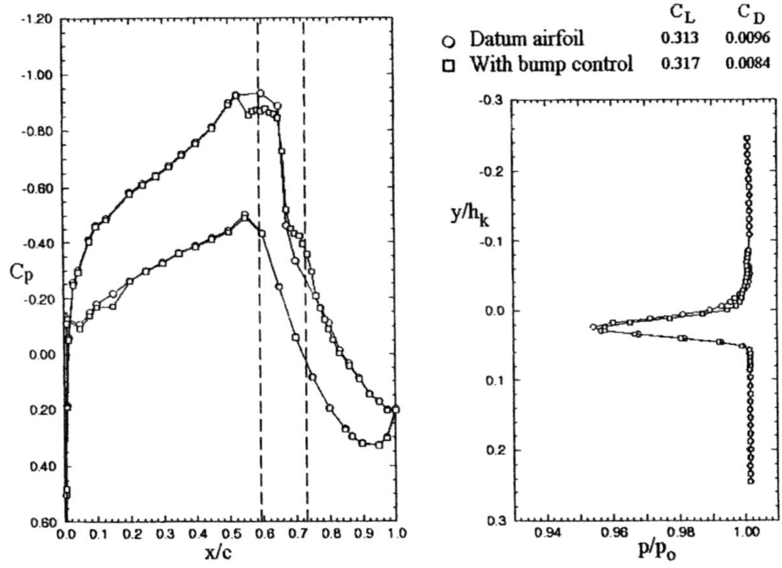

Figure 8 Surface and wake pressure distributions for the DA LVA-1A airfoil with and without bump control, $\alpha = 0.5°$, $M_\infty = 0.77$, $Re_c = 6 \times 10^6$, $x_{tr}/c = 48/48\%$

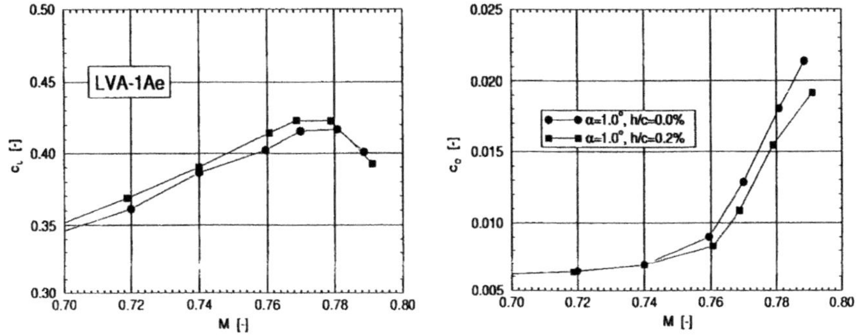

a. Lift coefficient

b. Drag coefficient

Figure 9 Variation of drag and lift coefficient with Mach number for the LVA-1A airfoil with and without shock control , $\alpha = 1.0°$, $Re_c = 6 \times 10^6$, $x_{tr}/c = 48/48\%$

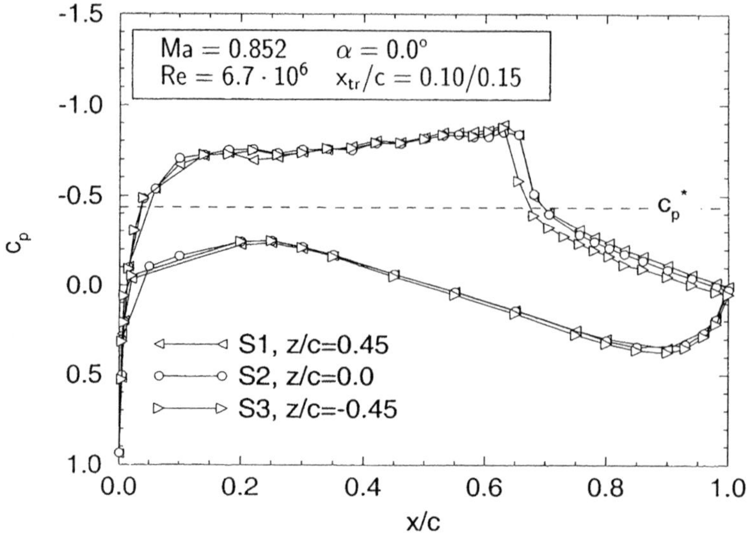

Figure 10 Effectiveness of contoured side walls at the design conditions, ADIF sheared wing, $\alpha = 0°$, $M_\infty = 0.852$, $Re_c = 6.7 \times 10^6$ [12]

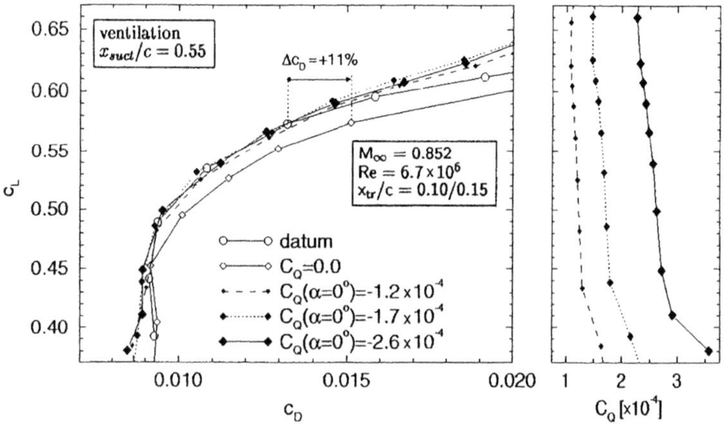

Figure 11 Control effects on the drag polars of the ADIF sheared-wing configuration with passive ventilation with and without upstream slot suction, $M_\infty = 0.852$, $Re_c = 6.7 \times 10^6$ [12]

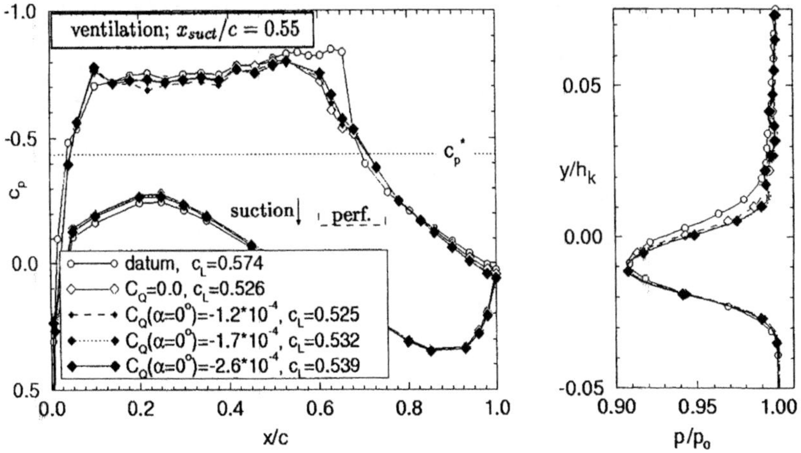

Figure 12 Surface and wake pressure distributions for the sheared wing with passive control with and without discrete upstream suction, $M_\infty = 0.852$, $\alpha = 0.0°$, $Re_c = 6.7 \times 10^6$, $x_{tr}/c = 10/15\%$ [12]

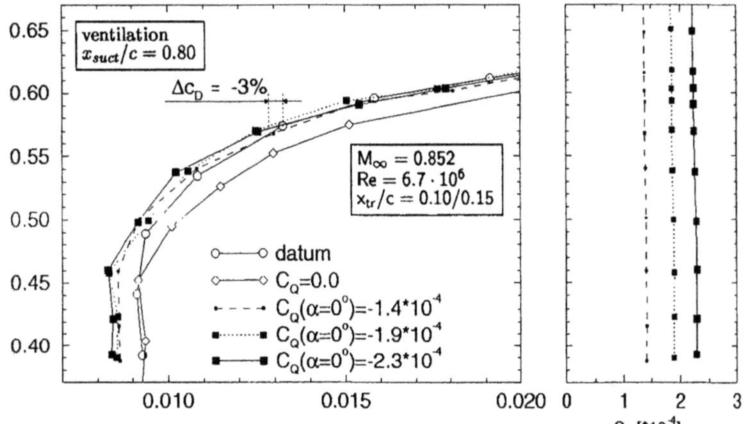

Figure 13 Control effects on the drag polars of the ADIF sheared-wing configuration with passive ventilation with and without downstream slot suction, $M_\infty = 0.852$, $Re_c = 6.7 \times 10^6$ [12]

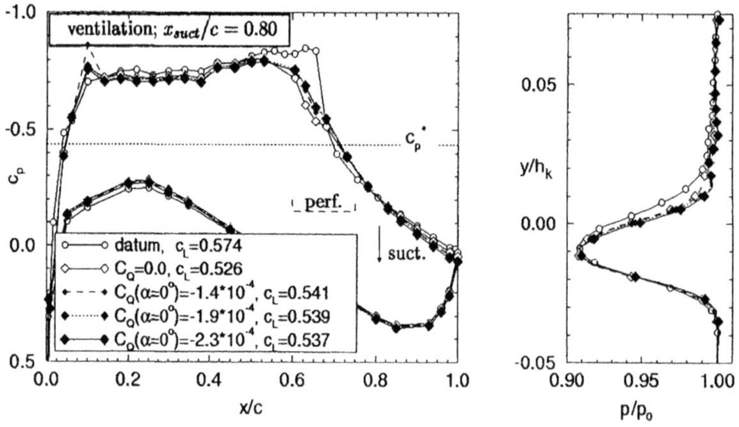

Figure 14 Surface and wake pressure distributions for the sheared wing with passive control with and without discrete downstream suction, $M_\infty = 0.852$, $\alpha = 0.0°$ $Re_c = 6.7 \times 10^6$, $x_{tr}/c = 10/15\%$ [12]

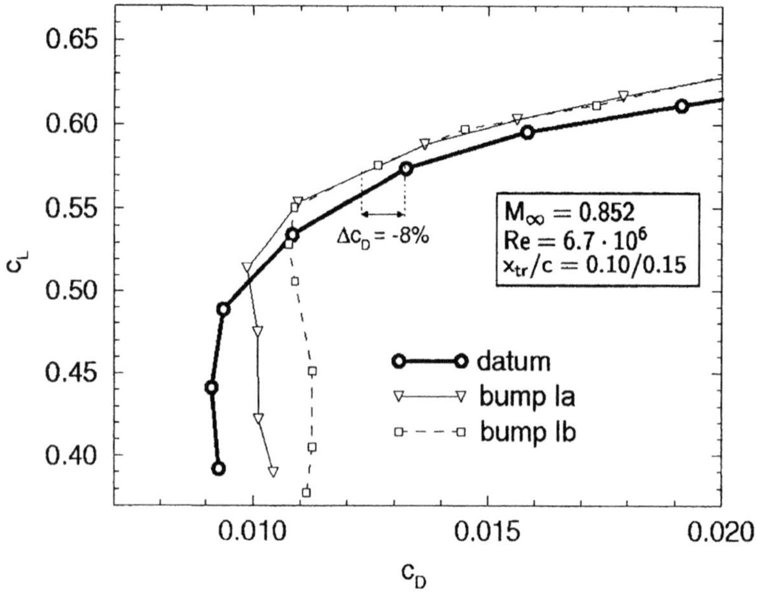

Figure 15 Drag polars for the ADIF swept wing without and with control by bumps, $M_\infty = 0.852$, $Re_c = 6.7 \times 10^6$, Ia: $h_{bump}/c = 0.1573\%$, Ib: $h_{bump}/c = 0.3146\%$ [12]

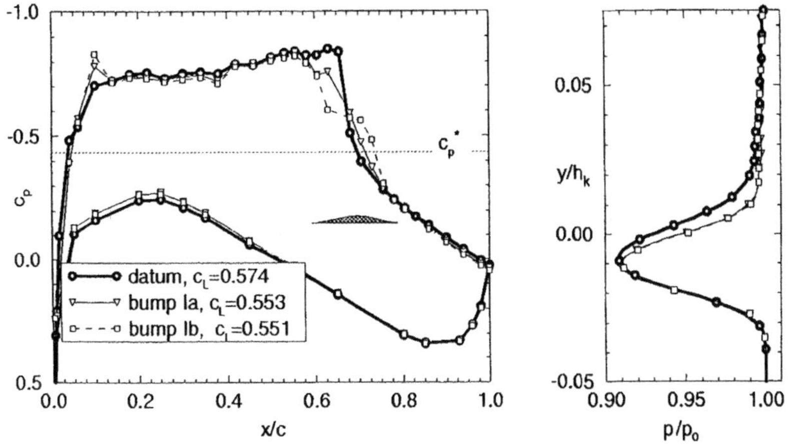

Figure 16 Surface and wake pressure distributions for the sheared wing with and without bumps, $M_\infty = 0.852$, $\alpha = 0.0°$, $Re_c = 6.7 \times 10^6$, $x_{tr}/c = 10/15\%$, [12] (for bump heights see Figure 15)

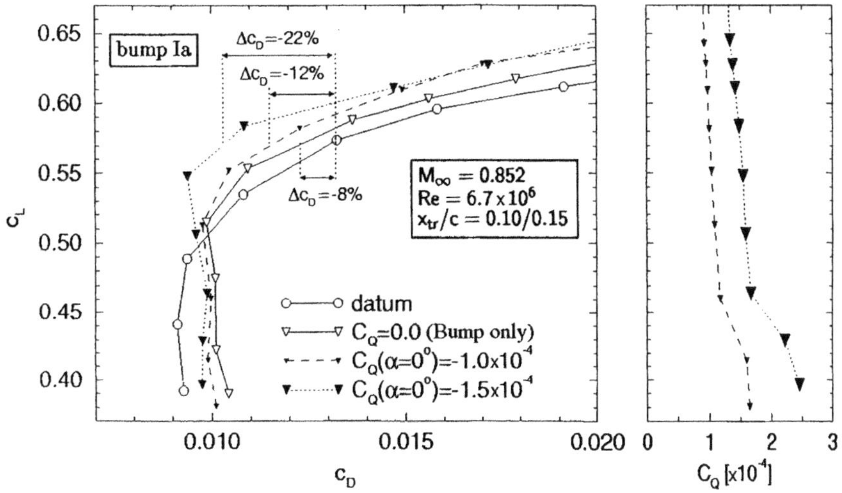

Figure 17 Drag polars for the ADIF swept wing with control by bump and bump plus upstream suction, $M_\infty = 0.852$, $Re_c = 6.7 \times 10^6$, $h_{bump} = 0.1573\%c$ [12]

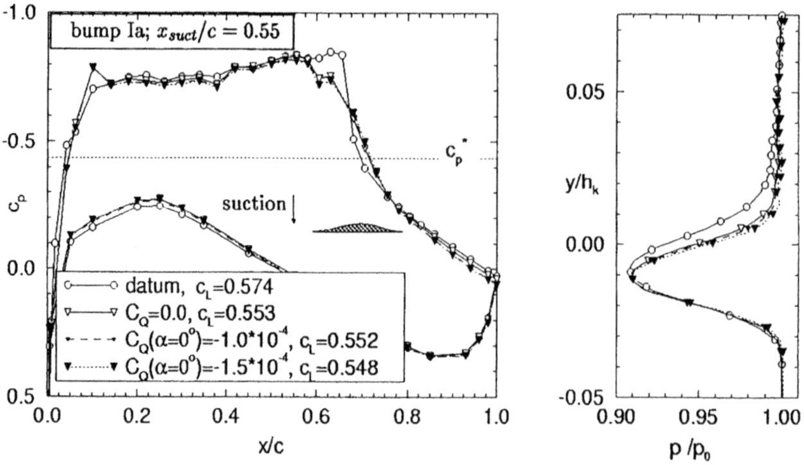

Figure 18 Surface and wake pressure distributions for the ADIF swept wing with control by bump and bump plus upstream suction, M_∞=0.852, $C_L \approx 0.55$, Re_c=6.7 x 10^6, x_{tr}/c = 10/15% [12]

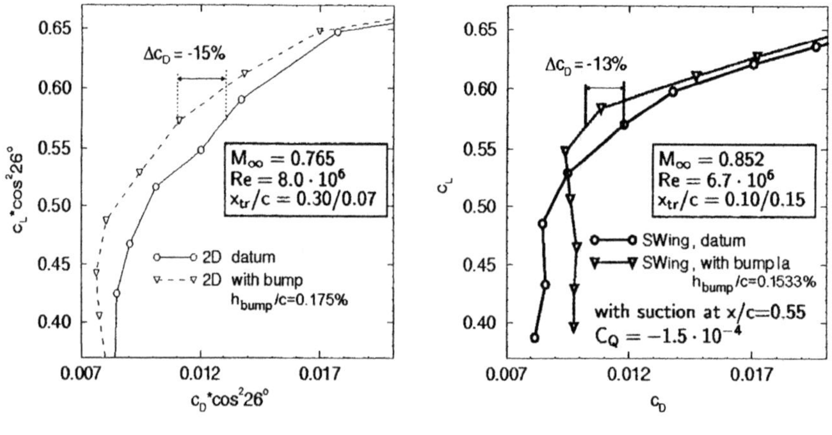

Figure 19 Drag polars for the ADIF sheared wing and the ADIF airfoil with and without bump at equivalent freestream and bump conditions [12]

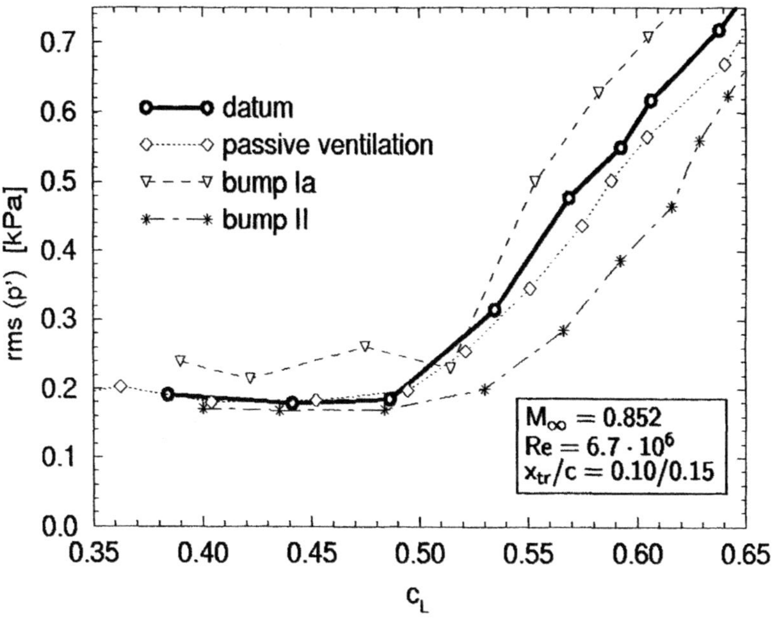

Figure 20 RMS-values of the upper-surface pressures at 90%-chord, ADIF swept wing without and with control by cavity ventilation and bumps, $M_\infty = 0.852$, $Re_c = 6.7 \times 10^6$ [12] (bump Ia: $x_{crest}/c = 0.715$, bump II: $x_{crest}/c = 0.76$, both bumps: $h_{bump}/c = 0.1573\%$)

21 Assessment of Shock and Boundary Layer Control Concepts for Hybrid Laminar Flow Wing Design

G. Dargel and P. Thiede
EADS Airbus GmbH
Hünefeldstr 1–5, 28183 Bremen,Germany

Summary

The contribution of Dasa Airbus to the EUROSHOCK II project in Task 2 and 4 is focused on the assessment of different control devices — active control by cavity/perforated plate arrangement, discrete suction, and bump— for the application to a hybrid laminar flow wing of a long–range transport aircraft.

The DA viscous inviscid interaction airfoil/swept wing code was improved and extented to treat various shock and boundary layer control devices within Task 2. Extensive computations were carried out for the laminar–type airfoil DRA 2303, the turbulent–type airfoils RAE 5225, MBB Va2 and ADIF and the sheared wing ADIF. The effect of flow control on the pressure distribution could be predicted with sufficient accuracy rendering the code suitable for the application in Task 4 for the control device assessment.

Shock control by using an adaptive bump has shown the greatest potential for drag reduction in the computations as well as in the experiments, while drag reduction by discrete slot suction although successful, is considerably reduced when pump drag is taken into account.

Based on the characteristic features of a laminar wing, the adaptive variable–height bump device seems to be the best means for drag reduction when shock waves are present. A bump parametric study was carried out leading to the establishment of limiting bump parameters and their sensitivity.

For a representative A340–type hybrid laminar flow wing section an adaptive bump design was performed. Its integration into the wing spoiler requires modifications of the A340 wing structure: modification of the wing box, enlargement of the spoiler and introduction of a variable camber flap system for roll control. Two structure and system concepts were suggested for a preliminary realization of the bump integration into the spoiler.

Using the wing section polars with the adaptive bump drag characteristics, the estimated drag balance indicates a drag reduction of up to 4% for the complete aircraft and fuel savings of up to 2.1% for a typical transatlantic mission.

21.1 Introduction

In EUROSHOCK I [1] it was found that the passive shock and boundary layer control by cavity ventilation, at least for laminar–type airfoils, always resulted in an increase in drag due to the wave drag reduction being offset by a viscous drag increase.

In order to overcome the penalties encountered with passive shock control, the following concepts of shock and boundary layer control (SC) were investigated in EUROSHOCK II [2], Figure 1:

- Active SC via a porous surface with a cavity underneath where partial suction was applied.
- Discrete suction by a single slot.
- SC by locally modifying the airfoil shape (contour bump).
- Hybrid SC by specific combinations of these concepts.

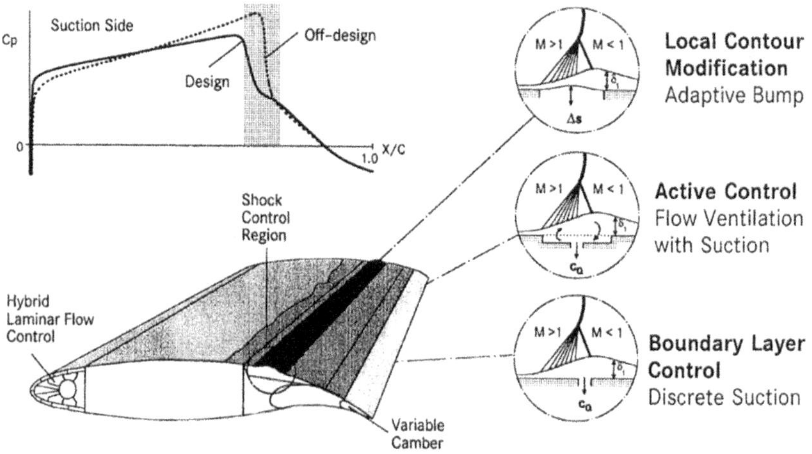

Figure 1 Shock and boundary layer control concepts investigated for an HLF wing

The passive control concept has the advantage of needing no additional power system, while the concepts investigated here not only require power input but also a higher structural and installation effort. Therefore, besides the aerodynamic improvement, the integration of the control device into a wing must be assessed under the aspects of weight, structures, systems, maintenance and associated costs.

The contribution of Dasa Airbus (DA) to the EUROSHOCK II project in Tasks 2 and 4 is focused on the application of various control concepts to a hybrid laminar flow (HLF) wing of a long–range transport aircraft [3,4]. After a short description of the basic computational method and its extension to shock and boundary layer control, the validation of the code using airfoil and sheared wing test cases measured in Task 3, is discussed with emphasis on laminar–type airfoils. After an aerodynamic assessment of control concepts, the most powerful control device, the bump, is further investigated by analyzing the influence of the bump parameters on drag reduction. Using an A340–type HLF wing section, the aerodynamic improvements are estimated and the consequences of the adaptive bump integration into a wing spoiler are outlined.

21.2 Extension of Numerical Code to Shock and Boundary Layer Control

21.2.1 Basic method

The basic prediction method has previously been developed for the computation of transonic airfoil and infinite swept wing flows by Dargel [5]. It is based on an inviscid–viscous coupling of a 2D potential flow solver with a 3D boundary layer finite difference method. Following the defect formulation concept of Le Balleur [6], the real viscous flow is split into a viscous defect flow and an equivalent inviscid flow which is extended to the wing surface and the wake streamline as the location of the viscous boundary conditions for the inviscid flow.

The equivalent inviscid flow with prescribed viscous boundary conditions is computed using the transonic 2D full potential method with an entropy correction in the shock region. The infinite swept wing (ISW) flow is accounted for by applying the inviscid sweep theory to the wing section shape and the freestream condition. After solving the inviscid flow in the plane normal to the sweep angle the solution is again transformed into the streamwise plane by sweep theory.

Following the defect formulation the displacement effect of the viscous layer is achieved by normal velocities on the surfaces of the transformed wing section and normal velocity jumps across the wake. In addition, 2nd order corrections for the curvature effects are introduced by a jump of the tangential velocities along the wake and a correction between the inviscid and viscous wall pressure.

Restricted to 1st order boundary layer theory, the viscous flow solution is obtained from the compressible boundary layer equation with an algebraic eddy viscosity formulation. Instead of solving the energy equation, Crocco's relationship is used for the density profile at adiabatic wall conditions.

In the case of ISW flow the 3D boundary layer equations are solved in a non–orthogonal curvilinear coordinate system with one coordinate in chordwise and one in spanwise direction under the assumption of infinite swept wing flow ,i.e., all spanwise flow derivatives are zero.

To perform calculations for flows with separation, the boundary layer equations are solved in an inverse mode by prescribing the displacement thickness for the chordwise direction and in a direct mode for the spanwise direction by prescribing the spanwise velocity component; the external velocity is computed as part of the solution by using Veldman's quasi–simultaneous coupling technique [7] and the extension for ISW flow by Cebeci [8]. To speed up the viscous–inviscid iteration, Carter's relaxation formula [9] for the displacement thickness is applied before the viscous boundary conditions are computed.

The turbulence model used is expressed in terms of the Cebeci and Smith (CS) eddy viscosity formulation [10]. Boundary layer transition is assumed to occur discontinuously.

Lift and pitching moment coefficients are obtained by appropriate integration around the wing section contour. The total drag is determined from its components: the wave drag from the entropy jump along the shock wave and the viscous drag from the momentum thickness in the far wake. This method for determining drag is found to be less sensitive to numerical errors than predicting drag from force integration.

21.2.2 Extension to Shock and Boundary Layer Control

The basic airfoil/swept wing flow prediction method was extended to account for different shock and boundary layer control schemes such as

- passive shock control by cavity ventilation
- active shock control by cavity ventilation with part suction, and
- single slot suction.

The shock control by contour bumps can be calculated without modifications of the basic method.

21.2.2.1 Control law procedure

In order to compute shock and boundary layer control by wall mass flow transfer a control law procedure is needed. Due to the pressure difference between the flow side and the cavity, mass flow transfer occurs through the holes of the perforated sheet or a slot and changes the boundary condition in the region of the interaction between the shock and the boundary layer.

A typical porous surface consists of n_h holes with a specified diameter d_h and pitch a_h, giving a porosity n_p of

$$n_p = \frac{n_h A_h}{A_p} \,. \qquad (1)$$

The characteristic of the porous surface, i.e., the relationship between the mass flow rate through and the pressure drop across the surface affects the ventilation velocity. A total mass flow rate $\mathcal{M}_p$ can be defined by

$$\mathcal{M}_p = n_h \mathcal{M}_h = \int_{L_p} v_p \rho_w b_p \, dx \qquad (2)$$

with length L_p and width b_p of the porous area A_p or in non–dimensional form as total mass flow rate coefficient

$$C_Q = \int_{L_p} \left(\frac{v_p}{a_\infty} \frac{\rho_w}{\rho_\infty} \right) \frac{b_p}{b} \, d\left(\frac{x}{c}\right). \qquad (3)$$

with b as reference span and c as reference chord of the airfoil or wing. The ventilation velocity v_p referenced to the stagnation conditions of the free–stream flow is related to the velocity within a hole of the perforation (hole velocity) by:

$$\frac{v_p}{a_o} = \frac{1}{\varrho_w/\varrho_o} n_p \frac{\mathcal{M}_h}{a_o \rho_o A_h} \tag{4}$$

$$\text{with} \quad \frac{\rho_w}{\rho_o} = \frac{1}{\left(1 + r\frac{2}{\kappa-1}M_e^2\right)} \left(\frac{p_w}{p_o}\right)^{\frac{1}{\kappa}} .$$

For the determination of the hole velocity as function of the pressure difference between the wall pressure p_w and the cavity pressure p_c Poll's calibration law [11] is implemented in the code; this law gives results simular to the 'standard' Doerffler/Bohning law [12] investigated in Task 1, when the correction factor in Poll's formulation is adjusted by calibration tests of the porous sheet.

21.2.2.2 Poll's calibration law

Under the assumption of a laminar and incompressible pipe flow, Poll developed a flow model for an ideal hole resulting in a quadratic relationship between the pressure drop $\Delta p = (p_w - p_c)$ and the mass flow rate $\mathcal{M}_h$ through a hole:

$$\frac{\Delta p}{p_o}\frac{\rho_h}{\rho_o}\frac{1}{\kappa} = \left[\frac{32}{R}\left(\frac{\mathcal{M}_h}{a_o\rho_o A_h}\right) + 1.203\left(\frac{\mathcal{M}_h}{a_o\rho_o A_h}\right)^2\right]\frac{1}{K} \tag{5}$$

$$\text{with} \quad R = \frac{a_o \varrho_o c}{\mu_o} \frac{d_h/c}{s_h/d_h \; \mu_h/\mu_o}$$

$$\text{and} \quad K = \left[\frac{\text{effective } d_h}{\text{measured } d_h}\right]^4 .$$

The quadratic equation is solved for the mass flow rate $\mathcal{M}_h$ by imposing $|\Delta p|$ for suction and blowing and taking the positive value of the root. Average values are taken for ϱ_h and μ_h using properties from the flow and cavity side of the perforation. The correction factor K is used to account for imperfections of the porous sheet and obtained from calibration tests.

21.2.2.3 Modifications of the solution procedure

The introduction of the wall mass flow transfer has necessitated a slight modification of the wall boundary condition in the inviscid and viscous flow solvers.

In addition, the wall mass flow yields a disturbance of the inner region of the turbulent boundary layer which is taken into account by a modification of the van Driest wall damping function [10].

The mesh size of the H–type grid of the inviscid flow solution had to be adapted in the control region as a compromise between accuracy and stability.

The grid used for the viscous flow solution contains 81 points in normal direction and follows in streamwise direction the grid point distribution of the inviscid grid but with an additional subgrid technique in the control region.

21.2.2.4 Prediction of pump drag

In case of suction the pump drag has to be included in the drag balance; pump drag, as used here, is given by the formula of Pankhurst and Gregory[13]:

$$c_{dQ} = \frac{1}{\eta_p}\left(\frac{p_o - p_\infty}{q_\infty} - c_{pcav} - \frac{\Delta p_{loss}}{q_\infty}\right) C_Q \tag{6}$$

with
C_Q = non–dimensional total mass flow coefficient
η_p = pump efficiency factor (=1)
p_o = total pressure
q_∞ = dynamic pressure
c_{pcav} = pressure coefficient in the cavity
Δp_{loss} = pressure losses in the ducts (= 0)

21.3 Code Validation for Shock and Boundary Layer Control

In Task 2 test cases with different control devices were established for code validation using experiments of Task 3.

Computations were carried out by DA for the following test cases

- airfoil DRA 2303 with SC bump
- airfoil RAE 5225 with SC bump
- airfoil DRA 2303 with suction slot
- airfoil DRA 2303 with SC bump and suction slot
- airfoil DRA 2303 with passive and active SC by cavity ventilation
- airfoil MBB Va 2 with passive and active SC by cavity ventilation
- airfoil ADIF with two SC bumps of different heights
- sheared wing ADIF with two SC bumps of different heights

Only a selection of test cases will be presented —in particular under consideration of the Task 4 objective to assess control devices for hybrid laminar wing flow application— since the code was, for certain control applications, already validated as described in [15].

21.3.1 Airfoil DRA 2303 with shock control bump device

The experimental data of this test case were distributed by DERA at the beginning of the project when experimental data of Task 3 were not yet available. Measurements with the laminar–type airfoil DRA 2303 were carried out in the DERA 8ftx8ft Subsonic–Supersonic Wind Tunnel [14]. The bump with circular arc shape had a length of l_B/c=20% and a height of h_B/c=0.25%.

A test case has been chosen with a lift coefficient of c_l=0.74, the design Mach number of M_∞=0.68 and a flight Reynolds number of $Re_\infty=19\cdot10^6$. Boundary layer transition was fixed at 5% chord on both surfaces.

The computed results and the DERA experimental data agree quite well concerning the pressure distribution without and with SC bump, Figure 2. The bump clearly reduces shock strength in the experiment as well as in the computation. The boundary layer parameters are also slightly improved by the presence of the bump. A reduction in total drag of about 10cts (Δc_d=0.0010) is predicted being of the same amount as the measured value after a correction due to a Mach number offset.

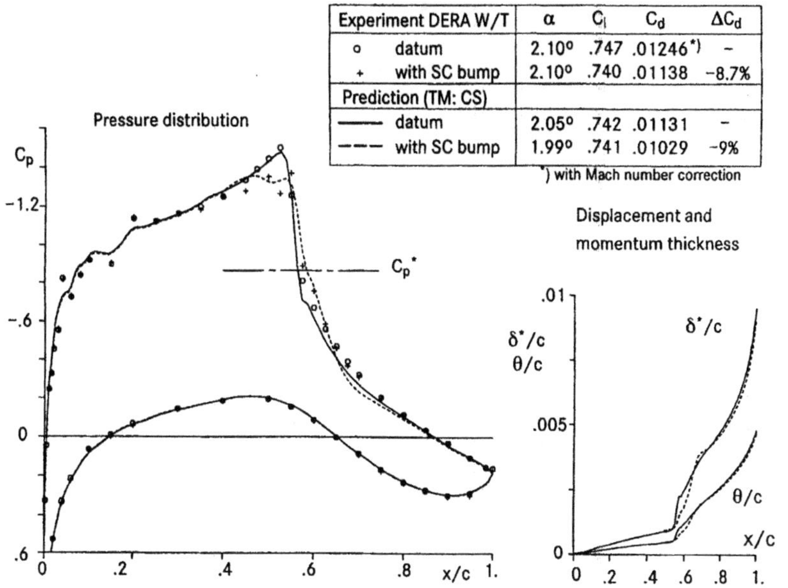

Figure 2 Effect of SC bump on flow parameters; airfoil DRA 2303 with bump (l_B/c=20%, h_B/c=0.25%), M_∞=0.68, $Re_\infty=19\cdot10^6$, transition 5%/5%, c_l=0.74

21.3.2 Airfoil DRA 2303 with discrete suction slot

This control case was experimentally investigated by Fulker et al. during the present project in Task 3. The airfoil DRA 2303 was equipped with a discrete suction slot in front of the shock location at the design Mach number of M_∞=0.68. The slot is located at x/c=0.45 and has a width of 1% chord. It is covered by a perforated sheet of 1mm thickness with a porosity of 4% formed by laser–drilled holes with a nominal diameter of 0.076mm. The prediction of the ventilation velocity is carried out by Poll's formulation using the nominal diameter and a correction factor of K=1 as no calibration data were available for the approximation of the characteristic of the porous sheet.

Due to the small size of the slot, the grid spacing was reduced up to Δx=0.14% giving 7 grid points over the slot. Figure 3 shows the inviscid grid used for the present computation.

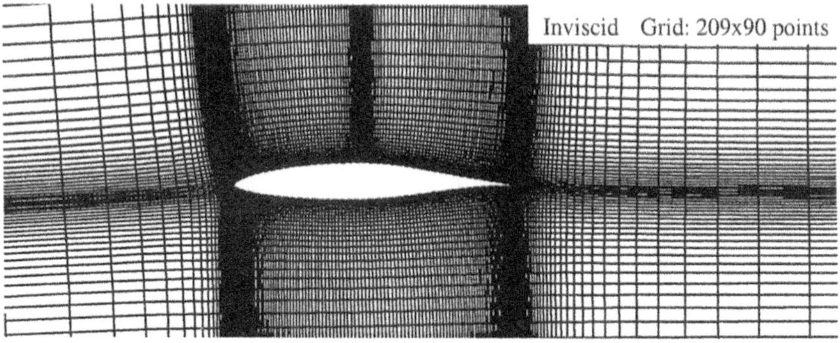

Figure 3 Grid of inviscid flow solution adapted to discrete suction slot; airfoil DERA 2303 with suction slot; grid size of $\Delta x/c$=0.14% (=7 grid points) within the slot between- x/c=0.45–.46

The predicted and measured pressure distributions are given in Figure 4 for the datum and the suction case at a lift coefficient of c_l=0.60. The total suction coefficient of c_Q=–0.00007 was prescribed in the calculations.

The drag reduction due to suction that is mainly caused by thinning of the boundary layer, as shown in Figure 4, is about 4.4% in the calculation compared to 7.5% in the experiment; the largest difference in drag occurs for the datum case. The pump drag is estimated to be about two drag counts for this test case.

21.3.3 Airfoil DRA 2303 with active and passive shock control

The experimental results of the active control concept have been obtained with the DRA 2303 airfoil by Fulker et al. in Task 3 (see Chapter 19 of this volume). The control device consists of a porous surface with a porosity of 4% and a length of 15% chord, and the same flow characteristic as in the previous suction case has been applied.

A grid spacing of $\Delta x/c$=0.75% has been chosen in the control region between x/c=47.5 – 62.5% since it was found to be favorable for dealing with the shock/ boundary layer interaction.

Experiment DERA W/T	α	C_l	C_d*)	ΔC_d
o datum	1.28°	.608	.00973	-
Δ with slot	1.28	.595	.00900	-7.5%
Prediction (TM: CS)				
—— datum	1.26°	.608	.00931	-
---- with slot	1.18°	.597	.00890	-4.4%

*) without pump drag

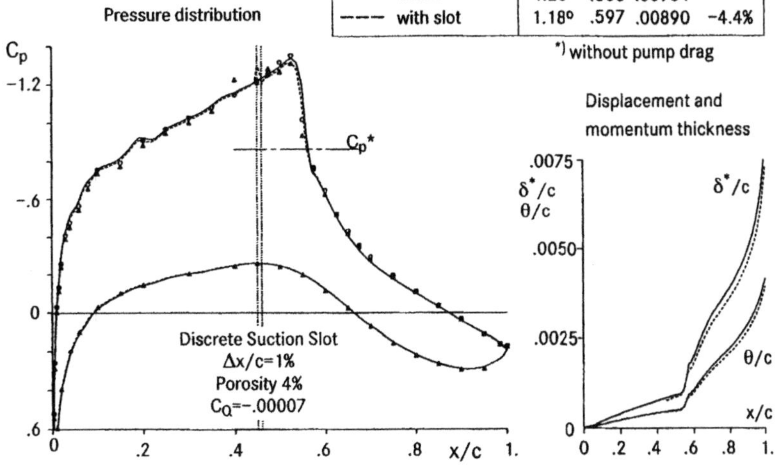

Figure 4 Effect of discrete suction on flow parameters; airfoil DRA 2303 with suction slot, $M_\infty=0.68$, $Re_\infty=19 \cdot 10^6$, transition 5%/5%, $c_l=0.60$

The passive control case $C_Q=0.0$ and the active one with a suction rate of $C_Q=-0.00009$ have been considered for code validation at a lift coefficient of $c_l=0.6$, Figure 5, but rather than prescribing the suction rate the measured cavity pressure is prescribed when using the Poll's formulation based on former experience [15].

The predicted pressure distributions for both control cases agree well with the measured ones, except for a somewhat stronger upstream influence of the control device in the experiment. The increase in drag due to passive control is caused by blowing in the front part of the control region and the associated increase in the boundary layer thickness parameters. This behaviour can be reduced by suction in the active case as shown by the boundary layer parameters in Figure 5.

In spite of the suction out of the cavity, an increase in drag is found for the active control case. The predicted drag balance is nearly half of the measured one. The actual suction value is too small to overcome the drag increase caused by passive control, but a higher suction rate needs a larger duct system with greater installation effort.

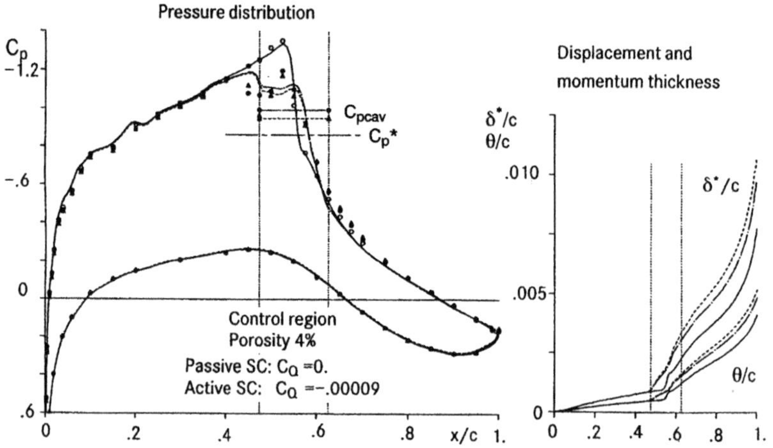

Experiment DERA W/T	α	C_l	C_d *)	ΔC_d
○ datum	1.28°	.608	.00973	-
△ with passive SC	1.28°	.586	.01196	22.9%
• with active SC	1.28°	.593	.01121	15.2%
Prediction (TM: CS)**)				
——— datum	1.26°	.607	.00932	-
– – – with passive SC	1.34°	.586	.01042	11.8%
——— with active SC	1.34°	.595	.00985	5.7%

*) without pump drag

**) measured C_{pcav} prescribed

Figure 5 Effect of passive and active control on flow parameters; airfoil DRA 2303 with active control device (x/c=0.475 – 0.625), M_∞=0.68, Re_∞=19·10^6, transition 5%/5%, c_l=0.60

21.3.4 Conclusions Concerning Code Application

A comprehensive study has been performed by DA to validate the present airfoil/swept wing code for shock and boundary layer control.

The computed results showed that the code seems to be able to assess different control devices, in qualitative agreement with the experimental data of Task 3. The effect of the control devices on the pressure distribution and the aerodynamic coefficients could be predicted with sufficient accuracy so that the code may be applied with confidence in Task 4.

Based on the computational and experimental results, the control devices could be assessed concerning their potential in drag reduction:

- Active SC by cavity ventilation with part suction decreases wave drag but increases viscous drag with the result of an increase of total drag (Δc_{dexp}=15% and Δc_{dcomp}=6%) when a moderate (but realistic for aircraft application) suction rate is used.
- Discrete suction in front of the shock decreases viscous drag at nearly unchanged wave drag for a moderate suction rate. Total drag reduction

(Δc_{dexp}=–7% and Δc_{dcomp}=–4%) is found to exist over the whole lift range investigated but may be partly offset by pump drag.

- SC bump device decreases wave drag and does not effect viscous drag. A substantial drag reduction of about $\Delta c_{dexp,comp} > 10\%$ is achievable within a certain lift range when the bump height is fixed.

The numerical investigation for validation purposes of the sweep effect on the adaptive variable–height SC bump suffered under the limited time of the project, but later computations showed that the type of pressure distribution seems to be unchanged and the drag reduction is still evident under sheared wing conditions as long as the flow is attached.

21.4 Assessment of Various Shock and Boundary Layer Control Concepts

The main assessment criterion of the control concepts for a hybrid laminar flow wing is linked to drag reduction at off–design conditions. Laminar wing design requires a favourable pressure gradient on the suction side. To maximize viscous drag reduction by laminarization, the shock should be located as far downstream as possible. In particular under off–design conditions, the rear shock position is associated with a rapid increase in shock strength and hence in wave drag thus possibly offsetting the benefit of laminarization.

Three control concepts — active control by cavity ventilation with part suction, discrete suction and bump — have been investigated, Figure 1, using the laminar–type airfoil DRA 2303. Combinations of these devices are also possible, but are not addressed within this study. The calculations are performed with the viscous inviscid interaction code described above and validated in Task 2.

21.4.1 Active shock control

The laminar–type airfoil DRA 2303, which was equipped with a passive SC device between x/c=0.5 to 0.6 in EUROSHOCK I [1], is used for the investigation of the active SC concept by applying partial suction in the cavity, Figure 6. The evaluation of the drag characteristics was studied for two porosities of the porous plate, 2% and 4%, by varying the total suction coefficient from C_Q=0 to 0.0006 at a constant lift coefficient of c_l=0.74. Pressure distributions are shown for the datum, the passive control and the active control case with a suction rate of C_Q =0.0003, respectively.

By reducing the viscous drag, the suction overcomes the drag increase found with passive shock control, i.e. C_Q =0. however, the wave drag reduction becomes smaller for increasing suction rates. For a porosity of 4% a maximum drag reduction of about Δc_d = 10cts is found, when a large suction rate of C_Q = 0.0005 is applied. Taking the pump drag (Eq.6) into account no substantial net drag reduction is predicted.

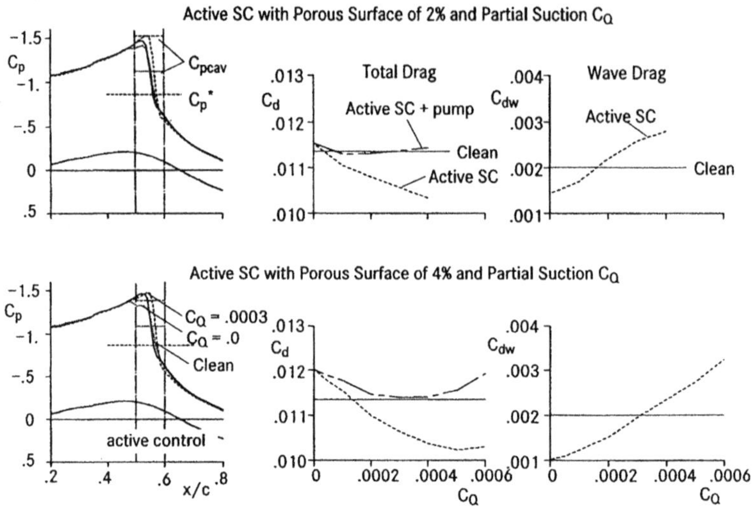

Figure 6 Drag balance for active SC; airfoil DRA 2303 with actice control device (x/c=0.5 – 0.6), M_∞=0.68, Re_∞=$19\cdot10^6$, transition 5%/5%, c_l=0.74

21.4.2 Discrete suction

For the same freestream conditions, different positions of a suction slot with a width of 2% chord were investigated by varying the suction coefficient, Figure 7. The pressure distributions in the shock region are shown for the datum and suction case with a suction rate of C_Q = 0.0003.

The main effects of applying suction in the shock region are a decrease of the viscous drag and the attendant increase in wave drag due to the thinner boundary layer. The highest drag reduction is about Δc_d = 10cts with the slot positioned in the subsonic part of the shock region and a large suction coefficient of C_Q = 0.0006.

When pump drag is added to the airfoil drag, the net drag reduction by a suction slot is less than 3cts. A slot location downstream of the shock needs less power due to a higher surface pressure, giving lower pump drag. The slot location in front of the shock is most suitable, because it is independent to shock movement. In this case the required suction coefficient to reduce drag has to be small (C_Q<0.0001), which is in agreement with the results of Task 2 and 3, see Chapter 21.3.2, althouigh the slot width is different.

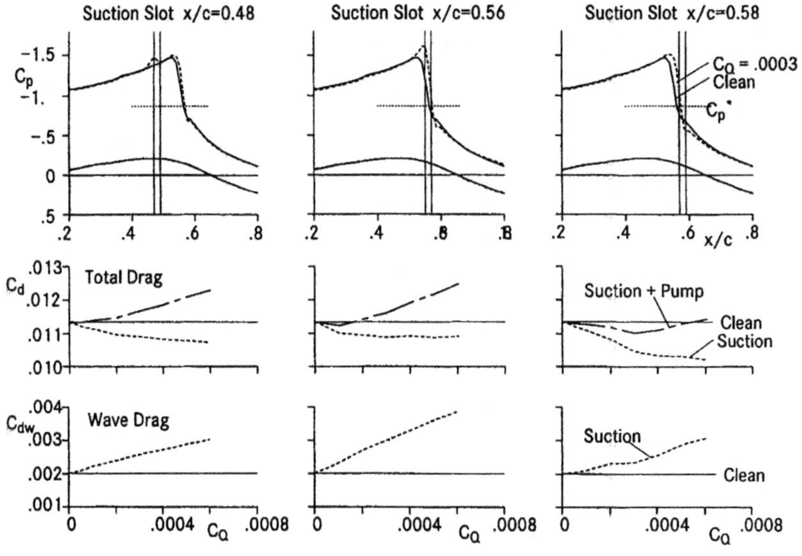

Figure 7 Drag balance for discrete suction; airfoil DRA 2303 with suction slot of width $\Delta x/c=2\%$, $M_\infty=0.68$, $Re_\infty=19\cdot10^6$, transition 5%/5%, $c_l=0.74$

Below $c_l=0.55$ the bump causes a weak double shock system with an increase of viscous drag, while for higher lift coefficients the pre–shock Mach number is reduced by the presence of the bump yielding a dominating reduction of the wave drag.

21.4.3 Shock control bump

As known from DERA experiments [14], used for code validation in Task 2, the SC bump device produces a substantial reduction in airfoil drag at off–design conditions, as shown in the measured and predicted drag polars in Figure 8.

21.4.4 Selection of a favorite control concept

Based on the calculations no substantial net drag reduction is predicted for a laminar–type airfoil with control devices using flow ventilation or suction. The experimental results of Task 3 and the computations in Task 2 support the previous conclusions concerning the active control by ventilation and the discrete suction slot. The suction rate used in the experiments was about $C_Q<0.0001$. In case of suction a net drag reduction of about 5cts was found with an upstream slot location amounting to 4% of the the airfoil drag.

Due to a drag reduction potential of more than $\Delta c_d = 10$cts for airfoils with a bump device, such devices are regarded as the only feasible shock control candidate to be used for laminar wing application, in particular since this amount of drag reduction is confirmed by experiments and predictions.

The drag behaviour due to a bump device with changing lift and Mach number suggests that two possible arrangements may be considered for a laminar wing:

- A fixed bump with a fixed, moderate height which does not lead to a significant drag increase in the basic design range of the airfoil but improves the off–design behaviour at higher lift and Mach number. This amounts to a local reshaping of a wing with nearly no penalties in structure, weight and costs.
- A bump with fixed location but variable height in order to realize the large potential in drag reduction. This requires an adaptive wing with strong modifications in the wing design, structure and system and implies adverse effects on weight, maintenance and costs.

The second alternative has been further investigated for the use on laminar wings.

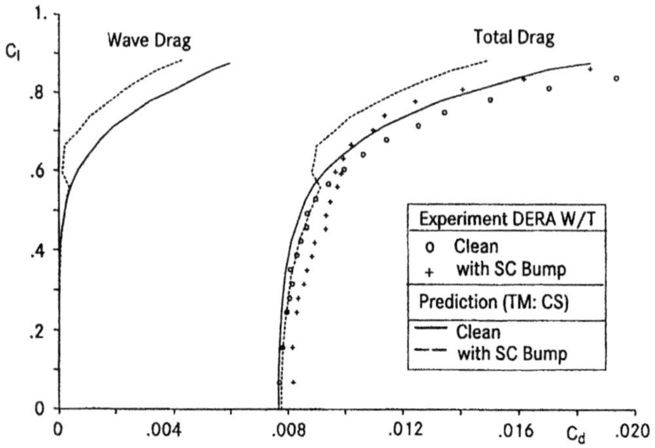

Figure 8 Drag polar with fixed bump device; airfoil DRA 2303, M_∞=0.68, Re_∞=19·10^6, fixed transition 5%/5%

21.5 Adaptive Shock Control Bump Parametric Study

Since the bump seems to be the most feasible control concept for laminar wing application, a parametric bump study has been carried out for the laminar–type airfoil DA LVA–1A [17].

The LVA–1A airfoil is a laminar–type airfoil designed by DA to achieve extended natural laminar flow up to 50% chord at M_∞=0.73 and Re_∞=20*10^6. The requirement for the design of a bump device is the presence of a certain amount of wave drag as is the case at the off–design condition with M_∞=0.76 and a lift coefficient of c_l =0.47. The Reynolds number of Re_∞=6*10^6 was chosen for the design in view of an experimental verification. For a laminar airfoil the chordwise shock movement relative to the bump design point is small with variations of lift and Mach number.

The calculations were performed with the MSES code [18], a viscous–inviscid interaction airfoil prediction method. The numerical formulation consists of a finite–volume discretization of the steady Euler equation on an intrinsic streamline grid coupled with integral boundary layer and wake flow formulations via the displacement thickness surface. This code has been chosen due to higher accuracy in wave drag prediction by the Euler formulation.

The SC bump is defined by its shape function $f_B(x)$, a length l_B/c, a height h_B/c and a crest c_B/l_B; it is added to airfoil surface relative to the shock position $(x_C - x_S)/c$, Figure 9.

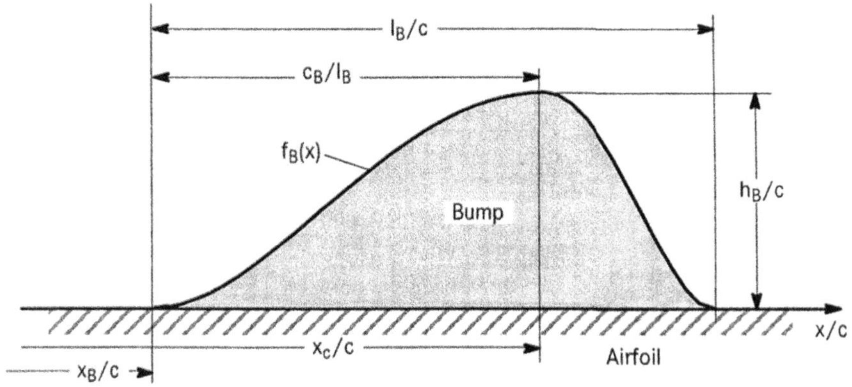

Figure 9 SC bump geometry parameters

21.5.1 Bump location

By adding a fixed bump shape normal to the airfoil surface, series of airfoils with the same bump geometry but different bump locations can be produced. In Figure 10 the pressure distributions and the drag–component coefficients are given dependent on the bump location, and compared with results for the datum airfoil.

If the crest is located in front of the shock, the bump shape forces first an isentropic compression followed by a strong expansion of the supersonic flow. The supersonic flow regime is terminated by a strong shock with increasing wave and viscous drag.

The bump is in an optimum position when the strength of the shock is weakened by an isentropic compression, giving a lower pre–shock Mach number. The resulting wave drag is reduced by about 50% compared to the wave drag coefficient of the datum airfoil and a total drag reduction of about $\Delta c_d = 16$ cts is produced. The location of the crest is a few percent chord downstream of the shock location of the clean airfoil flow.

If the bump is located in the subsonic flow region, i.e., in a rearward position, negligible effects of the bump on drag are observed.

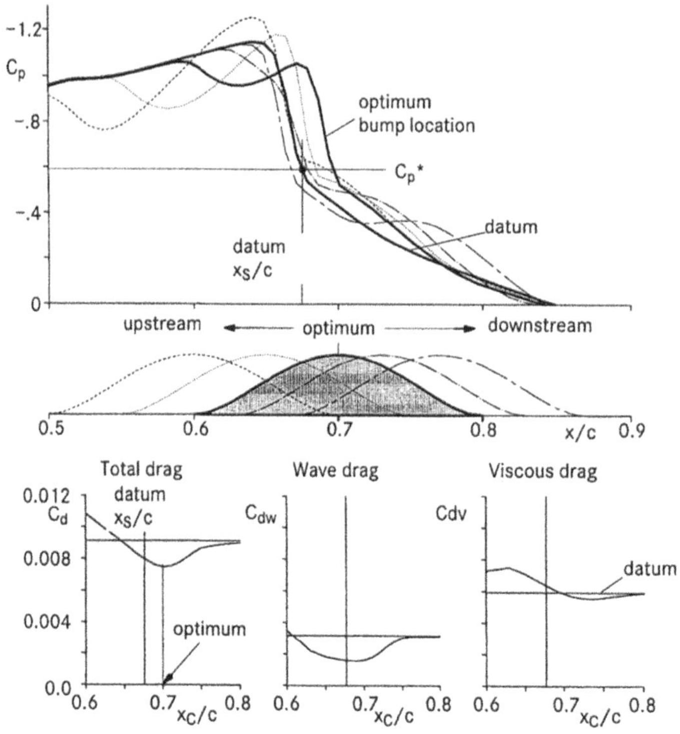

Figure 10 Effect of bump location on drag and pressure distribution; airfoil DA LVA–1A, M_∞=0.76, Re_∞=$6{\cdot}10^6$, fixed transition 50%/50%, c_l=0.47; bump geometry: $f_B(x)$=fixed beam, l_B/c=20%, h_B/c=0.3%, c_B/l_B=50%

21.5.2 Bump shape function

Four bump shapes — ramp, fixed beam, polynomial of 3rd order, sinus function — with a length of l_B/c=20% and a height of h_B/c=0.003 were investigated, Figure 11. The crest is set at c_B/l_B=50%, i.e., a symmetrical bump shape. The bumps are located to produce maximum drag reductions.

Although there are different shapes, the level of drag reduction is nevertheless quite similar. The difference in the drag reduction between the shapes is only about 3%. It has been found that the shape of the fixed beam gives the largest wave drag reduction amounting to about Δc_d=50%.

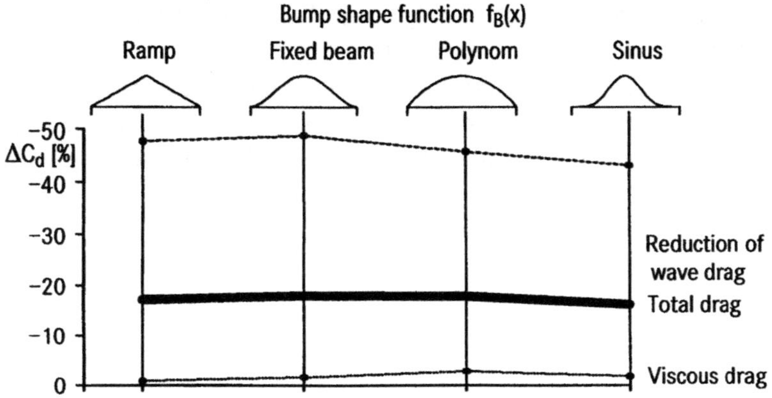

Figure 11 Effect of bump shape on drag reduction; airfoil DA LVA–1A, M_∞=0.76, Re_∞=6·10^6, fixed transition 50%/50%, c_l=0.47; bump geometry: l_B/c=20%, h_B/c=0.3%, c_B/l_B=50%, x_c/c=optimum

21.5.3 Bump length and crest location

To assess the effect of bump length and crest location on drag reduction, the bump shape 'fixed beam' is used with the bump height fixed at h_B/c=0.003. Three bump lengths were investigated:

- l_B/c = 12%, 15%, 20%

in combination with three to five crest locations, viz.,

- c_B/l_B= 30%, 50%, 70% and partly 20% and 80%.

Each bump configuration is moved along the airfoil surface giving a parameter array of predicted drag coefficients versus bump locations that is presented in Figure 12 referenced to the drag coefficient and shock location of the datum airfoil.

For a fixed crest an increase of the bump length produces an increase in drag reduction due to the widening of λ–shock system.

If the bump length is fixed, the maximum drag reduction is clearly a function of the crest location. Increasing the crest location up to c_B/l_B= 50% improves the maximum drag reduction, while a further increase beyond c_B/l_B= 50% slightly decreases it. The symmetrical bump produces the highest drag reduction while rearward asymmetrical shapes reduce drag in a wider chordwise range of the crest/shock positions.

The optimal bump location regarding maximum drag reduction clearly depends on the crest location and bump length. For a bump length of l_B/c = 20% the crest of the symmetrical bump should be located about 2.0% and of the rear asymmetrical bump about 5% downstream of the shock, respectively.

21.5.4 Bump height

For the investigation of the bump height, the bump location is now fixed at the corresponding optimal location. In Figure 13, the variation of the height has been performed for a fixed crest location of $c_B/l_B = 50\%$ but different bump lengths and for a fixed length of $l_B/c = 20\%$ but different crest locations.

Increasing the bump height up to a certain value improves the amount of drag reduction in all cases. Bumps of large extent ($l_B/c = 20\%$) and an asymmetrical shape ($c_B/l_B = 70\%$) give a drag reduction of up to $\Delta c_d = 25\%$ for the largest feasible height of $h_B/c = 0.005$.

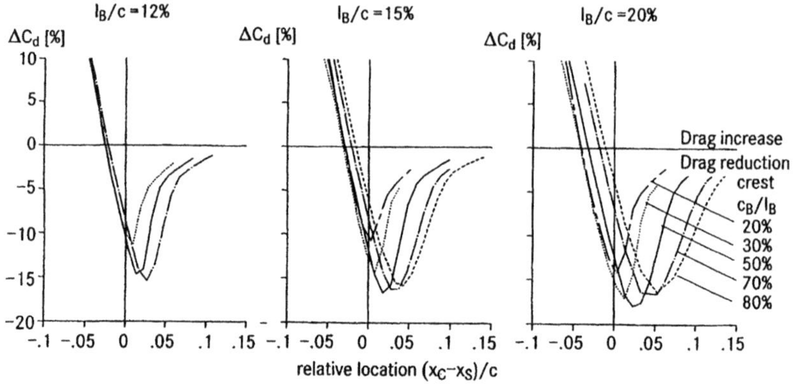

Figure 12 Effect of bump length and crest location on drag; airfoil DA LVA–1A, M_∞=0.76, Re_∞=6·10^6, fixed transition 50%/50%, c_l=0.47; bump geometry: $f_B(x)$=fixed beam, h_B/c=0.3%

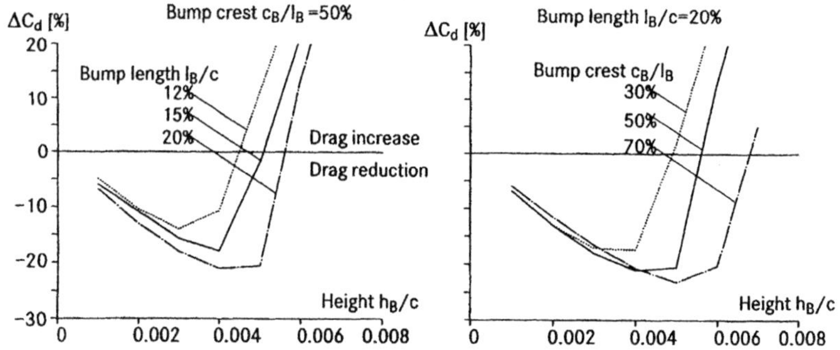

Figure 13 Effect of bump height on drag; airfoil DA LVA–1A, M_∞=0.76, Re_∞=6·10^6, fixed transition 50%/50%, c_l=0.47; bump geometry: $f_B(x)$=fixed beam, x_c/c=optimum

21.5.5 Off–design behaviour

While the previous results were achieved for a constant lift coefficient and Mach number, i.e., for the bump design point, the off–design behaviour has been investigated by predicting the drag polars for bumps with increasing bump height but fixed position at the design Mach number of $M_\infty = 0.76$.

Since the asymmetrical bump with a crest location of $c_B/l_B = 70\%$ and a length of $l_B/c = 20\%$ shows a large potential for drag reduction with varying of the bump height, it has been chosen for the prediction of drag polars with variation of the bump height h_B/c from 0.2% up to 0.5%, Figure 14.

Lower bumps reduce the airfoil drag at lower lift coefficients, while higher bumps produce larger drag reductions at higher lift coefficients. Therefore, the height can be used to adapt the drag reduction features of a bump with fixed location to the flow (or shock) characteristics of the clean airfoil.

In the meantime wind tunnel tests carried out in Task 3 by DLR and also tests by DA/ONERA[17] have confirmed the predicted drag behaviour of the adaptive bump. This feature can be particularly applied on laminar–type airfoils as the shock movement is small with variation of lift and Mach number.

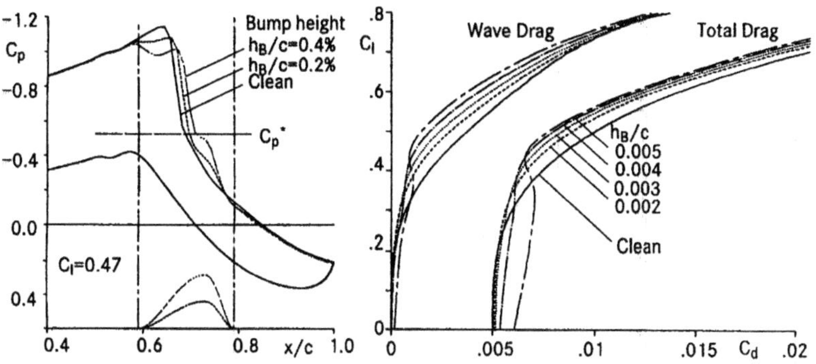

Figure 14 Effect of adaptive bump on drag under off–design conditions; airfoil DA LVA–1A, M_∞=0.76, $Re_\infty=6\cdot10^6$, fixed transition 50%/50%; bump geometry: $f_B(x)$=fixed beam, l_B/c=20%, c_B/l_B=70%, x_c/c=optimum

21.6 A340 HLF Shock Control Bump Device Integration

To achieve the largest benefit in drag reduction by the bump, the 'adaptive bump' approach is chosen for an integration into the A340 Hybird Laminar Flow (HLF) wing.

In order to proceed with the bump integration, a laminar wing based on the A340 planform should have been available, but could not be realized within the scope of the project. Therefore, all investigations were carried out on a wing section under sheared wing conditions and then scaled to the complete wing in 'handbook' manner. The sheared wing predictions are performed with the MSES code [18] by applying the inviscid sweep transformation. No 3D Navier Stokes computations were carried out to verify that certain assumptions are still fulfilled along the whole span.

21.6.1 Hybrid laminar wing section PHLF1

In a proposed successor project of the EU–project ELFIN [19] it was planned to design a HLF glove located between the engines of the A340 aircraft to demonstrate the drag reduction potential of laminar flow technology with a suction system in the wing nose under flight conditions.

For this purpose a laminar–type wing section PHLF1 [20] was designed to cover the original wing section at the spanwise location of η=0.50, Figure 15. During the design process it was established that at off–design conditions the PHLF1 airfoil led to an unacceptable drag rise due to strong shock waves so that the extent of the laminar flow was shortened by contour modifications.

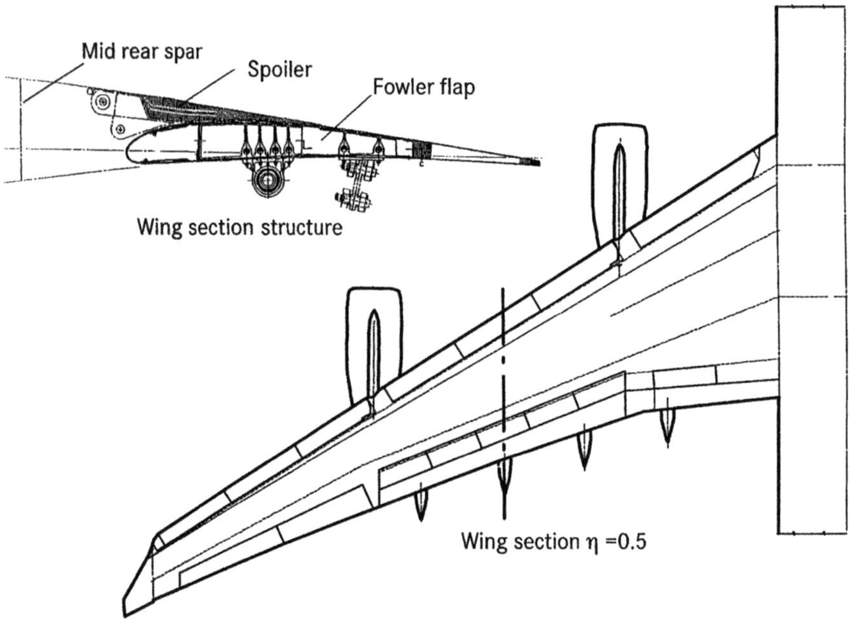

Figure 15 General arrangement of A340 wing

This representative A340 HLF wing section PHLF1 at 50% span is used for the application of the bump device at flight conditions assuming an extension of laminar flow up to 50% chord when suction is activated in the nose region.

The pressure distributions and the drag polars are shown for the design Mach number M_∞=0.82 of the laminar wing section and the off–design Mach numbers of M_∞=0.84 and 0.86 in Figure 16 and 17. A sweep angle of 26^0 is used for infinite swept wing calculations. The local Reynolds number of $Re_\infty=35\cdot10^6$ is chosen corresponding to an altitude of 37000ft. A strong increase of the wave drag and a little shock movement are clearly observed with increasing lift and Mach number.

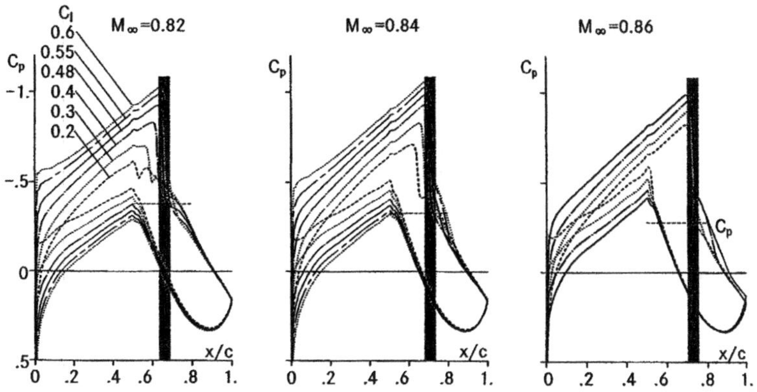

Figure 16 Pressure distribution of A340 wing section with HLF glove at flight conditions; wing section PHLF1, $Re_\infty=35\cdot10^6$, sweep $\Phi=26^o$, fixed transition 50%/50%

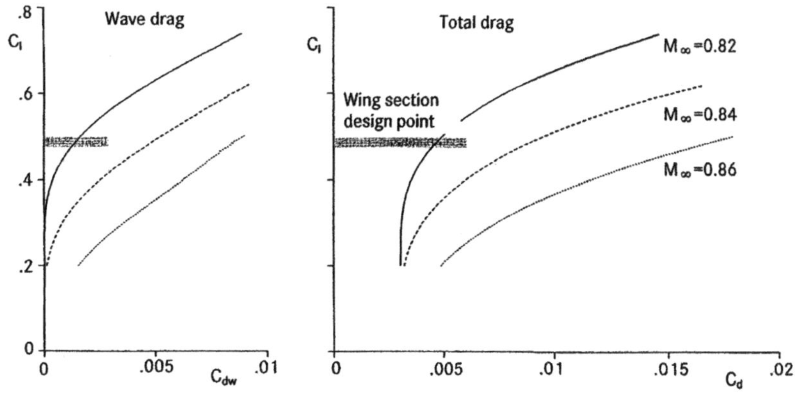

Figure 17 Drag polars of A340 HLF wing section; wing section PHLF1, $Re_\infty=35\cdot10^6$, sweep $\Phi=26^o$, fixed transition 50%/50%

21.6.2 Shock control bump design for laminar wing section PHLF1

The off–design Mach number of M_∞=0.84 and the local lift coefficient of c_l=0.48 were chosen for the bump design [21] in order to improve the off–design behaviour.

Based on the results of the bump parametric study, a bump with a length of $l_B/c = 20\%$ and an asymmetrical shape of $c_B/l_B = 70\%$ was chosen due to its favorite features concerning shock movement and drag reduction potential. In addition two symmetrical bumps of different lengths, l_B/c = 15% and 20%, were included for studying the bump integration into the present A340 wing structure. The shape function 'fixed beam' and a height of h_B/c = 0.3% have been fixed.

The evaluation of total drag with varying bump location is presented in Figure 18, indicating bump locations for maximum drag reduction:

- bump 1: l_B/c =20%,c_B/l_B =70% at x_B/c=63%
- bump 2: l_B/c =20%,c_B/l_B =50% at x_B/c=65%
- bump 3: l_B/c =15%,c_B/l_B =50% at x_B/c=66%

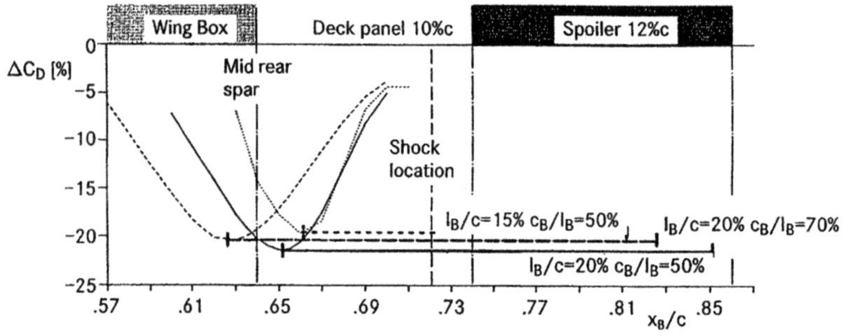

Figure 18 Three bump designs overlayed with A340 wing structure; bump geometry: $f_B(x)$=fixed beam, h_B/c=0.3%

In order to demonstrate the difficulty of integrating the bump into the present A340 wing structure, the positions of the mid rear spar (x/c=64%) and the spoiler leading and trailing edge (x/c=74% and 86%) are marked. As the shock is located in front of the spoiler, neither of the bumps can be integrated in the A340 spoiler as it was originally envisaged.

For the Mach numbers M_∞=0.82 and 0.84 the predicted drag polars show the effect of the asymmetrical bump (bump 1) with variation of the bump height, Figure 19. At the design point of the wing section at M_∞=0.82 and c_l=0.48 no drag reduction is produced by the bump, although the wave drag is about 10cts. To reduce the wave drag for the design point, the bump has to move further upstream and its wing integration becomes more difficult. For c_l>0.50, a decrease in drag can be generated by the bump. At the bump design point, M_∞=0.84 and c_l=0.48, a clear reduction in drag is found due to the bump, that can be further improved by adaptation of the bump height.

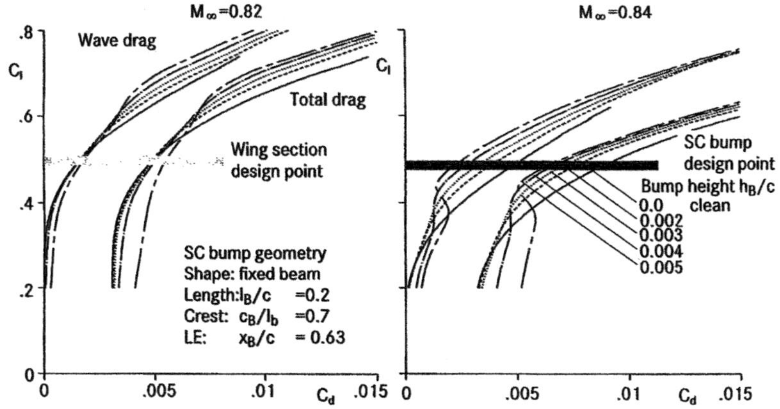

Figure 19 Drag polars for the A340 HLF wing section with adaptive bump device; wing section PHLF1, $Re_\infty=35\cdot10^6$, sweep $\Phi=26^o$, fixed transition 50%/50%

The predicted drag envelope of each adaptive bump configuration is presented in Figure 20 for M_∞=0.82 and 0.84. The asymmetrical bump is superior to the other bump designs concerning drag reduction and, in particular, the efficiency at low lift. Since all three bumps require a modification of the wing structure for their integration, the asymmetrical adaptive bump design was chosen for the further investigation.

21.6.3 Bump device integration into wing spoiler

The final positioning of the asymmetrical bump on the original A340 wing section structure is shown in Figure 21, again indicating the necessity of modifications of the rear part of the wing. For the integration of the bump into the spoiler, a shortening of the wing box and an enlargement of the spoiler up to 23% chord are needed, Figure 22. These modifications are associated with an increase in wing weight, but a larger height of the rear spar due to the laminar airfoil shape may partly compensate the weight increase.

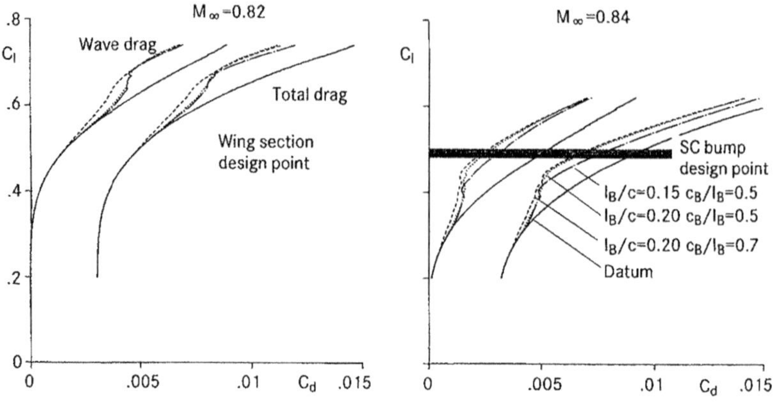

Fig. 20 Drag envelope of three bump designs; wing section PHLF1, Re_∞=35·10^6, sweep Φ=26°, fixed transition 50%/50%

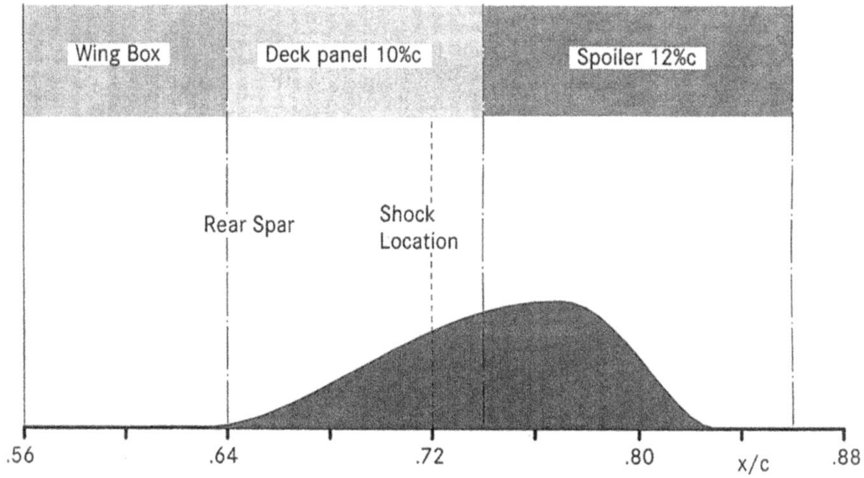

Figure 21 Selected bump design for integration in the A340 wing structure; bump geometry: l_B/c=20%, h_B/c=0.3%, c_B/l_B=70%

As strong shocks mainly appear on the outer wing, the spoiler has to be extended up to the wing tip, Figure 23. In order to maintain the capability of roll control, a variable camber (VC) flap system is introduced over the whole span replacing the aileron.

Different settings of the VC flaps in spanwise direction can be used to obtain the appropriate distance between the shock and the bump location. The VC flaps are also needed to ensure the favourable pressure gradients for laminar flow at off design flight conditions. It is assumed that laminar flow and shock control are effective over about 41% of the wing area.

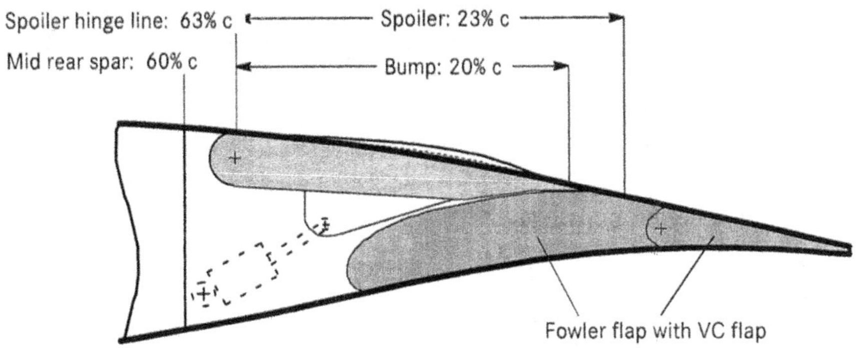

Figure 22 Integration of adaptive variable–height bump device into modified A340 wing spoiler

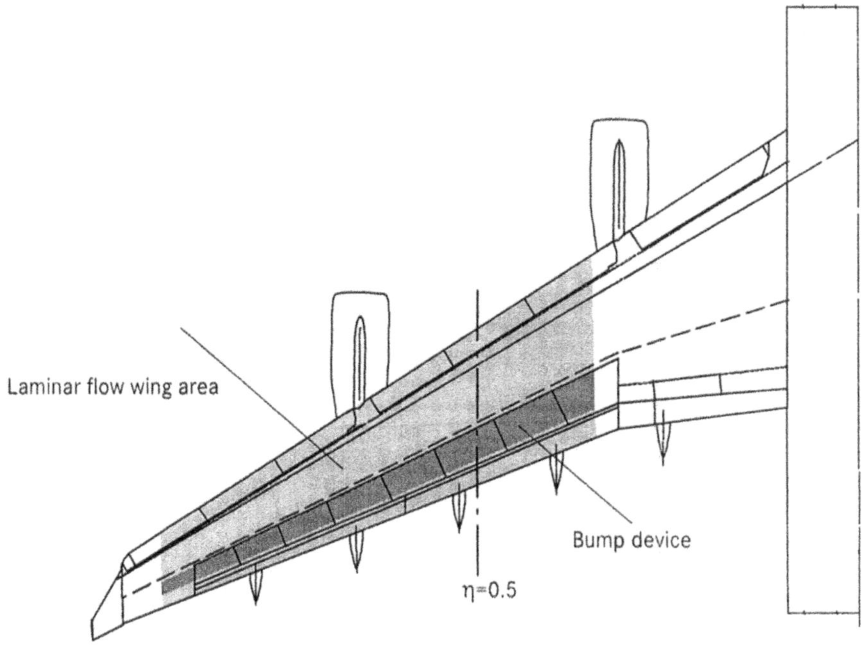

Figure 23 A340 HLF wing with integrated adaptive bump device

The modifications of the wing structure for shock control is restricted to the bump integration into the spoiler, while all other modifications are assumed to be linked to the HLF wing itself. An increase of the operating weight is estimated to be about 0.25tons for the bump integration into the spoiler.

21.6.4 Drag balance of HLF wing with adaptive bump device

Based on the wing section drag characteristic at 50% span, the laminar wing profile drag is corrected by the drag reduction due to the adaptive variable–height bump device. The corresponding flight polar and drag balance are estimated for the cruise Mach number of M_∞ =0.82 and the off–design Mach number of M_∞ =0.84 and shown in Figure 24, indicating reductions in drag for higher lift coefficients at cruise and over the whole lift range at off–design compared to basic A340–HLF polar. The highest drag reduction of about 4% is obtained for c_L=0.38 at off–design and for c_L=0.58 at cruise, respectively.

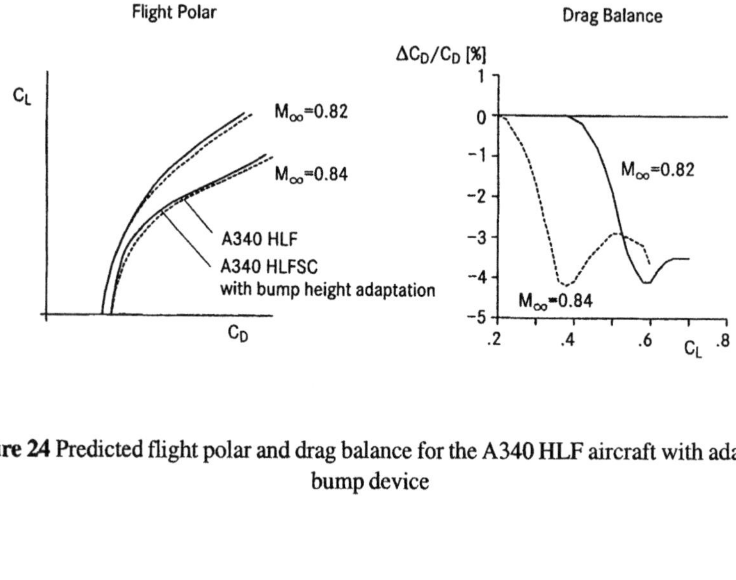

Figure 24 Predicted flight polar and drag balance for the A340 HLF aircraft with adaptive bump device

21.6.5 Flight mission, fuel prediction and operating costs

With the given flight polar and weight penalty a standard northatlantic flight mission with a range of 3500 nm was investigated, Figure 25, assuming 600 trips per year at the cruise Mach number M_∞ =0.82 and 620 trips per year at the off–design Mach number M_∞ =0.84. The flight profile has been predicted by the standard procedure with steps of 4000ft for optimal flight altitude. At these conditions, a reduction in fuel consumption per year due to shock control of about 353 tons (1.23%) at M_∞ =0.82 and of about 729 tons (2.11%) at M_∞ =0.84 is estimated by flight mission calculation.

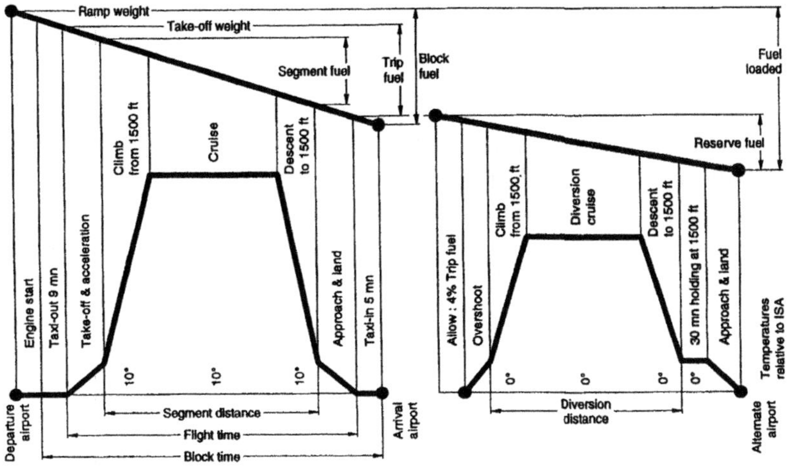

Range 3500 NM

Mach No 0,82	Utilization	Blocktime	Blockfuel	Fuel/Year	Gain in Fuel/Year	Gain in Fuel
	Trips/Year	hr	kg	to	to	%
HLF	600	7,9	47998	28799		
HLFSC	600	7,9	47410	28446	353	1,23
Mach No 0,84						
HLF	620	7,6	55797	34594		
HLFSC	620	7,6	54622	33866	729	2,11

Figure 25 Profile of flight mission for the A340 HLF aircraft and the A340 HLF aircraft with shock control by the bump device (HLFSC)

Using the formula for Cash Operating Costs (CoC) provided by BAe for Task 4

$$\Delta CoC/CoC = 0.49\ \Delta C_D/C_D + 1.9 \cdot 10^{-3}\ \Delta W + 0.113\ \Delta mc/mc$$

a decrease of $\Delta CoC/CoC$ of about 1.3% is achieved assuming an average drag reduction of $\Delta C_D/C_D = 3\%$, a weight penalty of $\Delta W = 0.25$tons and an increase in maintenance costs of $\Delta mc/mc = 0.5\%$.

21.7 Adaptive Shock Control Bump Structure and System Concepts

It is difficult to finally assess the bump benefits/penalties due to the lack of a sound integration study that is outside the scope of the present project. However, two preliminary structure and system concepts of an adaptive variable–height bump device are presented, showing the realization of the bump integration into the spoiler.

Due to the thickness of the spoiler only a small space is available for the integration. A compact actuator system is, therefore, required with an efficient mechanism to deform the skin and the ability to contribute to the stiffness of the spoiler.

21.7.1 Pneumatic bump concept of DLR

The DLR Institute of Structural Mechanics proposed a pneumatic bump concept, Figure 26, consisting of a supporting lower spoiler layer, a layer with actuators and a flexible upper skin, [22] . The actuator consists of a pipe–spring made of carbonfibre (CFK), forming two mirror–image hollow spaces. The pipe–spring is deformed by pressuring the inner liners made out of glasfibre and rubber. The open ends of the pipe–spring can be displaced by up to 2/3 of their height by changing the inner pressure . A hinged connection with the stringer of the spoiler skin ensures that the displacement is transformed into a linear movement normal to the skin. The pipe–spring is placed into the supporting structure of the spoiler and improves the stiffness of the spoiler in spanwise direction.

A bump with a length of l_B=2000mm (l_B/c=20%), integrated into the spoiler with a length of 2200mm, can be deployed up to a height of h_B=55mm (h_B/c=0.55%) with this bump concept assuming 21 actuators equally distributed over the spoiler.

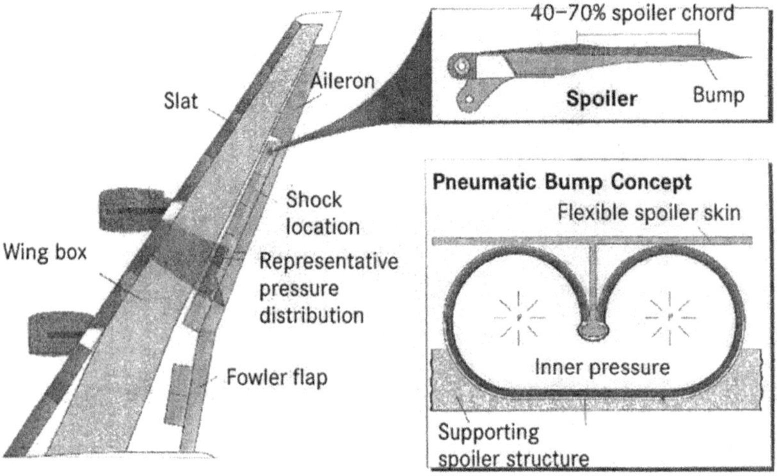

Figure 26 Pneumatic bump concept of DLR

21.7.2 Kinematic bump concept of DaimlerChrysler F&T

The realization of a bump integrated into a spoiler was also investigated by DaimlerChrysler Forschung & Technologie [23] using a kinematic concept, Figure

27. Due to the small volume of the spoiler the ribs are replaced by a special design containing a trapezoidal kinematic system on two levels. A two–bar linkage system driven by pushrods is employed to achieve the deformation of the flexible skin. This special rib design gives a high stiffness of the whole structure, and the variation of the bump height is achieved by only one actuator.

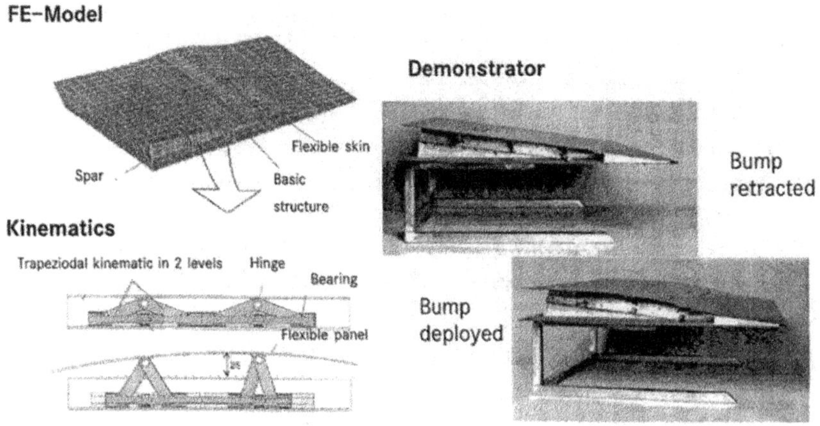

Figure 27 Kinematic bump concept of DaimlerChrysler F&T

21.8 Conclusions

As agreed upon at the start of EUROSHOCK II, DA was to look at the benefits/penalties of control for an A340–type aircraft with a hybrid laminar flow wing design.

Different shock and boundary layer control concepts – active shock control by cavity ventilation with part suction, discrete suction and contour bump – have been evaluated for the application on an HLF wing of a long–range transport–type aircraft. Substantial net drag reductions could only be achieved by using a bump device, a fact also confirmed by the computational results in Task 2 and experimental data of Task 3. Based on the characteristic features of a laminar wing, such as relatively fixed shock positions and a strong wave drag increase at near–design and off–design conditions, the introduction of an adaptive bump seemed to be the best means for cruise performance improvement at changing flight conditions.

A parametric bump study was carried out leading to the establishment of limiting bump parameters and their effectiveness in drag reduction. For a fixed bump location the adaptation of the bump height was found to be the most feasible adaptation concept for laminar–type wings to reduce wave drag at near–design and off–design conditions.

A representative A340 HLF wing section, designed to produce laminar flow on a wing glove mounted between the engines of an A340 wing when suction is applied at the nose, was used for the study of the integration of a bump into wing and the consequences for the wing design. Although the shock is, at near–design and off–design conditions, located downstream the A340 rear wing spar, the installation of an optimized asymmetrical adaptive bump with a length of $l_B/c= 20\%$ requires modifications of the wing structure: an enlargement of the spoiler if the bump is to be integrated into the spoiler, a modification of the wing box, and the addition of a VC flap system needed for roll control, possibly shock positioning and the adjustment of the pressure gradient for maintaining laminar flow.

Wing section polars for two Mach numbers — design and slightly off design — and a variable bump height were the basis for predicting the corresponding drag balance of the aircraft assuming that 41% of the total wing area will be laminar and the shock accessible to bump control. This resulted in aircraft drag reductions of up to 4%.

Considering only the weight penalty for the bump installation into the spoiler (0.25tons), the estimates of a typical transatlantic mission gave fuel reductions of 1.23% at the design Mach number of M_∞=0.82 and 2.11% at the off–design Mach number of M_∞=0.84. Under the given assumptions, a reduction in cash operating costs (CoC) of 1.3% seems to be achievable.

The realization of the adaptive bump installation into the spoiler was exemplified by two preliminary structure and system concepts: the pneumatic bump concept of DLR and kinematic bump concept of DaimlerChrysler F&T.

The performed investigation of the adaptive variable–height bump control device, integrated in a long–range HLF–wing transport aircraft, showed a benefit in aircraft performance and operating costs; however, further studies are necessary to prove the spanwise efficiency of the bump device at "true" three–dimensional conditions and to determine the system and structural integration with respect to weight, maintenance and costs penalties in more detail.

21.9 References

[1] Stanewsky,E., Delery,J.,Fulker,J.,Geißler,W.:"EUROSHOCK–Drag Reduction by Passive Shock Control
Results of the Project EUROSHOCK, AER2–CT92–0049, Supported by the European Union, 1993–1995, Notes on numerical Fluid Mechanics, Vol. 56, Vieweg Verlag, Braunschweig 1997.

[2] Stanewsky,E.:"EUROSHOCK II – Drag Reduction by Shock and Boundary Layer Control",Technical Annex, Project BE–1316, IMT, Aera–3 Aeronautics, 1995–1998.

[3] Dargel,G.:" Validation of DA Airfoil/Swept Wing Code for Shock and Boundary Layer Control – EUROSHOCK II–Final Technical Report of DA Contribution to Task 2 ", Euroshock TR–BRPR–95–0076, DA–Report EF–068/99, 1999

[4] Thiede,P.,Dargel,G.:" Assessment of Shock and Boundary Layer Control Concepts for Hybrid Laminar Flow Wing Design – EUROSHOCK II–Final Technical Report of DA Contribution to Task 4 ", Euroshock TR–BRPR–95–0076, DA–Report EF–069/99, 1999

[5] Dargel,G.:"Ein Programmsystem für die Berechnung transsonischer Profil– und konischer Fluegelstroemungen auf der Basis gekoppelter Potential– und Grenzschichtloesungen", DGLR–Bericht 92–07,1992.

[6] Le Balleur,J.C.: "Strong matching method for computing transonic viscous flows including wakes and separations. Lifting airfoils", La Rech. Aerosp. 1981–3, 1981.

[7] Veldman,A.E.P.:"New, quasi–simultaneous method to calculate interacting boundary layers", AIAA Journal, Vol. 154, 1981.

[8] Cebeci,T.,Khattab,A.A.,Chen,H.H.,Chen,L.T .:"An approach to the design of wings; the role of mathematics, physics and economics",AIAA Paper No.92–0286,Reno,1992.

[9] Carter,J.E.:"A new boundary–layer inviscid iteration technique for separated flow", AIAA Paper No. 79–1450,1979.

[10] Cebeci,T.,Smith,A.M.O.:"Analysis of turbulent boundary layers", Academic Press, New York, 1974.

[11] Poll,D.I.A.,Danks,M.,Humphreys,B.E.: "The aerodynamic performance of the laser drilled sheets", In Proceedings 'First European Forum on Laminar Flow Technology', DGLR–Bericht 92–06,1992.

[12] Bohning,R., Doerffler,P.:"Hybrid and Active Control of the Shock Wave–Turbulent Boundary Layer Interaction and Porous Plate Transpiration Flow", Final Technical Report, Contribution of Karlsruhe University to the project EUROSHOCK II, Contract BRPR–CT95–0076, 1999. (also see Chapter 12 of the present Volume)

[13] Pankhurst,R.C.,Gregory,N.:"Power requirements for distributed suction for increasing maximum lift", ARC CP 82, 1952.

[14] Ashill,P.R,Fulker,J.,L., Simmons,M.,J.:"Simulated active control of shock waves in experiments on aerofoil models", Proceedings ICEFM,Turin,July 1994.

[15] Dargel,G.:"Extension, Validation and Application of the DA VII Transonic Airfoil Code with Passive Shock Control", Chap. 14, p.195–220, in Notes on numerical Fluid Mechanics, Vol. 56, Vieweg Verlag, Braunschweig 1997.

[16] de Matteis,P.,Dima,C.:"Assessment of CFD Capabilities and Concepts for Shock and Boundary Layer Active Control on Transonic Airfoils", EURO-SHOCK–TASK 2, Cira TR–97–097/Euroshock TR–BRPR–95–76/2.2, 1997.

[17] Dargel,G.,Rodde,A.M.,Archambaud,J.P.:"Assessment of the Capability of Drag Reduction of the Shock Control Device 'SC Bump' on Airfoil Flows and Applications Aspects on Wings", in IUTAM Symposium on Mechanics of Passive and Active Flow Control, Editor: G.E.A.Meier and P.R. Viswanath, Kluwer Academic Publishers,1999.

[18] Drela,M.,Giles,M.B.:"ISES: A two–dimensional viscous aerodynamic design and analysis code",AIAA Paper 87–0424, 1987.

[19] Dziomba,B.:"ELFIN– Final Technical Report",BRITE/EURAM Area5 Aeronautics, Contract No AERO–0013–C(MB), 1992

[20] Seumenicht,L.:"Untersuchung zum Potential der Hybridlaminarisierung von Verkehrsflugzeugen",Studienarbeit,Institut für Aerodynamik und Gasdynamik,Universität Stuttgart, 1992.

[21] Mertol,S.,Dargel,G.:"SC Bump Parameterstudie und Entwurf eines Bumps am A340–Hybridlaminarprofil PHLF1", DA Report EF–034/97,1997

[22] Monner,H.P.,Breitbach,E.,Bein,T.,Hanselka,H.:"Strukturkonzepte für den adaptiven Flügel", Aeroelastik–Tagung der DGLR, DLR Göttingen,29.–30.Juni,1998

[23] Martin,W.: Private Communication, DaimlerChrysler Forschung&Technologie, 1999.

22 Assessment of Bump Control and its Application to a Long-range Turbulent-wing Aircraft

R. Doe

BAE SYSTEMS Airbus UK, Wing Aerodynamics Engineering
Filton, Bristol BS99 7AR, United Kingdom

Note: The full contribution was, unfortunately, not available for inclusion into the present volume. The reader is requested to accept our apologies and to refer to Chapter 6 for an overview and essential results of the work performed. A brief summary is given below.

Summary

The objective of the present study was, similar to the work of EADS-Airbus D described in the preceding chapter, to apply shock and boundary layer control to an A340-type aircraft wing of existing turbulent design and to determine benefits and penalties of control implementation on the operational aircraft. The control mechanism considered was foremost the contour bump. (Originally, also a new turbulent wing design with integrated bump should have been considered; however, this task was not performed due to time limitations.)

After initial two-dimensional and sheared-wing computations for a characteristic A340-type wing section, computations were repeated and further computations carried out for the complete wing with and without bump control, employing the BAE SYSTEMS-Airbus UK 3D Euler VII code, since the predicted 2D/sheared-wing shock location did not match the shock location of the three-dimensional wing. The latter was considered important for subsequent structural considerations.

In optimizing bump control for a turbulent wing, characterized by a rather flat, slowly increasing upper-surface pressure distribution, difficulties arise from the rapid movement of the shock with changing lift coefficient and/or Mach number. This shock behavior suggests that a fixed bump will give an effective benefit only over a narrow lift and Mach number range centered around the design point. It was, for instance, determined that a bump with a height of 0.2% chord and a length of 12.5% chord, located 2% chord downstream of the shock at the design condition M_∞=0.82, C_L=0.59, is, at this Mach number, only effective in a lift range between C_L=0.565 and 0.655, with the highest effectiveness at off-design, i.e., outside the operating range of the aircraft.

In an attempt to improve the effectiveness of the bump in general and, in particular, over the required operating range, a study was performed varying the bump shape parameters and its position. Here, the largest effect was achieved by increasing the bump height from 0.2% chord to 0.3% chord with the maximum drag reduction in the lift range of interest, i.e., C_L=0.5 to 0.6, being about 5%. Other parameters, such as the detailed bump shape, including the crest position with respect to the bump chord, had little effect on drag in this C_L-range, any positive changes tending to occur at higher values of C_L. Moving the bump location on the wing shifted the maximum drag reduction to a different lift coefficient. The relatively low drag reductions near design for the (existing) turbulent wing — in comparison to a laminar-type wing — is mainly due to the inherently low shock strength, hence low wave drag, associated with the turbulent wing design.

Finally, assessments have been made of the benefits attainable over typical mission profiles taking into account the aerodynamic benefits over the operating C_L-range as well as the weight penalties associated with the bump installation, here assumed to be ΔW=0.5 tons for a variable-height bump and ΔW=0.2 tons for a fixed bump. For the variable-geometry bump, a reduction in Cash Operating Costs (CoC) of ΔCoC = -0.4% was determined for typical long-range missions, while for the fixed bump only a ΔCoC = -0.1% was realized. The CoC could, of course, be further reduced by also making the bump adaptive in chord-wise direction and, furthermore, by considering aircraft operation at off-design conditions, e.g., within the drag rise, which can, generally, not be avoided and was, therefore, also considered when treating the laminar-wing aircraft mission in the preceding chapter.

23 Drag Reduction and Buffet Damping by a Contour Bump Control Device and Regional-jet Application

N.Catino, N. Ceresola
ALENIA Aerospazio - Divisione Aeronautica
C.so Marche, 41 - 10142 Torino, Italy

Summary

In the present work Alenia (ALN) has performed an evaluation of shock control by contour bumps in order to explore their capability to reduce the wing drag in transonic flow and to allow damping of aerodynamic buffet. Evaluations on 2D airfoils have been performed using the ALN Navier-Stokes code. Indications have been obtained that a local, well designed, upper surface deformation (bump) can sensibly reduce airfoil drag in transonic flow in a large range of lift coefficients. The bump, when effective in drag reduction, also dampens buffet. At off-design conditions, the bump effectiveness vanishes very rapidly and, in some cases, the bump may increase drag and/or induce buffet. ALN has also generated a simplified configuration of a regional jet for which numerical evaluations have been made. The results obtained indicate that benefits can be gained only at the design point, and that even a reduction in efficiency may occur at off-design conditions. It is concluded that, to obtain a true benefit from the application of bump devices, the adoption of more advanced techniques is needed, such as generating a bump shape through multipoint optimisation or developing an adaptive bump device.

23.1 Introduction

In the present document the ALENIA contribution to the EUROSHOCK II project is described. The objectives of the research were the evaluation of the capabilities of a contour bump in reducing aerodynamic drag and delaying buffet, the optimisation of a bump shape, and its application on a simplified Regional-jet aircraft configuration.

After an assessment phase of the ALN Navier-Stokes code, an evaluation of the bump effects on drag was made for two airfoils, viz., the turbulent RAE-5225 and the laminar-type DRA-2303. Then, the effect on buffet has been assessed using an unsteady Navier-Stokes code, which was employed to simulate the actual buffet process on the airfoil DRA-2303.

In a joined effort with Dassault, a business-type aircraft wing was considered. An optimisation of the bump shape was performed for a 2D-airfoil geometry derived from the 3D Falcon wing at the scaled freestream conditions.

To evaluate the real potential of the bump technique, a simplified complete Regional-jet configuration was generated. For this hypothetical aircraft, the effect of an optimised bump in terms of lift and drag benefits at design and off-design conditions was assessed.

23.2 Assessment of the ALENIA Navier-Stokes Code

The numerical code employed at ALENIA solves the Reynolds-averaged full Navier-Stokes equations with a finite differences technique. Centred space discretization is used, with added non-linear second and fourth-order numerical damping. Implicit relaxation is employed to drive the solution to a converged result. A two-equations, fully point-wise k-Rt turbulence model is implemented for turbulent flow computations. Up to third-order accuracy can be attained in time.

Computations at transonic conditions were performed for two airfoils, the turbulent RAE-5225 and the laminar-type DRA-2303, with and without bump. In order to make a comparison with experiment meaningful, the angle of attack in both cases was modified until the shock fitted the experimental shock position for the "clean" configuration. The pressure distribution for the RAE-5225 airfoil with and without bump is shown in Fig. 1a and Fig.1 b, respectively. The effect of the presence of the bump appears to be well represented by the numerical simulation. Similar results for the DRA-2303 airfoil are presented in Figs. 2a and b.

The computed lift and drag coefficients are reported in the following tables together with the DERA experimental data. The agreement is quite good after applying a correction in α between 0.06 and 0.41 degrees to match the shock positions in case of the datum airfoils as mentioned above.

Table 1 RAE-5225 w/o bump (Data Point 288)

	M	Re	α[deg]	CL	CM	CD
ALENIA	0.7266	$18.8\text{x}10^6$	2.900	0.756290	-0.08980	0.011907
DERA EXP	0.7271	$18.8\text{x}10^6$	2.959	0.755392	-0.09180	0.011920

Table 2 RAE-5225 with Bump (Data Point 715)

	M	Re	α[deg]	CL	CM	CD
ALENIA	0.7266	$18.8\text{x}10^6$	2.900	0.758657	-0.08991	0.010817
DERA EXP.	0.7255	$18.7\text{x}10^6$	3.016	0.757207	-0.09030	0.010290

Table 3 DRA-2303 w/o Bump (Data Point 276)

	M	Re	α[deg]	CL	CM	CD
ALENIA	0.6800	$18.9\text{x}10^6$	2.500	0.732168	-0.08612	0.013236
DERA EXP.	0.6829	$18.9\text{x}10^6$	2.097	0.747208	-0.10060	0.013460

Table 4 DRA-2303 with Bump (Data Point 1639)

	M	Re	α[deg]	CL	CM	CD
ALENIA	0.6800	$18.9\text{x}10^6$	2.50000	0.731232	-0.09067	0.011651
DERA EXP.	0.6801	$18.9\text{x}10^6$	2.09800	0.739879	-0.09890	0.011380

23.3 Bump Evaluation and Optimisation

23.3.1 DRA-2303 Airfoil

23.3.1.1 Bump effect on drag

The first approach to optimise a bump was performed for the DRA-2303 airfoil for which the existing bump allowed a drag reduction only for C_L>0.5. The present target was to design a new symmetrical bump with a design point of C_L=0.4.

In order to reduce costs, the optimisation cycle was based on a 2D full-potential code coupled with a boundary layer simulation. The parameters investigated were bump length, thickness and maximum thickness position. After the optimisation cycle, the bump shape attained was tested with the Navier-Stokes code.

Fig. 3a shows, respectively, the drag polars for the basic airfoil (w/o bump), for the airfoil with the original bump, and for the airfoil with the new bump design. The drag variation obtained with the two different bumps is depicted in Fig. 3b: the new bump gives a benefit only in a very narrow range between C_L=0.3 and 0.5, confirming that an optimised shape is able to generate a drag reduction only around its design point.

A comparison between full-potential and Navier-Stokes results was carried out to understand the reason why the gain obtained with the optimised bump was only marginal. Results of this comparison indicated that a small discrepancy existed in the shock position leading in the case of the Navier-Stokes computations to the reduced effectiveness.

23.3.1.2 Bump effect on buffet

An interesting question is whether a "passive" device can control buffet onset on a wing and dampen the shock oscillations inside the buffet regime, limiting the extent of the shock-induced separation that is at the origin of the phenomenon. To answer this question, a preliminary study has been performed, simulating the flow past the DRA-2303 airfoil with and without bump, at a Mach number of M=0.68 and an angle of incidence of $\alpha=5^{\circ}$. The ALN Navier-Stokes solver was used for the computations.

The results, in terms of time histories of the lift coefficient and the position of the shock, are depicted in Figs. 4a and b: the presence of the bump has clearly the effect to dampen the oscillations of the shock wave. Noteworthy is that no attempt has been made to optimise bump shape and position on the airfoil to control buffet, the bump parameters being the same used to reduce drag at a much lower angle of incidence.

23.3.1.3 DRA-2303 airfoil: a brief conclusion

On the basis of the numerical evaluations for the DRA-2303 airfoil, it is possible to conclude that a bump device may have the dual role of reducing drag at cruise incidence and of reducing buffet oscillations at off-design conditions.

23.3.2 DASSAULT Airfoil

Within the ALENIA-DASSAULT joint effort, ALN designed a symmetrical bump shape to be placed on an airfoil derived from a Falcon aircraft outer-wing section (also see Chapter 24).

The bump optimisation has been performed for two freestream conditions:

- Design condition: M=0.720, α=2.60°
- Off-design condition: M=0.765, α=2.00°.

On the basis of the preceding experience with the DRA-2303 airfoil, and to further evaluate the possibility of using a low cost 2D full-potential code in the optimisation cycle, ALN has made preliminary calculations for the DASSAULT airfoil at the same freestream conditions with the full-potential and with the Navier-Stokes code, respectively. It was found that for this airfoil the full potential code, coupled with a boundary layer method, calculates with sufficient accuracy the shock position in case of fully attached flow.

The bump effect at the design condition on the pressure distribution is shown in Fig. 5 indicating the weakening of the shock by the bump. In Fig. 6, the drag polars are presented (6a) together with the drag balance (6b). The drag reduction appears to be sensible and present over the entire C_L-range explored.

At the specified off-design condition, the evaluations have been performed by DASSAULT and the results (see Chapter 24) indicated that the bump does not generate any major drag reduction, suggesting that the bump effectiveness diminishes very rapidly at off-design.

23.4 Bump Evaluation for a Simplified Regional-Jet Configuration

23.4.1 Generation of a Regional-Jet Configuration

As shown in Fig. 7, a complete Regional-jet configuration, although simplified, was generated deriving its characteristics from an average of data associated with existing jets: the wing is swept 26° at the leading edge, having no twist, and a constant airfoil section in span-wise direction. The airfoil used is the DRA-2303 airfoil, extensively analysed within EUROSHOCK II project, with and without bump. The horizontal and vertical tails have a constant symmetric NACA-0010 airfoil and the engine nacelles are axially symmetric. The bump, when present, is fitted over the full wing span.

23.4.2 Numerical Evaluations

The code used to evaluate the bump effect on the Regional-jet configuration is a 3D Full-Potential code coupled with a boundary layer method.

23.4.2.1 Performance at the design condition

A drag polar was computed at M=0.73 and lift coefficients between C_L=0.15 and C_L=0.68 in order to evaluate the bump effect at design and off-design cruise conditions. In order to properly compare the results for the bump-off / bump-on configurations, calculations for the bump-on configuration have been performed by varying the incidence until the same lift coefficient as in the case without bump was achieved. To properly locate the shock position, since indications obtained from 2D calculations showed that this parameter is of paramount importance concerning the effectiveness of the bump, a local mesh refinement was made.

The pressure distributions at the same lift coefficient, C_L=0.54, Fig. 7, shows that the bump reduces the shock strength. The analysis of the drag components shows not only a wave drag reduction but also friction and pressure drag reductions. It is worth outlining that, when the bump is present, the same C_L is obtained at a lower angle of attack.

In Fig. 8 the computed drag polars for the complete aircraft are presented: it appears that the bump is effective over the whole range explored. The gain in terms of total drag reduction ranges from 1.7% to 5.9% when changing lift from 0.3 to about 0.70. Since these values are typical of cruise conditions, one may conclude that the bump technique, when applied over the full span of the wing, satisfies the requirement given by DASSAULT and BAE-SYSTEMS (Airbus-UK), namely, that the "Minimum Significant Total Drag Reduction" for the whole aircraft should at least be 2%.

The above results, therefore, show that significant total drag reductions for a commercial transport configuration are possible by bump control.

23.4.2.2 Performance at off-design Mach number conditions

In order to complete the evaluation of the influence of a bump on the performance of a typical regional-jet configuration, two off-design conditions have been considered, the first one being M=0.8, C_L=0.45 (Reynolds=$6.7x10^6 m^{-1}$); at this flight speed the bump gives only a negligible 1% drag reduction compared to the 5% obtained at the design condition (M=0.73). In case of subsonic flow, M=0.50, C_L=0.60 (Re=$6.7x10^6 m^{-1}$), the bump has, as can be expected in the case of weak shocks, no positive effect on drag.

It seems, therefore, as also found by other investigators, that the bump effectiveness vanishes rapidly at off-design Mach number conditions. It is important to underline that these last results confirm findings obtained for the 2D DASSAULT airfoil.

23.5 Conclusions

Within the EUROSHOCK II project, ALENIA has performed an evaluation of bump control devices (BUMP) acting on shock-boundary layer interaction in order to explore the capability of this type of device to reduce drag on a wing in transonic flow and to dampen aerodynamic buffet. The final objective was to apply bump control to a Regional-jet aircraft assessing benefits and penalties associated with bump control.

Evaluations for 2D airfoils (RAE-5225, DRA-2303, DASSAULT Falcon-type airfoil), performed with the ALN Navier-Stokes code, (jointly with and similar to other Euroshock II partners) allow the following conclusions:

1. A local, well designed upper surface deformation (bump) can effectively reduce airfoil drag in transonic flow.
2. At the same design Mach number, it is possible to reduce drag over a large range of lift coefficients.
3. The bump, when effective in reducing drag, may also dampen buffet.
4. At off-design Mach number conditions the bump effectiveness vanishes very rapidly and, in some cases, the bump may increase drag and/or induce buffet.

In order to evaluate bump effectiveness for a new commercial aircraft, ALN has generated a simplified complete regional-jet aircraft configuration. Numerical evaluations have then been made. (Similarly BAE-SYSTEMS and Dassault have evaluated the possibility of using bump control on existing turbulent-wing aircraft types while EADS-Airbus (DASA)has applied bump control to an A340-type aircraft with a laminar wing design.)

The target drag reduction was defined by Dassault estimating that it is necessary to have at least a 2% total drag reduction for the complete aircraft in order to obtain a real benefit from the application of this technique.

The conclusions are here that:

1. Bump application does not seem to be effective for existing turbulent-wing aircraft (retrofit).
2. Attractive drag reductions for new aircraft with a fixed bump, applied over the full wing span, seem to be possible, but only at the design Mach number condition.

The need is to have a bump which is effective over, at least, a part of the Mach number range of interest, and not only for one Mach number. It would also be necessary to avoid the current penalties in terms of drag increase and buffet deterioration, at off-design Mach number conditions. ALENIA's final comment is, therefore, that more advanced techniques should be applied to the design and application of bump devices, such as:

1. Generating a bump shape through a multi-point optimisation cycle.
2. Having an adaptive bump device, i.e., to have a device able to change its shape according to the prevailing flight condition.

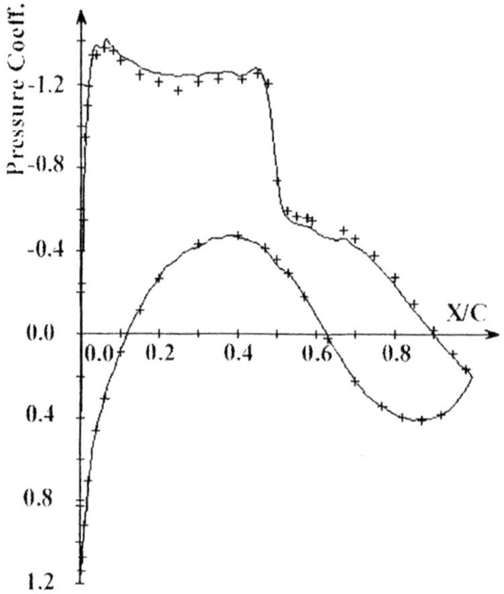

a. Without bump control

+ DERA EXPERIMENT

- ALN Navier-Stokes Calculations

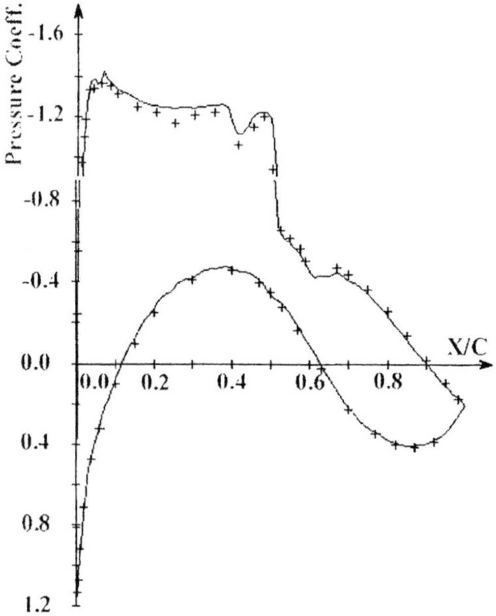

a. With bump control

+ DERA EXPERIMENT

- ALN Navier-Stokes Calculations

Figure 1 Pressure distributions for the turbulent RAE-5225 airfoil with and w/o bump control, $M = 0.72$, $\alpha = 2.9°$, $Re = 19x10^6$

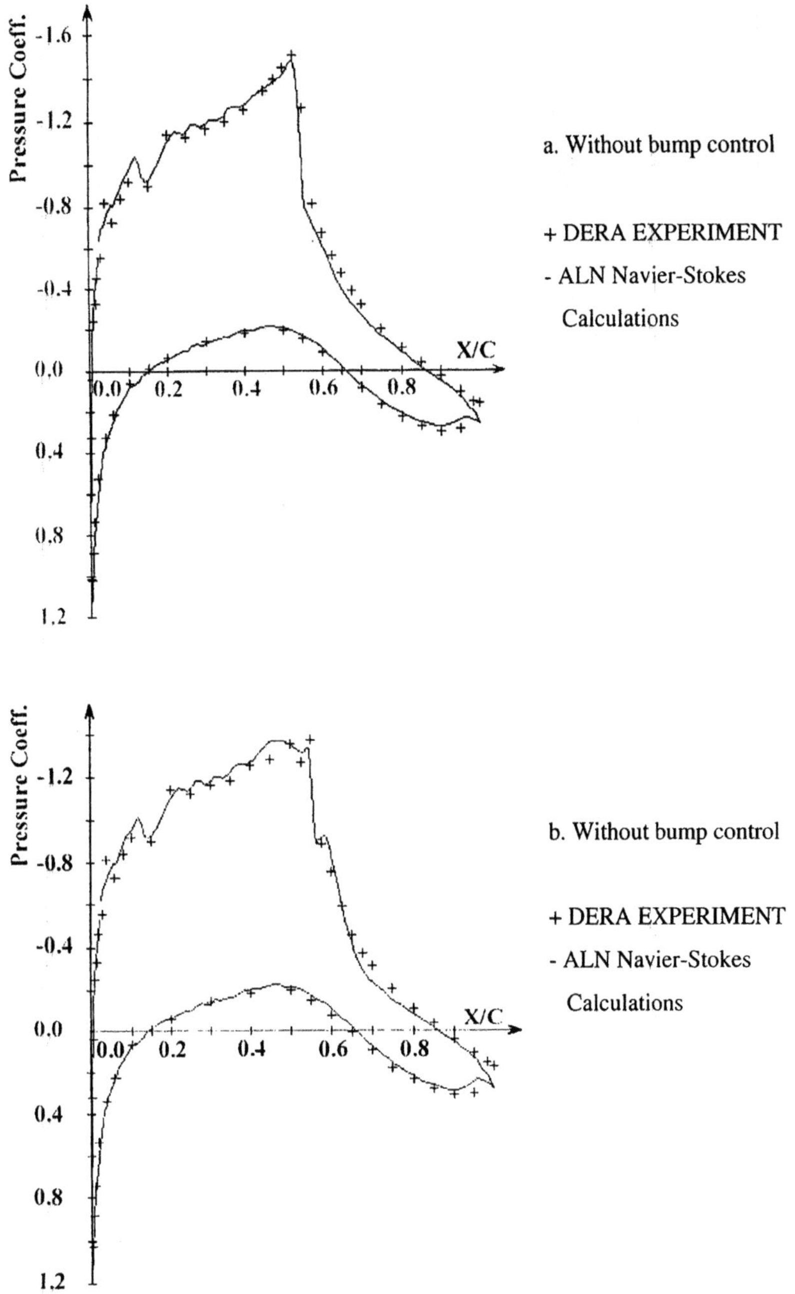

Figure 2 Pressure distributions for the laminar airfoil DRA-2303 with and w/o bump control, $M = 0.68$, $\alpha = 2.5^{\circ}$, $Re = 19 \times 10^{6}$

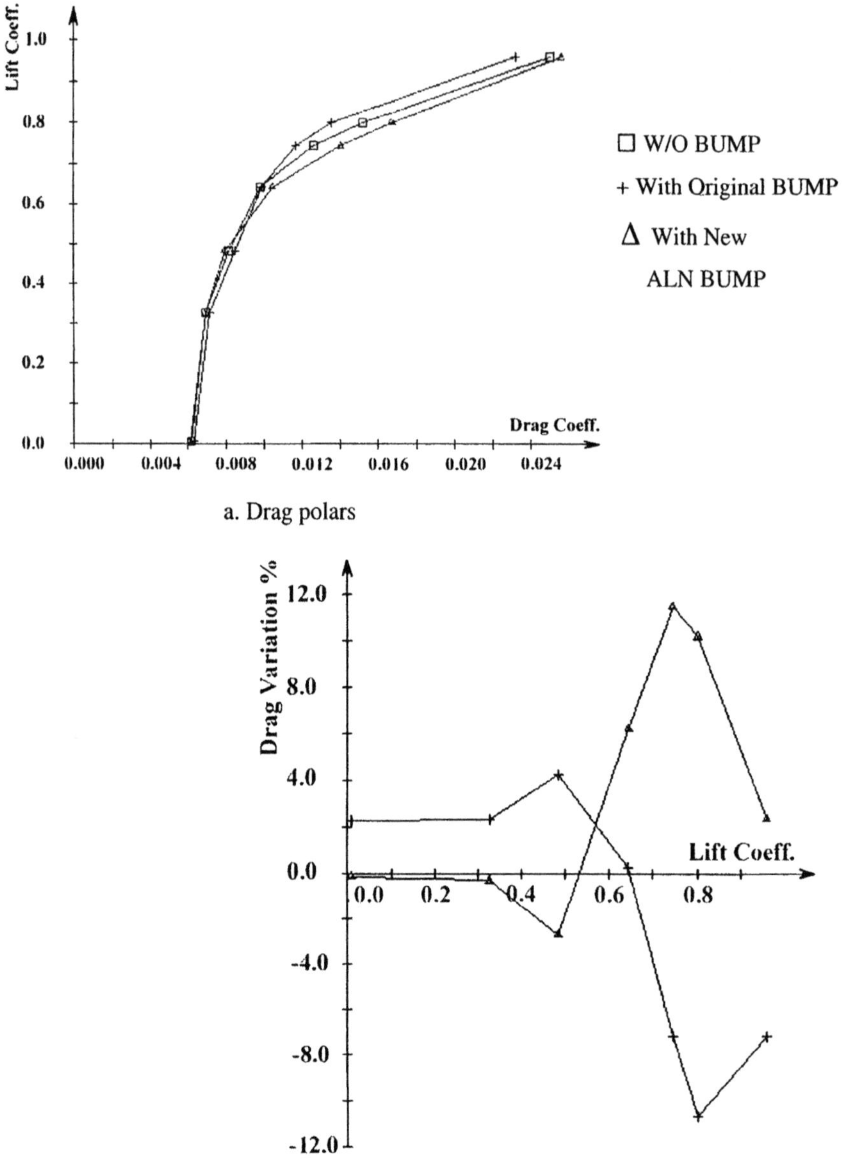

a. Drag polars

b. Drag balance (drag reductions negative)

Figure 3 Drag polars and drag balance for the DRA-2303 airfoil with two different bump designs, ALN Navier-Stokes calculation, M=0.68, Re = 19×10^6

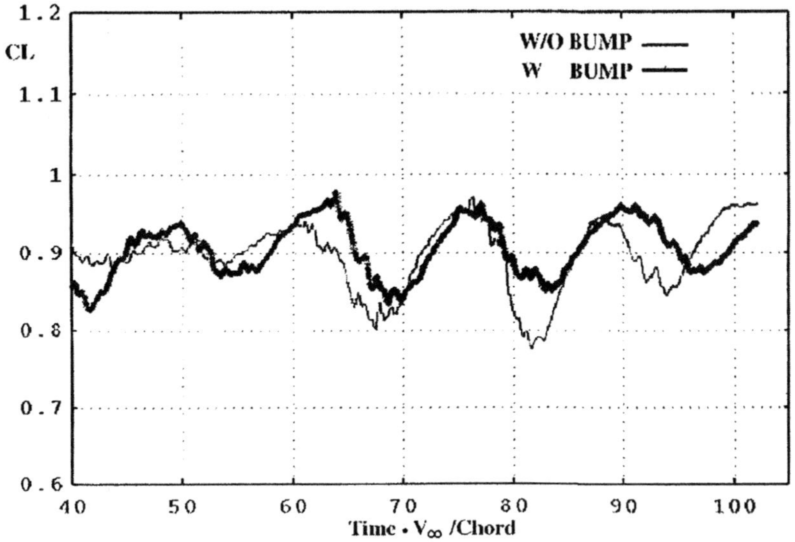

a. Time history of the lift coefficient

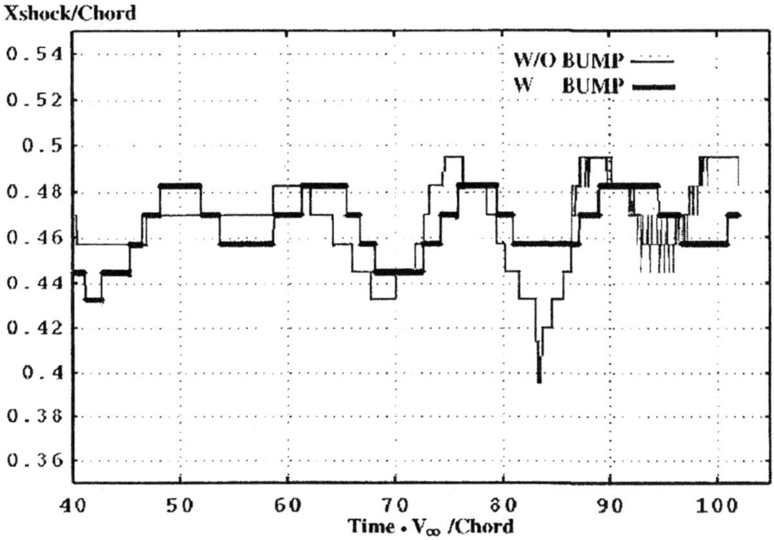

b Shock oscillations

Figure 4 Time histories of lift and shock location with and w/o bump: Unsteady Navier-Stokes calculations, airfoil DRA-2303, M=0.68, α = 5.0°, Re=19×10^6

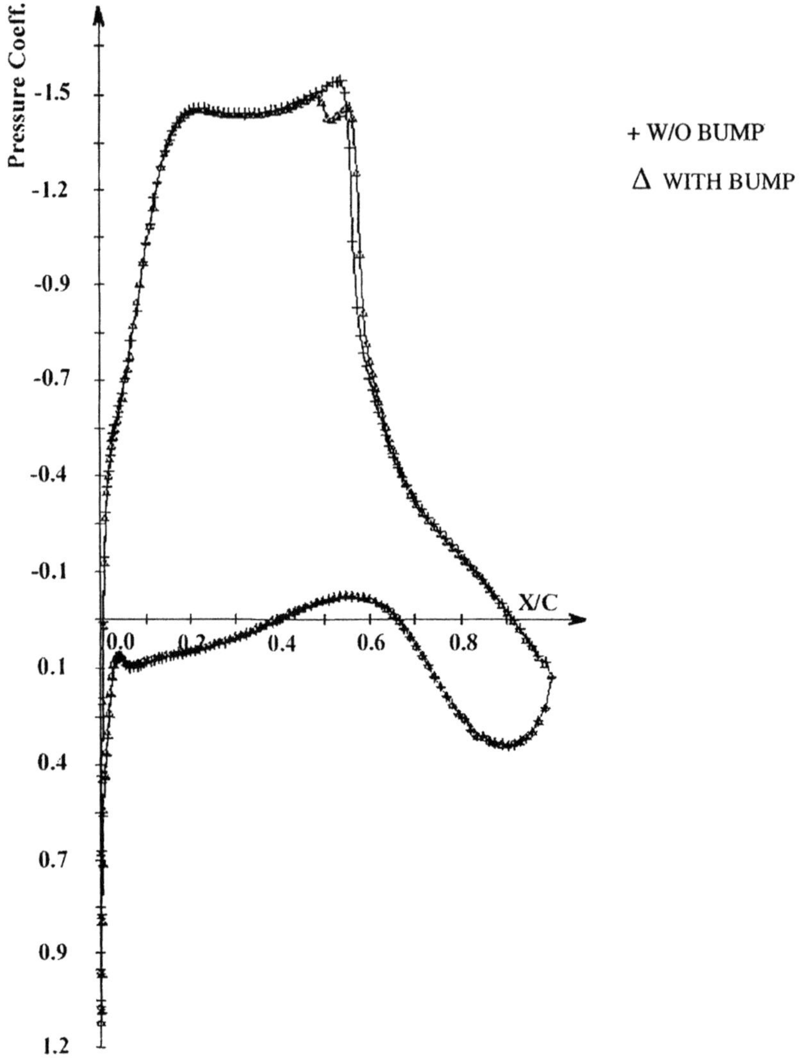

Figure 5 Pressure distributions for the Dassault Falcon-type airfoil with and w/o bump control, ALN Navier-Stoke calculations, M=0.72, $\alpha = 2.6^{o}$, Re = $8{\cdot}10^{6}$

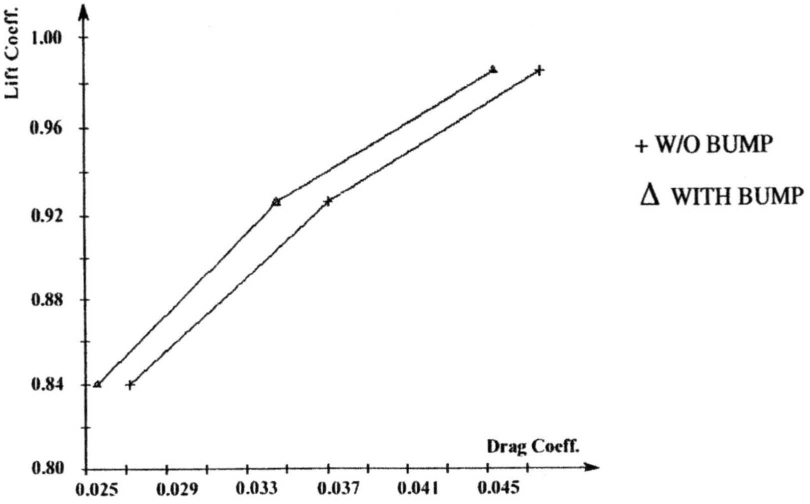

a. Drag polars

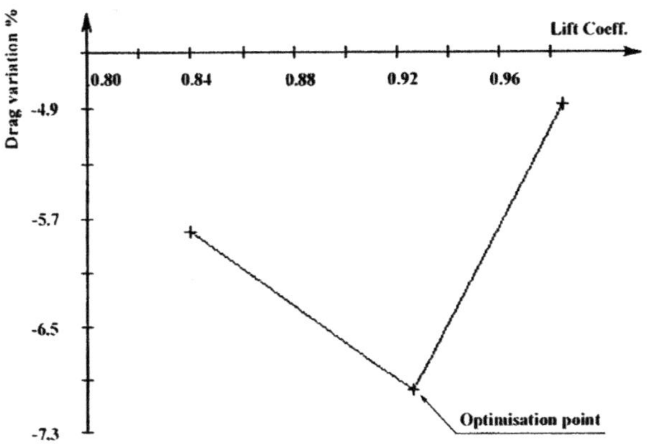

b. Drag balance

Figure 6 Drag polars and drag balance for the Dassault Falcon-type airfoil with and w/o bump control, ALN Navier-Stoke calculation, M=0.72, $\alpha = 2.6^{\circ}$, Re = $8\text{x}10^6$

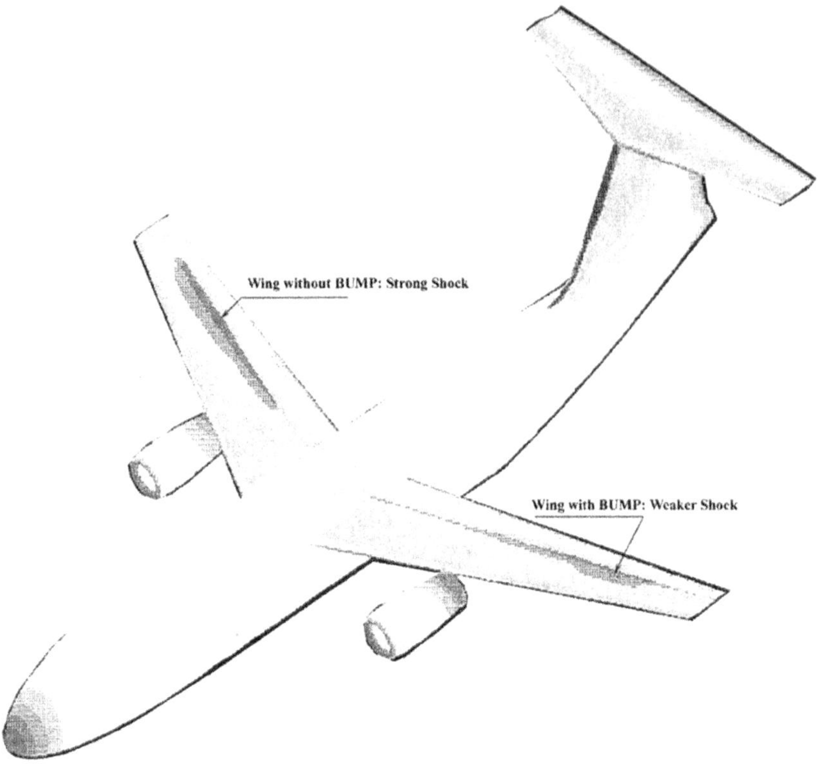

Figure 7 Bump effect at constant lift coefficient for a 100-seat Regional-jet, $M = 0.73$, $Re = 6.7x10^6\ m^{-1}$, free transition, $C_L = 0.54$

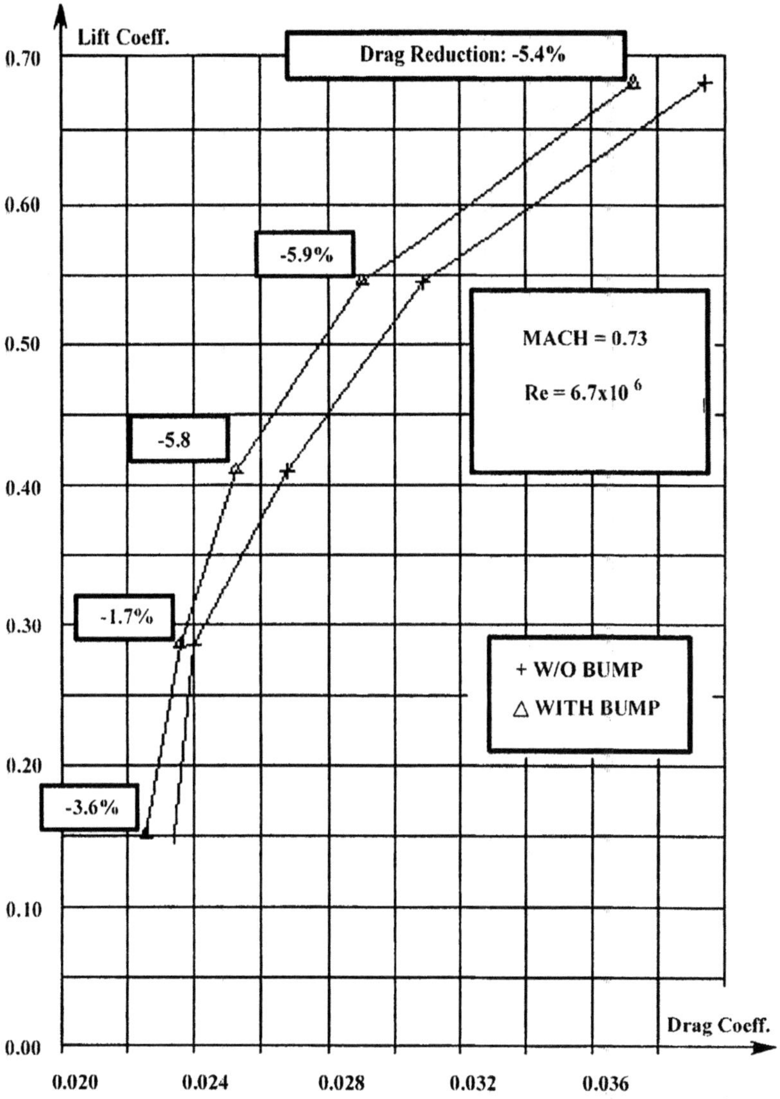

Figure 8 Evaluation of bump effect on drag for a 100-seat Regional-jet in transonic flight, $M = 0.73$, $Re = 6.7 \times 10^6\ m^{-1}$, free transition

24 Application of Shock and Boundary Layer Control to a Business-jet Aircraft

J.J. Vallee
DASSAULT Aviation, 78, Quai Marcel Dassault, Saint-Cloud, France

Summary

The present paper describes the application of shock and boundary layer control to a business-jet aircraft similar to the Dassault Falcon. It is comprised of the aerodynamic assessment of the effect of a bump on the flow development about a Falcon-type airfoil at design and off-design conditions, the technological integration of control by an adaptive bump and slot suction, respectively, into a Falcon-type wing, and the assessment of minimum benefits making an introduction of control worthwhile. It was found that for the turbulent Falcon wing an adaptive bump is required. Installation penalties for such a bump are assessed to be around 3% of total drag, so the bump does not show sufficient performance to be introduced into the present turbulent wing and, as a consequence, a new wing design is needed whose aerodynamic behavior in the transonic regime must include a very constant shock position. For slot suction, an estimation of the suction power accessible on the FALCON allows the operation of a suction system at rates compatible to the ones studied in Task 3, however, the costs of implementation correspond to more than 6% in total drag with gains being limited to 2% so that such a system is not a viable tool for drag reduction for the present aircraft.

24.1 Introduction

The Dassault Aviation contribution to EUROSHOCK II is concerned with Task 4: Assessment of Shock and Boundary Layer Control Application Aspects. The main goal of control is to reduce wave drag on transonic wings, however, the possible effect of control on the occurrence of buffet is also being considered.

The work performed during the three years of the contract covers:

- The evaluation of the minimum performance necessary to make the control device worthwhile for a business jet like the FALCON, and establishment of a criterion to verify this also taking into account the cost for operating the control device.
- The aerodynamic assessment of a bump device on a 2D airfoil representative of the pressure distribution on the FALCON outer wing at flight conditions.
- A technological study related to the integration into an existing business-jet wing of two different control devices, viz., a moveable bump and a suction

slot, and the assessment of the penalties induced and the deduction of the net benefits.

The main result concerning the implementation of a bump on the turbulent business jet wing is that, due to the large motion of the shock over the transonic flight domain, this bump must be variable in position and height to remain efficient. This has important consequences for the penalties associated with bump installation and operation: looking at the minimum performance benefits necessary to counteract these penalties (nearly 3% of the total drag), there is little additional gain for the aircraft. Moreover, the geometrical extension of the bump leads to several strong interactions with the existing control surfaces of the actual FALCON wing making its use impossible for a retrofit. The integration of a bump into a turbulent wing must be seen in conjunction with the development of a new wing.

Concerning the suction slot, aerodynamic performance benefits considered are taken from the work carried out in Tasks 1 and 3 of EUROSHOCK II. The technological study shows that the FALCON engines can reasonably well provide the energy needed for the suction system, but considering the penalties due to the integration and operation of the suction device (over 6% of total drag), the performance gains due to the system are insufficient.

24.2 Evaluation Criterion and Aircraft Balance

To evaluate the benefits due to shock control on the aircraft, two problems have to be considered, firstly, the global impact of a local drag reduction which determines whether it is worthwhile to use such a device, and, secondly, the penalties due to the integration of the control device into the wing.

For the first problem, a global drag balance is made for the aircraft to quantify the drag reduction due to the device and a minimum weighted efficiency is defined. For the second problem, technological solutions for integrating the control device are studied and drag penalties are assessed using a criterion which takes into account the different impacts for the aircraft such as weight, fuel (volume) loss or power taken from the engines. The device is worth considering only if its effect is at least equal to the sum of the drag penalties and the minimum efficiency increase expected for the aircraft.

24.2.1 Transposition Local/Global Drag Reduction

Considering the FALCON business jet, the Shock and Boundary Layer Control (SaBLC) device should be used on the part of the outer wing where the wave drag is a maximum, Figure 1. This part represents 35% of the wing span and is, at a typical cruise flight condition, responsible for nearly 10% of the total drag. Then, in order to reduce drag for the aircraft by 1%, the local airfoil drag reduction must be at least of 10%.

Industrially speaking, the order of magnitude of total drag reduction that may motivate the use of a SaBLC device is around 2%, i.e., the minimum local (airfoil) drag reduction expected is about 20%, not including penalties.

24.2.2 Evaluation Criterion

Integration of a SaBLC device into the wing will generate some penalties due to volume and energy needs to install and operate a device, as well as due to a weight increase.

Thus the following criterion (defined for a mid-size jet and based on the maximization of the range) has been proposed to classify control concepts, considering their efficiency and complexity, by a parameter "C" which must be maximized:

$$C = -10^{-5}\ \Delta C_d - 0.068\ \Delta M - 9.27\ \Delta V - 241.5\ m \tag{1}$$

where ΔC_d is equal to the drag change ($\Delta Cd < 0$) for the average cruise lift coefficient, ΔM is the weight change (lb) ($\Delta M > 0$ for weight penalties), ΔV is the fuel capacity change (ft^3) due to system installation inside the wing box ($\Delta V > 0$ for a fuel capacity penalty), and m is the total suction mass-flow rate through the suction slot (lb/s), if applicable.

The above formula (1) takes into account the efficiency of the jet pump, fed by the engine bleed, which is the equipment expected to be installed on a mid-size business jet (low weight and easy installation). In order to keep the suction solution feasible, the mass flow rate taken from the engines has to be lower than 0.35 lb/s for the aircraft (acceptable engine cycle modification).

24.3 Aerodynamic Assessment of the Bump

24.3.1 Calculation Conditions

To study the effect of the bump by 2D Navier-Stokes computations, a typical pressure distribution on the outer wing of the FALCON is taken from a 3D calculation. An airfoil geometry is then extracted from the wing in the direction normal to the leading edge and modified to reproduce as close as possible the 3D pressure distribution, Figure 1b.

Once this 2D geometry was defined, two typical flight points were transformed into the corresponding 2D conditions for the computations:

- Cruise point $M_{2D} = 0.72$ $\alpha_{2D} = 2.6^\circ$ and
- Off-design point $M_{2D} = 0.765$ $\alpha_{2D} = 2^\circ$

The off-design point corresponds to buffet conditions, so this point will also be useful to assess the ability of a bump to delay buffet onset.

In the first section, we will present results obtained for the symmetrical bump proposed by ALENIA (also see Chapter 23). The numerical simulation is performed with the DASSAULT-AVIATION VIRGINI Navier-Stokes code using a $k-\varepsilon$ turbulence model. For the study of the symmetrical bump, this turbulence model is used under the assumption of wall laws for the region close to the body.

The second section presents results for an asymmetric bump shape, already studied by EADS-Airbus (also see Chapter 21). In this second set of calculations, the wall laws are replaced by a two-layer approach since results at cruise conditions obtained in previous calculations shed some doubts on the use of the wall laws, especially with regard to separation as will be seen below.

For the asymmetric bump, a parametric study of the bump height is carried out as indicated in Table 1.

Table 1 Characteristics of the bumps studied (dimensions relative to airfoil chord c)

	L/c (length) %	X_{deb}/c (L.E.-position) %	X_H/c (crest position) %	h/c (height) %
ALENIA symmetrical bump	20	50	60	0.2
DASA asymmetrical bump	20	48	64	0.2
	20	56	72	0.2
	20	48	64	0.4
	20	56	72	0.4

24.3.2 Results for the Symmetric Bump of ALENIA

Taking the airfoil at cruise conditions, ALENIA has optimized a symmetrical bump in height and position. Characteristics of this bump are given in Table 1.

In order to verify the effect of this bump, calculations were made for cruise and off-design Mach numbers at two angles of attack, the latter to obtain a segment of a polar which allows to estimate drag reductions for the same lift coefficient. Results are presented on Figures 2, 3, and 4.

Cruise Mach number Concerning the pressure distributions at cruise, Figures 2a and 3a, the effect of the bump on shock strength is clearly indicated; lift is increased due to the downstream displacement of the shock. Considering the polar curves for the present conditions, Figure 4, one observes that the influence on lift really constitutes a major effect. Indeed, the result for the angle of attack of 2.6° shows that drag is increased due to the bump, but that lift is also sufficiently increased to produce, at the same lift level (linear interpolation between the two calculated points), a reduction in drag of about 30 cts (ΔC_d=0.0030), i.e., a reduction of almost 10%.

Looking at the skin friction coefficients, Figures 2b and 3b), we can see that the bump has no significant effect on separation. This seems to indicate that the use of the present bump to postpone problems linked to separation, such as buffet, might not be effective.

It should be noted that the relatively large extent of the shock-induced separation is believed to be due to the modeling of the wall laws used in the turbulence model. The overestimation of separation leads to the preference of the two layer model for further calculations involving the asymmetric bumps.

Off-design Mach number For the off-design Mach number, the associated polars, Figure 4, show that no benefits are achieved by the bump at the higher angle of attack, while at the lower incidence, the drag reduction is similar to the one obtained at the design Mach number. The ineffectiveness of the bump in case of the former is essentially due to the misplacement of the bump with respect to the shock location.

Conclusion: As a first conclusion concerning the present symmetric bump and the given freestream conditions, we can say that with a bump correctly placed with respect to the shock position, a drag reduction of about 10% is certainly possible, but that the performance falls off quickly, as is well known, with shock movement.

24.3.3 Results for the Asymmetric Bumps

The same aerodynamic assessment has been made for the asymmetric bump. This kind of bump, studied by DASA (EADS-Airbus) for a laminar-type A340-type aircraft wing, shows positive results for this class of aircraft wing. Moreover, it appears that for a laminar wing the bump can be fixed in location since the shock does, by design, not move significantly in chordwise direction within the transonic flight domain.

The bump geometries studied are described in Table 1: the bump has a crest placed at 80% of its length and its position on the airfoil is considered to be correct when the "outer inviscid" shock is located roughly at 40% of the bump length. The parametric study covers bump location and height.

As for the symmetrical bump, the bump location optimized for cruise conditions does not result in noticeable benefits for the off-design Mach number (results not presented), so for the off-design conditions and corresponding computations the bump was moved downstream by 8%-chord. For both Mach numbers, i.e., $M_{design} = 0.72$ and $M_{off\text{-}design} = 0.765$, three angles of attack were considered.

24.3.3.1 Cruise Mach number

For $\alpha = 2.6^{\circ}$, as an example, Figure 5, the bump is correctly placed and shock strength is reduced due to the bump. The shock shifts somewhat downstream with this effect not being amplified by the higher bump. The same behavior also holds for $\alpha = 2^{\circ}$. For $\alpha = 1^{\circ}$, the shock is too far upstream, with the bump increasing this tendency, and an expansion zone appears behind followed by a second somewhat weaker shock. Concerning separation, only the $\alpha = 2.6^{\circ}$ case, Figure 5b, shows a small separation bubble just behind the shock for the datum airfoil. In case of the lower bump, this bubble disappears, but it reappears for the higher bump. Moreover, the recompression at the end of this bump generates conditions closer to separation.

Concerning global effects, Figure 7, the bump is efficient for the three angles of attack investigated and achieves, at the same lift coefficient, a drag reduction close to 10%. The higher bump gives some additional drag reduction at lift coefficients above 0.8.

24.3.3.2 Off-design Mach number

As shown by the polars in Figure 7 at $M = 0.765$ and the three angles of attack, there is a major effect on lift due to the shock displacement as indicated in Figure 6a for the (representative) angle of attack of $\alpha = 1^{\circ}$. Simultaneously there is a weakening of the shock leading to a drag reduction, at the same lift coefficient, of close to 10% for the lower bump and 20% for the higher one. At these conditions, the optimum bump height is certainly greater than 0.4%.

Concerning separation, Figure 6b, there exists a large separated zone, also present at $\alpha = 2^{\circ}$, which is not influenced by the bump. This result confirms the conclusion already drawn in conjunction with the symmetrical bump.

24.3.4 Conclusion

Symmetric and asymmetric bumps of various positions and heights have been evaluated for a FALCON-type airfoil. The assessment was made for two flight conditions, viz., cruise and off-design (buffet).

At cruise conditions, a reduction in airfoil drag of about 10% is achieved with the lower bump height, while the higher bump gives only minor improvements. At off-design, the drag reduction reaches about 20% and is obtained with the higher bump.

The effect of the bump is twofold: a shock-strength reduction (drag effect) and a shock displacement (lift effect). Depending on conditions, these two effects combine differently with emphasis more on drag or lift. To obtain an optimum in performance improvement with this device, it seems clear that its parameters must be adapted to each new situation, which means that the bump must be adaptable to flight conditions.

The bump influence on separation seems very weak, especially for large separation zones as they occur in the case of buffet conditions. So there is no hope to delay the occurrence of buffet substantially by means of a bump. This is, however, contrary to the findings of other investigations within EUROSHOCK II and might possibly be related to the present airfoil.

24.4 Technological Integration of Shock and Boundary Layer Control Devices into a Falcon Wing

24.4.1 Adaptive Bump

The fixed bump is a highly interesting solution as its cost is close to zero, but the aerodynamic assessment made above shows that, for the present turbulent wing, it is impossible to apply this kind of device since it is efficient only in a very narrow range of flight conditions. An adaptive bump is then necessary and a precise estimation of its cost must be made. For the technical integration of the bump into the FALCON wing, specifications are taken from the asymmetric-bump study.

24.4.1.1 Bump area

In Figure 8 a concept for the adaptive bump is displayed. The area of the wing covered by the bump is cross-hatched. In spanwise direction it extends from 45% to 80% of the wing span. To cover the range of positions pointed out in the previous section, the area reserved for the bump in chordwise direction extends from 50% to 80% of the chord.

If we consider a retrofit solution on an existing wing, the following conclusions can be drawn: ① For cruise conditions only the wing box is affected by the modification, but for buffet conditions the bump will extend over the wing box and the air-brake system, and, partially, the aileron, which is, of course, not viable. ② It follows that the bump solution, fixed or retractable, is, for the present aircraft, not a (retrofit) solution.

24.4.1.2 Technological concept

A distortable skin controlled by actuators located underneath is possible due to the low amplitude required for the bump, Figure 9. The skin is rigid in the spanwise direction and somewhat flexible in the chordwise direction. It is actuated by rollers moving in helical ramps machined into special crank levers. The whole system is actuated by an actuator located in an adjoining wing box.

One part of the total movement of the actuator is used for deforming the skin and the second, opposite part is used for moving the rollers in order to be able to carry out the skin assembly which closes the deformable part of the wing box. It is also mandatory to have a wing box bottom which is very stiff and which contributes to the leeward skin continuity.

The zone concerned is located partly in the fuel wing box and partly downstream of the rear spar; this has an impact on the fuel capacity and on the spar height. This solution leads to a weight penalty of 264 lbs. and to a fuel capacity penalty of 4.6 ft^3.

By applying Equation 1, the change in drag should be

$$10^5 \, \Delta C_d = -0.068 \times 264 - 9.27 \times 4.6 = -60.5$$

which implies a minimum drag reduction of 3% of the total drag in order to balance the fuel and weight penalties. As previously seen, the drag reduction obtained with a bump is, at best, 20% locally, i.e., about 2% for the aircraft. This is just sufficient to cover the penalties without any net gain. One should, however, remember that the numbers used are for a specific aircraft with a turbulent wing and applying a very special solution to the control action. Here, the reader is also referred to Chapter 23 where bump control is applied to a regional jet aircraft with a laminar-type wing.

24.4.2 Discrete Slot Suction

An integrated solution for discrete slot suction has also been studied, the solution consisting of a row of spanwise suction slots, located at 50% chord, with the slots being 10 mm wide and having a total length of 3 m in the spanwise direction, Figure 10; there are 6 suction slots separated by the wing ribs. The idea is to manufacture a special leeward panel for which the skin is lower than the normal one with a minimum at the center of two subsequent ribs in order to create a plenum chamber closed above by trap doors perforated in the desired zone where suction is to be applied.

Each plenum chamber has its own suction system which runs inside the wing to an ejector system providing the desired suction rate.

Taking into account that for the aircraft the maximum allowable mass flow rate is 0.35 lb/s, this leads for a total slot length of 6 m (for two wings) to Cq = 1.5×10^{-4} (flow rate in kg/m/s divided by $\rho_\infty V_\infty$ c) which is compatible with mass flow rates applied in Tasks 1 and 3 for such a device.

The wing integration study leads to a weight penalty of 140 lbs. and to a loss of fuel capacity of 4.7 ft^3. By again applying Equation 1, the drag change is

$$10^5 \, \Delta C_d = -0.068 \times 140 - 9.27 \times 4.7 - 241.5 \times 0.35 = -137.6$$

which implies a minimum drag reduction of 6.8% of the total drag in order to balance the fuel-mass and suction penalties. But we have seen, for instance in the investigations within Task 3 (see, e.g., Chapter 19), that slot suction is able to achieve a total (aircraft) drag reduction of about 2% which is well below the present requirements.

24.5 Conclusion

The work performed by DASSAULT-Aviation within the EUROSHOCK II project was mainly concerned with the assessment of Shock and Boundary Layer Control (SaBLC) applied to a mid-size business jet like the FALCON.

The part of the FALCON wing directly affected by a high wave drag contribution, and hence susceptible to SaBLC, represents about a third of the span in the outer wing region. The drag contribution of this area is about 10% of the total drag. Correspondingly, in order to gain 1% drag reduction for the aircraft, the local airfoil gain must be about 10%.

Since the integration of SaBLC devices into the wing involves some penalties for the aircraft (weight, fuel loss, engine power needs), a criterion is used to quantify these penalties in terms of "minimum drag reduction" required. The devices must at least compensate the penalties associated with their implementation and operation and provide extra benefits of sufficient magnitude to motivate their integration; this extra "benefit" is generally about 2% of drag reduction for the aircraft.

Concerning the bump, local airfoil drag reductions of 10% have been numerically determined at cruise conditions and 20% for an off-design point. This is the minimum local gain of interest for the aircraft, so the bump must show very low penalties. A fixed bump will meet these requirements since it generates almost no costs.

The variation of the shock position with flight conditions on the present turbulent wing renders a fixed bump efficient only for one flight point. So it is necessary to introduce a bump adaptive in location and height, the latter since it has also been shown that the optimum bump height must increase with shock strength. The penalties due to the wing integration of such an adaptive bump are assessed to be around 3% of total drag, so the bump does not show sufficient performance to be introduced on the present turbulent wing; in addition, a retrofit for the present wing would involve strong interference with existing control surfaces.

To conclude: for a bump on a business-jet wing like the FALCON wing to be of interest, the performance of the device must be such that its operational costs are negligible. This is only met by a fixed bump, and, as a consequence, the wing aerodynamic behavior in the transonic regime must be altered to provide a very constant shock position since the bump efficiency falls quickly when not optimally placed with respect to the shock; this would, of course, require a new wing.

For slot suction, an estimation of the suction power available on the FALCON (taken from the engines) allows to control the same part of the wing as above at suction rates studied in Task 3 (C_Q=1.5x 10^{-4}). But the costs of the implementation of slot suction are much higher than the ones for the adaptive bump, being equivalent to more than a 6% penalty in total drag, at equal performance.

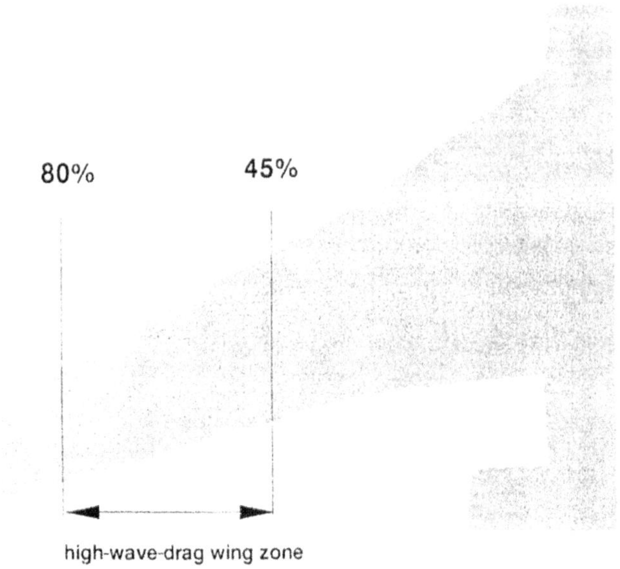

a. Wing plan form with area of high wave drag

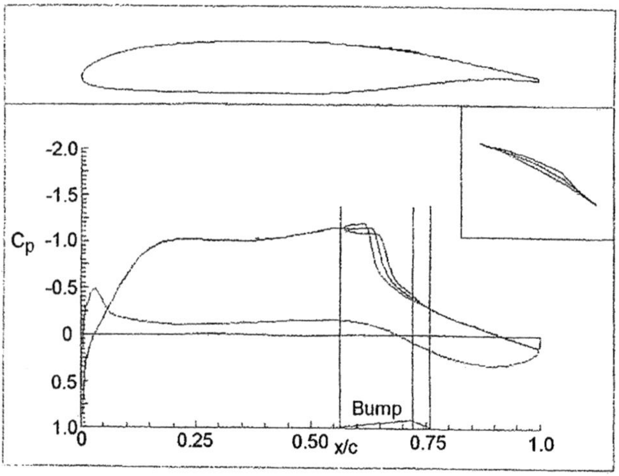

b. Falcon-equivalent outer-wing airfoil shape and characteristic pressure distribution

Figure 1 Falcon wing planform and equivalent outer-wing airfoil shape

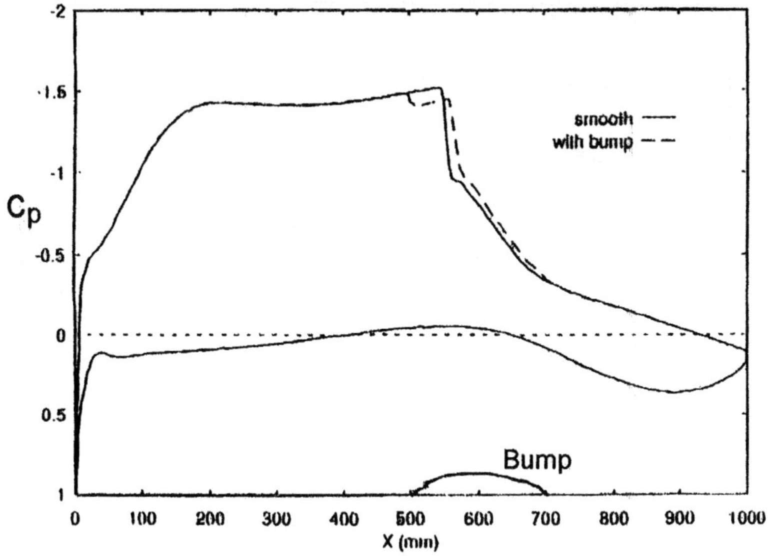

a. Airfoil pressure distribution

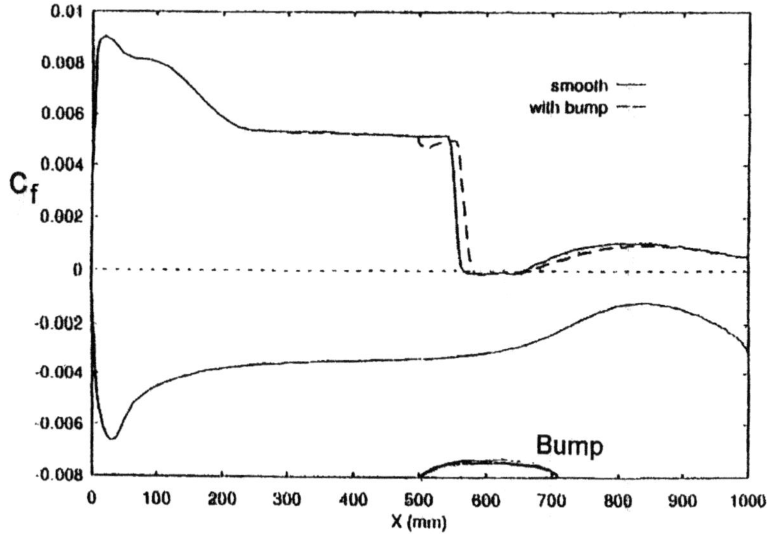

b. Airfoil skin-friction distribution

Figure 2 Pressure and skin-friction distributions with and w/o symmetrical bump
$M = 0.72,\ \alpha = 2^\circ$

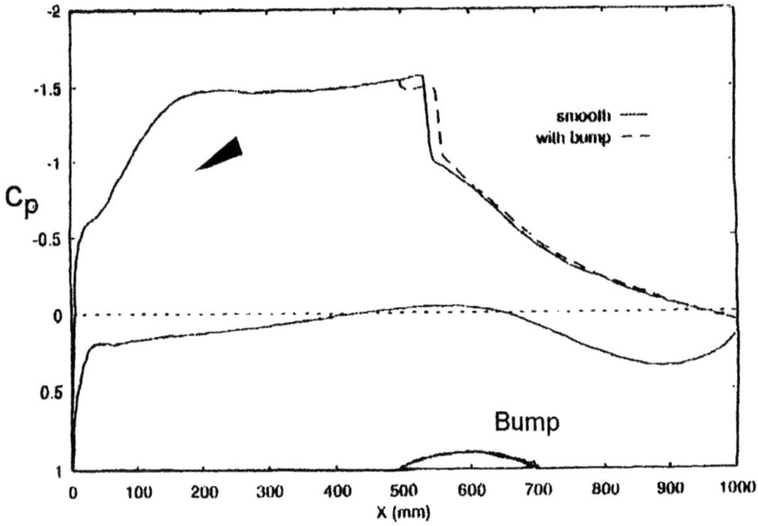

a. Airfoil pressure distribution

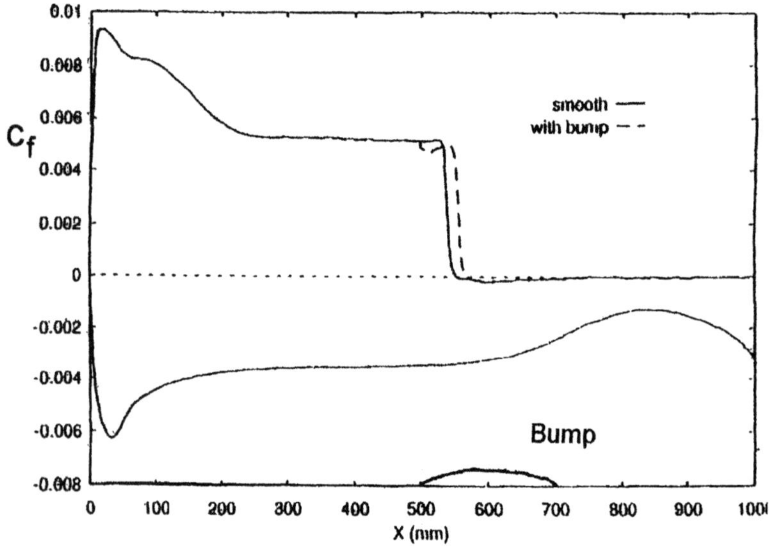

b. Airfoil skin-friction distribution

Figure 3 Pressure and skin-friction distributions with and w/o symmetrical bump $M = 0.72,\ \alpha = 2.6^{\circ}$

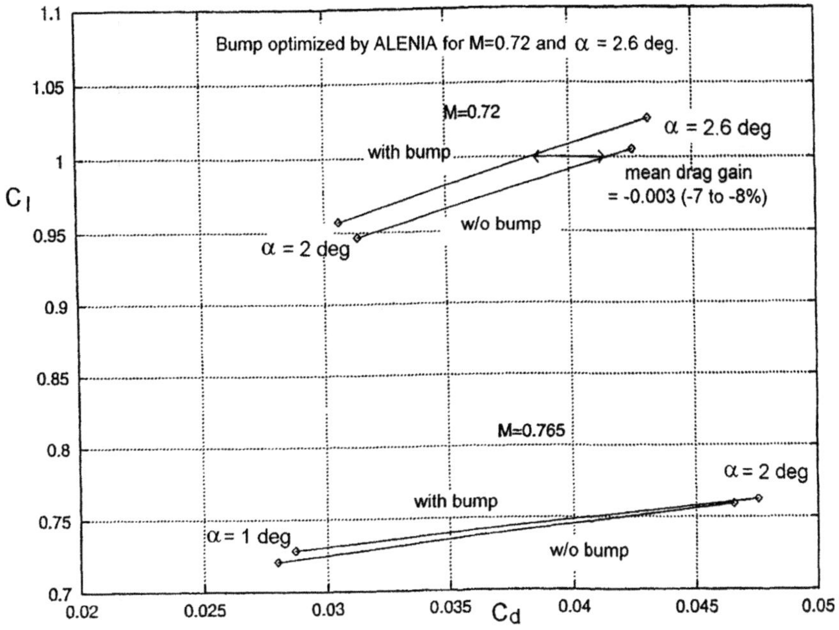

Figure 4 Drag polars for the Dassault airfoil with and without symmetrical bump

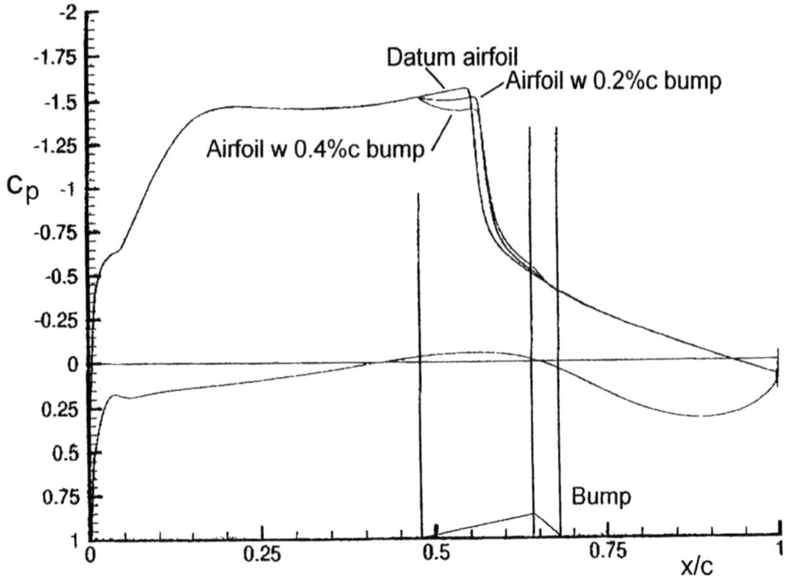

a. Pressure distribution

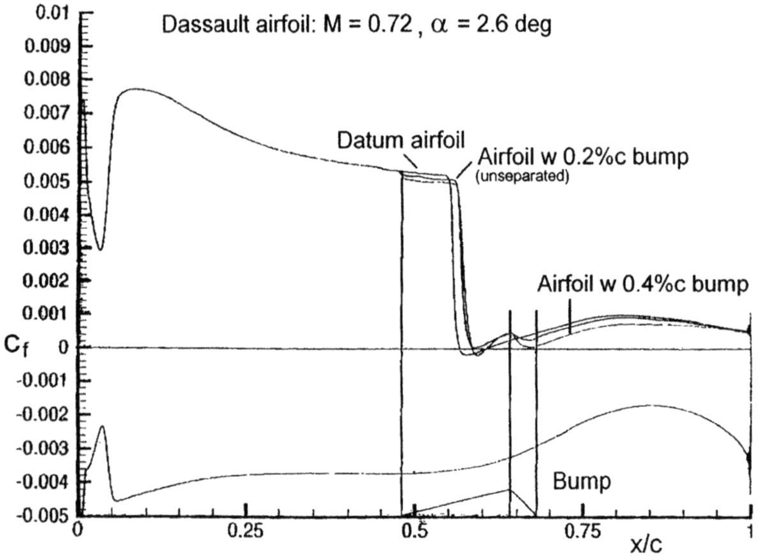

b. Skin-friction distribution

Figure 5 Pressure and skin-friction distributions with and w/o the asymmetrical bump
Design M = 0.72, $\alpha = 2.6^{\circ}$

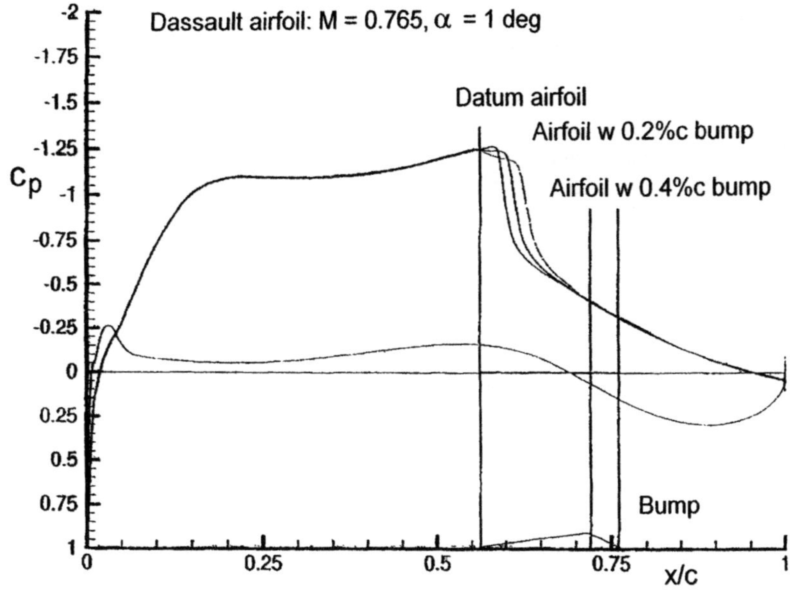

a. Pressure distribution

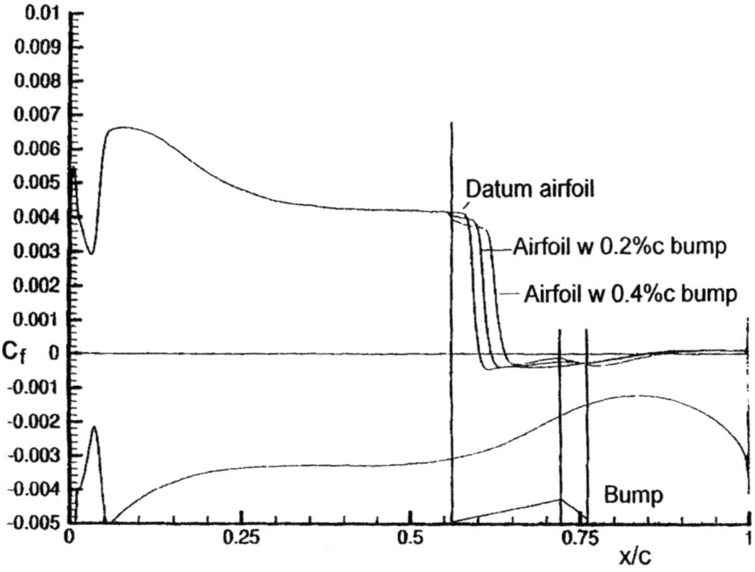

b. Skin-friction distribution

Figure 6 Pressure and skin-friction distributions with and w/o the asymmetrical bump
Off-design M = 0.765, $\alpha = 1.0^{\circ}$

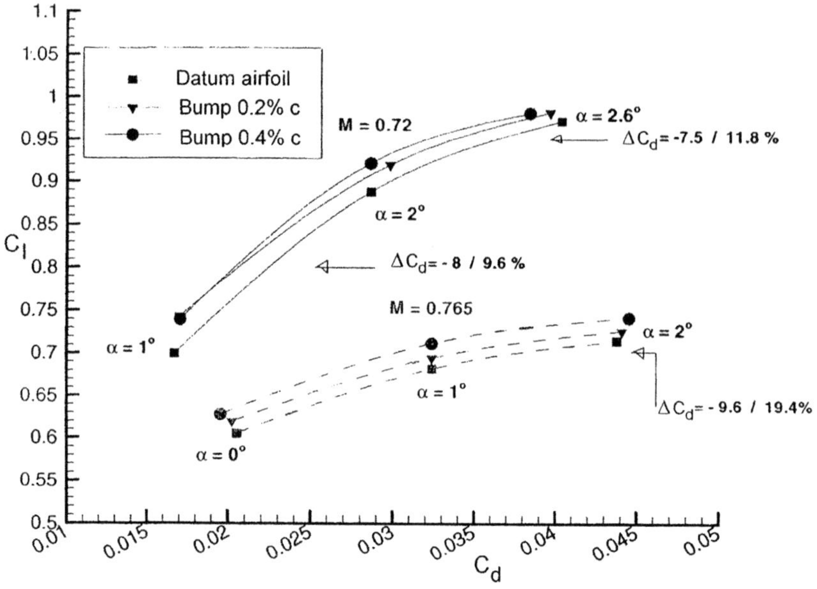

Figure 7 Drag polars for the Dassault airfoil with and without asymmetrical bump

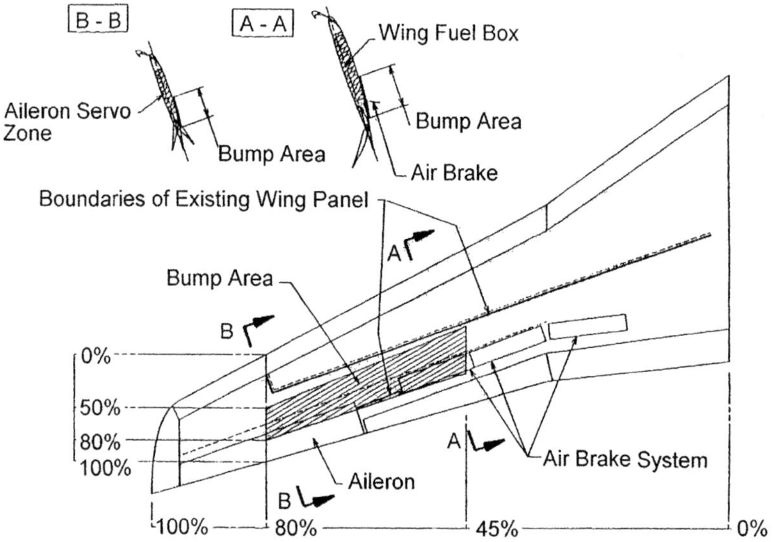

Figure 8 Bump location on a Falcon wing

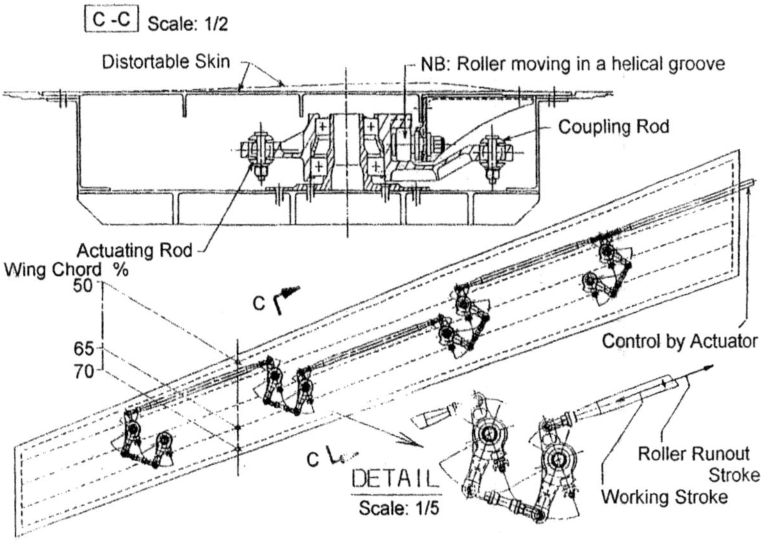

Figure 9 Wing-bump mechanism concept

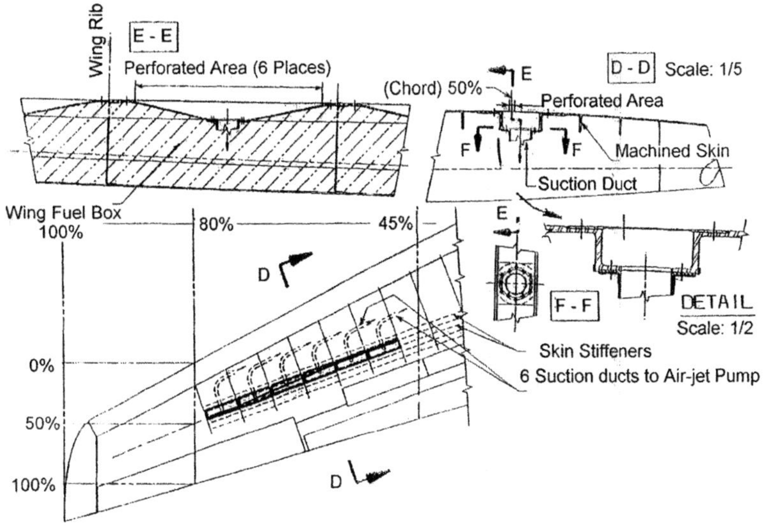

Figure 10 Boundary layer suction concept on a Falcon wing

Notes on Numerical Fluid Mechanics and Multidisciplinary Design

Available volumes

Volume 79: B. Schulte-Werning, R. Grégoire, A. Malfatti, G. Matschke (eds.): TRANSAERO – A European Initiative on Transient Aerodynamics for Railway System Optimisation. ISBN 3-540-43316-3

Volume 78: M. Hafez, K. Morinishi, J. Periaux (eds.): Computational Fluid Dynamics for the 21st Century. Proceedings of a Symposium Honoring Prof. Satofuka on the Occasion of his 60th Birthday, Kyoto, Japan, 15-17 July 2000. ISBN 3-540-42053-3

Volume 77: S. Wagner, U. Rist, H.-J. Heinemann, R. Hilbig (eds.): New Results in Numerical and Experimental Fluid Mechanics III. Contributions to the 12th STAB/DGLR Symposium Stuttgart, Germany 2000. ISBN 3-540-42696-5

Volume 76: P. Thiede (ed.): Aerodynamic Drag Reduction Technologies. Proceedings of the CEAS/DragNet European Drag Reduction Conference, 19-21 June 2000, Potsdam, Germany. ISBN 3-540-41911-x

Volume 75: E.H. Hirschel (ed.): Numerical Flow Simulation II. CNRS-DFG Collaborative Research Programme, Results 1998–2000. ISBN 3-540-41608-0

Volume 74: D. Henry, A. Bergeon (eds.): Continuation Methods in Fluid Dynamics. Contributions to the ERCOFTAC/EUROMECH Colloquium 383, Aussois, France, 6.–9. September 1998. ISBN 3-540-41556-4

Volume 73: M. Griebel, S. Margenov, P. Yalamov (eds.): Large-Scale Scientific Computations of Engineering and Environmental Problems II. Proceedings of the Second Workshop on "Large-Scale Scientific Computations" Sozopol, Bulgaria, June 2–6, 1999. ISBN 3-540-41551-3

Volume 72: W. Nitsche, H.-J. Heinemann, R. Hilbig (eds.): New Results in Numerical and Experimental Fluid Mechanics II. Contributions to the 11th AG STAB/DGLR Symposium Berlin, Germany 1998. ISBN 3-540-41552-1

Volume 70: W. Hackbusch, G. Wittum (eds.): Numerical Treatment of Multi-Scale Problems. Proceedings of the 13th GAMM-Seminar, Kiel, January 24–26, 1997. ISBN 3-540-41562-9

Volume 69: D. P. Hills, M. R. Morris, M. J. Marchant, P. Guillen (eds.): Computational Mesh Adaptation. ECARP – European Computational Aerodynamics Research Project. ISBN 3-540-41548-3

Volume 67: C. H. Sieverding, G. Cicatelli, J. M. Desse, M. Meinke, P. Zunino: Experimental and Numerical Investigation of Time Varying Wakes behind Turbine Blades. Results of the Industrial and Material Technology Aeronautics Research Project AER2-CT92-0048, 1992–1996. ISBN 3-540-41546-7

Volume 66: E. H. Hirschel (ed.): Numerical Flow Simulation I. CNRS-DFG Collaborative Research Programme. Results 1996–1998. ISBN 3-540-41540-8

Volume 64: R. Friedrich, P. Bontoux (eds.): Computation and Visualization of Three-Dimensional Vortical and Turbulent Flows. Proceedings of the Fifth CNRS-DFG Workshop on Numerical Flow Simulation, München, Germany, December 6 and 7, 1996. ISBN 3-540-41557-2

Volume 63: M. Fiebig, N. K. Mitra (eds.): Vortices and Heat Transfer. Results of a DFG-Supported Research Group. ISBN 3-540-41558-0

Volume 62: M. Griebel, O. Iliev, S. D. Margenov, P. S. Vassilevski (eds.): Large-Scale Scientific Computations of Engineering and Environmental Problems. Proceedings of the First Workshop on "Large-Scale Scientific Computations" Varna, Bulgaria, June 7–11, 1997. ISBN 3-540-41560-2

Volume 57: H. Deconinck, B. Koren (eds.): Euler and Navier-Stokes Solvers Using Multi-Dimensional Upwind Schemes and Multigrid Acceleration. Results of the BRITE/EURAM Projects AERO-CT89-0003 and AER2-CT92-00040, 1989–1995. ISBN 3-540-41535-1

Printing: Mercedes-Druck, Berlin
Binding: Buchbinderei Lüderitz & Bauer, Berlin

GPSR Compliance

The European Union's (EU) General Product Safety Regulation (GPSR) is a set of rules that requires consumer products to be safe and our obligations to ensure this.

If you have any concerns about our products, you can contact us on ProductSafety@springernature.com

In case Publisher is established outside the EU, the EU authorized representative is:

Springer Nature Customer Service Center GmbH
Europaplatz 3
69115 Heidelberg, Germany

Batch number: 10755284

Printed by Printforce, the Netherlands